Implications of the Blood–Brain Barrier and Its Manipulation

Volume 1
Basic Science Aspects

Implications of the Blood–Brain Barrier and Its Manipulation

Volume 1
Basic Science Aspects

Edited by
Edward A. Neuwelt, M.D.

The Oregon Health Sciences University School of Medicine
Portland, Oregon

Plenum Medical Book Company
New York and London

Library of Congress Cataloging in Publication Data

Implications of the blood–brain barrier and its manipulation.

Includes bibliographies.
Contents: v. 1. Basic science aspects.
1. Blood–brain barrier. I. Neuwelt, Edward A., 1948–   . [DNLM: 1. Blood–Brain
Barrier. WL 200 I34]
QP375.5.I46   1988   616.8'046   88-9845

ISBN-13: 978-1-4612-8039-2   e-ISBN-13: 978-1-4613-0701-3
DOI: 10.1007/978-1-4613-0701-3

Softcover reprint of the hardcover 1st edition 1989

© 1989 Plenum Publishing Corporation
233 Spring Street, New York, N.Y. 10013

Plenum Medical Book Company is an imprint of Plenum Publishing Corporation

To my parents, Matthew and Mariam Neuwelt, whose devotion to my education so greatly impacted my career;

To my wife, Jill, whose love and support persisted despite the distractions of my academic neurosurgical career; and

To my three children, Jennifer, Simone, and Sasha, whose welcome distractions allowed me to maintain a balanced perspective during the preparation of this monograph

Contributors

Michael W. Bradbury • Department of Physiology, King's College, London WC2R 2LS, England

Milton W. Brightman • National Institute of Neurological Communicative Diseases and Stroke, National Institutes of Health, Bethesda, Maryland 20892

Hugh Davson • Department of Physiology, King's College, and St. Thomas' Hospital Medical School, University of London, London WC2R 2LS, England

Joseph D. Fenstermacher • Departments of Neurological Surgery, and Physiology and Biophysics, State University of New York, Stony Brook, New York 11794

Eugene P. Frenkel • Department of Internal Medicine, School of Medicine, University of Texas Health Science Center at Dallas, Dallas, Texas 75235

Nigel H. Greig • Laboratory of Neurosciences, National Institute on Aging, National Institutes of Health, Bethesda, Maryland 20892

Conrad E. Johanson • Department of Clinical Neurosciences, Brown University and Rhode Island Hospital, Providence, Rhode Island 02902

Marianne Juhler • Department of Neurology, Rigshospitalet, Copenhagen, Denmark

Mark J. Kupersmith • Departments of Neurology and Ophthalmology, New York University, New York, New York 10016

Robert R. Myers • Departments of Anesthesiology and Neurosciences, Veterans Administration Medical Center, La Jolla, California 92093

Edward A. Neuwelt • Divisions of Neurosurgery and Biochemistry, School of Medicine, Oregon Health Sciences University, and Neurosurgery Section, Veterans Administration Medical Center, Portland, Oregon 97201

Hanna M. Pappius • The Goad Unit of The Donner Laboratory of Experimental Neurochemistry, Montreal Neurological Institute, McGill University, Montreal, Quebec H3A 2B4, Canada

Henry C. Powell • Department of Pathology, University of California, San Diego, La Jolla, California 92093

Manoucher Shakib • Department of Ophthalmology, New York University, New York, New York 10016

Quentin R. Smith • Laboratory of Neurosciences, National Institute on Aging, National Institutes of Health, Bethesda, Maryland 20892

Foreword

Understanding the structure and function of the blood–brain barrier (BBB) and recognizing its clinical relevance require a concert of scientific disciplines applied from a viewpoint of integrative physiology rather than from only molecular or analytical approaches. It is this broad scope that is emphasized in this book.

In my opinion, four original contributions define the field as it exists today. The first, a monograph by Broman,[1] entitled *The Permeability of the Cerebrospinal Vessels in Normal and Pathological Conditions,* was the model for many subsequent clinical and experimental studies on BBB pathology. Second, experiments by Davson,[3] summarized in his book entitled *Physiology of the Ocular and Cerebrospinal Fluids,* indicated that passive entry of nonelectrolytes into brain from blood is governed largely by their lipid solubility. This research supported the original suggestion by Gesell and Hertzman[4] that cerebral membranes have the semipermeability properties of cell membranes. The modern era of the barrier was introduced with the 1965 paper by Crone,[2] entitled "Facilitated transfer of glucose from blood to brain tissue." This paper identified stereospecific, facilitated transport of glucose as part of a system of regulatory barrier properties at a time when only a barrier to passive diffusion had been contemplated. Finally, the 1967 paper by Reese and Karnovsky,[11] entitled "Fine structural localization of a blood–brain barrier to exogenous peroxidase," sited the barrier at the continuous layer of cerebrovascular endothelial cells, which are connected by tight junctions.

Consider the complexity of the problem when it was first approached: to characterize, understand the function of, and manipulate the interfaces among the blood, brain, and cerebrospinal fluid compartments. These interfaces are the 180 cm²/g brain of cerebral capillary surface, the choroid plexus (which elaborates cerebrospinal fluid) and the arachnoid membranes. It is no surprise that progress in the field has been related so closely to new methodology, both experimental and theoretical.

Quantitative approaches began with two-compartment models (e.g., blood–brain, blood–cerebrospinal fluid) that were used to interpret exchange following programmed infusions of radioisotopes to maintain steady-state blood concentrations and that helped to characterize the selective permeability of the barrier for lipid-soluble agents.[3] The indicator dilution technique later distinguished facilitated, stereospecific transport of D-glucose into the brain,[2] and could be used to measure brain uptake of moderately permeant substances. With the introduction of the Brain Uptake Index technique by Oldendorf,[8] the entire spectrum of stereospecific carrier systems for transport into the brain was elaborated. The intravenous bolus technique[7] made it then possible to measure brain and spinal fluid uptakes of poorly permeant substances, such as [^{14}C]sucrose and ^{36}Cl, and to start to distinguish the contribution of the capillary pathway to regional brain uptake, as compared with the cerebrospinal fluid pathway.[12] Finally, the in vivo brain perfusion technique,[13] by allowing absolute control of intravascular perfusate composition, provided accurate measurements of maximum velocities and half-saturation constants for transport of individual amino acids and other substances into brain, and could be used to examine the role of protein binding on uptake.[5] We do not as of yet have complete compartmental-diffusion models to analyze local transport relations among brain, blood, and cerebrospinal fluid.[10,12] Such models would help to evaluate continuous in vivo measurements of isotope concentrations in small regions of the human brain, which are made available by positron emission tomography.

Distinctions between basic and clinical research do not often exist when studying the BBB. Frequently, fundamental experiments in transport and cell structure are soon carried over into the clinic. This has occurred in predicting drug entry into the brain and designing new drugs, in understanding the potentially lethal process of brain edema, and in characterizing the pathophysiology of brain tumors. Animal observations on the effect of hypertonic solutions on barrier permeability, which contributed to our understanding of the lability of the interendothelial tight junction,[9] were transferred within a short period of time to the clinic to enhance chemotherapy of brain tumors.[6]

The contributions in this book critically present the extensive current data on the BBB and on its clinical relevance. The chapters on the blood–nerve and blood–ocular barriers remind us that many principles applying to the barrier in the central nervous system are also valid for the eye and peripheral nerve. In the next 10 years, it is likely that molecular biologic techniques will provide information about how the barrier is programmed during development and in genetic disorders and how specific transport proteins operate at barrier membranes. More understanding will surely be forthcoming about the immunologic role of the barrier, and about barrier enzymes and their influence on brain function. We still know little about how divalent ions are transported at BBB sites and about how ion homeostasis is regulated in the central nervous system. Finally, future needs for treatment of cerebral dysfunction in Alzheimer disease, Huntington disease, brain tumors, or AIDS with drugs, antibodies, immune enhancers, or gene replacement will constrain researchers to develop creative approaches that will bypass current barrier limitations to these agents. Many of the new directions of this research are outlined in this book.

Stanley I. Rapoport, M.D.

Chief, Laboratory of Neurosciences
National Institute on Aging
National Institutes of Health
Bethesda, Maryland

References

1. Broman T: *The Permeability of the Cerebrospinal Vessels in Normal and Pathological Conditions.* Munksgaard, Copenhagen, 1949.
2. Crone C: Facilitated transfer of glucose from blood to brain tissue. *J Physiol (Lond)* 181:103–113, 1965.
3. Davson H: *Physiology of the Ocular and Cerebrospinal Fluids.* Churchill, London, 1956.

4. Gesell R, Hertzman AB: The regulation of respiration. IV. Tissue acidity, blood acidity and pulmonary ventilation: A study of the effects of semipermeability of membranes and the buffering action of tissues with the continuous method of recording changes in acidity. *Am J Physiol* 78:610–629, 1926.
5. Levitan H, Ziylan Z, Smith QR, et al: Brain uptake of a food dye, erythrosin B, prevented by plasma protein binding. *Brain Res* 323:131–134, 1984.
6. Neuwelt EA, Frenkel EP, Diehl J, et al: Reversible osmotic blood–brain barrier disruption in humans: Implications for the chemotherapy of malignant brain tumors. *Neurosurgery* 7:44–52, 1980.
7. Ohno K, Pettigrew KD, Rapoport, SI: Lower limits of cerebrovascular permeability to nonelectrolytes in the conscious rat. *Am J Physiol* 235:H299–H307, 1978.
8. Oldendorf WH: Brain uptake of radiolabeled amino acids, amines and hexoses after arterial injection. *Am J Physiol* 221:1629–1639, 1971.
9. Rapoport SI, Hori M, Klatzo I: Testing of a hypothesis for osmotic opening of the blood–brain barrier. *Am J Physiol* 223:323–331, 1972.
10. Rapoport SI, Fitzhugh R, Pettigrew KD, et al: Drug entry into and distribution within brain and cerebrospinal fluid: [^{14}C]urea pharmacokinetics. *Am J Physiol* 242:R339– R348, 1982.
11. Reese TS, Karnovsky MJ: Fine structural localization of a blood–brain barrier to exogenous peroxidase. *J Cell Biol* 34:207–217, 1967.
12. Smith QR, Rapoport SI: Cerebrovascular permeability coefficients to sodium, potassium and chloride. *J Neurochem* 46:1732–1742, 1986.
13. Takasato Y, Rapoport SI, Smith QR: An in situ brain perfusion technique to study cerebrovascular transport in the rat. *Am J Physiol* 247:H484–H493, 1984.

Preface

During the past ten years, two excellent books were published by basic scientists reviewing the concept of the blood–brain barrier (BBB), its anatomy, and its physiology. More recently, this concept of the BBB has attained clinical significance in the treatment of a wide variety of central nervous system (CNS) disorders. However, there has yet to appear a volume that takes our knowledge of the BBB one step further into the clinical arena.

Clearly, the demonstration in Parkinson disease that dopamine deficiency of the basal ganglia could be remedied by the administration not of dopamine itself (a substance that does not cross the BBB) but of its precursor (which does cross) was a major breakthrough in the treatment of this neurologic disease. Furthermore, it highlighted the importance of the BBB in clinical medicine. This clinical advance, as well as the basic research as summarized in the books of Dr. Rapoport and Dr. Bradbury, has fueled major advances in the treatment of a number of neurologic disorders based primarily on an understanding and manipulation of the BBB. It is therefore the purpose of this monograph to bridge the gap from the laboratory to the clinical setting.

Clearly, the BBB has an impact on the therapy of movement disorders. More recently, significant advances in the treatment of CNS infection and brain tumors through the delivery of drugs across the BBB have been made. Evidence is beginning to accumulate that the manipulation of the BBB may be important even in the treatment of CNS genetic disorders. In neuroradiology, the relationship of the various contrast agents to the BBB has become a major area of investigation. Over the past 5 years, patient studies regarding the therapeutic potential of reversible osmotic BBB opening have progressed. These studies relating the clinical impact of the BBB in neurologic disease have been published in widely diverse journals by investigators in a number of medical specialties. It is the purpose of this volume to condense this mass of information and provide a substantial reference source as well as a readable text.

Edward A. Neuwelt

Portland, Oregon

Acknowledgments

We would like to thank the following colleagues for their editorial expertise in reviewing various chapters in this monograph. Their assistance was indispensable in the realization of this project and is much appreciated. We offer thanks to James H. Austin, Louis Bakay, John A. Barranger, Mitch Berger, Michael W. Bradbury, Donald B. Calne, Anthony D'Agostino, N. deTribolet, Burton Drayer, Dieter Enzmann, Joseph D. Fenstermacher, Scott Goodnight, Jr., Nigel H. Greig, Robert G. Grossman, Mary K. Gumerlock, Phillip Gutin, Julian T. Hoff, Conrad E. Johanson, Richard Johnson, Toshihiko Kasuroiwa, Michael L. Klein, Kenneth Krohn, Victor Levin, Peter Lipsky, Don M. Long, Phillip A. Low, William G. Mahyan, David A. McCarron, Leena Mela-Riker, Gajanan Nilaver, George A. Ojemann, Hanna M. Pappius, William Pardridge, Michael E. Phelps, Michael Pollay, Stanley I. Rapoport, John A. Resko, William K. Riker, James T. Rosenbaum, David A. Rottenberg, Merle A. Sande, Jay P. Sanford, W. Michael Scheld, Steven Shenker, Thomas Shults, Quentin R. Smith, Reynold Spector, Wendell C. Stevens, Larry J. Strausbaugh, Jay Tureen, E. Michael VanBuskirk, William Wara, and Keasley Welch.

Special thanks to Ms. Patricia Butler, whose unique editorial and secretarial expertise was so important to this monograph.

Contents

3. The Anatomic Basis of the Blood–Brain Barrier

Milton W. Brightman

4. Quantitation of Blood–Brain Barrier Permeability

Quentin R. Smith

5. Transport across the Blood–Brain Barrier

Michael W. Bradbury

6. Pharmacology of the Blood–Brain Barrier

Joseph D. Fenstermacher

7. Ontogeny and Phylogeny of the Blood–Brain Barrier

Conrad E. Johanson

10. The Blood–Brain Barrier and the Immune System

Marianne Juhler and Edward A. Neuwelt

11. Cerebral Edema and the Blood–Brain Barrier

Hanna M. Pappius

12. Drug Delivery to the Brain by Blood–Brain Barrier Circumvention and Drug Modification

Nigel H. Greig

13. The Blood–Ocular Barrier

Mark J. Kupersmith and Manoucher Shakib

Contents of Volume 2

Color Plates

Chapter 1

Figure 9. CNS lymphoma. Radionuclide brain scans, vertex view. (A) Control scan 5 months after initiation of protocol. (B) 24 hr later, after BBB disruption via right internal carotid artery. (The patient's right is on the reader's left.) The radionuclide uptake is color coded from low uptake (blue) to high uptake (white) as follows: blue, green, yellow, orange, red, white. (From Neuwelt et al.[40])

Chapter 4

Figure 2. Computerized images of phenylalanine influx into rat brain. L-[14C]phenylalanine was infused into the femoral vein for 90 sec, after which time the rat was killed by decapitation. Regional brain 14C content was determined by a computer-driven densitometer from autoradiographs prepared from coronal sections of the brain. The readings were converted into values of phenylalanine influx and displayed by colors corresponding to discrete levels of transport. Influx was calculated as K_{in} times the plasma phenylalanine concentration. The sections were taken at the level of the following structures: (1) caudate, (2) lateral ventricles, (3) anterior thalamic nuclei, (4) lateral geniculate, (5) medial geniculate, (6) inferior colliculus, and (7) superior olive. The color scale (8) is in units of nanomoles per minute per gram. (From Mans et al.[69])

Chapter 11

Figure 5. Cerebral glucograms prepared from selected [14C]deoxyglucose autoradiographs. LCGU in micromoles/100 g per min. Sections through the lesion area are shown from a normal animal and representative animals at 24, 72, and 120 hr after a freezing lesion. (Data from Pappius.[62])

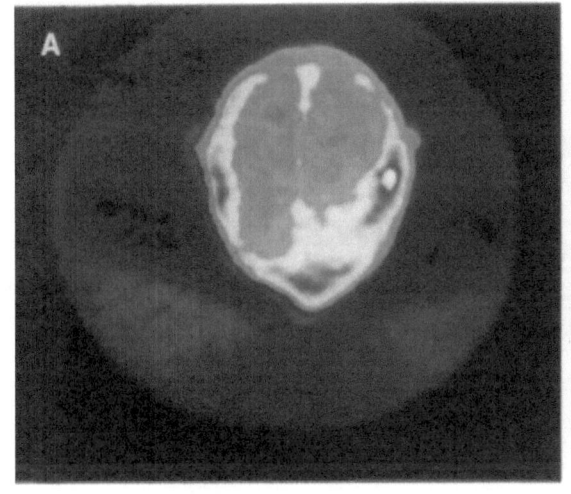
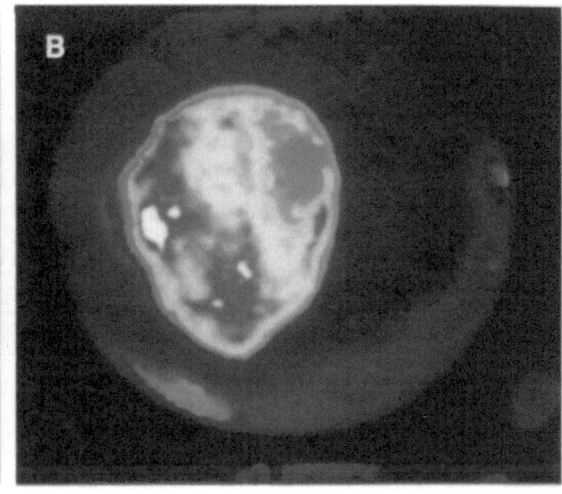

Chapter 1: Figure 9

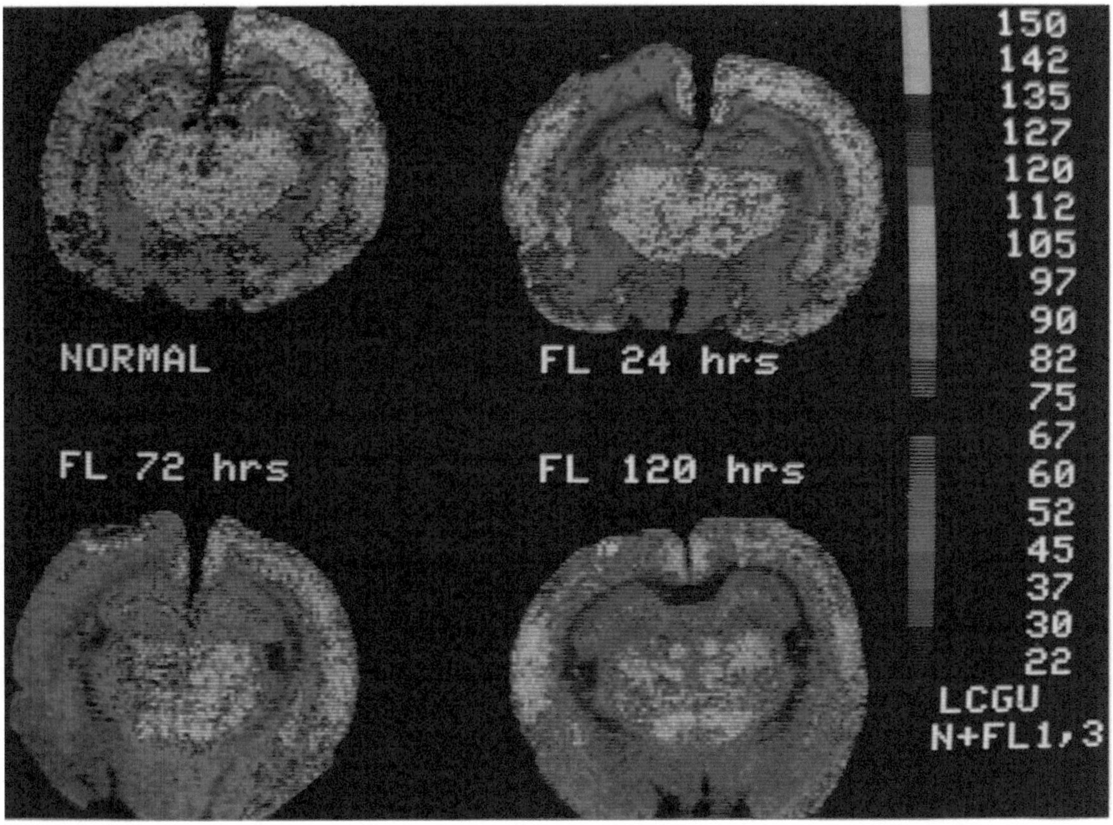

NORMAL   FL 24 hrs

FL 72 hrs   FL 120 hrs

150
142
135
127
120
112
105
97
90
82
75
67
60
52
45
37
30
22
LCGU
N+FL1,3

Chapter 11: Figure 5

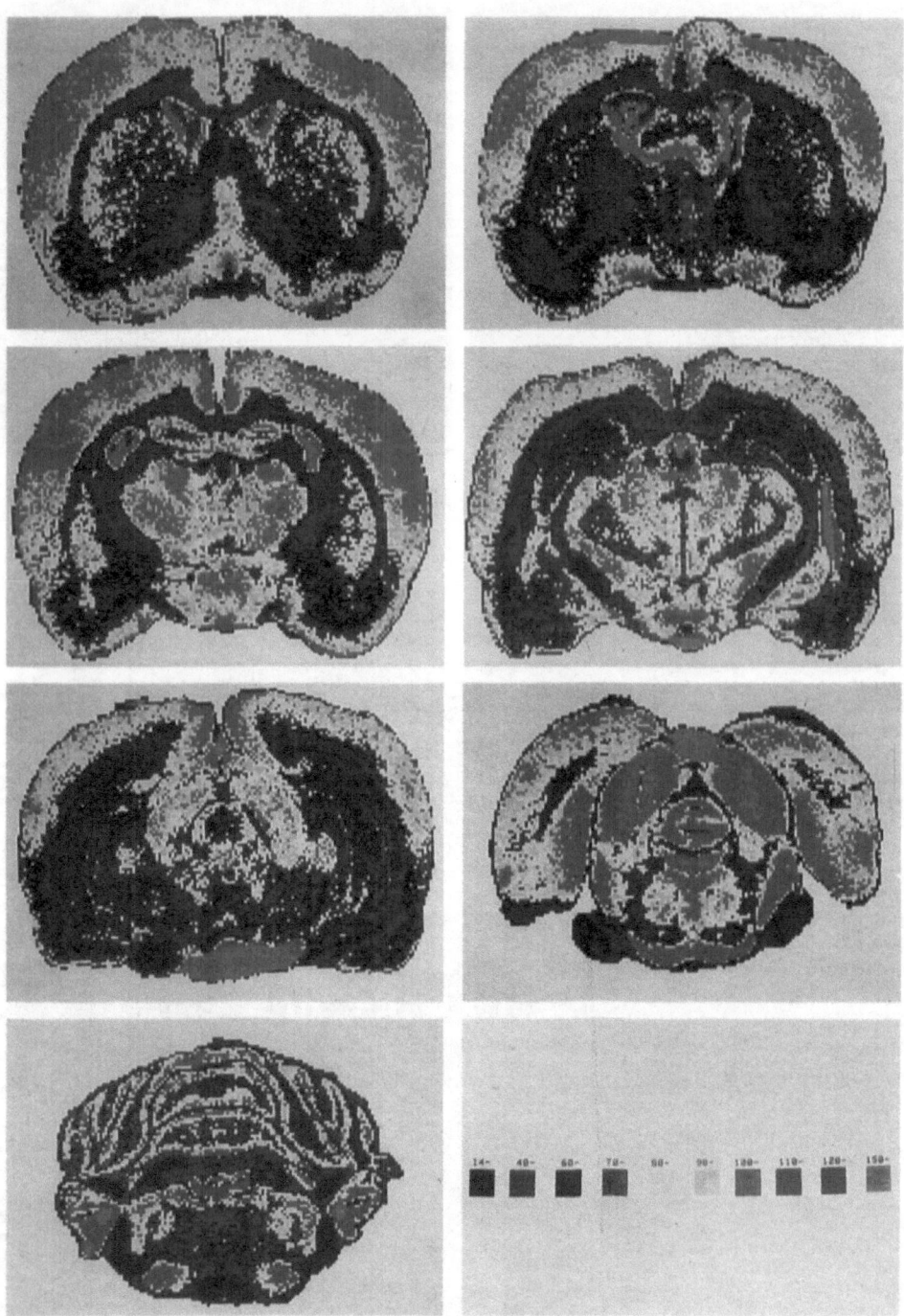

Chapter 4: Figure 2

1

The Challenge of the Blood–Brain Barrier

Edward A. Neuwelt and Eugene P. Frenkel

1. INTRODUCTION

Contemporary understanding of the blood–brain barrier (BBB) has been highlighted by Davson,[5] Rapoport,[50] and Bradbury[3]; their reviews have provided the critical basis for understanding the physiologic and pathophysiologic functioning of the BBB as it relates to clinical medicine. Human diseases and experiments of nature have helped illustrate the functioning of the BBB. This chapter presents examples in illustrative case reports.

2. DO MONOCYTES STREAM ACROSS THE BLOOD–BRAIN BARRIER IN RESPONSE TO BRAIN INJURY?

Lymphatic channels and lymphoid tissue are not found in the normal central nervous system (CNS), although an occasional mononuclear cell, structurally similar to a lymphocyte, can be seen in the perivascular spaces and small numbers of lymphocytes (0–3/μl) may be found in the cerebrospinal fluid (CSF). The rod, or microglial cell, is the major lymphoreticular cell in the CNS (Fig. 1). This cell was initially described in 1919–1921 by del Rio-Hortega,[52,53] using a silver carbonate staining technique. He subsequently demonstrated that this cell could assume a variety of morphologic configurations ranging from rod-shaped cells to swollen fat-granule cells. These observations led to the concept that the microglial cell is the macrophage of the CNS. This cell population is still a mystery in terms of origin, properties, and function. As a macrophage, it appears to have a role in both afferent and efferent limbs of the immune response, and it may be the cell of origin of the tumors termed microglioma, now more commonly and correctly termed primary CNS lymphoma.

Edward A. Neuwelt • Divisions of Neurosurgery and Biochemistry, School of Medicine, Oregon Health Sciences University, and Neurosurgery Section, Veterans Administration Medical Center, Portland, Oregon 97201.   *Eugene P. Frenkel* • Department of Internal Medicine, School of Medicine, University of Texas Health Science Center at Dallas, Dallas, Texas 75235.

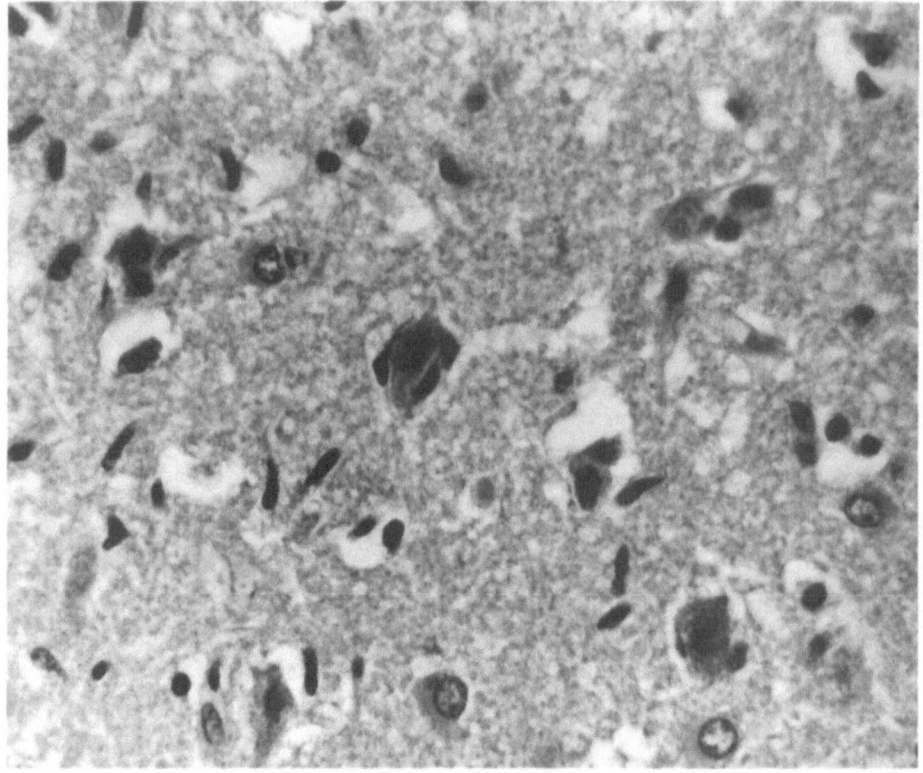

Figure 1. Hippocampal cortex demonstrating shrunken neurons and increased numbers of rod cells 12 days after a prolonged hypoxic episode. Hematoxylin and eosin (H & E). (×480) (From Neuwelt and Clark.[25])

Whether the microglial cell originates primarily in the CNS or from the systemic circulation is uncertain.[25] For instance, Hain[12] carried out an experiment that supported the view of a CNS origin for the microglial cell. Cerebral stab wounds were made after either radiating the head of an animal while shielding the body or radiating the body while shielding the head. When only the head was radiated, no macrophages were found around the cerebral stab wounds; when only the body was radiated, a normal microglial response was seen. Hain concluded that the microglial response was not dependent on a cellular origin from the systemic circulation.[12]

Subsequently, Konigsmark and Sidman[18] came to the opposite conclusion. They labeled circulating monocytes with [3H]thymidine in vivo and then made cerebral stab wounds. Since most macrophages around these lesions were radiolabeled, they concluded that the systemic circulation was the primary source of CNS macrophages.

Even less certain than the findings in animal models is the origin of the microglial cell or macrophage in humans. In humans, an interesting working concept was proposed by Feigin,[7] who stated:

> If large numbers of macrophages were streaming into the tissues from the vessels, one might anticipate that there would be a centrifugal distribution of these cells centered around vessels, with large numbers adjacent to the vessels and lesser numbers more distantly located. . . . This does not occur. . . . Instead, the distribution of macrophages, as viewed with hematoxylin and eosin preparations, or microglia as viewed with silver carbonate techniques, is remarkably uniform and homogeneous.

The following clinical case[26] emphasizes these descriptive findings.

Case Report: Microglial Response Following Cerebral Ischemia in an Aplastic Patient

A 34-year-old black man was transferred to the Baltimore Cancer Research Center with a 2-week history of fever. On admission, he had sinusitis, a perirectal abscess, an infiltrate in the right upper lobe of the lung, and *Pseudomonas* septicemia. The peripheral blood white blood count (WBC) was 2400/μl, of which less than 10% were mature polymorphonuclear leukocytes. The patient was confused, and this appeared to be secondary to the *Pseudomonas* sepsis; he was otherwise neurologically normal. He had no history of either seizures or hypoxia. An examination of the bone marrow provided the diagnosis of acute myelomonocytic leukemia. The patient was started on antibiotics for sepsis and on antileukemic therapy with daunorubicin and cytosine arabinoside. Sepsis continued unabated. The total WBC count dropped to <100 WBC/μl and the marrow was severely hypoplastic. On the fourth hospital day, he had a prolonged hypoxic episode but was resuscitated. He was comatose thereafter. Over the next several days, he remained septic and his neurologic status continued to deteriorate. He died 10 days following cardiopulmonary arrest.

At autopsy, there was a perirectal abscess and a pulmonary infarct in the patient's right upper lobe. No gross abnormalities of the brain were seen. On microscopic examination of the brain, there was diffuse and marked neuronal shrinkage and neuronal eosinophilia. There was diffuse rod cell (i.e., microglial) hyperplasia consistent with global cerebral ischemia 10 days prior to death (Fig. 1). No leukemic or other cellular infiltrate seen.

Comment. The pattern of distribution of microglia diffusely and homogeneously throughout the CNS conforms to the conceptual pattern described by Feigin.[7] It is noteworthy that the microglial proliferation was consistent with global cerebral ischemia. Since this event occurred when the patient was essentially aplastic, it is particularly difficult to implicate circulating WBCs as the origin of these microglia. The microglia in this patient were not of neoplastic origin.

In summary, this case provides clinical support for the view that the macrophage–microglial cell can arise in the CNS. However, a dual origin (i.e., systemic and CNS) for such cells may exist.

3. EFFECTS OF INJURY TO THE BLOOD–BRAIN BARRIER BY MASS LESIONS IN THE CNS: NEURORADIOLOGIC CHARACTERISTICS

The introduction of enhanced computerized tomographic (CT) scanning of the brain has improved the rate of detection of mass lesions to greater than 90%.[16] The enhancement primarily results from water-soluble iodinated contrast agents that are able to cross a damaged BBB. The following cases, in which the presumptive diagnosis was tumor, aptly demonstrate the remarkable sensitivity but lack of specificity of CT in identifying mass lesions in the CNS when the barrier is altered.

Case Report: Tuberculous Brain Abscess

R.M. had multiple intracranial enhancing lesions due to BBB damage in both the cerebral and posterior fossa compartments as shown by CT (Fig. 2) and an epidural block in the cervical spine at myelography.[16] Systemic workup for a primary tumor or other lesion was negative. The patient was begun on empirical radiotherapy; after 2000 rad, his brain herniated and he died. At autopsy, multiple tuberculous abscesses were found and there was no evidence of tumor.

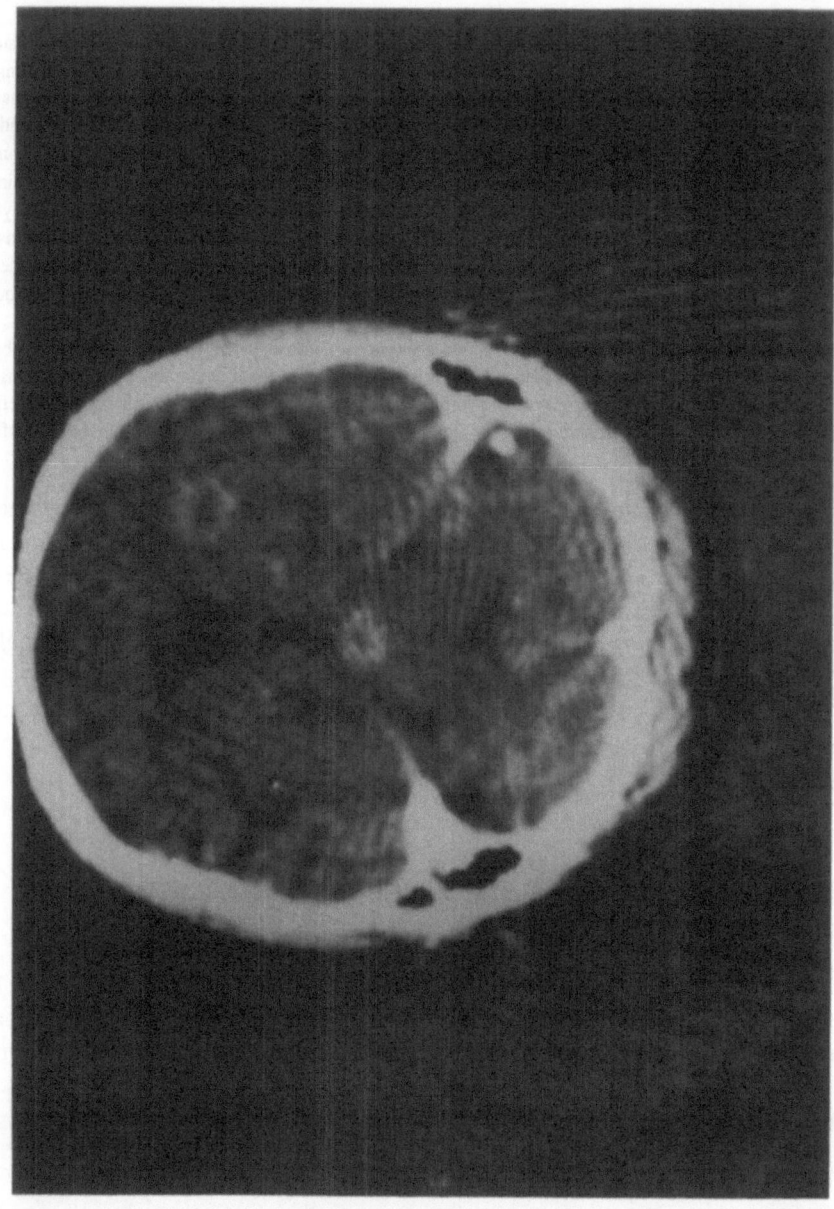

A

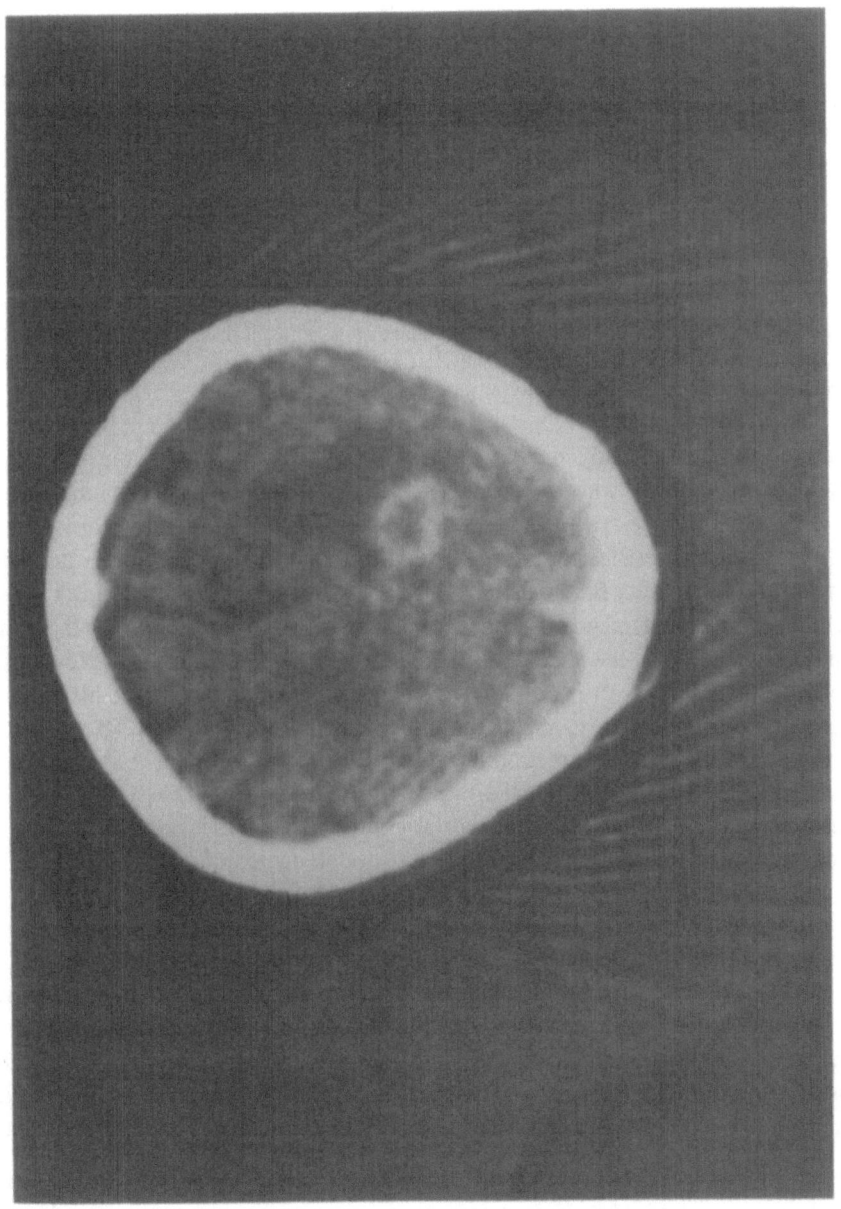

Figure 2. (A, B) Enhanced computed tomographic scans showing three separate enhancing ring lesions with edema. At autopsy, all proved to be tuberculous abscesses. (From Kandalaft et al.[16])

Case Report: Cryptococcal Brain Abscess

R.H. had multiple uniformly enhancing cerebral lesions as seen on CT scan (Fig. 3). Angiography revealed multiple avascular mass lesions.[16] There was no radiologic or other evidence of systemic tumor or disease. The presumptive preoperative diagnosis was metastases to the CNS from an unknown primary neoplasm. At craniotomy multiple cryptococcal abscesses were found.

Case Report: Traumatic Cerebral Hematoma

J.M. had a several-week history of severe temporal headaches after a fall from his chair.[16] Computed tomography revealed a 3-cm left frontal enhancing ring lesion with edema and ventricular shift (Fig. 4). Angiography revealed abnormal vascularity with early filling veins. All neurologic findings were firmly in support of a diagnosis of tumor. At craniotomy, however, the lesion was an intracerebral hematoma, presumably traumatic. Multiple biopsies from the hematoma cavity wall showed no evidence of tumor.

Case Report: Giant Aneurysm

C.H. presented with multiple lower cranial nerve palsies on the left,[16] and a CT scan revealed a large enhancing lesion with areas of low density to the left of midline in the posterior fossa (Fig. 5). Angiography revealed evidence of an avascular mass. The presumptive preoperative diagnosis was glioma. At operation, a partially thrombosed giant vertebral artery aneurysm was identified and clipped.

Comment. These cases aptly emphasize the remarkable sensitivity of contemporary neuroradiologic technology, which is capable of eliciting important information after damage to the BBB. There is little question that presumptive preoperative histologic diagnosis of intracranial mass lesions has improved markedly, largely due to the pattern of enhancement on CT scan after BBB damage. However, despite the exquisite sensitivity of these techniques including magnetic resonance imaging (MRI) and positron emission tomography imaging (PET), the problem of specificity remains. The four cases presented help emphasize the fallibility of neuroradiologic conclusions in predicting the histologic nature of a given mass lesion in the brain. Most studies report at least 5% of neuroradiologic diagnoses to be incorrect on the basis of subsequent histologic studies.

Kendall et al.,[17] from the National Hospital for Neurological Diseases in London, reviewed radiologically identified supratentorial gliomas. Of 311 gliomas diagnosed by the degree and pattern of enhancement as well as by the degree of mass effect on CT scanning, 30 cases (9.7%) later proved to have different structural diagnoses. Obviously, the misdiagnosis of a glioma for a malignant tumor of different histology is not as serious as the error of calling a benign lesion a glioma. That clinical interpretations have been made on the basis of these findings is supported by one series in which three meningioma patients were sent home, operation was delayed in one abscess case, and an unnecessary resection was done in one infarct patient.[17] In addition, giant aneurysms filled with thrombus and mistaken for neoplasm have been described by Pinto et al.[48] Similarly, Zimmerman et al.[62] showed a "ring blush" due to BBB damage with a resolving intracerebral hematoma.

Multiple intracranial cryptococcomas and tuberculomas (in the absence of evidence of systemic disease), an intracerebral hematoma, and thrombosis of a giant aneurysm, have each been called a malignant neoplasm by virtue of the clinical and neuroradiologic diagnostic criteria, much of which results from the pattern of BBB damage.

An important consequence of incorrect neuroradiologic presumptive diagnoses is that patients with benign tumors, inflammatory lesions, or vascular lesions can be subjected to unnecessary radiation and/or chemotherapy. Since damage to the BBB is not specific, these findings emphasize the indispensable need for a specific tissue diagnosis prior to the initiation of treatment. The availability of both microsurgery and CT- or MRI-guided stereotactic biopsy has meant that most lesions can be safely subjected to biopsy.[9]

4. ENCEPHALOPATHY ASSOCIATED WITH CRYOPRECIPITABLE AUSTRALIA ANTIGEN

The hepatitis B-associated antigen (HAA) has been indirectly implicated as a cause of encephalopathy. The patient with antigen-positive hepatitis who develops hepatic encephalopathy is the most common and important example of this indirect association.[1,6,57] The following case suggests, on a more direct immunologic basis, that the HAA can cause the development of an acute encephalopathy.

Case Report

A 19-year-old white woman who was comatose and decerebrate was seen in the emergency room.[54] Her sister indicated that she had lived with a heroin addict who was jaundiced. She had left him 2 months before admission. She had complained of arthralgias and urticaria over the previous 6 weeks. She had sought medical attention and an assay for the hepatitis B surface antigen (HB$_s$Ag) at that time had been positive.[21,23,56] Two weeks prior to admission, she had complained of nausea, vomiting, and headaches. She was described as irritable, sleeping most of the day, and complaining of joint pains. On the morning of admission the patient was found comatose on the floor.

On examination her pulse was 140 beats/min, temperature 105°F (40.5°C), and she had Cheyne–Stokes respirations. She had mild scleral icterus and a normal-sized liver. Rectal examination revealed guaiac-positive stool. On neurologic examination her disc margins were blurred bilaterally, and she had downbeat nystagmus. She alternated between decorticate and decerebrate posturing. Her hemoglobin was 11.7 g/dl, WBC count 6800/μl, and platelet count 470,000/μl. Her serum chemistries, arterial blood gas (ABG), and clotting studies were normal except for a direct/indirect bilirubin concentration of 7.2/5.0 mg/dl, serum glutamic oxalacetic transaminase (SGOT) of >500 units, and serum alkaline phosphatase of 160 units. Arterial and CSF ammonia levels were only minimally elevated. Her HB$_s$Ag in serum remained positive. Serum was also positive for anti-HAA antibody by radioimmunoassay (RIA).[21,23,56] Evaluation for collagen vascular disease (antinuclear antibody titer, lupus erythematous cell preparations, serum protein electrophoresis) was negative. Screens for toxic substances and a salicylate level were both normal. Urinalysis revealed trace proteinuria.

Changes observed in a four-vessel intracranial arteriogram were consistent with mild cerebral edema. There were no masses or shifts of major vessels. Cerebrospinal fluid obtained via a frontal twist drill ventriculostomy contained 138 red blood cells (RBCs) and no white cells/μl, protein of 6 mg/dl, and glucose of 187 mg/dl (with a plasma glucose of 207 mg/dl). There was no detectable Australia antigen in the CSF. A portable electroencephalogram (EEG) revealed diffuse slowing and absence of any triphasic waves, suggestive of hepatic coma.

The patient was intubated in the emergency room. She was treated with dexamethasone, mannitol, and furosemide to reduce the cerebral edema. She developed generalized major motor seizures on the second hospital day which were controlled with anticonvulsants. Subsequently, she began to improve, and by the fourth hospital day she moved all extremities in response to pain without decerebrate posturing, and the nystagmus resolved. However, she then developed an interstitial pneumonia and died of progressive respiratory failure. Cultures of tracheal secretions, serum, and CSF did not grow viruses.

The findings at autopsy included mild cerebral edema, mild resolving hepatitis, interstitial and bronchial pneumonia, and acute gastric ulcers. There was no evidence of vasculitis in the brain, spinal cord, meninges, or choroid plexus by either light or electron microscopy. The kidneys were normal.

Immunologic studies showed that the patient's serum produced a visible cryoprecipitate after 72 hr at 4°C; the HB$_s$Ag was markedly concentrated in the cryoprecipitate as measured by RIA. Total serum hemolytic complement was borderline low at 33 CH$_{50}$ U/ml (normal, 22–61 CH$_{50}$ U/ml). The serum C'3 value was borderline low at 0.81 mg/ml (normal, 0.82–1.7 mg/ml), with C'4, C'5, and C'1 normal (0.281, 0.77, and 0.223 mg/ml, respectively). The CSF hemolytic C'4 was low at 13 CH$_{50}$ U/ml (normal, 43–99 CH$_{50}$ U/ml), but

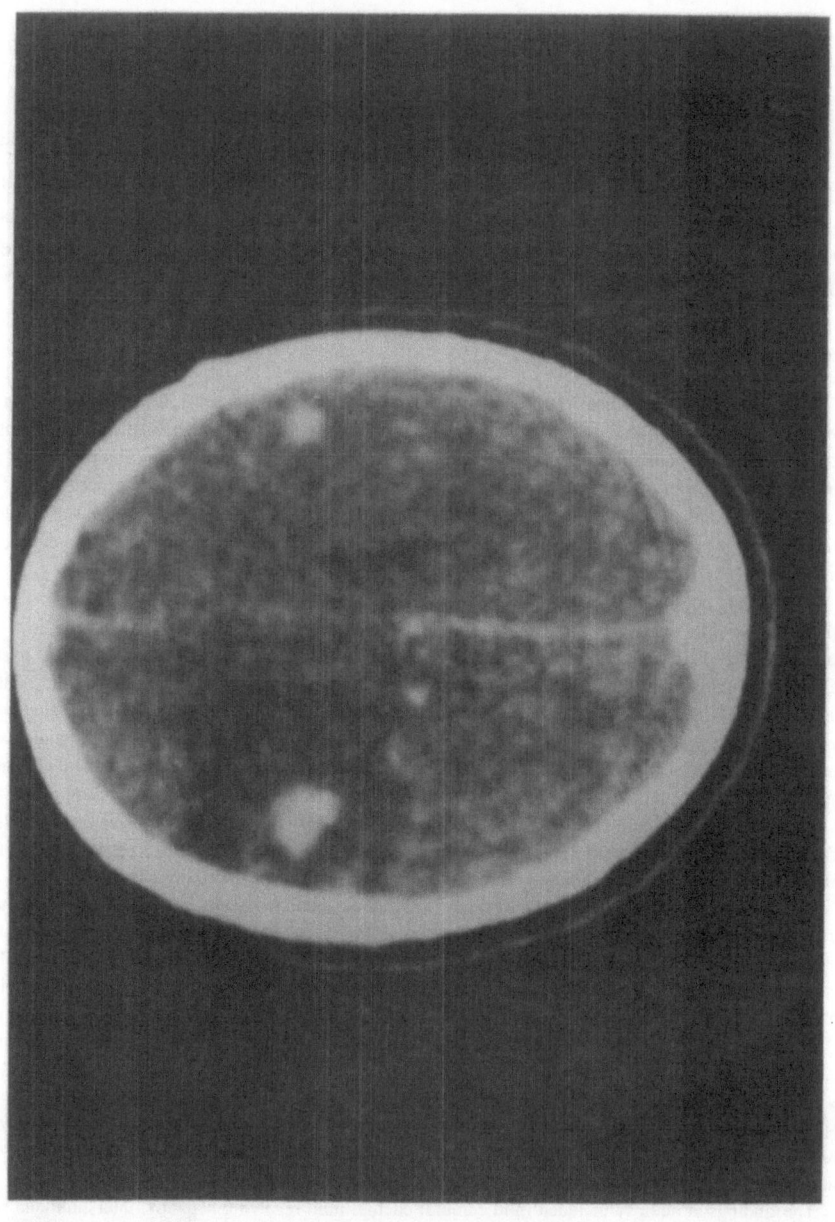

A

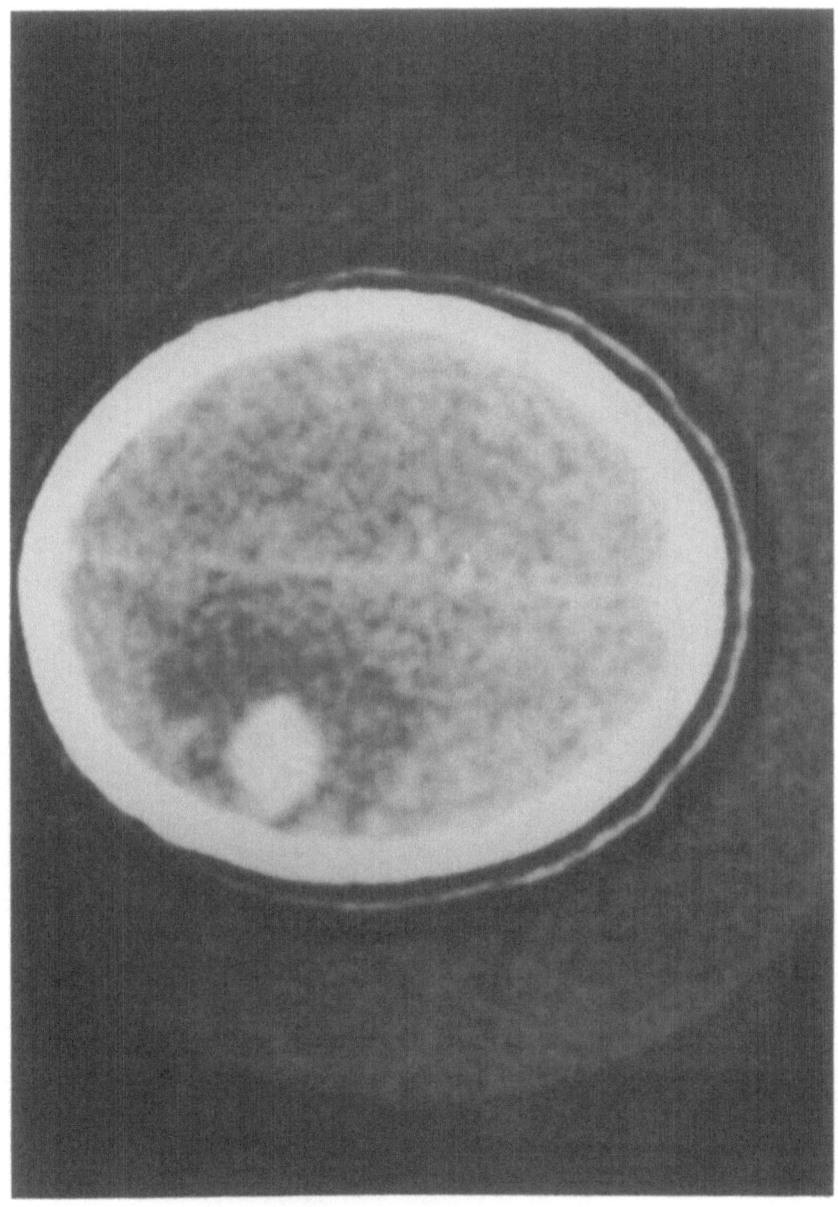

Figure 3. (A, B) Enhanced computed tomographic scans showing two uniformly enhancing round lesions with associated edema. At surgery, these proved to be cryptococcomas. (From Kandalaft et al.[16])

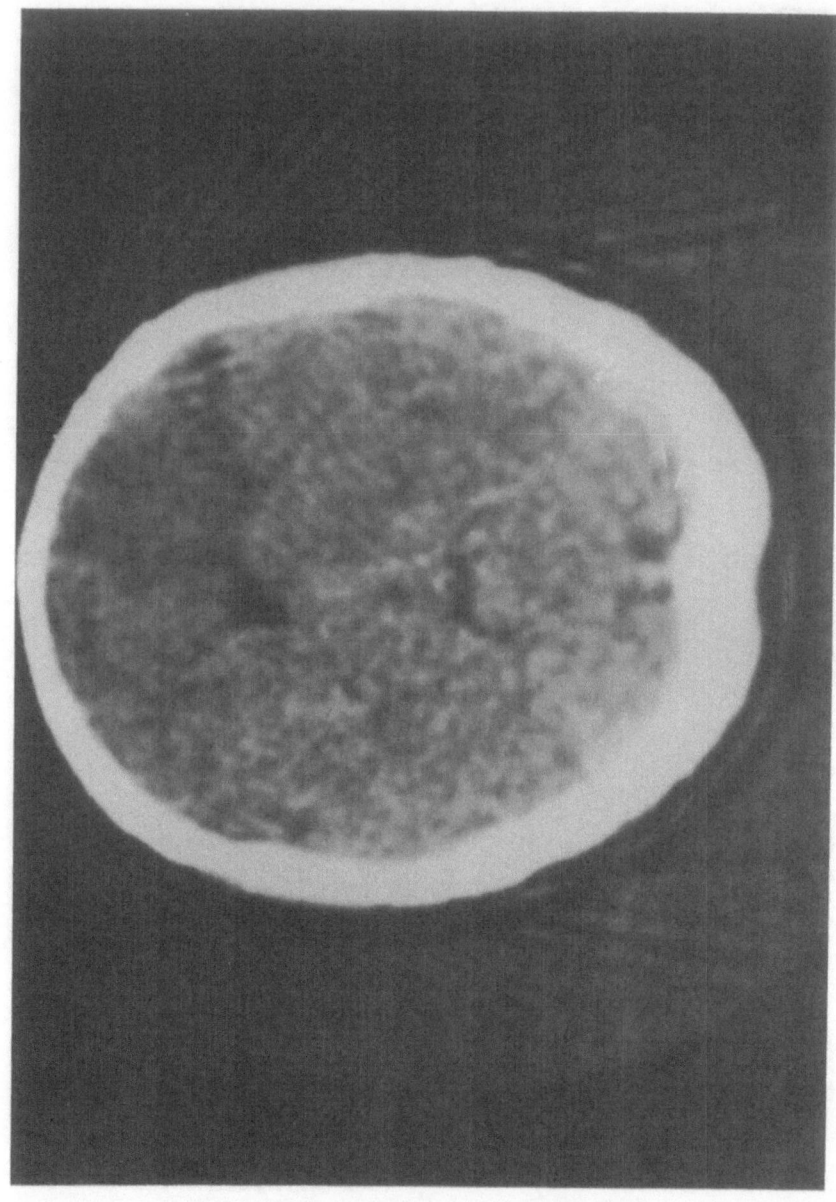

A

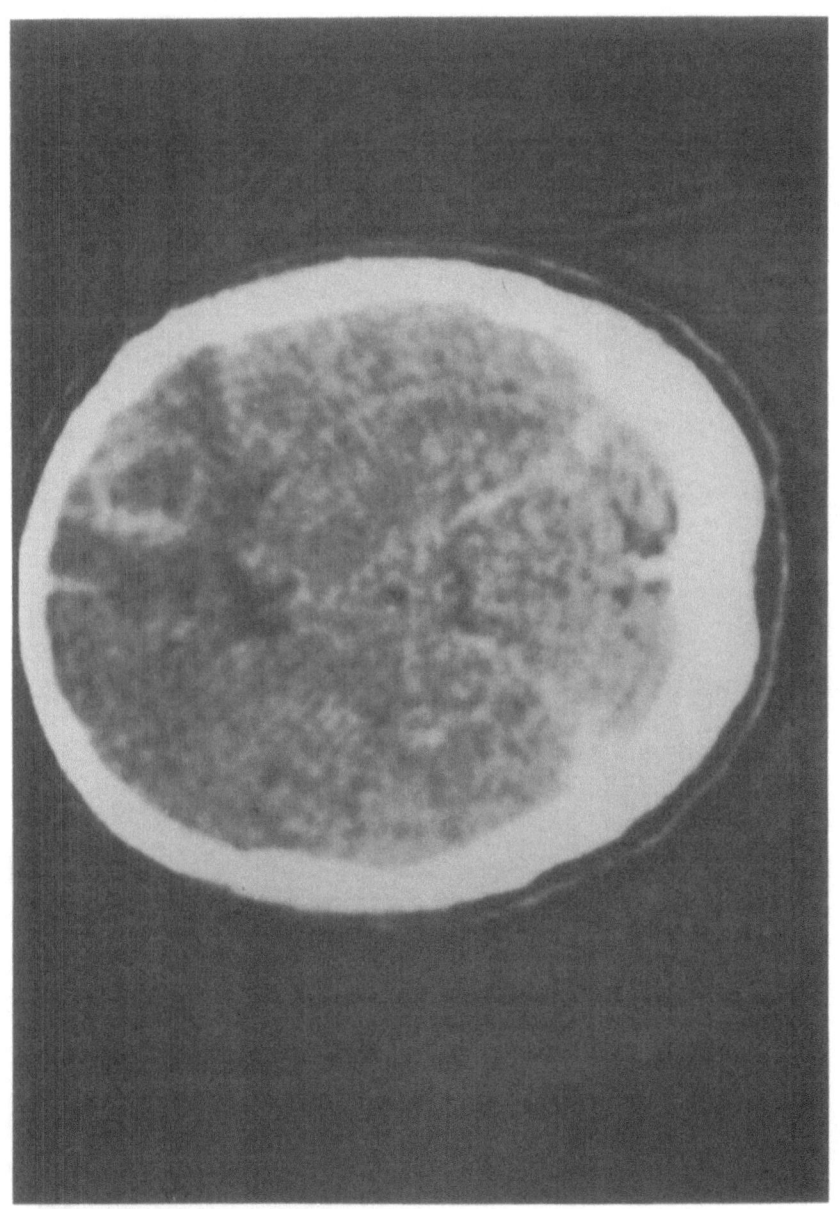

B

Figure 4. (A) Unenhanced computed tomographic (CT) scan shows low density in left frontal region with mass effect. (B) CT scan with enhancement demonstrates ring lesion that, at surgery, proved to be an old hematoma. (From Kandalaft et al.[16])

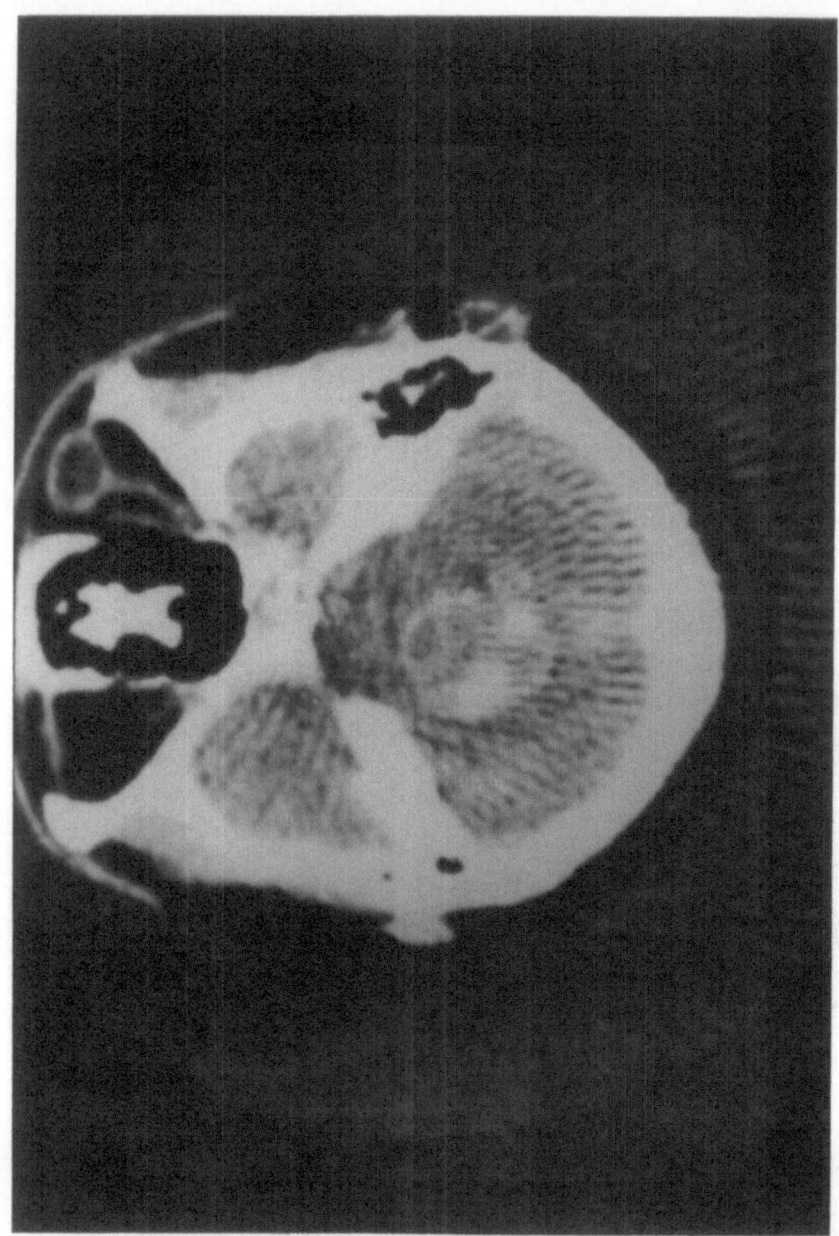

Figure 5. Enhanced computed tomographic scan shows a complex ring lesion in cerebellar area with localized zones of both increased and diminished density. At surgery, the lesion proved to be a partially thrombosed aneurysm. (From Kandalaft et al.[16])

the CSF was stored at 20°F (6.6°C), a temperature at which CSF complement, unlike serum complement, is unstable.[47] The results of indirect immunofluorescent studies using fluorescein-conjugated anti-IgM, anti-IgA, anti-IgG, anti-IgG'3, and antifibrinogen were all negative when incubated with cerebral cortex or choroid plexus.

Comment. In this young woman who developed symptoms of circulating immune complex disease and subsequently an encephalopathy, the presence of cryoprecipitable HB_sAg-antibody complexes was demonstrated in the serum at the time of her maximal illness. No mass was demonstrated by arteriography, and resolving cerebral edema was the only abnormal postmortem finding in the brain. A search for a metabolic or infectious cause of the encephalopathy revealed only mild hepatic disease. The findings suggest that her encephalopathy and cerebral edema were due to an underlying immunologic disorder. The findings are similar to those seen in patients with systemic lupus erythematosus (SLE), who often develop a severe encephalopathy with minimal histologic changes.[8,15] The mechanism whereby the encephalopathy was produced is only speculative. It is very likely that the cryoprecipitable HB_sAg–anti-HB_sAg complexes present in her serum may have produced the encephalopathy in the manner seen in SLE.

5. ADOPTIVE IMMUNOTHERAPY AS A MEANS TO CIRCUMVENT THE BLOOD–BRAIN BARRIER AND THE IMMUNOLOGIC PRIVILEGE OF THE CNS

Autologous lymphoid cell infusions into the lumbar subarachnoid space in patients with glioma have been studied as a form of antitumor immunotherapy.[26,31] The rationale for this route of administration was to reduce the risk of creating a localized mass effect with lumbar rather than intracranial infusion and to avoid the risk of infection with a reservoir of the Ommaya type. In addition, animal studies had demonstrated that cells infused into the cisterna magna migrated throughout the intracranial subarachnoid space and then appeared to return to the systemic circulation via the superior sagittal sinus.[27] Cells infused intrathecally could conceivably enhance systemic antitumor immunity by first being exposed to tumor that may be in contact with the subarachnoid space and then by returning to the systemic circulation, enhancing sensitization of the immune system. Thus, on serial subsequent infusions, the lymphocytotoxicity of the infused cells may be further increased. Phase I (i.e., toxicity) studies of such autologous lymphoid cell infusions demonstrated neither infections nor other complications.[31]

Case Report

K.R., a 28-year-old right-handed woman, had a left prefrontal lobectomy for an anaplastic astrocytoma.[32] Following her operation, a low-density cavity with minimal surrounding enhancement and edema was seen in the left frontal region on CT examination. A metrizamide cisternogram showed good filling of the entire subarachnoid space but no entrance of the metrizamide into the tumor cavity (Fig. 6A). Because previous studies have emphasized the critical requirement for communication between the tumor cavity and the subarachnoid space for effective lymphocyte administration, further resection of the tumor was performed and a defect in the bone flap was developed to facilitate percutaneous lymphocyte infusions into the tumor cavity. Whole head radiation was begun; 1 week later, her first intratumor lymphocyte infusion was administered. Intratumor metrizamide infused prior to the lymphocytes showed a large cavity in the frontal area with no communication with either the ventricles or the subarachnoid space (Fig. 6B). The lumbar CSF WBC count did not rise after intracranial lymphoid cell infusion. The patient completed her radiation therapy and subsequently had 12 courses

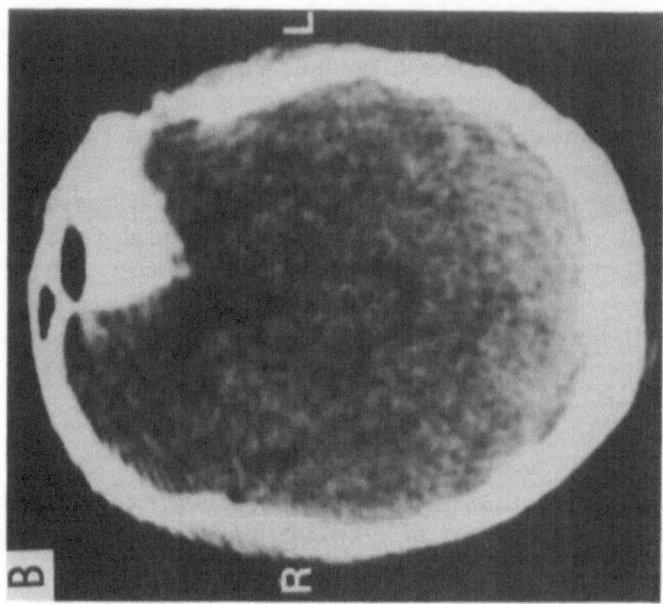

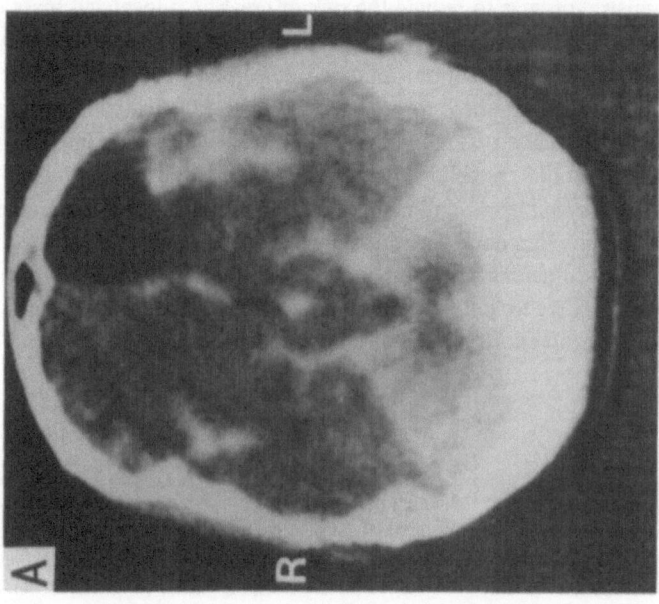

Figure 6. Case report 4. (A) CT scan 1 hr after intrathecal administration of metrizamide. The entire intracranial subarachnoid space is filled with metrizamide, in contrast to the tumor cavity in the left frontal region. The contrast agent is unable to diffuse beyond the pial margins of that tumor cavity. (B) CT scan after percutaneous intracranial infusion of metrizamide into the tumor cavity. The contrast agent is unable to escape from the tumor cavity into either the subarachnoid space or the ventricular system. (From Neuwelt et al.[32])

of lymphocyte infusion. Metrizamide insertion before lymphocyte infusion was repeated with no evidence of communication with the ventricles or subarachnoid space, and pre- and postlymphocyte infusion lumbar punctures revealed no increase in WBCs in the CSF, as might have been expected had a communication existed. The patient did well for 6 years until recurrent tumor was identified.

Comment. Neuwelt and Hill[29] extensively studied the kinetics of lymphocytes introduced into the subarachnoid space in animals and in phase I clinical trials as exemplified in the above case.[32] Although the problem of the BBB was circumvented by adoptive transfer of lymphocytes directly into the CSF (or tumor cavity), clinical efficacy was not identified. This failure may have related to inadequate activation of the lymphocytes. In in vitro studies, mitogen-induced lymphocyte activation was inadequate when concentrated CSF replaced fetal calf serum (FCS) even with CSF protein concentrations equivalent to those of serum (i.e., 6 g%). Recent studies with interleukin-2 (IL-2) have provided an alternative means of lymphokine-activated killer (LAK)-cell activation. Such an approach may provide adequate activation with the potential for continued maintenance of activity in LAK cells by subsequent administration of IL-2 by intrathecal or intratumor infusion.[14]

6. THE DRAMATIC EFFECT OF CORTICOSTEROIDS ON VASOGENIC CEREBRAL EDEMA

Adrenal corticosteroids have often produced marked decreases in mass effect and degree of enhancement, as seen on CT scan, in a variety of clinical circumstances. That this is not an artifact of the imaging system is proved by finding similar effects with radionuclide brain scanning,[24] angiography,[58,61] and echoencephalography.[19]

Case Report

J.M. is a 40-year-old right-handed man[32] who presented with a left temporal mass on CT scan (Fig. 7A). Dexamethasone was begun (96 mg/day), and 1 week later a repeat scan showed a marked decrease in ventricular shift (Fig. 7B). The patient underwent a craniotomy for subtotal excision of a left temporal glioblastoma multiforme.

Comment. In normal patients, the intravenous administration of iodinated contrast agents does not provide enhancement on CT scan evaluation; that is, these agents do not cross the BBB under normal circumstances. However, when there is a lesion in the brain, whether it be due to a tumor, a cerebrovascular accident (CVA), infection, or sometimes even demyelinating disease, the BBB may be damaged, thereby permitting these iodinated contrast agents to cross the BBB. Such injury allows not only contrast agents but other intravascular substances to cross the barrier. The result of such entry is edema, termed vasogenic cerebral edema.

Corticosteroids have long been known to have a dramatic effect on the vasogenic cerebral edema due to tumor. Inexplicably, steroids affect edema associated with some types of lesions (tumors) considerably more than that associated with other kinds of lesions (trauma). Furthermore, although steroids can dramatically decrease the enhancement due to iodinated contrast agents, a similar effect is not seen with drug delivery. Thus, steroids have only a modest effect in decreasing water-soluble antibiotic delivery across the BBB in brain abscesses[41] and similarly a very slight effect on delivery of chemotherapeutic agents to malignant intracerebral brain tumors.[39]

The reasons for such paradoxical effects are unknown. The mechanisms by which

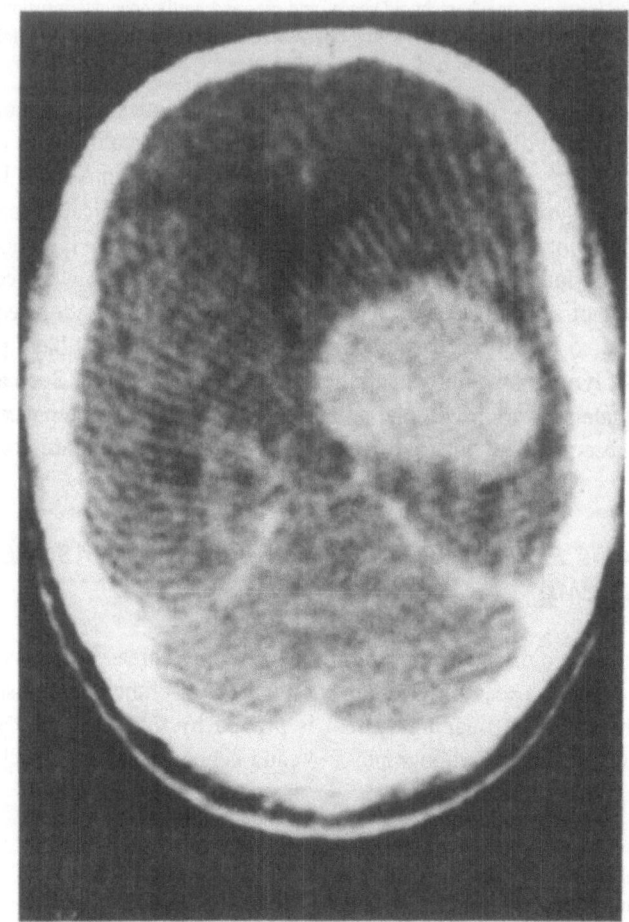

Figure 7. (A) Enhanced computed tomographic (CT) scan showing a large enhancing lesion in the left middle fossa with marked ventricular shift. (B) Enhanced CT scan 1 week after high-dose dexamethasone therapy (96 mg/day). Compared with (A), the tumor is significantly smaller, there is significantly less enhancement, and there is significantly less ventricular shift. The tumor is a glioblastoma. (From Neuwelt et al.[32])

steroids affect cerebral edema are unclear, and they may have other related effects beyond those on the BBB. For instance, it appears that steroids may have an effect on intravascular volume and/or the size of the extracellular space in CNS lesions.

7. THE ROLE OF BLOOD–BRAIN BARRIER MODIFICATION IN THE TREATMENT OF PRIMARY BRAIN TUMORS

The inability of chemotherapy to produce significant prolongation of life in the treatment of CNS malignancies has led to a consideration of the possible factors responsible for this limited success.[13,55] One important factor is the potential intrinsic resistance of glioma cells to currently available anticancer drugs. Such a concept of resistance is

B

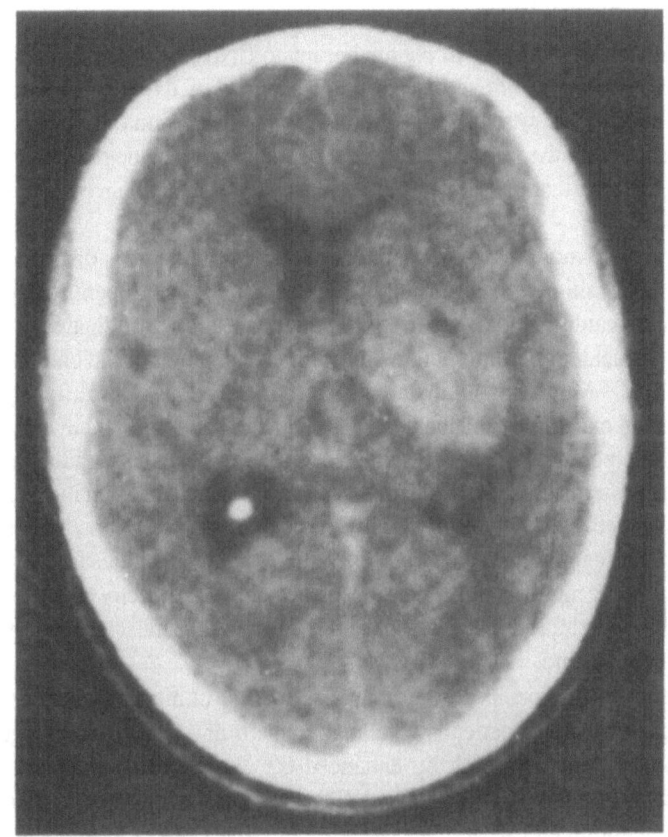

Figure 7. (Continued)

weakened by the observation that with a variety of systemic cancers (i.e., lymphoma, breast cancer, small cell undifferentiated cancer of lung, and testicular germ cell tumors), chemotherapy causes clinical regression or even complete remission of the tumor at non-neural sites, yet does not in the CNS.[2,42,49]

This dichotomy of response inside and outside the BBB suggests that delivery of drug to the tumor is critical. Neoplasms in the brain are subject to the classic pharmacologic issues of drug delivery (e.g., blood flow, drug concentration, time of exposure); drug entry is also affected by the BBB.[10,11,28,50,51] Although an older view expressed the concept that malignant tumors in the CNS lack an effective barrier,[59] this has now been shown to be incorrect. It is clear that an intact barrier, albeit variable in degree, is present within malignant tumors in the CNS.[10,11,42] In addition, there is heterogeneity of barrier integrity among individual neoplastic nodules in a given patient. Even the well-vascularized actively proliferating edge of the tumor, the brain adjacent to tumor (BAT), is known to have a variable and complex barrier integrity.[22]

Successful attempts to modulate or open the BBB have permitted examination of this variability relative to the response to chemotherapy by tumors in the CNS. Transient reversible BBB modification by intraarterial infusion of hypertonic mannitol has been documented and evaluated in a variety of models.[39,43]

Osmotic BBB opening has now developed to a level at which this technique can be used to increase the delivery to the CNS of a variety of drugs, proteins, and perhaps even DNA (genetic material). Delivery of chemotherapy to CNS malignancies and surrounding tumor-infiltrated brain is facilitated by osmotic opening of the tight junctions between capillary endothelial cells, which is accomplished by the infusion of hypertonic mannitol into the artery supplying the tumor-bearing part of the brain. The degree, extent, and time course of barrier modification can be monitored by contrast-enhanced CT and/or radionuclide scanning.

For instance, CT monitoring has demonstrated that the degree, distribution, extent, and temporal relationships of the reversibility of osmotic barrier opening can be quantitated.[30,33,38] In the canine model, the timing of administration of iodinated contrast agent was shown to be crucial in order to obtain optimum visualization of the brain in the area of the disrupted BBB.[36] Meglumine iothalamate given intravenously just after osmotic BBB opening resulted in excellent enhancement on CT scan. The importance and relevance of this CT monitoring potential are emphasized by canine studies, showing that methotrexate given systemically after osmotic BBB disruption results in increased brain methotrexate levels in areas that matched the regions in which CT scan enhancement has been seen.[33,35,36] Radionuclide brain scanning has also been examined in animal studies to evaluate its ability to document BBB modification; these scans proved unsatisfactory because of the increased absorption of the radionuclide substance in the large muscle mass overlying the cranial vault in the canine.

After careful evaluation in various animal models, clinical evaluation of osmotic BBB disruption in terminal patients with malignant brain tumors was begun. Barrier modification was documented both by enhanced CT and by radionuclide scanning, although the greater sensitivity of enhanced CT compared with nuclear brain scanning resulted in the predominant use of enhanced CT.[34,38] These studies provided documentation that osmotic BBB modification resulted in increased enhancement in the tumor as well as in the surrounding brain (i.e., BAT). This increased enhancement effect has provided important new clinical data, as it identified tumor nodules seen after barrier modification that had not been evident on the routine enhanced CT scan.[34] With region-of-interest analysis, the measurement of change in CT number permitted documentation and quantitation of the changes in enhancement in the tumor and surrounding brain after barrier modification.

Case Report: Primary CNS Lymphoma

A 37-year-old man presented with hemianopsia and multifocal mass lesions. At craniotomy these lesions were shown to be primary CNS lymphomas. He was treated with multiagent chemotherapeutic agents delivered with osmotic barrier modification.[40] After five such treatments, remarkable tumor regression was documented. Serial follow-up evaluation of the areas exposed to the therapy was available, because the barrier opening achieved was clearly delineated by the enhanced CT scans (Fig. 8). Iodinated contrast material was used for these enhanced CT studies. The patient experienced a focal motor seizure after his third therapy procedure and a grand mal seizure after his fourth BBB modification. Because of the known epileptogenic potential of contrast media, for the fifth barrier modification a radionuclide brain scan (Fig. 9) was used instead of contrast-enhanced CT scan; no seizures occurred.

Comment. Of special interest is the fact that the use of these contrast agents to help document the degree to which barrier modification is achieved with the osmotic procedure is associated with an increased seizure frequency. As one would expect, radionuclide imaging can be done following BBB opening with a significantly lower risk of seizures. Although this approach does permit evaluation of the status of barrier modification (Fig.

8), it is neither as sensitive nor as informative as the data obtained by CT scan. For example, after three courses of chemotherapy, this patient's tumor in the right frontal lobe could be seen by enhanced CT imaging but not by radionuclide scans. The effectiveness of chemotherapy delivery enhanced by BBB modification is emphasized by the fact that this patient is functionally normal with no evidence of residual tumor 5 years following completion of therapy—and he has never been irradiated.

8. COMPETITION AT THE BLOOD–BRAIN BARRIER BY DIETARY AMINO ACIDS CAUSING FLUCTUATIONS IN THE MOTOR PERFORMANCE OF PATIENTS WITH PARKINSON DISEASE WHO ARE RECEIVING LEVODOPA THERAPY

As reviewed by Pollay,[49] in experimental animals and to a lesser extent in clinical studies, eight separate transport systems have been identified and characterized. These carrier systems include: (1) a hexose transport system that regulates the supply of D-glu-

A

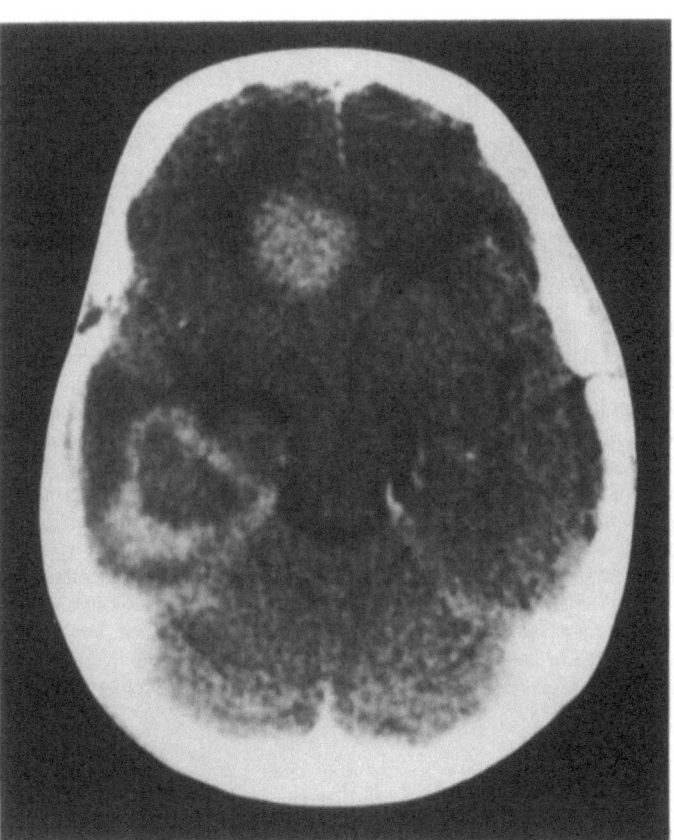

Figure 8. Serial enhanced computed tomographic (CT) scans in a patient with CNS lymphoma. (A) CT scan after the patient's second craniotomy while on dexamethasone PO (decadron), 24 mg/day, just before entering the experimental protocol. Three lesions (arrowheads) are present on this transverse cut CT scan at the level of the lateral ventricles: a right frontal, uniformly enhancing mass, a ring lesion in the right posterior temporal

B

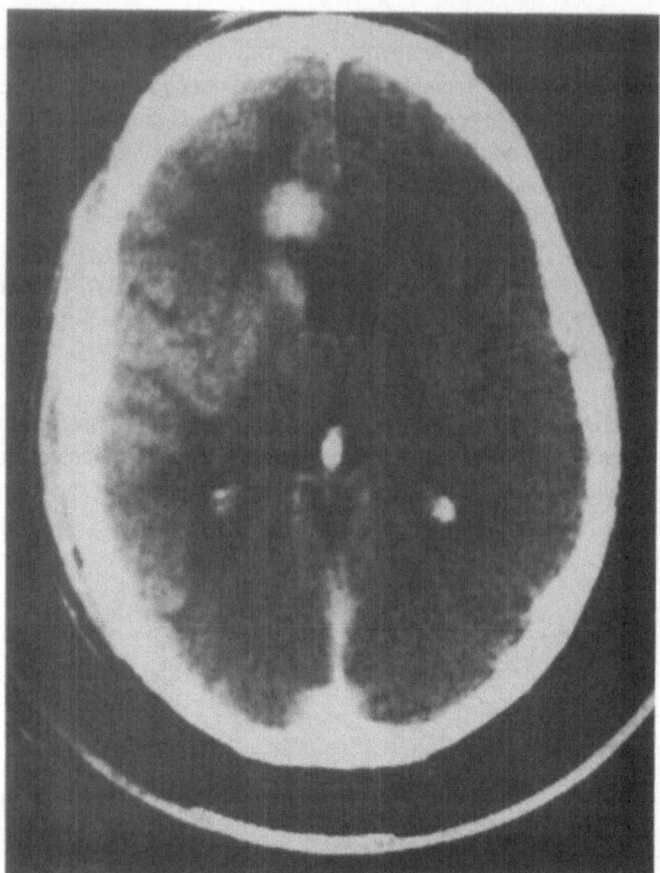

Figure 8. (*Continued*) region at the site of the second craniotomy, and a low-density lesion in the left frontal region, which was the site of the initial tumor excision. (The white square in the left cerebral hemisphere is a CT marker and should be disregarded.) (B) Transverse cut enhanced CT scan at the level of the lateral ventricles 30 min after the third BBB modification. The white areas depict the areas of iodinated contrast agent penetrating the BBB in the distribution of the anterior and middle cerebral arteries. Barrier modification also occurred in the anterior cerebral distribution on the left side (arrowheads) due to collateral flow via the anterior communicating artery. There is barrier modification in and around all three known tumor regions. The CT number in the right frontal tumor increased from 54 on a control enhanced CT scan 24 hr before BBB opening to 75 after BBB opening. The time between IV contrast administration and scanning and the contrast dose were the same for both scans. Both the illustrated and the control scan (not shown) were obtained 5 days after the discontinuation of steroids. (C) CT scan at the level of the lateral ventricles just before the eighth course of therapy showing disappearance of both the right posterior temporal lesion and the right frontal lesion. The low-density area that does not enhance at the site of the previous operation in the left frontal region persists and is probably a postoperative cavity. Thus, there is no evidence of residual tumor. (From Neuwelt et al.[40])

cose; (2) a monocarboxylic transport system that is activated during starvation or other conditions leading to increased blood ketones in order to provide other nutrients (acetoacetic acid and β-hydroxybutyric acid) in place of glucose and thereby maintain cerebral metabolism; (3) the choline transport system, which has not been completely characterized but seems to be important in brain neurotransmitter metabolism; (4) the adenine transport system, which also regulates the entry of neurotransmitters into the brain, since it transports purine bases such as guanine and adenine; (5) the purine nucleosides and

C

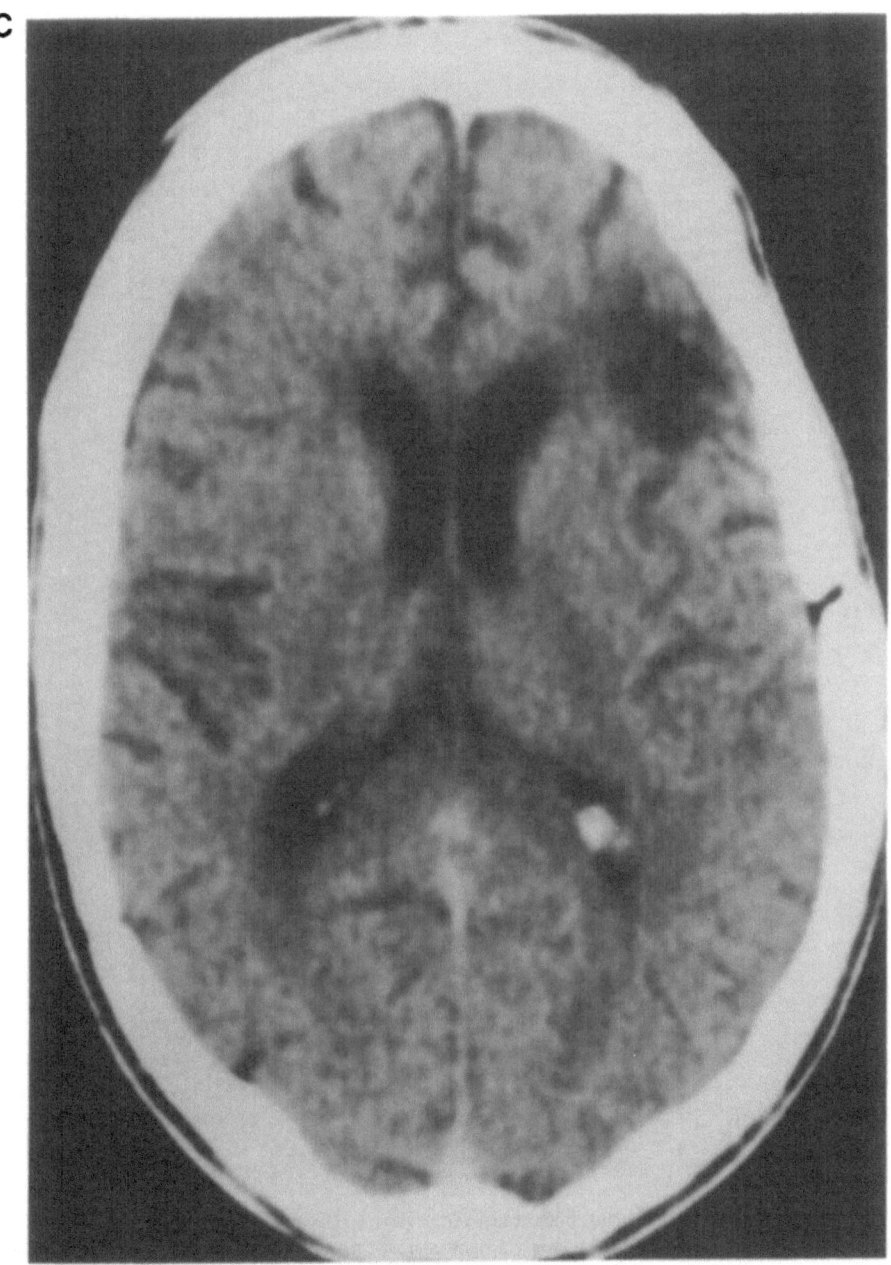

Figure 8. (*Continued*)

uridine that enter the brain by the nucleoside transport system; (6) the amino acids arginine, lysine, and orthinine that are transported by the basic amino acid transport system, an important system in normal protein metabolism; (7) the acidic amino acid transport system that, unlike the above carrier transport systems that operate in both directions across the BBB, actively pumps excitatory neurotransmitters out of the extra-cellular space of the brain into the blood; and (8) the neutral amino acid transport system

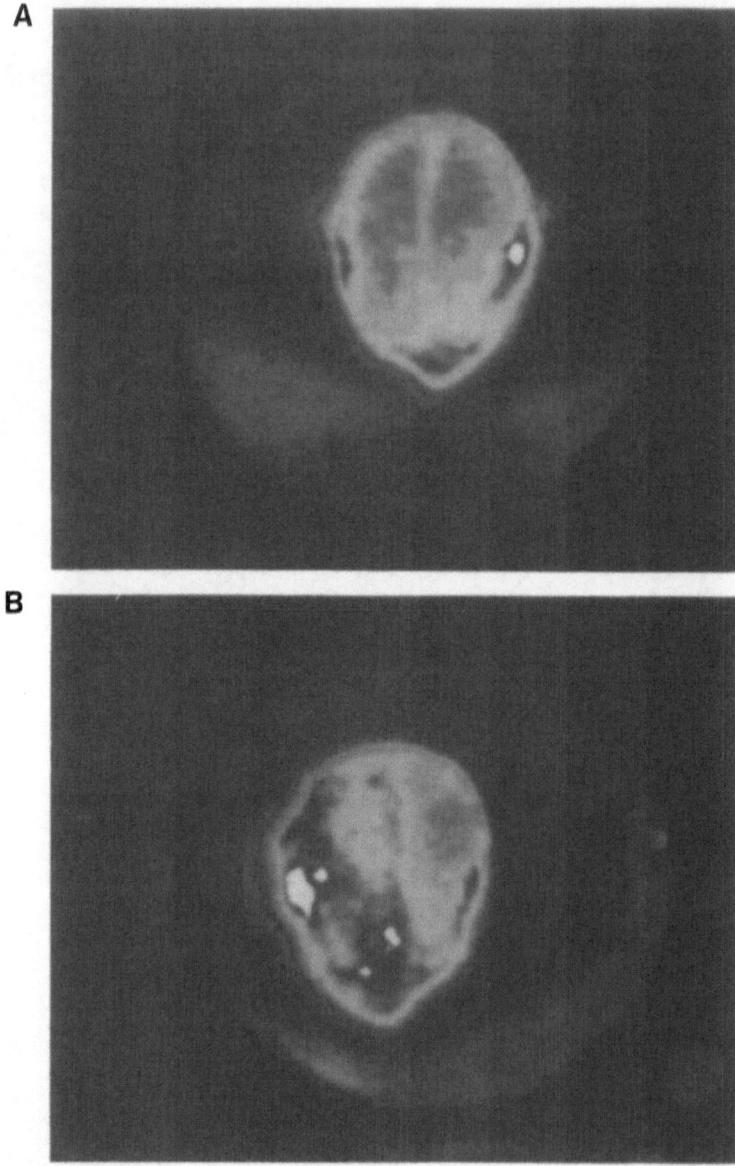

Figure 9. CNS lymphoma. Radionuclide brain scans, vertex view. (A) Control scan 5 months after initiation of protocol. (B) 24 hr later, after BBB disruption via right internal carotid artery. (The patient's right is on the reader's left.) From Neuwelt et al.[40] A color reproduction of this figure appears following p. xxviii.

that moves 14 neutral amino acids from blood to brain. Separate and distinct carrier mechanisms also exist for the basic and acidic amino acids. These are important in maintaining protein and neurotransmitter synthesis within the CNS. For example, in hepatic encephalopathy, this carrier system becomes deranged, with an abnormal accumulation of tyrosine, phenylalanine, and trytophan in the brain that, in turn, leads to an imbalance of neurotransmitters. Based upon this knowledge, therapeutic strategies involving dietary restriction of these amino acids can be instituted so that normal neurologic function may be restored. As in the case below, it is also upon this transport system that L-dopa therapy in parkinsonism is based.[4]

Case Report

W.F., a 56-year-old retired logger, developed parkinsonism at age 40.[46] His disease was initially treated with anticholinergics and amantadine, but his disability progressed sufficiently so that by age 44 he was begun on carbidopa–levodopa (Sinemet). His initial response was gratifying, but over the ensuing 8 years he began to develop marked fluctuations in response and had painful cramps in his legs when the drug effects were at their nadir (so-called off-dystonia). By age 54, his life was dominated by swings between almost normal mobility when he was able to perform such chores as chopping wood, and painful dystonic cramps, that left him unable to walk or perform the simple activities of daily living. Of note, he and his wife had found that the medication was less effective after a high-protein meal and that the reduced medication response lasted 4 to 6 hr following meals.

Because of this history of a marked effect by dietary protein, the patient was given L-dopa by constant intravenous infusion to determine whether dietary protein interfered with his clinical response by decreasing absorption of levodopa. Infusion of L-dopa at a constant rate produced a constant mobile state, but when the patient was fed a meal high in protein, the clinical effect of the infused L-dopa was diminished. A high-protein meal on the second day of the infusion had the same effect.

Comment. The results of this study suggested that transport at the BBB may be an important step in the clinical response to L-dopa (Fig. 10). This observation initiated studies demonstrating that dietary amino acids (from the large neutral amino acid group) could inhibit the effect of infused L-dopa without altering the plasma L-dopa concentra-

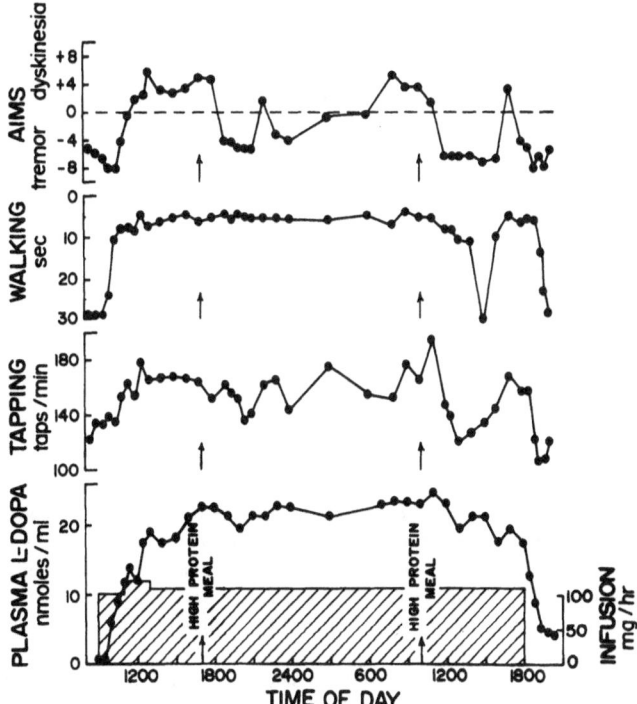

Figure 10. Effects of high-protein meals on the plasma L-dopa concentration and the clinical response during intravenous infusion of L-dopa. The patient was also receiving carbidopa, 25 mg PO q2h. Algebraic sum of involuntary-movement scores (AIMS): positive values indicate dyskinesia, and negative values indicate tremor. (From Nutt et al.[46])

tion.[46] Dietary amino acids from the acidic, basic, and small neutral amino acid groups did not antagonize the effect of L-dopa. These observations are consistent with animal studies demonstrating that L-dopa enters the brain by the saturable large neutral amino acid transport system.[60] The entry of the [18F]fluorodopa into brain, as detected by PET, can also be antagonized by raising plasma large neutral amino acid concentrations.[20]

9. CONCLUSION

These studies emphasize the clinical significance of the blood–brain barrier. The BBB appears to be an important factor in the immunologically privileged state of the CNS.[26] The barrier appears to be significant in affecting circumstances of encephalopathy from a wide variety of metabolic derangements. Similarly, it is now clear that effective chemotherapy of malignant neoplasms in the CNS is impeded by the existence of the BBB, although to different degrees in different areas of even the same tumor.[39,43–45] The ability to modulate (open) the barrier transiently in animal models and in humans provides a new powerful tool with which to examine pathophysiologic and neuropharmacologic events heretofore restricted from study. The anecdotal cases presented exemplify the clinical significance and challenge of the BBB.

REFERENCES

1. Adams RD, Foley JM: The neurological disorder associated with liver disease. *Proc Assoc Res Nerv Ment Dis* 32:198–237, 1952.
2. Benjamin RS, Wiernik PH, Bachur NR: Adriamycin in chemotherapy efficacy, safety, and pharmacologic basis of intermittent single high dosage schedule. *Cancer* 33:19–27, 1974.
3. Bradbury MWB: *The Concept of a Blood–Brain Barrier.* Wiley, New York, 1979.
4. Cotzias GC, Van Woert MH, Schiffer LM, et al: Modification of Parkinsonism. *N Engl J Med* 276:375–379, 1967.
5. Davson H: *The Cerebrospinal Fluid.* Ciba Foundation Symposium, Churchill, London, 1958.
6. Denny-Brown D: The cerebral control of movement. In *No. 8: Cerebral Factors in Movement, Liverpool University Sherrington Lectures, Liverpool, England,* Liverpool University Press, Liverpool, 1966, pp. 110–123.
7. Feigin I: Mesenchymal tissues of the nervous system: The indigenous origin of brain macrophages in hypoxic states and in multiple sclerosis. *J Neuropathol Exp Neurol* 28:6–24, 1969.
8. Glasner H: Barrier impairment and immune reaction in the cerebrospinal fluid. *Eur Neurol* 13:304–314, 1975.
9. Goldstein S, Gumerlock MK, Neuwelt EA: Diagnostic CT-guided needle biopsy: Replaced by stereotactic biopsy? *J Neurosurg,* 1986 (in press).
10. Groothius DR, Fischer JM, Lapin G, et al: Permeability of different experimental brain tumor models to horseradish peroxidase. *J Neuropathol Exp Neurol* 41:164–185, 1982.
11. Groothius DR, Molnar P, Blasberg RG: Regional blood flow and blood.to. tissue transport in five brain tumor models: Implications for chemotherapy. *Prog Exp Tumor Res* 27:132–153, 1984.
12. Hain RF: In discussion of Konigsmark BW and Sidman RL: Origin of gitter cells in the mouse brain. *J Neuropathol Exp Neurol* 22:327–328, 1963.
13. Hildebrand J: Current status of chemotherapy of brain tumors. In *Prog Exp Tumor Res* 29:152–166, 1985.
14. Jacobs SK, Wilson DJ, Kornblith PL, et al: Interleukin-2 or autologous lymphokine-activated killer cell treatment of malignant glioma: Phase I trial. *Cancer Res* 46:2101–2104, 1986.
15. Johnson R, Richardson E: The neurological manifestations of systemic lupus erythematous. *Medicine (Baltimore)* 47:337–369, 1968.
16. Kandalaft N, Diehl J, Neuwelt EA: Nonneoplastic intracranial lesions simulating neoplasms on computed tomographic scan. Excellent sensitivity with limited specificity. *JAMA* 248:2166–2168, 1982.

17. Kendall BE, Jakubowski J, Pullicion P, et al: Difficulties in diagnosis of supratentorial gliomas by CAT scan. *J Neurol Neurosurg Psychiatry* 42:485–492, 1979.
18. Konigsmark BW, Sidman RL: Origin of brain macrophages in the mouse. *J Neuropathol Exp Neurol* 22:643–676, 1963.
19. Kullberg G, West K: Influence of corticosteroids on the ventricular fluid pressure. *Acta Neurol Scand* 13:445–452, 1965.
20. Leenders KL, Poewe WH, Palmer AJ, et al: Inhibition of [^{18}F]fluorodopa uptake into human brain by amino acids demonstrated by positron emission tomography. *Ann Neurol* 20:258–262, 1986.
21. Leers WD, Kouroupis GM: Comparison of the reverse passive hemagglutination with radioimmune assay methods for hepatitis B antigen. *J Clin Microbiol* 2:8–10, 1975.
22. Levin VA, Freeman-Dove M, Landahl HD: Permeability characteristics of brain adjacent to tumors in rats. *Arch Neurol* 32:785–791, 1975.
23. Ling CM, Overby LR: Prevalence of hepatitis B virus antigen as revealed by direct radioimmune assay with I-antibody. *J Immunol* 109:834–841, 1972.
24. Marty R, Cain ML: Effects of corticosteroid (dexamethasone) administration on the brain scan. *Radiology* 107:117–121, 1973.
25. Neuwelt EA, Clark WK: Unique aspects of central nervous system immunology. *Neurosurgery* 3:419–430, 1978.
26. Neuwelt EA, Clark WK: *Clinical Aspects of Neuroimmunology*. Williams & Wilkins, Baltimore, 1978.
27. Neuwelt E, Doherty D: Toxicity kinetics and clinical potential of subarachnoid lymphocyte infusions. *J Neurosurg* 47:205–217, 1977.
28. Neuwelt EA, Frenkel EP: Is there a therapeutic role for blood–brain barrier disruption? *Ann Intern Med* 93:137–139, 1980.
29. Neuwelt EA, Hill S: Intrathecal lymphocyte infusions: Clinical and animal studies. In Wood JH (ed.): *Neurobiology of Cerebrospinal Fluid.* Plenum, New York, 1980, pp. 525–548.
30. Neuwelt EA, Rapoport SI: Modification of blood–brain barrier in the chemotherapy of malignant brain tumors. *Fed Proc* 43:214–219, 1984.
31. Neuwelt EA, Clark K, Kirkpatrick JB, et al: Clinical studies of intrathecal autologous lymphocyte infusions in patients with malignant glioma: A toxicity study. *Ann Neurol* 4:307–314, 1978.
32. Neuwelt EA, Diehl JT, Hill SA, et al: Use of metrizamide computerized tomographic cisternography in the evaluation of patients with malignant glioma for immunotherapy. *Neurosurgery* 5:576–582, 1979.
33. Neuwelt EA, Maravilla KR, Frenkel EP, et al: Osmotic blood–brain barrier disruption: Computerized tomographic monitoring of chemotherapeutic agent delivery. *J Clin Invest* 64:684–688, 1979.
34. Neuwelt EA, Frenkel EP, Diehl JT, et al: Reversible osmotic blood–brain barrier disruption in humans: Implications for the chemotherapy malignant brain tumors. *Neurosurgery* 7:44–52, 1980.
35. Neuwelt EA, Frenkel EP, Rapoport SI, et al: Effect of osmotic blood–brain barrier disruption on methotrexate pharmacokinetics in the dog. *Neurosurgery* 7:36–43, 1980.
36. Neuwelt EA, Maravilla RK, Frenkel EP, et al: Use of enhanced computerized tomography to evaluate osmotic blood–brain barrier disruption. *Neurosurgery* 6:49–56, 1980.
37. Neuwelt EA, Diehl JT, Frenkel EP, et al: Osmotic blood–brain barrier disruption in posterior fossa of the dog. *J Neurosurg* 55:742–748, 1981.
38. Neuwelt EA, Frenkel EP, Diehl JT, et al: Monitoring of methotrexate delivery in patients with malignant brain tumors after osmotic blood–brain barrier disruption. *Ann Intern Med* 94:449–454, 1981.
39. Neuwelt EA, Barnett PA, Bigner DD, et al: Effects of adrenal cortical steroids and osmotic blood–brain barrier opening on methotrexate delivery to gliomas in the rodent: The factor of the blood–brain barrier. *Proc Natl Acad Sci USA* 79:4420–4423, 1982.
40. Neuwelt EA, Specht HD, Howieson J, et al: Osmotic blood–brain barrier modification: Clinical documentation by enhanced CT scanning and/or radionuclide brain scanning. *Am J Neuroradiol* 4:907–913, 1983; and *Am J Radiol* 141:829–836, 1983.
41. Neuwelt EA, Baker DE, Pagel MA, et al: The cerebrovascular permeability and delivery of gentamicin to normal brain and experimental brain abscess in rodents. *J Neurosurg* 61:430–439, 1984.
42. Neuwelt EA, Hill SA, Frenkel EP: Osmotic blood–brain barrier modification in areas of barrier opening and progression in brain regions distant to barrier opening. *Neurosurgery* 15:362–366, 1984.
43. Neuwelt EA, Frenkel E, D'Agostino AN, et al: Growth of human lung tumor in the brain of the nude rat as a model to evaluate antitumor agent delivery across the blood–brain barrier. *Cancer Res* 45:2827–2833, 1985.
44. Neuwelt EA, Frenkel EP, Gumerlock MK, et al: Developments in the diagnosis and treatment of primary CNS lymphoma: A prospective series. *Cancer* 58:1609–1620, 1986.

45. Neuwelt EA, Howieson J, Frenkel EP, et al: Therapeutic efficacy of multiagent chemotherapy with drug delivery enhancement by blood–brain barrier modification in glioblastoma. *Neurosurgery* 19:573–582, 1986.
46. Nutt JG, Woodward WR, Hammerstad JP, et al: The "on–off" phenomenon in Parkinson's disease. Relation to levodopa absorption and transport. *N Engl J Med* 310:483–488, 1984.
47. Petz L, Sharp G, Cooper N, et al: Serum and cerebral spinal fluid complement and serum autoantibodies in systemic lupus erythematosus. *Medicine (Baltimore)* 50:259–275, 1971.
48. Pinto RS, Kricheff II, Butler AR, et al: Correlation of computed tomographic, angiographic, and neuropathological changes in giant cerebral aneurysms. *Radiology* 132:85–92, 1979.
49. Pollay M: Blood–brain barrier: Review of clinical aspects. *Contemp Neurosurg* 9:1–6, 1987.
50. Rapoport SI: *Blood–Brain Barrier in Physiology and Medicine.* Raven, New York, 1976.
51. Rapoport SI, Fredericks WR, Ohno K, et al: Quantitative aspects of reversible osmotic opening of the blood–brain barrier. *Am J Physiol* 238:R421–R431, 1980.
52. Rio-Hortega P del: El tercer elemento de los centros nerviosos. *Bol Soc Esp Biol* 9:69–120, 1919.
53. Rio-Hortega P del: Microglia. In Penfield W (ed.): *Cytology and Cellular Pathology of the Nervous System.* Vol. 2. Hoeber, New York. 1932, pp. 483–534.
54. Rosenberg RN, Neuwelt EA, Kirkpatrick J, et al: Encephalopathy associated with cryoprecipitable Australia antigen. *Ann Neurol* 1:298–300, 1977.
55. Shapiro WR: Therapy of adult malignant brain tumors: What have the clinical trials taught us? *Semin Oncol* 13:38–45, 1986.
56. Shorey J, Bombes B: Radioimmunoassay studies of Ab to Australia antigen. *Gastroenterology* 62(abstr.):881, 1972.
57. Stokes JF, Owen JR, Holmes EG: Neurological complication of infective hepatitis. *Br Med J* 2:642–644, 1945.
58. Verdura J, Brown H, White RJ: Use of adrenal steroids in cerebral metastasis: Case report of improvement documented by angiography. *Ohio State Med J* 50:693–694, 1963.
59. Vick NA, Khandeka VD, Bigner DD: Chemotherapy of brain tumors: The "blood–brain barrier" is not a factor. *Arch Neurol* 34:523–526, 1977.
60. Wade LA, Katzman R: Synthetic amino acids and the nature of L-DOPA transport at the blood–brain barrier. *J Neurochem* 25:837–842, 1975.
61. Weinstein JD, Toy FJ, Jaffe ME, et al: The effect of dexamethasone on brain edema in patients with metastatic brain tumors. *Neurology (NY)* 23:121–129, 1973.
62. Zimmerman RD, Leeds NE, Naidich TP: Ring blush associated with intracerebral hematoma. *Radiology* 122:708–711, 1977.

2

History of the Blood–Brain Barrier Concept

Hugh Davson

1. STUDIES WITH DYESTUFFS

The first studies that led to the concept of the blood–brain barrier (BBB) were performed by Ehrlich at the beginning of this century. His studies on vital staining were carried out in the search for chemotherapeutic agents and eventually culminated in the discovery of the sulfonamides. Ehrlich[59] observed that many intravenously injected dyes stained the tissues of practically the whole body but not the brain. Later, Lewandowsky[77] showed that the Prussian blue reagents did not pass from blood to brain, and he formulated the concept of the BBB (Bluthirnschranke). Goldmann's experiments with trypan blue demonstrated very distinctly the existence of this BBB. After intravenous dye injection, the brain was unstained and the dye was not found in the cerebrospinal fluid (CSF), although the choroid plexuses and meninges were stained.[66]

In a second type of experiment, the importance of which was generally ignored for more than 50 years, Goldmann[67] injected trypan blue into the CSF and found that the whole of the brain became heavily stained with granular accumulations of the dye in all cell types. The demonstrated barrier between blood and nervous tissue seemed to be almost absolute as far as trypan blue (and other acidic dyestuffs) were concerned. Thus, the dye could not leave the vascular bed of the nervous tissue, although it could leave the blood vessels of the choroid plexuses, as indicated by staining of the connective tissue of these bodies. By contrast, the barrier could be circumvented by direct injection into CSF. Goldmann was actually responsible for the view, later taken up by L. Stern and V. Monakow, that the mode of entry into the brain was by way of the choroid plexuses *Die Weg über den Liquor* ("the way through the CSF"). Thus, according to Goldmann, the choroid plexuses could be viewed as analogous to the placenta; trypan blue failed to pass from the maternal circulation into the fetus of a pregnant dog.

Hugh Davson • Department of Physiology, King's College and St. Thomas' Hospital Medical School, University of London, London WC2R 2LS, England.

2. THE STERN-GAUTIER HYPOTHESIS

Stern and Gautier[100] made the first systematic study of the movement of a variety of substances from the blood into the CSF. They used solutes that could be estimated by chemical methods instead of relying on dyestuffs alone. A given substance was injected into the blood of the nephrectomized animal. After a time, CSF was withdrawn, and a qualitative test was applied to the fluid to determine whether the substance had penetrated. Among those that did penetrate the blood–CSF barrier were bromide, thiocyanate, strychnine, morphine, atropine, and bile salts, whereas iodide, ferrocyanide, salicylate, curare, epinephrine, bile pigments, eosin, and fluorescein were invariably absent from the fluid.

In a second experiment, Stern and Gautier[101] studied the effects of a variety of substances on the central nervous system (CNS) and were able to establish a correlation between penetration into the CSF and influence on the nervous system. For example, intravenous bromide had a depressant effect on rabbits and could be found in both CSF and nervous tissue; thiocyanate increased the excitability of cats and could also be found in both CSF and nervous tissue. Iodide and ferrocyanide, on the other hand, had no nervous influence after intravenous injection and were absent from the CSF and nervous tissue. Stern and Gautier[101] then injected these substances into the subarachnoid space and found that, in general, they all passed into the brain tissue, producing characteristic neurologic effects. Their general conclusions were as follows:

1. There is a barrier between blood and brain, in the sense that certain substances may be excluded from access. After intravenous injection, these substances are prevented from exerting their characteristic effects. To this barrier, Stern and Gautier gave the general name, *barrière hématoencéphalique*.
2. Substances that cannot reach the brain tissue from the blood fail to enter the CSF; also, substances that reach the brain tissue appear in the CSF.
3. Passage from CSF to brain tissue is possible for those substances that fail to pass the *barrière hématoencéphalique* (e.g., curare, iodide).

From these facts, the conclusion was drawn that passage from blood to brain tissue took place with the CSF as an intermediary, and the CSF was therefore regarded as the sole nutrient medium for the central nervous tissue. The theoretical objections to this conclusion, attributing no function to the large capillary bed in vertebrate nervous tissue, were adequately expressed by Walter,[107] and the hypothesis need not be considered further.

However, these hypotheses did help distinguish two separate barriers, which we now call the BBB and blood–CSF barrier, and emphasize their similarity and the importance of interchange between CSF and brain. Thus, if a BBB were necessary for the economy of the brain, a blood–CSF barrier was also a necessity because of the CSF–brain interchange.

3. SELECTIVITY OF THE BARRIER

The choice of the term *barrier* to describe the experimental exclusion of some dyestuffs from the brain tissue may have been unfortunate, suggesting as it does an absolute exclusion of solutes from access by way of the blood. Nevertheless, the early workers, especially Stern and Gautier, recognized a definite selectivity: morphine had access to the brain, whereas curare did not. Similarly, when dyestuffs were the sole solutes examined, it seemed that there was some ionic selectivity. Dyes that were anions in solution—the acidic dyes—were excluded, whereas basic dyes were apparently not excluded, at any rate according to Friedmann and Elkeles.[65]

An excellent study by Becker and Quadbeck[11] showed that the situation is complex. They studied the penetration of a number of basic and acidic dyes into the brain tissue after intravenous injection. The basic dye, triphenyltetrazolium chloride (TTC), apparently did not penetrate the BBB; however, within 1 and one-half min of the death of the animal into which TTC had been injected intravenously while alive, the brain became red, the gray matter more rapidly than the white. Most of the basic dyestuffs commonly used in biologic studies become colorless on reduction, being converted into the leuco base; TTC, however, behaves in the opposite manner, being converted into a red substance on reduction. The postmortem coloring of the brain could therefore be interpreted as a reduction of the TTC that had already crossed the BBB. Thus, this basic dye did conform to the rule of ionic selectivity.

Becker and Quadbeck[11] then proceeded to study several basic and acidic dyes and found one basic dye, Astra violet, that failed to cross the BBB. This dye was definitely not reduced in vivo to a colorless compound. Becker and Quadbeck had demonstrated the falsity of the generalization that basic dyes penetrate the barrier whereas acidic dyes do not. The question thus arose as to what characteristic of the dye determined its ability to cross the BBB, and Becker and Quadbeck showed that an important factor was lipid solubility. Acridine orange, Nile blue, neutral red, and toluidine blue, all lipid-soluble, crossed the barrier. Methylene blue, Janus green, and malachite green were apparent exceptions in that they could cross the BBB but were not lipid-soluble. If these dyes were previously treated with a reducing agent, however, they became lipid-soluble. Apparently, they penetrated the BBB after a preliminary in vivo conversion to the leuco base. In the nervous tissue, they accumulated in this form but also, by oxidation, in the colored form, in which condition they were recognized by the experimenter.

4. LIPID SOLUBILITY

Subsequent work carried out using quantitative methods, often with suitable mathematical analysis of rate, demonstrated the importance of lipid solubility in determining the ease with which the CSF and BBB are crossed. Anticipating more accurate quantitative studies, Krogh[75] had already drawn attention to the similarity between these barriers and the selective membrane that separates the individual cell from its environment; the characteristics of this latter cellular barrier were formulated at the same time in Davson and Danielli's monograph.[47]

Figure 1 illustrates the application of adequate quantitative experimental methods to the study of selective permeability of the blood–CSF barrier to some nonelectrolytes of increasing lipid solubility. The influence of lipid solubility on penetration is obvious; creatinine, with low lipid solubility, penetrates slowly and, interestingly, shows no sign of reaching equality of concentration with creatinine in the blood. In these experiments, the plasma concentration was established by intravenous infusion to maintain a steady level (equal to 100 in the graph) over a period of up to 5 hr.

Kinetic analysis showed that penetration could be described in terms of a two-

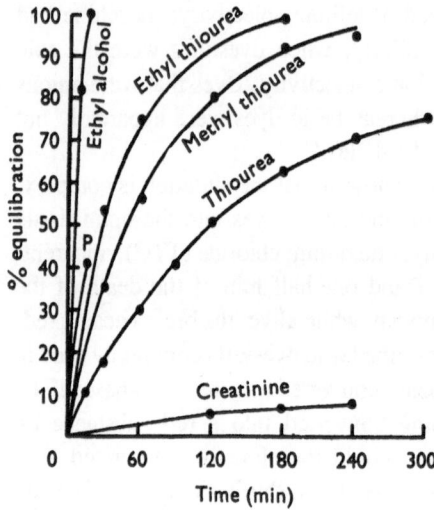

Figure 1. The blood–CSF barrier. A steady level of the solute (=100) was maintained in the blood plasma, and at intervals the animal was anesthetized and its CSF was removed for analysis (Davson[43]). The relative values of various substances in the CSF are shown.

compartment model following an equation of the form

$$dC_{CSF}/dt = K_{in}C_{pl} - K_{out}C_{CSF}$$

With substances that penetrate relatively rapidly, the two transfer coefficients, K_{in} and K_{out} could be equated, giving a value of unity for the ratio $C_{CSF} : C_{Pl}$ at infinite time, i.e., equality of CSF and plasma concentrations. Computed values of K_{out} are shown in Table I, which includes a value for ^{24}Na, the penetration of which into CSF is illustrated by Fig. 2. If creatinine penetrated into CSF in a low concentration compared with that which remained in plasma and left the CSF compartment by flow through nonselective channels, such as the arachnoid villi, we would have a situation in which K_{out} was greater than K_{in} to give a steady-state ratio of $C_{CSF} : C_{Pl}$ (equal to $K_{in} : K_{out}$).

Figure 3 illustrates a more extensive study conducted by Brodie et al.[32] in which the results were plotted semilogarithmically; values of K_{out} derived from these plots using comparable coefficients are shown in Table II. The same type of study can be used to examine the BBB. The animal is killed at the end of appropriate periods, during which a steady level of the solute has been maintained in the plasma. Figure 4 illustrates simulta-

Table I. Values of K_{out} Deduced from Experiments on the Penetration of Different Solutes from Blood to CSF[a]

Solute	K_{out} (min)
Thiourea	0.0057
Methyl thiourea	0.0135
Ethyl thiourea	0.021
Propyl thiourea	0.035
Ethyl alcohol	0.225
^{24}Na	0.004

[a]From Davson.[45a]

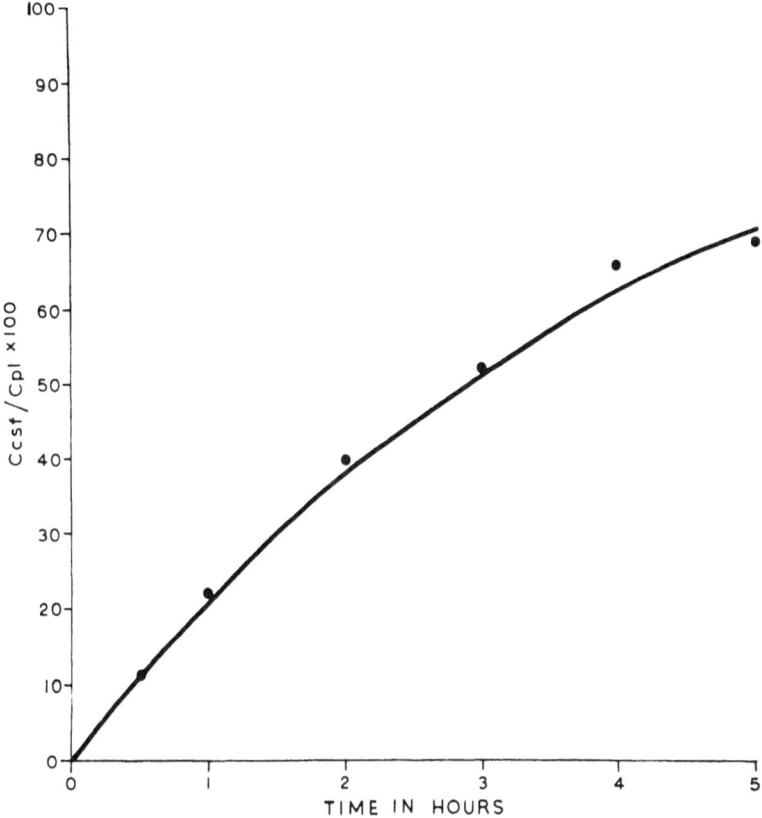

Figure 2. Penetration of [24]Na into rabbit CSF; the technique was similar to that for Fig. 1 (Davson[43]).

neous penetration of thiourea and ethyl thiourea into CSF and brain of the rabbit. Figure 5 shows penetration of [24]Na into brain and CSF; in this case, the curve represents penetration into CSF and the plotted points represent penetration into the chloride space of the brain because, with this solute, uptake into cellular water is likely to be negligible. Remarkably, the two separate compartments—CSF and the chloride space (or extracellular fluid)—come into equilibrium with plasma at the same rate, so the interchanges between these two compartments during the approach to equilibrium seem to be insignificant, justifying in this case the use of a two-compartment model for both the blood–CSF and the BBB. With the more rapidly penetrating lipid-soluble solutes, the two-compartment model applies for each system, although there is no doubt that interchanges between CSF and brain water must take place during the approach to equilibrium.

As far as these lipid-soluble solutes are concerned, Fig. 4 shows that net movement is from brain to CSF because the brain equilibrates with plasma faster, giving a gradient in favor of the CSF; thus, K_{in} for CSF is composed of a component of transport in the primary CSF secretion and another of transport from brain tissue to CSF. With solutes that are preferentially accumulated by brain cells, such as K^+, the gradient will be in the opposite direction, giving the curve of Fig. 6, which shows that accumulation in the CSF tends to slow at later stages because of uptake by the cells.[22,43] In this case, a simple two-compartmental analysis is not feasible and the more complex treatment of Welch[109] and Davson and Welch[49] is necessary.

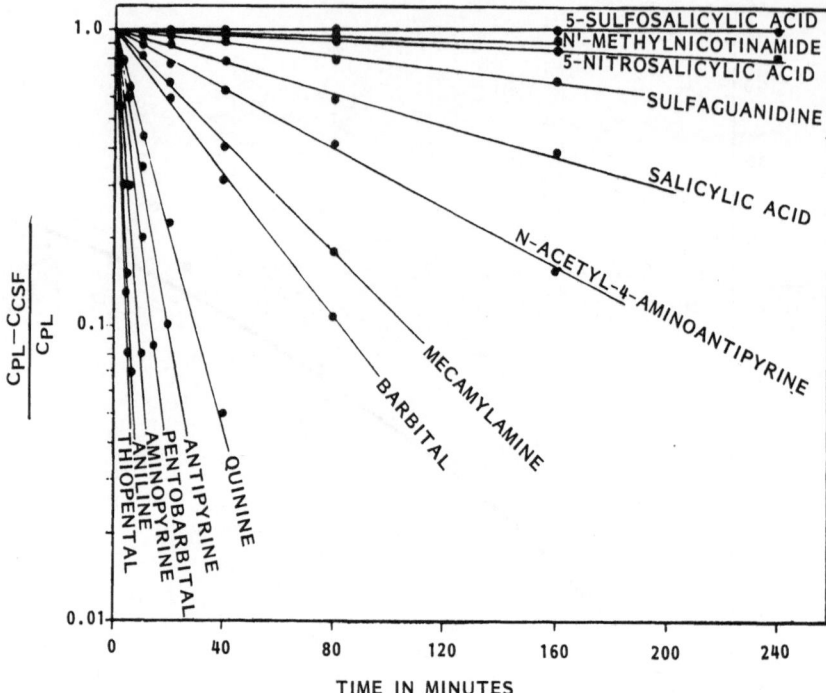

Figure 3. Penetration of various drugs into dog CSF. Note the logarithmic scale of the ordinate. (From Brodie et al.[30])

Table II. Heptane–Water Partition Coefficients and Transfer Coefficients for Penetration across the Blood–CSF Barrier of the Dog[a]

Substance	Heptane–water partition coefficient (concentration in heptane: concentration in water)	K_{out} (min)
Thiopental	3.3	0.50
Aniline	1.1	0.40
Aminopyrine	0.21	0.25
Pentobarbital	0.05	0.17
Antipyrine	0.005	0.12
Barbital	0.002	0.026
n-Acetyl-4-aminoantipyrine	0.001	0.012
Sulfaguanidine	0.001	0.003

[a]From Brodie et al.[32]

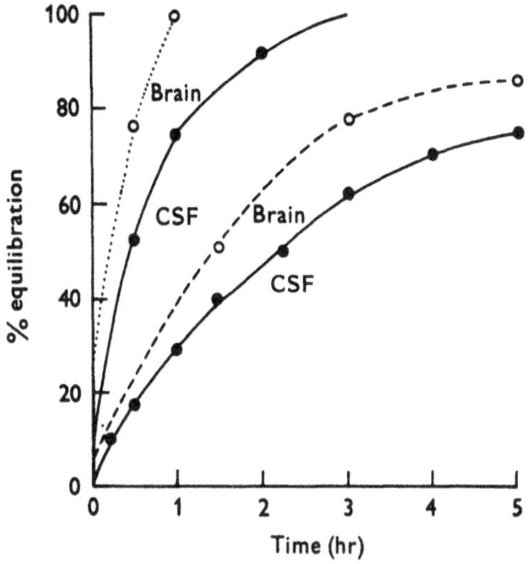

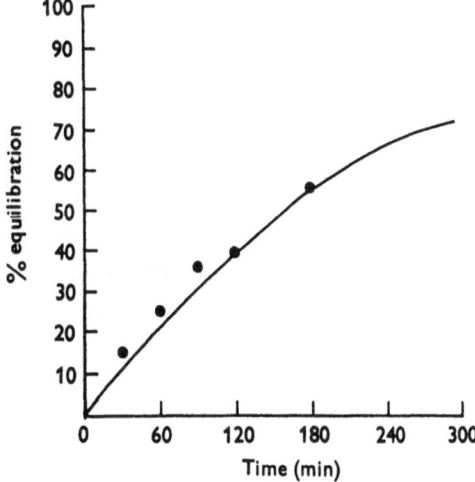

Figure 4. Comparison of the blood–CSF and BBB. The procedure was similar to that described for Figs. 1 and 2, but brain was analyzed after homogenization. Upper curves show values for ethyl thiourea, and lower curves show thiourea values. (-----) penetration into brain; (——) CSF values. (From Davson et al.[51])

Figure 5. Blood–CSF and BBB to ^{24}Na. The procedure was similar to that described for Fig. 4. The continuous curve represents penetration into the CSF, and the points represent penetration into the extracellular space of the brain, computed as the chloride space. (From Davson.[43])

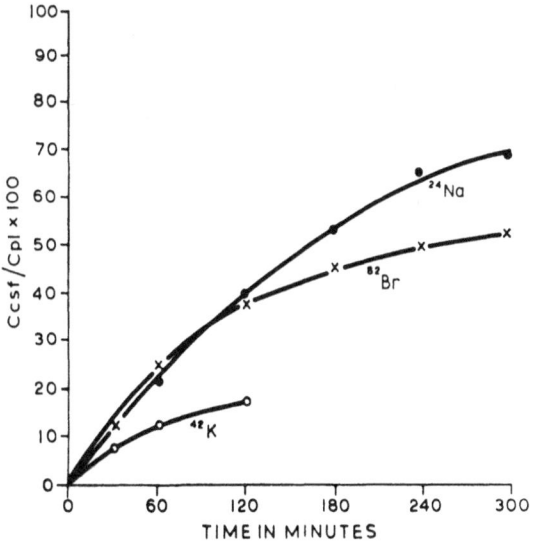

Figure 6. Penetration of isotopic ions into rabbit CSF. (From Davson.[43])

5. SINK ACTION OF THE CEREBROSPINAL FLUID

Quantitative studies on penetration of lipid-insoluble substances into the brain tissue have shown that equilibration between plasma and CSF is analogous with equilibration between plasma and brain. The system tends to a steady state, with the concentration in brain much less than that in plasma. If we consider that these substances are probably excluded from the brain cells, we must also consider equilibration with the extracellular fluid, which makes up about 15% of the brain weight. The steady-state level in this fluid shows no sign of ever achieving a value of unity for the ratio of extracellular fluid : plasma concentrations.

Figure 7 illustrates the penetration into both CSF and brain extracellular fluid of $^{35}SO_4$. As with the lipid-soluble substances, the brain fluid tends to equilibrate more rapidly than the CSF, so that there is a gradient favoring passage from brain to CSF. Because the CSF circulates and the brain fluid does not, the CSF may act as a perpetual drain or sink to the brain, accounting for the low steady-state levels reached in the brain extracellular fluid.[44,45] Figure 8 illustrates the process graphically. This sink action may be important in the economy of the brain, especially when undesirable solutes appear in the tissue, such as neurotransmitters in excessive amounts. The flowing CSF is able to prevent accumulation in the brain tissue, especially if the choroid plexuses actively remove the solute in question, which does occur for a number of neurotransmitters,[60,97,104,105] prostaglandins,[19] and iodide and thiocyanate.[10,18,87,88,108] Thus, the very low levels of iodide and thiocyanate found in the brain after lengthy intravenous infusions are not due exclusively to a slow penetration of the BBB, but are also due to a highly efficient sink action of the CSF favored by the active removal of these ions from the CSF by the choroid

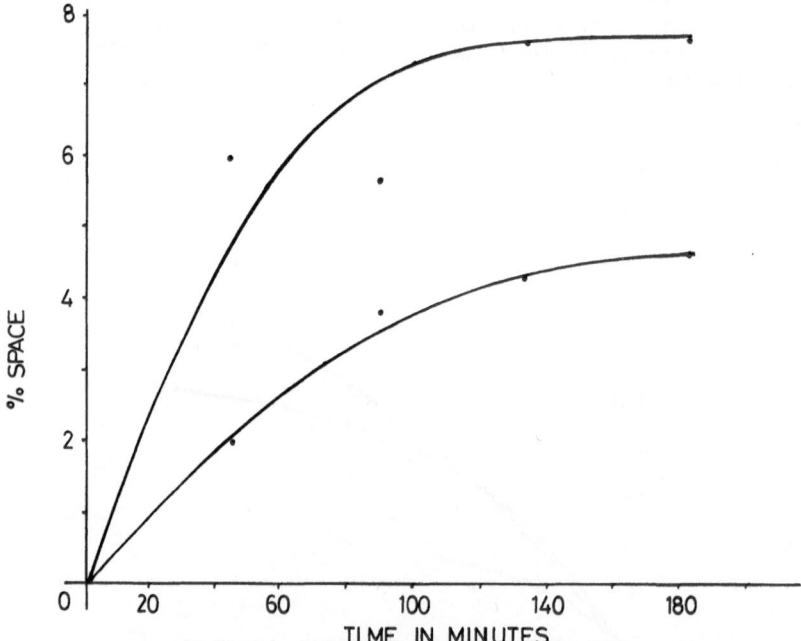

Figure 7. Penetration of $^{35}SO_4$ from blood into brain extracellular fluid (upper curve) and CSF (lower curve) of the rabbit. Ordinate: concentration in fluid: concentration in plasma. ($\times 100$) (After Hollingsworth and Davson.[70])

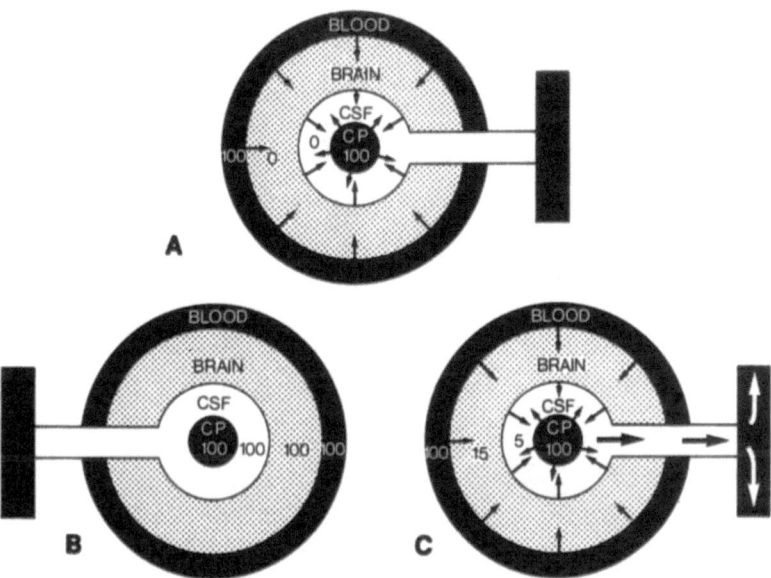

Figure 8. The sink action of the CSF. The black compartments represent blood. That in the peripheral annulus corresponds with blood in brain capillaries; that in the central spot corresponds with blood in the choroid plexuses (CP). The dotted area represents brain extracellular fluid, and the white area is CSF. (A) Initial state with the concentration of the extracellular marker equal to 100 in blood plasma and 0 in the other compartments. (B) Infinite time, assuming no flow of CSF. (C) Infinite time with a flow of CSF. The bulk flow of CSF into the venous blood prevents the establishment of diffusional equilibrium between the choroid plexus blood and the CSF; the low concentration produced in this way acts as a sink for material in the extracellular fluid of the brain. (From Davson.[46])

plexuses. The practical importance of this sink action in the experimental determination of the extracellular space of the brain is discussed in Section 7.

6. MORPHOLOGY OF THE BLOOD–BRAIN BARRIER

The idea that the capillary of the brain tissue is itself impermeable to the dyestuffs studied by earlier investigators seemed repugnant to those who maintained that the adventitial layer surrounding the blood vessels (made up of the invaginating pia and arachnoid as the vessels plunged from the subarachnoid space into the nervous tissue) was responsible for the low permeability of the barrier to dyes. It was assumed that this sheath extended as far as the capillaries. This view of a pia–glial sheath acting as a significant barrier to dyestuffs was contested vigorously by Spatz,[98] who pointed out that Goldmann's second type of experiment, in which trypan blue was injected into the subarachnoid space or ventricles, showed that the dye passed from these regions directly into the nervous tissue, i.e., they crossed the pia–glia without difficulty. Spatz showed, moreover, that the passage was not along the so-called Virchow–Robin spaces (the spaces between the blood vessels and their pia–glial covering). The dye progressed along a broad front not limited to definite spaces and thus could not have circumvented the supposed barrier presented by the pia–glia. Spatz concluded that the site of the barrier was the capillary endothelium itself, and a similar conclusion was reached by Riser.[93]

Figure 9. The astrocytic covering of a brain capillary. (Drawing after Wolff.[110])

However, the possibility that the cerebral capillary required some limiting sheath to permit it to function as a barrier was not abandoned easily because of an apparently complete similarity between cerebral and connective tissue capillaries, with the exception of the glial covering. According to the excellent study of Wolff,[110] the cerebral capillary, when viewed as an endothelial tube, is little different from a capillary of the skeletal muscle. As described by Wolff, the capillary in the cerebral cortex of the rabbit is a tube some 3 μm in diameter made up of a single layer of endothelial cells 1000–1500 Å thick; pericytes are attached (Fig. 9), enclosed by the basement membrane, which splits to contain them. The cytoplasm appears very dense, and there are relatively few vesicles. The junctions between the endothelial cells appear similar to those described in other parts of the body. These junctions are 100–150 Å wide and are interrupted by characteristic terminal bars similar to the occluding junctions between endothelial cells in other capillaries and also, apparently, in epithelia.[86]

6.1. Brain Capillary Endothelial Cell End-Feet

The special feature of the brain capillary recognized by light microscopists is the attachment of astrocytic processes, or end-feet, to the basement membrane of the endothelial cell, so that the whole endothelial tube seems to be covered by a protoplasmic layer. In muscle, the capillary endothelium is surrounded by various connective tissue elements, such as collagen fibrils. The capillaries of the CNS lack such elements. The endothelial cells (of mesenchymal origin) are apparently separated from direct contact with neurons by the protoplasm of astrocytes (of ectodermal origin). As summarized by Wolff, the brain capillaries are characterized by the high electron density of the endothelial cytoplasm, the thickness of the basement membrane, the absence of perivascular connective tissue, the complete covering of the endothelial surface by astrocytic processes, and the small number or absence of cytoplasmic vesicles in the endothelial cells. Because there seemed to be no difference in the nature of the junctional apparatus of cerebral and muscle capillaries, the astroglial covering seemed to the impartial enquirer to be the only possible morphologic correlate of the BBB. This view was fortified by Clemente and Holst's observation that massive doses of X-rays applied to the brain of the monkey caused the escape of trypan blue from the blood vessels associated with degeneration of the astrocytic end-feet.[36] Finally, electron microscopic examination showed that the junctions between astrocytic processes were apparently of the occluding-type, similar to those seen in epithelia. Occluding-type junctions are considered to be the basis of the low permeability of epithelia (including that of the choroid plexus) to lipid-insoluble substances (including trypan blue).

Against this view, however, was the study by Rodriguez, who found that aminoacridine dyes stained the nuclei of all cells of the body except those of the CNS.[94] Moreover, the nuclei of the brain capillary endothelium were also not stained. It was argued that if the barrier were to escape the astrocytes, the endothelium would have taken up the dye; it was concluded that the barrier was at the luminal side of the capillary endothelial cell. If the acridine dyes were given intraventricularly, the endothelial cells were stained, showing that the nuclei could indeed take up the dye, and incidentally, that the abluminal surface of the capillary was permeable to acridine dye. Finally, the concept of the astrocytic lining as a barrier was completely disposed of by the observation of Brightman that ferritin particles, injected into the CSF, passed out of the ventricles across the ependymal lining and found their way between the astroglial end-feet, penetrating as far as the endothelial basement membrane.[30] Penetration of an electron-dense marker with such a large diameter (some 100 Å) between the end-feet meant that the astrocytic sheath could not act as a barrier to relatively small molecules such as sucrose, inulin, or trypan blue. To complete the picture concerning identification of the locus of the barrier, Karnovsky's discovery[72] that horseradish peroxidase could be employed as an electron microscopic marker to determine whether a cleft between cells was closed, as in the conventional tight junction, was applied by Reese and Karnovsky[92] to show that the junctions between brain capillary endothelial cells are in fact tight, thus demonstrating that the capillary endothelium has the ultrastructure necessary for a low-permeability type of membrane.

6.2. Electron Microscopists' Heresy

When histologic examination of the brain was limited to the use of the light microscope, the difficulties in staining all the very fine processes of both neurons and glial cells led to histologic preparations that revealed a picture of cells embedded in a homogeneous medium described by Nissl, for example, as an *Intrazellulares Grau*.[79] This view was upset by the first electron micrographs made by Wyckoff and Young, which showed the lack of any large intercellular spaces filled with ground substance,[111] a finding supported by subsequent studies. Thus, when thin sections of brain cortex were studied, all intervening spaces between neuronal, glial, and vascular elements were found to be filled with submicroscopic structures that could be recognized as cellular extensions. All these elements were encased in cell membranes separated only by the characteristic 140–200-Å-wide gaps. It was, therefore, argued that Goldmann's failure to observe transport of trypan blue from the blood to the brain tissue was due to the absence of sufficient extracellular space to accommodate the dye. The capillaries of the brain were permeable to the dye, but there was nowhere for it to go.

The answer to this argument was already in the literature, namely Goldmann's second experiment, in which he showed that the brain was capable of taking up considerable quantities of trypan blue if the dye were presented by way of the CSF, i.e., circumventing the BBB. However, as with so many faulty hypotheses, it required more than an argument based on a classical study to reestablish the concept of the BBB as an ineluctable fact. The BBB provided the only logical basis for so many pharmacologic studies showing CNS effects produced by injections into the CSF or brain tissue of drugs that were completely ineffective when presented intravascularly.

7. THE EXTRACELLULAR SPACE OF THE BRAIN

To the physiologist, the first task was to obtain an accurate measure of the extracellular space of the brain and spinal cord. With non-nervous tissue, measurement of the extracellular space presented no serious problem. A steady level of an extracellular tag was maintained in the plasma until diffusion equilibrium between plasma and extracellular fluid took place; the tissue was excised and the amounts of the tag in unit volumes of tissue and plasma determined. By contrast, with nervous tissue, equilibration of such solutes with brain extracellular fluid is apparently never achieved, owing to the sink action of the CSF. By presenting the extracellular tag by way of the CSF, this sink action may be avoided; alternatively, the tissue may be placed in a medium containing the tag in a known concentration in vitro. Both techniques have been used by Davson and Spaziani,[48] Davson et al.,[50] Zadunaisky and Curran,[113] and Rall et al.[90]

Rall et al.[90] perfused dogs via the ventricle with an artificial CSF containing [^{14}C]in-

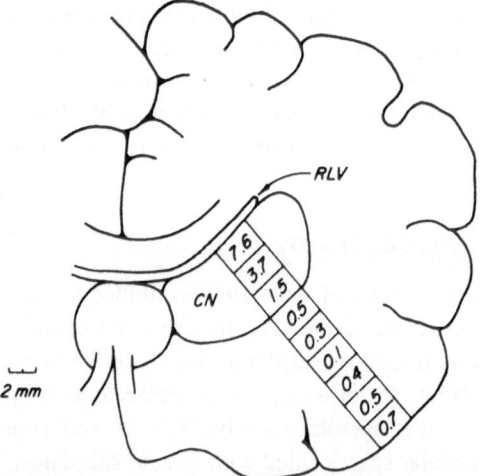

Figure 10. Inulin concentration as the percentage of perfusate in tissue blocks cut from the caudate nucleus (CN) outward. RLV, right lateral ventricle. (From Rall et al.[90])

ulin; after 3–5 hr of perfusion, the brain was removed and coronal sections some 5 mm thick were cut through the caudate nucleus, thalamus, and pons. Small sections were cut from these areas at successively greater distances from the ventricular surface, as indicated in Fig. 10. The blocks were analyzed and, by an elegant mathematical analysis of diffusion process, the concentration in the extracellular fluid in the tissue immediately adjacent to the ventricle was obtained by a linear extrapolation. At this point, the brain tissue could be assumed to be in equilibrium with the perfusion fluid, and thus an extracellular space could be computed. Rall et al. obtained an average value of 12%. Subsequent improvements of technique indicated a rather larger space: 16–17% for cerebral cortex and 18% for spinal cord gray matter; the extracellular space for white matter was rather less, at 12–14%.[62,68] An in vitro study of the frog brain by Bradbury et al.[28] indicated a space of some 20%. Cold-blooded tissue is much more amenable to in vitro study than mammalian tissue.[48] These studies provided accurate measurements of an important brain parameter necessary for most kinetic studies of the BBB and gave an unequivocal answer to the problem of the existence of a barrier in the first place. However, the answer had already been provided by Goldmann's second experiment.

8. THE BARRIERS AND HOMEOSTASIS

We have looked at the BBB simply as a restraint on the rate of exchange of solutes between blood and nervous tissue. As such, it would act as a damping mechanism that reduces the effects of fluctuations in the blood plasma levels of metabolites and other constituents. The mechanism would help to maintain homeostasis of the composition of the fluid environment of the nerve cells. However, a mere slowing of exchanges would not provide an effective homeostasis when long-continued changes of plasma concentration were maintained. A more positive contribution to homeostasis would be desirable.

Bekaert and Demeester's early studies[14,15] showed that the CSF is well insulated from changes in the plasma concentration of K^+ brought about in various ways, such as by injections of isotonic KCl or insulin. Studies on human subjects with metabolic disorders revealed the same picture.[38] The highest serum K^+ observed pathologically was 7.16 mEq/liter when the CSF concentration was 3.55 mEq/liter; the lowest was 1.97 mEq/liter with a CSF concentration of 2.43 mEq/liter. Although there was a small response to chronically altered plasma K^+, the response was never great enough to bring the concentration in the CSF outside the normal range of variation which, for human subjects, is 2.33–4.59 mEq/liter (mean, 2.96 ± 0.45 [SD]). The range for hypokalemic subjects was 2.43–2.74 and that for hyperkalemic subjects was 2.65–3.55 mEq/liter. An essentially similar insulation of the CSF was observed when the effects of fluctuation of plasma Ca^{2+} in different species and Mg^{2+} were examined.[25,37,69]

By studying the effects of altered plasma levels of K^+ on the movement of ^{42}K into and out of rabbit CSF, Bradbury and Stulcova[27] concluded that there is virtually perfect homeostasis of the CSF K^+ concentration, which was achieved by a carrier-mediated type of transport. When the plasma level was raised, the influx tended to be inhibited through saturation of a carrier; when the CSF level was raised, Bradbury and Stulcova observed an interesting facilitation of transport outward into blood, as illustrated in Fig. 11. A combination of the two processes would favor maintenance of a steady level in CSF in the face of a fluctuating plasma concentration.[22,23]

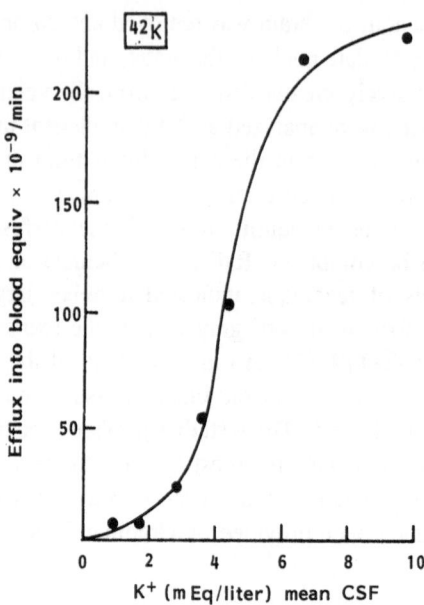

Figure 11. Computed efflux of K^+ into the blood during ventriculocisternal perfusion as a function of the concentration of K^+ in the perfusion fluid. (From Bradbury and Stulcova.[27])

8.1. Brain Extracellular Fluid

There is good experimental evidence that the concentrations of most solutes in the CSF are similar to those in the extracellular fluid of the brain. The likelihood of this was pointed out by Davson who indicated that the concentrations of glucose and some other solutes in the CSF were the same whether the fluid was sampled from ventricles, cisterna magna, or lumbar subarachnoid space.[44] It was argued that if, for example, the brain extracellular fluid had a concentration of glucose equal to that in plasma, the CSF would tend to gain glucose from the brain as it passed from ventricles to the subarachnoid spaces. A similar argument would apply to any of the CSF solutes with a concentration in the freshly secreted CSF that was different from that of plasma; the fluid would tend to gain or lose solute as it passed through the ventriculosubarachnoid system unless it were formed with the same concentration as that in the extracellular fluid. This virtual identity in concentration of CSF samples from different localities was further established by Bito and Davson[16] for K^+ and by Bito and Myers[17] for Ca^{2+} and Mg^{2+}.

Bradbury and Davson[24] perfused artificial CSF with different concentrations of K^+ through the ventricles of the rabbit. Only when the concentration of K^+ in this artificial CSF was equal to that in normal CSF (i.e., 2.9 mEq/liter) did the perfusion fluid neither lose nor gain K^+ to or from the brain. A similar situation was found by Bradbury[21] with respect to Mg^{2+}, as illustrated in Fig. 12.

Therefore, if active processes are involved in maintaining a fixed composition of CSF in relation to fluctuations of plasma composition, the same forces must be operative in maintaining the composition of the extracellular fluid of the brain. It is agreed that the epithelium of the choroid plexuses is responsible for producing CSF; moreover, when freshly secreted CSF is analyzed, it has the approximate composition of that found in the ventricles, as shown by Ames et al.[7] for K^+ and some other ions and by Deane and

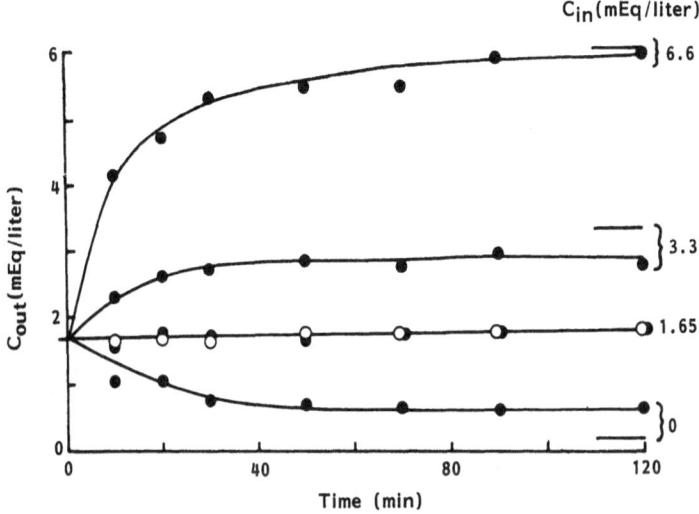

Figure 12. The concentration of magnesium in the effluent (C_{Out}) during the perfusion of fluids of different magnesium contents from the lateral ventricle to the cisterna magna of the rabbit. The mark on the left of the ordinate represents the normal concentration of magnesium in rabbit CSF. The horizontal marks at the right indicate the expected (C_{Out}) for magnesium if there had been no exchange among the fluid, blood, and brain and if the newly formed fluid contained 1.65 mEq/liter. (From Bradbury.[21])

Segal[52] with respect to glucose. The extracellular fluid of the brain is presumably produced by its capillaries. This capillary endothelium must have the power to produce a fluid with characteristic levels of K^+ glucose, and so forth, similar to the power possessed by the choroidal epithelium. Thus, we must attribute to the brain capillaries (i.e., to the BBB) a much more positive role than that of merely slowing down exchanges of these solutes with plasma. The BBB not only restrains rapid exchanges but also controls them so that steady-state levels different from those expected of a simple equilibrium distribution are maintained. The BBB must be capable of active transport and, as the glial astrocytic lining of the capillaries has been excluded as the site of the barrier, we are led to attribute this power to the capillary endothelium, a power not imagined for the capillary endothelium of muscle. Recognition of this capability of the brain capillary endothelium is an important step in the evolution of our concepts of the BBB in general. Bradbury and Segal[26] demonstrated experimentally that the carrier-mediated processes operating to create homeostasis of the CSF K^+ concentrations also operate across the BBB, which lends powerful support to our view of the BBB as being capable of active transport.

8.2. Cerebral Edema

Recognition of the positive role of the barrier in the economy of brain fluids made way for better understanding of the phenomenon of brain edema. If the brain extracellular fluid is formed by a positive act of secretion rather than as a result of the interplay of hydrostatic and osmotic forces, as has been demonstrated with muscle and other connective tissue, edema in brain tissue must be due to a disturbance of the active transport mechanisms that control the composition and amount of the extracellular fluid. In connective tissue, a breakdown in capillary permeability leading to rapid losses of protein is a sufficient cause for edema; in brain, this need not be true because a requisite for the escape of fluid along with protein is that the capillaries be highly permeable to the noncolloidal solutes of plasma (e.g., Na$^+$, Cl$^-$). Conversely, an increased capillary permeability to noncolloidal solutes may well cause edema in brain tissue, but not in connective tissue. An increased capillary permeability to noncolloidal solutes is probably the basis of many of the toxic brain edemas observed both experimentally and pathologically.

9. ONTOGENY OF THE BLOOD–BRAIN BARRIER

The immaturity of the BBB at birth was recognized by Behnsen.[12,13] He observed that trypan blue, injected into young mice, accumulated extensively in nervous tissue until the animals were about 5 weeks old, when staining remained localized to special regions—regions that, in the normal adult, lack a barrier (e.g., the area postrema in medulla). Stern and Peyrot[102] similarly demonstrated a more rapid penetration of ferrocyanide into the CSF of the newborn rabbit, rat, mouse, cat, and dog but not that of the guinea pig, which is born at a more mature stage in development. In a study of fetal guinea pigs, it was only when their eyelids were still sealed and their hair was undeveloped that penetration from blood to CSF and nervous tissue would occur.

For the neurologist, this delayed development of the BBB explained such congenital defects as kernicterus, i.e., jaundice of the brain nuclei associated with hemolytic and other forms of jaundice in the newborn.[98] More extensive studies using the classic trypan blue technique in general failed to confirm this immaturity of the barrier at birth. Grontoft[68] profited by Broman's observation that the barrier remains apparently intact in the dead animal for at least 12 hr[33] by studying previable fetuses and prematurely deceased infants with a perfusion technique. Human fetuses exhibited a barrier to trypan blue at 5–30 cm in length, and in human infants the barrier was as complete as in the adult. Millen and Hess[78] found that the barrier to trypan blue in 2–8-day-old rats is just as complete as that in adults. Subsequent studies were based on quantitative measurements of solutes that could be determined accurately, such as [^{14}C]inulin[106] and [^{14}C]sucrose,[54,61] and took into account the kinetics of transport into both CSF and brain and the relations between brain and CSF. Generally, these confirmed that during early fetal stages and, in some species such as the rat, during early postnatal stages, the permeability of the barriers to these lipid-insoluble substances is greater than in the adult. With lipid-soluble solutes, on the other hand, the permeability of the BBB in the early postnatal rat is not significantly different from that in the adult.[39] According to Saunders,[95] the change from the fetal to the adult blood–CSF barrier consists of a reduction in the number of highly permeable channels or leaks that permit free exchange between blood and CSF.

9.1. Rate of Protein Exchange

Of special interest are the changes that occur during development in the rate of exchange of proteins between blood and brain and in the character of the plasma and, thus, the CSF and brain proteins, a subject being actively pursued by Saunders and colleagues. The very low concentrations of plasma proteins in adult CSF are a manifestation of the blood–CSF barrier and BBB to these large molecules. An immature blood–CSF barrier is revealed by a higher value of the CSF : plasma concentration ratio (R_{CSF}). This higher value is found in newborn and juvenile rats. Amtorp and Sørenson[9] found the R_{CSF} for total proteins to fall from about 0.9 at birth to about 0.01 in the adult rat; a similar trend for serum albumin was found by Ramey and Birge.[91]

As far as quantitative measurements of the barriers are concerned, it seems that immaturity is a factor of greater importance in the blood–CSF barrier than in the BBB. Thus, Amtorp[8] found the rate constants for the uptake of labeled albumin by the brains of newborn, 5-day-old, and 30-day-old rats to be the same. Dziegielewska et al.[55,56] measured the transport of a number of labeled proteins from blood into the CSF and brain of fetal sheep and concluded that the blood–CSF barrier in the fetal animal is generally more permeable than that in the adult, although the differences in penetration of the proteins suggested the operation of specific factors rather than just the presence of large pores. By contrast, the BBB, on the other hand, seemed to be absolute in the fetal animal. The plasma proteins probably penetrated into brain through a secondary route from the CSF. Such a route would involve crossing the ependymal linings of the ventricles. In the adult, this is possible, but a recent study by Fossan et al.[64] suggests that the fetal ependyma differs from that of the adult in that it has intercellular tight junctions that would provide an impediment to the passage of proteins from CSF into brain tissue. Dziegielewska et al.[56] concluded that the route from CSF into brain might involve uptake by the neuroepithelial cells lining the ventricular system; these cells would give rise to immature neurons that would migrate to the cortical plate. Such a process would be consistent with the immunohistochemically demonstrated distribution of plasma proteins in the developing sheep brain.

9.2. Concentrations of Individual Proteins

Albumin is by no means the most quantitatively significant protein in fetal plasma or CSF. Such proteins as fetuin, β-fetoprotein, and transferrin represent a much larger proportion of the whole than in the adult. Table III gives the concentration of albumin, β-fetoprotein, fetuin, β-antitrypsin, and transferrin in fetal CSF during the period of peak concentration of total proteins. To some extent, but not completely, the concentrations of such proteins as β-fetoprotein in CSF are a reflection of their high concentrations in plasma, but the parallelism is not always valid. Specific factors control the concentrations of some of these proteins in CSF, notably the actual synthesis of protein by the choroid plexuses and brain tissue. Ali et al.[6] demonstrated the synthesis of β-fetoprotein by newborn rat cells in tissue culture, and Dziegielewska et al.[57] demonstrated the incorporation of labeled leucine into albumin, transferrin, and β-fetoprotein by newborn rat cells in

Table III. Concentrations of Individual Proteins (mg/dl) in the Fetal CSF of Four Species at the Peak Total Protein Concentration[a]

Protein	Sheep (31 days)	Pig (31 days)	Rat (22 days)	Human (19 weeks)
Albumin	121	—	85	143
β-Fetoprotein	532	406	76	72
Fetuin	229	113	+	+
β-Antitrypsin	427	++++	++	++
Transferrin	51	195	62	6
Total protein (Lowry)	1143	961	317	375

[a]From Saunders.[96]

tissue culture. A considerable amount of the labeled protein escaped into the incubation medium. Finally, several laboratories demonstrated, in the developing brain, a messenger RNA (mRNA) capable of directing the synthesis of, and therefore determining, the brain concentration and presumably the CSF concentration, of proteins that subserve specific functions in developing nervous tissue.[53,76] This local synthesis, combined with the immaturity of the blood–CSF barrier to proteins in the fetus, might suffice to maintain adequate supplies to the tissue.[58]

10. COMPARATIVE ASPECTS

The physiology of the BBB is intimately interwoven with that of the blood–CSF barrier. Early investigators were thus concerned with the existence of the CSF, its production, and its drainage in forms other than mammal. Serious comparative studies on the physiology of the CSF were largely carried out by those investigators of the mammalian system who had opportunities to study the same phenomena in cold-blooded vertebrates, especially in the dogfish, *Squalus acanthias*, a favorite species of those who spent their summers at Mount Desert Island, Maine.

10.1. Elasmobranchs

It was early established that the rate of production of CSF in the dogfish is less than that in mammals when represented as a percentage of the renewal rate. The existence of a BBB was established by Fenstermacher and Patlak[62,63]; this barrier was found to be much more resistant to trauma than is the mammalian barrier.[73]

The most interesting aspect of the comparative studies on vertebrates was the location of the barrier. Recognizing that the site of the barrier in the mammal is the brain capillary, investigators asked whether this site was a consistent feature of the nonmammalian barrier. As it turned out, the favorite species for study (the dogfish and some other elasmobranchs, such as the skate) were an exception to the mammalian rule. The barrier, as studied with horseradish peroxidase and the electron microscope, proved to be at the astrocytic end-feet layer rather than at the capillary endothelium. Thus, horseradish peroxidase escaped from the blood capillaries of the elasmobranch, but was prevented from diffusing throughout the nervous tissue by the layer of astrocytes with intercellular clefts sealed by tight junctions.[31]

10.2. Holocephalans

The other class of cartilaginous fishes is the holocephalans, such as the ratfish, *Chimaera monstrosa*. According to Bundgaard,[35] the capillary endothelium of this species is radically different from that of the shark and skate. Elaborate tight junctions seal the intercellular clefts, but the glial end-feet do not have tight junctions sealing their clefts.

10.3. Cyclostomes

Cyclostomes, such as the hagfish, *Myxine glutinosa*, and the lamprey, *Petromyzon*, have typical warm-blooded BBB features. The clefts between capillary endothelial cells are long and tortuous and sealed by tight junctions.[34,35] This demonstration of a BBB in the cyclostomes emphasizes the great age of the feature. Lampreys and hagfish are the only extant representatives of a very ancient class of ostracoderms (*Agnatha*), being derived from the armored ostracoderms, which lived in Silurian and Devonian times and developed independently of other vertebrate classes during the past 300–400 million years. Exceptions to the basic mechanism, such as the absence of a choroid plexus in the hagfish and the glial nature of the barrier in elasmobranchs, must be regarded as secondary modifications of this primary BBB principle.

10.4. The Invertebrate Barrier

Cephalopods have a closed circulation; Abbott et al.[5] showed that horseradish peroxidase failed to pass out of the blood into the neuropil of the cuttlefish, *Sepia*, being held up outside the capillaries by a glial barrier. In contrast to the leaky blood–brain interfaces in other molluscan groups, this cephalopod possesses a BBB. The situation in crustaceans was examined by Abbott and colleagues[1,3,4] and by Kristensson et al.[74] The vascular arrangements of the decapod are shown in Fig. 13. The CNS consists of a series of ganglia joined by interganglionic connectives; branches from the major arterial vessels enter the ganglia, penetrating only a short distance and opening into glia-lined blood channels bounded by basement membrane, but having no endothelial cells. The blood channels discharge into the hemocele cavity in which ganglia and connectives lie. The superficial blood–brain interface is formed by an epithelium-like layer of glial cells (the perineurium) and an associated basal lamina (neural lamella). The perineurium and perivascular glia seem to be structurally similar and together constitute the cellular layer of the blood–brain interface. Gap junctions are found between the glial cells in the perineural and paravascular layers, but tight junctions are rarely seen.

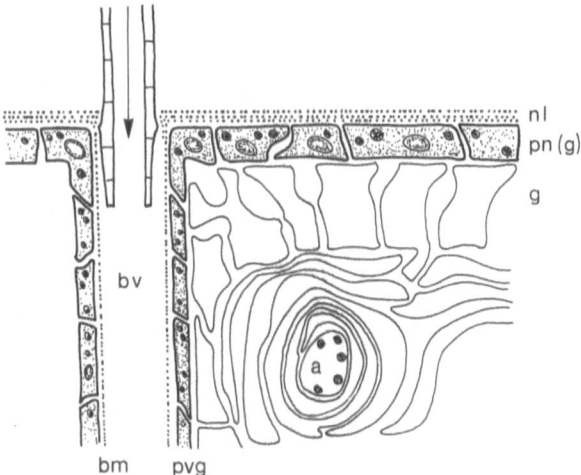

Figure 13. Diagrammatic representation of the fine structure of the decapod crustacean CNS. The intracerebral blood channels (bv) have no endothelium, but are lined by perivascular glia (pvg) and their basal laminae (bm). The superficial blood–brain interface is formed by a layer of modified glia, the perineurium (pn), with its associated basal lamina, the neural lamella (nl), axon (a), and glia (g). (From Abbott et al.[4])

Abbott injected tracers of sizes varying from erythrocytes (6–19 μm in diameter) through dextran (16.7 nm) to inulin (3.0 nm) and sodium (0.5 nm) into the vascular circulation of *Carcinus maenas* and estimated their volumes of distribution in brain. The volumes of distribution of inulin and sucrose were found to be large and probably equal to the volumes in the extracellular space. The view that the passage of small molecules across the BBB interface is unrestricted was tested electrophysiologically by Abbott et al.[3] by assessing the movements of K$^+$ as revealed by changes in the potential across various interfaces. With central nervous tissue, the responses to an altered external K$^+$ concentration suggested the existence of a definite barrier across the perineural layer of the outer surface of the connectives to give an extraneuronal potential between the outside medium and the space surrounding the nervous elements.

10.5. Value of Comparative Studies

As Abbott and Zlokovic emphasized when describing their preparation of the dog-fish, which enabled them to study its BBB by the Oldendorf technique,[80] the presence of a glial barrier provides a unique opportunity for studying the permeability and active transport features of a glial sheet.[2] As far as the function of the CSF is concerned, the finding of species of fish with ventricular volumes that vary from only a fraction of that of the brain to as much as 123% (in the lungfish) provides a means of evaluating the role of the CSF in blood–brain interchanges. Experimentally, moreover, the availability of fish with virtually no CSF, such as the hagfish, permits the study of the BBB and homeostasis of the brain extracellular fluid in the absence of complications due to exchanges between CSF and brain extracellular fluid.

11. THE PRESENT ERA

I have been able to discuss the physiology of the BBB and the CSF in the light of experimental studies that are now largely outdated insofar as technique is concerned. Single-pass techniques now permit examination of the permeability of the BBB to a large variety of solutes in experiments of very short duration. A large number of quantitative measurements of the permeability of the barrier under different conditions have been made.

In Oldendorf's brain uptake index (BUI) technique,[80] a rat is given a bolus injection through the common carotid artery of a solution containing the solute to be studied together with an internal standard, i.e., a solute such as tritiated water or [^{14}C]butanol with a rate of penetration determined by blood flow rather than by its permeability across the BBB. The BUI measures the uptake within 15 sec (the time between injection and decapitation of the animal) of a given solute; this uptake index is independent of variations in blood flow. With this technique, large numbers of solutes may be studied in the time required for the study of a single solute by the earlier methods.

In general, modern studies have done rather more than confirm the ideas derived from earlier studies because they have permitted quantitative evaluation of the parameters of facilitated transport using the classic Michaelis–Menten analysis of the penetration of the barrier.[84,85] This evaluation of the transport parameters has been of great value in assessing the quantitative importance of transport across the BBB in relationship to utilization by the brain of sugars, amino acids, and fatty acids[40,84,85] and in relationship to the transport of peptide hormone,[114] vitamins,[99] and steroid hormones[82,83] within the CNS.

The other type of single pass experiment was first reported by Crone[41,42] and was developed by Yudilevitch and De Rose.[112] In this approach, the solute is administered as an intraarterial injection and the brain uptake is calculated from the extraction from the blood, as deduced from the concentration in the venous blood drained by the torcula. This technique has the advantage of eliminating the need to destroy the animal for brain analysis. As with the BUI technique, the method has permitted the analysis of sugar and amino acid transport in terms of Michaelis–Menten kinetics.

11.1. Perfused Brain

Both single-pass techniques suffer from the limited time of exposure of the investigated solutes to the BBB. Solutes such as ^{24}Na or thiourea would not penetrate the brain in significant quantities during this period. The technique is limited to the study of those solutes that are either lipid soluble or exhibit the phenomenon of facilitated transport, such as sugar and amino acids. A virtue of single-pass techniques, on the other hand, is the limited exposure to metabolic change, so that the use of isotopically labeled solutes is feasible. A method that allows more lengthy exposure of the solute to the blood–brain interface, is provided by the perfused head preparation, examples of which have been described by Takasato et al. using the rat,[103] by Bradbury et al. using the mouse,[29] and most recently by Zlokovic et al. using the guinea pig.[115] When applied to the study of labeled metabolites, hormones, and neurotransmitters, these techniques should yield a rich reward in the form of accurate quantitative measurements of transport parameters.

11.2. Isolated Choroid Plexus

Insofar as transport across the choroid plexuses may be regarded as complementary and similar to transport across the brain capillaries, preparations of viable perfused choroid plexus should provide valuable information uncomplicated by metabolic influences of the brain; the most successful of these are those developed in the laboratories of Pollay[89] and Segal.[20,52] When combined with the paired tracer technique of Crone[41,42] the perfused isolated plexus is an exceptionally valuable tool.[116]

11.3. Isolated Brain Microvessels

Finally, a technique that opens the way to the study of transport characteristics of the brain capillaries is that involving the isolation of these microvessels by ultracentrifugation. The results of such studies were recently reviewed by Joó.[71]

12. SUMMARY

Figure 4 briefly traces the history of the development of the concept and experimental study of the BBB, as illustrated by my colleague, Professor M. W. B. Bradbury, at a recent symposium at Salisbury Cove, Maine.

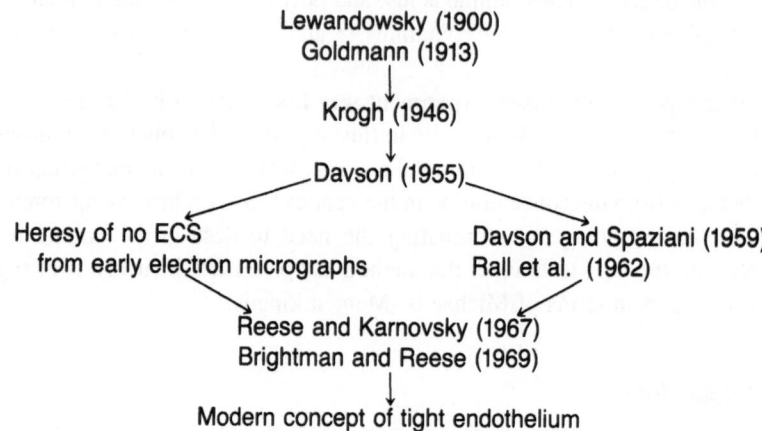

Figure 14. Steps in the history of the BBB. (Courtesy of Professor M.W.B. Bradbury.)

REFERENCES

1. Abbott NJ: Absence of blood–brain barrier in a crustacean, *Carcinus maenas. Nature (Lond)* 225:291–293, 1970.
2. Abbott NJ, Zlokovic BV: A preparation for the study of glial transport and blood–brain barrier permeability—the elasmobranch *Scyliorhinus canicula. J Physiol (Lond)* (in press), 1988.
3. Abbott NJ, Moreton RB, Pichon Y: Electrophysiological analysis of potassium and sodium movements in crustacean nervous system. *J Exp Biol* 63:85–115, 1975.
4. Abbott NJ, Pichon Y, Lane NJ: Primitive forms of potassium homeostasis: Observations on crustacean central nervous system. *Exp Eye Res* 25:259–271, 1977.
5. Abbott NJ, Bundgaard M, Cserr HF: Fine structural evidence for a glial blood–brain barrier to protein in the cuttle fish *Sepia officinalis. J Physiol (Lond)* 316:52–53P, 1981.
6. Ali M, Jujoo K, Sahib MK: Synthesis and secretion of alpha-fetoprotein and albumin by newborn rat brain cells in culture. *Dev Brain Res* 6:47–55, 1983.
7. Ames A, Sakanoue M, Endo S: Na, K, Ca, Mg and Cl concentrations in choroid plexus fluid compared with plasma ultrafiltrate. *J Neurophysiol* 27:672–681, 1964.
8. Amtorp O: Transfer of ^{125}I-albumin from blood into brain and cerebrospinal fluid in newborn and juvenile rats. *Acta Physiol Scand* 96:399–406, 1976.
9. Amtorp O, Sørensen SC: The ontogenetic development of concentration differences of proteins and ions between plasma and cerebrospinal fluid in rabbits and rats. *J Physiol (Lond)* 243:387–400, 1974.
10. Becker B: Cerebrospinal fluid iodide. *Am J Physiol* 201:1149–1151, 1961.
11. Becker H, Quadbeck G: Tierexperimentelle tersuchungen uber die Funktsionswise der Blut-Hirnschranke. *Z Naturforsch* 7B:493, 1952.
12. Behnsen G: Farbstoffversuche mit Trypanblau an der Schranke zwischen Blut und Zentralnervensystem der wachsenden Maus. *MMW* 73:1143–1147, 1926.
13. Behnsen G: Uber die Farbstoffspeicherung im Zentralnervensystem der weissen Maus in verschiedenen Alterszuständen. *Z Zellforsch* 4:515–572, 1927.
14. Bekaert J, Demeester G: The influence of the infusion of potassium chloride on the cerebrospinal fluid concentration of potassium. *Arch Int Physiol* 59:393–394, 1951.
15. Bekaert J, Demeester G: The influence of glucose and insulin upon the potassium concentration of serum and cerebrospinal fluid. *Arch Int Physiol* 59:262–264, 1951.
16. Bito LZ, Davson H: Local variations in cerebrospinal fluid composition and its relationship to the composition of the extracellular fluid of the cortex. *Exp Neurol* 14:264–280, 1966.
17. Bito LZ, Myers RE: The ontogenesis of haematoencephalic cation transport processes in the rhesus monkey. *J Physiol (Lond)* 208:153–170, 1970.

18. Bito LZ, Bradbury MWB, Davson H: Factors affecting the distribution of iodide and bromide in the central nervous system. *J Physiol (Lond)* 184:322–2543, 1966.

19. Bito LZ, Davson H, Hillingsworth J: Facilitated transport of prostaglandins across the blood-cerebrospinal fluid and blood–brain barriers. *J Physiol (Lond)* 256:273–285, 1976.

20. Blount R, Foreman P, Harding M, et al: The perfusion of the isolated choroid plexus of the sheep. *J Physiol (Lond)* 232:12–13, 1973.

21. Bradbury MWB: Magnesium and calcium in cerebrospinal fluid and in the extracellular fluid of the brain. *J Physiol (Lond)* 179:67–68, 1965.

22. Bradbury MWB: Potassium homeostasis in cerebrospinal fluid. In Siesjö BK, Sørensen SC (eds): *Ion Homeostasis of the Brain*. Munksgaard, Copenhagen, 1971, pp. 138–153.

23. Bradbury MWB: Discussion. In Siesjö BK, Sørensen SC (eds): *Ion Homeostasis of the Brain*. Munksgaard, Copenhagen 1971, p. 171.

24. Bradbury MWB, Davson H: The transport of potassium between blood, cerebrospinal fluid and brain. *J Physiol (Lond)* 181:151–174, 1965.

25. Bradbury MWB, Kleeman CR: Stability of the potassium content of cerebrospinal fluid and brain. *Am J Physiol (Lond)* 213:519–528, 1967.

26. Bradbury MWB, Segal MB: Transport of potassium at the blood–brain barrier. In *Proceedings of the Wates Foundation, Symposium on the Blood–Brain Barrier*, Truex Press, Oxford, 1970, pp. 143–149.

27. Bradbury MWB, Stulcova B: Efflux mechanism contributing to the stability of the potassium concentration in cerebrospinal fluid. *J Physiol (Lond)* 208:415–430, 1970.

28. Bradbury MWB, Villamil M, Kleeman CR: Extracellular fluid, ionic distribution and exchange in isolated frog brain. *Am J Physiol* 214:643–651, 1968.

29. Bradbury MWB, Deane R, Rosenberg G: Regional blood flow, EEG and electrolytes in mouse brain perfused with the perfluorchemica, FC 43. *J Physiol (Lond)* 355:31, 1984.

30. Brightman MW: The distribution within the brain of feitin injected into cerebrospinal fluid compartments. *J Cell Biol* 26:99–123, 1965.

31. Brightman MW, Reese TS, Olsson Y, et al: Morphologic aspects of the blood–brain barrier to peroxidase in elasmobranchs. *Prog Neuropathol* 1:146–161, 1971.

32. Brodie BB, Kurz H, Schanker LS: The importance of dissociation constant and lipid-solubility in influencing the passage of drugs into the cerebrospinal fluid. *J Pharmacol* 130:20–25, 1960.

33. Broman T: Supravital analysis of disorders in the cerebrovascular permeability. *Acta Pshychiatry (Kbh)* 25:19–31, 1950.

34. Bundgaard M: The blood–brain barrier in the lamprey. *Acta Physiol Scand (suppl)* 440:1–83, 1976.

35. Bundgaard M: The ultrastructure of cerebral blood capillaries in the ratfish, chimaera monstrosa cell. *Tissue Res* 226:145–154, 1982.

36. Clemente CD, Holst EA: Pathological changes in neurons, neuroglia and blood–brain barrier induced by x-irradiation of the heads of monkeys. *Acta Neurol Psychiatry Scand* 71:66–79, 1954.

37. Cohen H: The magnesium content of the cerebrospinal and other body fluids. *Q J Med* 20:173–186, 1927.

38. Cooper ES, Lechner E, Bellet S: Relations between serum and cerebrospinal fluid electrolytes under normal and abnormal conditions. *Am J Med* 18:613–621, 1955.

39. Cornford EM, Braun LD, Oldendorf WH, et al: Comparison of lipid-mediated blood–brain barrier permeability in neonates and adults. *Am J Physiol* 243:C161–C168, 1982.

40. Cremer JE, Braun L, Oldendorf WH: Changes during development in transport processes of the blood–brain barrier. *Biochim Biophys Acta* 448:633–637, 1976.

41. Crone C: *Om diffusionen af nogle organiske non-elektrolyter fra bold til hjernavaev*. Munksgaard, Copenhagen, 1961.

42. Crone C: The permeability of brain capillaries to non-electrolytes. *Acta Physiol Scand* 64:407–417, 1965.

43. Davson H: A comparative study of the aqueous humor and cerebrospinal fluid in the rabbit. *J Physiol (Lond)* 129:111–133, 1955.

44. Davson H: Some aspects of the relationship between the cerebrospinal fluid and the central nervous system. In Wolstenholme GEW, O'Conner CM (eds): *The Cerebrospinal Fluid*. Ciba Foundation Symposium. Churchill, London, 1958, pp. 189–203.

45. Davson H: The cerebrospinal fluid. *Erg Physiol* 52:21–73, 1963.

45a. Davson H: *Physiology of the Cerebrospinal Fluid*. Churchill, London, 1967.

46. Davson H: The environment of the neurone. *Trends Neurosci* 1:39–41, 1978.

47. Davson H, Danielli JF: *The Permeability of Natural Membranes*. Cambridge University Press, Cambridge, 1942.

48. Davson H, Spaziani E: The blood–brain barrier. *J Physiol (Lond)* 149:135–143, 1959.
49. Davson H, Welch K: The permeation of several materials into the fluids of the rabbit's brain. *J Physiol (Lond)* 218:37–351, 1971.
50. Davson H, Kleeman CR, Levin E: Blood–brain barrier and extracellular space. *J Physiol (Lond)* 159:67–*68P, 1961.
51. Davson H, Kleeman CR, Levin E: The blood–brain barrier. In Hogben CAM (ed): *Proceedings of the First International Congress of Pharmacology.* Vol. 4. Pergamon, Oxford, 1963, pp. 71–94.
52. Deane R, Segal MB: The transport of sugar across the perfused choroid plexus of the sheep. *J Physiol (Lond)* 362:245–260, 1985.
53. Dickson PW, Alred AP, Marten PD, et al: High prealbumin and transferrin mRNA levels in the choroid plexus of the rat brain. *Biochem Biophys Res Commun* 127:890–895, 1985.
54. Dziegielewska KM, Evans CAN, Malinowska D, et al: Studies of the development of blood–brain systems to lipid insoluble substances in fetal sheep. *J Physiol (Lond)* 292:207–231, 1979.
55. Dziegielewska KM, Evans CAN, Fossan G, et al: Proteins in cerebrospinal fluid and plasma of fetal sheep during development. *J Physiol (Lond)* 300:441–455, 1980.
56. Dziegielewska KM, Evans CAN, Lorscheider FL, et al: Plasma proteins in fetal sheep brain: Blood–brain barrier and intravertebral distribution. *J Physiol (Lond)* 318:229–250, 1981.
57. Dziegielewska KM, Bock E, Cornelis MEP, et al: Identification of fetuin in human and rat fetuses and in other species. *Comp Biochem Physiol* 76A:241–245, 1983.
58. Dziegielewska KM, Saunders NR, Soreq Q: Messenger ribonucleic acid (mRNA) from developing rat cerebellum directs in vitro synthesis of plasma proteins. *Dev Brain Res* 355:259–267, 1985.
59. Ehrlich P: *Uber die Beziehungen von chemische Constitution, Vertheilung, und Pharmakologischer Wirkung. Collected Studies in Immunity.* Repr and Transl Wiley, New York, 1906, pp. 567–595.
60. Eriksson K-H, Winbladh B: Choroid plexus uptake of atropine and methylatropine in vitro. *Acta Physiol Scand* 83:300–308, 1971.
61. Evans CAN, Reynolds JM, Reynolds ML, et al: The development of a blood–brain barrier mechanism in foetal sheep. *J Physiol (Lond)* 238:371–386, 1974.
62. Fenstermacher JD, Patlak CS: The movements of water and solutes in the brains of mammals. In Pappius HM, Feindel W (eds): *Dynamics of Brain Edema.* Springer-Verlag, Heidelberg, 1976, pp. 87–97.
63. Fenstermacher JD, Patlak CS: CNS, CSF, and extradural fluid uptake of various hydrophilic materials in the dogfish. *Am J Physiol* 232:R45–R53, 1977.
64. Fossan G, Cavanagh ME, Evans CAN, et al: CSF-brain permeability in the immature sheep fetus: A CSF-brain barrier. *Dev Brain Res* 18:113–124, 1984.
65. Friedmann U, Elkeles A: Kann die Lehre von der Bluthirnschranke in ihrer heutigen Form aufrechterhalten werden? *Dtsch Med Wochenschr* 57:1934–1935, 1931.
66. Goldmann EE: Die äussere und innere sekretion des genden und gekranken Organismus im Licht der vitalen Färbung. *Beitr Klin Chir* 64:192–265, 1909.
67. Goldmann EE: Vitalfärbung am Zentralnervensystem. *Abh Preuss Akad Wiss Phys-Math* 1:1–60, 1913.
68. Grontoft O: Intracranial haemorrhage and blood–brain barrier problems in the newborn. *Acta Pathol Microbiol Scand (Suppl C)* 1–109, 1964.
69. Herbert FK: The total and diffusible calcium of serum and the calcium of cerebrospinal fluid in human cases of hypocalcaemia and hypercalcaemia. *Biochem J* 27:1979–1991, 1933.
70. Hollingsworth JG, Davson H: Transport of sulfate in the rabbit's brain. *J Neurobiol* 4:389–396, 1973.
71. Joó F: The blood–brain barrier in vitro: Ten years of research on microvessels isolated from the brain. *Neurochem Int* 7:1–25, 1985.
72. Karnovsky MJ: The ultrastructural basic of capillary permeability studied with peroxidase as a tracer. *J Cell Biol* 35:213–236, 1967.
73. Klatzo I, Steinwall O: Observations on cerebrospinal fluid pathways and behavior of the blood–brain barrier in sharks. *Acta Neuropathol* 5:161–175, 1965.
74. Kristensson K, Strömberg E, Eloffon R, et al: Distribution of protein tracers in the nervous system of crayfish (*Astacus astacus L*) following systemic and local application. *J Neurocytol* 1:35–47, 1972.
75. Krogh A: The active and passive exchanges of inorganic ions through the surfaces of living cells and through living membranes generally. *Proc R Soc Lond B* 133:140–200, 1946.
76. Levin MJ, Tuil D, Uzan G, et al: Expression of the transferring gene during development of non-hepatic tissues. *Biochem Biophys Res Commun* 122:212–217, 1984.
77. Lewandowsky M: Zur Lehre der Cerebrospinalflüssigkeit. *Z Klin Med* 40:480–494, 1900.

78. Millen JW, Hess A: The blood–brain barrier: An experimental study with vital dyes. *Brain* 81:248–257, 1958.

79. Nissl F: Nervenzellen und graue substanz. *MMW* 45:988, 1889.

80. Oldendorf WH: Brain uptake of radiolabeled amino acids, amines and hexoses after arterial injection. *Am J Physiol* 221:1629–1639, 1971.

81. Oppelt WW, Patlak CS, Zubrod CG, et al: Ventricular fluid production rates and turnover in elasmobranch. *Comp Biochem Physiol* 12:171–177, 1964.

82. Pardridge WM, Mietus LJ: Transport of steroid hormones through rat blood–brain barrier. Primary role of albumin-bound hormone. *J Clin Invest* 64:145–154, 1979.

83. Pardridge WM, Mietus LJ: Effects of progesterone-binding globulin versus progesterone antiserum on steroid hormone transport through the blood–brain barrier. *Endocrinology* 106:1137–1141, 1980.

84. Pardridge WM, Oldendorf WH: Kinetics of blood–brain barrier transport of hexoses. *Biochim Biophys Acta* 382:377–392, 1977.

85. Pardridge WM, Oldendorf WH: Kinetic analysis of blood–brain barrier transport of amino acids. *Biochim Biophys Acta* 401:128–136, 1975.

86. Peters A: Plasma membrane contacts in the central nervous system. *J Anat* 96:237–248, 1962.

87. Pollay M: Cerebrospinal fluid transport and the thiocyanate space of the brain. *Am J Physiol* 210:275–279, 1966.

88. Pollay M, Davson H: The passage of certain substances out of the cerebrospinal fluid. *Brain* 86:137–150, 1963.

89. Pollay M, Stevens A, Estrada E, et al: Extracorporeal perfusion of choroid plexus. *J Appl Physiol* 32:612–617, 1972.

90. Rall DP, Oppelt WW, Patlak CS: Extracellular space of brain as determined by diffusion of inulin from the ventricular system. *Life Sci* 2:43–48, 1962.

91. Ramey BA, Birge WJ: Development of cerebrospinal fluid and the blood–cerebrospinal fluid barrier in rabbits. *Dev Biol* 68:292–298, 1979.

92. Reese TS, Karnovsky MJ: Fine structural localization of a blood–brain barrier to exogenous peroxidase. *J Cell Biol* 34:208–217, 1967.

93. Riser: Le Liquide Céphalo-rachidien. Masson, Paris, 1920.

94. Rodriguez LA: Experiments on the histologic locus of the hemato-encephalic barrier. *J Comp Neurol* 102:27–45, 1955.

95. Saunders NR: Ontogeny of the blood–brain barrier. *Exp Eye Res* (suppl) 25:523–550, 1977.

96. Saunders NR: Plasma proteins and fetal brain development. In Dupont AM, Kato AC, Weser M (eds): *The Role of Cell Interactions in Early Neurogenesis.* Raven, New York, 1984, pp. 191–199.

97. Schanker LS, Prockop LD, Schou J, et al: Rapid efflux of some quaternary ammonium compounds from cerebrospinal fluid. *Life Sci* 10:515–521, 1962.

98. Spatz H: Die Bedeutung der vitalen Färbund für die Lehre vom Stoffaustausch zwischen dem Zentralnervensystem und dem übrigen Körper. *Arch Psychiat Nervenheilk* 101:267–358, 1934.

99. Spector R: Vitamin homeostasis in the central nervous system. *N Engl J Med* 296:1393–1398, 1977.

100. Stern L, Gautier R: Rapports entre le liquide céphalo-rachidien et la circulation sanguine. *Arch Int Physiol* 17:138–192, 1921.

101. Stern L, Gautier R: Les rapports entre le liquide céphalo-rachidien et les éléments nerveux de l'axe cérébrospinal. *Arch Intern Physiol* 17:391–448, 1922.

102. Stern L, Peyrot R: Le fonctionnement de la barrière hémato-encéphalique aux divers stades de développement chez diverses espèces animales. *C R Soc Biol Paris* 96:1124–1126, 1927.

103. Takasato Y, Rapoport SI, Smith QR: An in situ brain perfusion technique to study cerebrovascular transport in the rat. *Am J Physiol* 247:H484–H493, 1984.

104. Tochino Y, Schanker LS: Active transport of quaternary ammonium compounds by the choroid plexus in vitro. *Am J Physiol* 208:666–673, 1965.

105. Tochino Y, Schanker LS: Transport of serotonin and norepinephrine by the rabbit choroid plexus in vitro. *Biochem Pharmacol* 14:1557–1566, 1965.

106. Vernadakis A, Woodbury DM: Cellular and extracellular spaces in developing rat brain. *Arch Neurol* 12:284–293, 1965.

107. Walter FK: *Die Blut-Liquorschranke. Eine Physiologische und Klinische Studie.* Thieme, Leipzig, 1929.

108. Welch K: Active transport of iodide by choroid plexus of the rabbit in vitro. *Am J Physiol* 202:757–760, 1962.

109. Welch K: A model for the distribution of materials in the fluids of the central nervous system. *Brain Res* 16:453–468, 1969.
110. Wolff J: Beiträge zur Ultrastruktur der Kapillaren im Zentralnervensystem. *Z Zellforsch* 60:409–431, 1963.
111. Wyckoff RWG, Young JZ: The motoneuron surface. *Proc R Soc Lond B* 144:440–450, 1956.
112. Yudilevich DL, De Rose N: Blood–brain transfer of glucose and other molecules measured by rapid indicator dilution. *Am J Physiol* 220:841–846, 1971.
113. Zadunaisky JA, Curran PF: Sodium fluxes in isolated frog brain. *Am J Physiol* 205:949–956, 1963.
114. Zlokovic BV, Begley DJ, Chain-Eliash DG: Blood–brain barrier permeability to leucine encephalin, D-alanine[2]-D-leucine[5]-encephalin and their N-terminal amino acid (tyrosine). *Brain Res* 336:125–132, 1985.
115. Zlokovic BV, Begley DJ, Djuricic BM, et al: Measurement of solute transport across the blood–brain barrier in the perfused guinea pig brain: Method and application to N-methyl-alpha-aminoisobutyric acid. *J Neurochem* 46:1444–1451, 1986.
116. Zlokovic BV, Segal MB, Begley DJ, et al: Permeability of the blood-cerebrospinal fluid and blood–brain barriers to thyrotrophin releasing hormone (TRH). *Brain Res* 358:191–199, 1985.

3

The Anatomic Basis of the Blood–Brain Barrier

Milton W. Brightman

The morphologic features of cerebral endothelium that prevent large and small solutes from being transferred from blood to extravascular, interstitial fluid, are the complete belts of tight junctions between them and the low number of endocytotic invaginations in the luminal part of their cell membrane.[15,108,131] When the capillaries of the central nervous system are perturbed in a number of unrelated ways, solutes can move from blood to brain tissue. The mechanisms of this transmural passage are still disputed. It is the intention of this chapter to review some firm and some preliminary evidence of (1) the ways in which endothelium may interact with the astrocyte in the establishment or maintenance of the barrier, (2) how the intact endothelium may selectively transfer specific ligands by receptor-mediated means, (3) the influence of electric charge on solute uptake by endothelium, (4) how the barrier may be circumvented, and (5) where extravasated solutes are disseminated interstitially.

1. INTERCELLULAR TIGHT JUNCTIONS

The capillaries of skeletal muscle and other tissues act as molecular sieves that enable hydrophilic substances of relatively low molecular weight (10,000–40,000) to move passively across the endothelium while restricting the passage of larger molecules ($>$40,000 M_r).[72,97] Dextrans of different sizes are restricted from traversing peripheral vessels in proportion to the dimensions of the molecules, up to a 4-nm radius, after which the transfer remains constant.[3] It has been inferred from such data that the capillary is transversed by small and large pores, in the form of water-filled channels of uniform width. These channels could be either cylinders 9.0 nm in diameter[97] or slits 5.0–5.5 nm wide.[75] The pores or channels would need occupy only a very small fraction, about 0.02%, of the capillary surface area to account for solute passage.[69] Their dimensions

Milton W. Brightman • National Institute of Neurological Communicative Diseases and Stroke, National Institutes of Health, Bethesda, Maryland 20892.

have been well within the resolving power of the electron microscope for decades, yet such pores, channels, or slits do not appear to exist in these forms.

There is some evidence that the morphologic equivalent of the small pores is the junction between contiguous endothelial cells.[69,146] The structure of the junction in epithelia and endothelia and how it may be maintained or altered in cerebral endothelium will now be discussed.

The permeability of tight junctions has been defined by electrical measurements made primarily across epithelia.[77] In general, the greater the electrical resistance across the epithelium, the tighter or more restrictive the intercellular junctions to the passage of ions. Since almost all the direct physiologic measurements of permeability have been made on epithelia and since their tight junctions are far more amenable to freeze-fracture analysis, we shall first consider the junctions between epithelial cells.

1.1. Structure and Permeability

Attempts have been made to correlate the paracellular permeability of epithelia and the structure of their tight junctions. When epithelia are fixed in aldehydes and are frozen and cleaved, and the fractured surfaces are replicated with platinum and carbon, the tight junctions appear as rows of anastomotic strands or fibrils on their inner protoplasmic (P) face (Fig. 1). The external (E) face of the junction is marked by furrows or grooves complementary to the fibrils. The composition of the fibrils is still in question; they may consist of micellar lipids rather than protein.[66] It has been proposed that epithelia with relatively high electrical resistance or low conductance from their mucosal to serosal sides have tight junctions with a greater number of strands and depth of strand rows than do epithelia that are relatively leaky to ions.[31] This proposal is supported by the finding that the addition of a hyperosmotic (600-mOsm) solution of mannitol to the mucosal surface of isolated jejunum causes a rapid increase in electrical resistance across its epithelium and a concomitant change in the structure of its tight junctions. Within 20 min, the electrical resistance rises by 64% with a statistically significant increase in the number of fibrils and the total depth occupied.[76]

These observations notwithstanding, there is very good evidence that the paracellular flow of ions is not directly related to junctional structures. In the toad urinary bladder, the continuous epithelium is regarded as tight because of its high restriction to the intercellular diffusion of electrolytes. When this epithelium is exposed to hypertonic solutions of lysine, the transepithelial electrical resistance or degree of tightness falls sharply, yet the number of strands and their total depth do not change appreciably.[77] During development, the choroid plexus epithelium in 40-day-old fetal sheep permits small lipophobic molecules such as erythritol, sucrose, and inulin to leave the blood and enter the cerebrospinal fluid (CSF) at a rate and amount that suggest a passive unrestricted diffusion. Here too, however, the number of fibrils and their overall depth are the same as in 125-day-old fetuses, when the blood–CSF barrier to these molecules has become established.[80] Although the number and depth of junctional strands may not be directly related to paracellular permeability, the continuity of strands may be. Endothelial cells of organs other than the central nervous system (CNS) are linked by tight junctions, but they are fascial rather than zonular in extent. Thus, the endothelial cells of capillaries supplying skeletal muscle[69] and lung[114] are connected by tight junctions that are not circumferential and therefore do not occlude the paracellular clefts completely. Moreover, the tight junctions

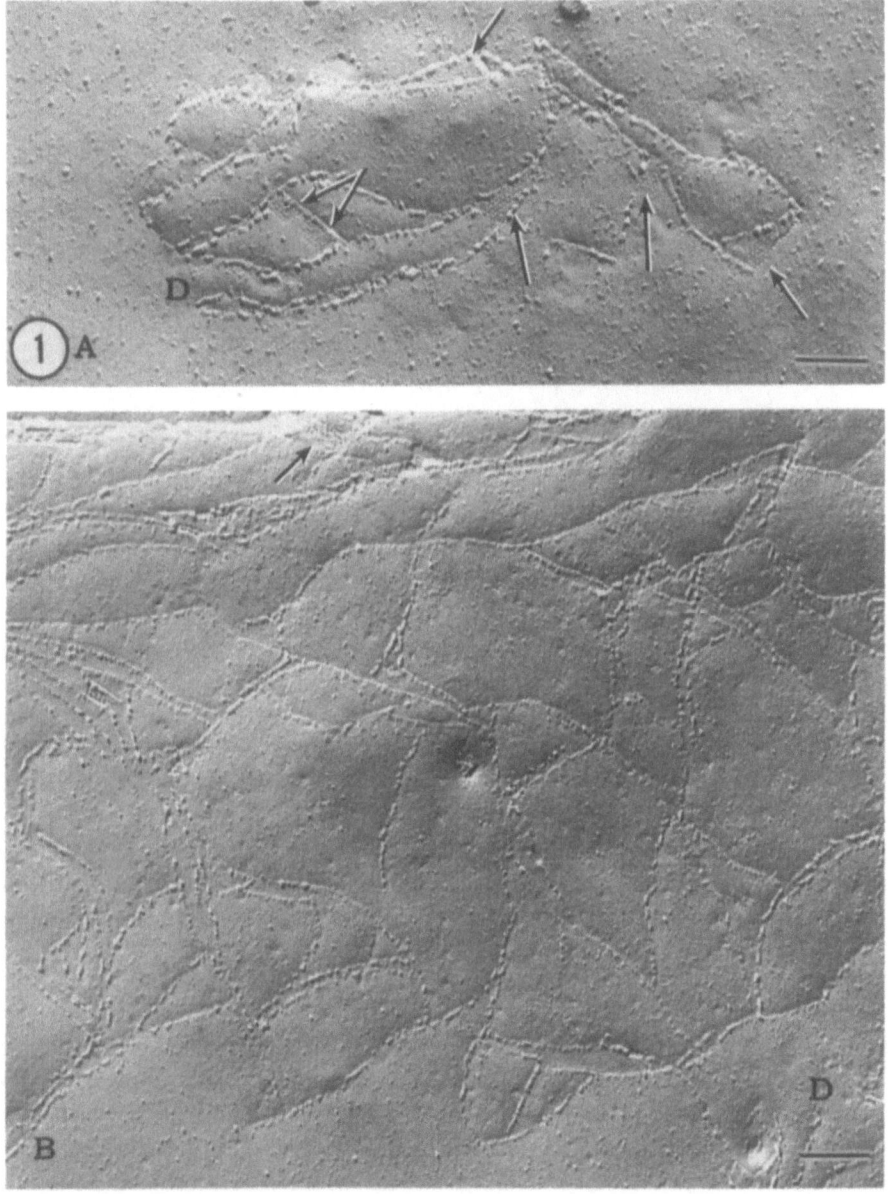

Figure 1. (A) Tight junctions between bovine brain endothelial cells maintained 6 days in vitro, then fixed, and freeze-fractured. Junctional strands are short; some are discontinuous (D). Five gap junctions (arrows) in outer face of cell membrane lie between but do not interrupt the strands. Bar = 145 nm. (B) Coculture of rat glial cells and bovine brain endothelial cells. Glial bed was 6 days old when endothelial cells were seeded onto it. Glial cells are now 12 days and endothelial cells are 6 days in vitro. Many more and longer junctional strands with very few D and only 1 small gap junction (arrow). Bar = 200 nm.

themselves may be interrupted by gap junctions, as in the carotid artery and aorta[60] and alveolar capillaries of the lung.[114]

The tight junctions between endothelial cells of cerebral capillaries do not appear to be interrupted by gap junctions. The plane of fracture usually yields disappointingly small portions of junction strands, but what published illustrations of en face replicas there are,[44,88,118] have not revealed a gap junction between, or adjacent to, the fibrils of the tight junction in cerebral capillaries. The absence of gap junctions is concordant with the impermeability of the tight junctions to protein from blood to brain[108] or brain to blood[10] and to ions.[11,41] This critical difference—the zonularity of the tight junctions, without the presence of gap junctions—sets the cerebral capillary apart from that of other organs. The tight seal provided by these junctions would be expected to impart a high electrical resistance. This expectation is borne out by the high average resistance of $1870\Omega\cdot cm^2$ in the frog's pial vessels; this value is similar to that of a tight epithelium.[33]

1.2. Glial Modulation

The highly occlusive junctions between cerebral endothelial cells may be established during development and subsequently maintained through some interactions with the surrounding cells. It has been suggested that barrier properties may be the consequence of perivascular astrocytes acting on the endothelium they ensheath.[35,93] The elegant experiments of Stewart and Wiley[124] give credence to this notion. Tissue from quail and chick can be distinguished by their different nucleolar morphology. Fragments of embryonic quail brain that had not yet become vascularized were transplanted to the chorioallantoic membrane of chick hosts. The regenerating vessels from the membrane are normally permeable but, upon entering the grafted brain tissue, assumed the barrier features of CNS vessels. Conversely, when embryonic quail somites were transplanted to the brain, the invading vessels that came from the surrounding brain developed characteristics of permeable, peripheral vessels.[124] Some of the grafts were intraventricular and could have been vascularized by the normally permeable vessels of the choroid plexus.[129] However, the grafts within the brain substance were vascularized by cerebral vessels that were induced by the somites to develop the permeable features of peripheral vessels.

The source of the blood supply to grafts of tissue containing vessels that are normally permeable in situ is not as readily apparent. Are all the intrinsic vessels of the graft replaced by those from the surrounding brain? Within 9–12 hr following transplantation of superior cervical ganglion to the surface of the fourth ventricle, anastomoses are formed between the vessels of graft and surrounding brain.[130] The rapidity with which these anastomoses are made suggests that some of the graft vessels had survived the trauma of transplantation. The permeability of the graft vessels could be accounted for by the surviving ones rather than by alterations imposed on invading, central vessels by their new environment.

Enzymic expression, in addition to structural characteristics, may also be determined by the milieu of a vessel. γ-glutamyltranspeptidase activity, normally present in cerebral endothelium in situ, is lost in vitro. When cerebral endothelial cells are cocultured with an astrocytic type of cell derived from a C6 glioma cell line, this enzymic activity reappears.[36] However, the specificity of this induction was not tested by the substitution of the astrocytic cell with another cell type (e.g., pericyte, smooth muscle, or fibroblast). The proximity in vitro of astrocytes also enhances the uptake by endothelial cells of neutral amino acids.[27]

Possible interactions that may determine membrane structure are currently being examined in our laboratory. The question we have asked is whether the coexistence of endothelial cells and astrocytes in culture affects the structure of the endothelial tight junction. Endothelial cells, dissociated from bovine brain enzymically and isolated by differential centrifugation, can be maintained in vitro.[49] In primary cultures, such cells, grown to confluence, have tight junctions[118] that are sufficiently well developed to exclude ionic lanthanum.[42] When the cultured endothelial cells are exposed to hyperosmotic solutions, they behave as they do in situ presumably shrinking so as to render the junctions permeable to hydrophilic solutes.[107]

The exclusion of ionic lanthanum by the tight junctions between cultured bovine brain endothelial cells maintained in isotonic medium,[42] is incompatible with the junction's appearance after freeze fracture. When, instead of primary cultures, several passages of endothelial cells from beef brain are made to enrich the number of these cells, we have found that the tight junctions become few, short, and discontinuous. Even when grown to confluence, the endothelial cells are separated by extracellular spaces that are much wider and more irregular than the usual 20-nm clefts in situ. In order for the intercellular flow of tracers to be halted in vitro, contiguous endothelial cells must, as in vivo, be tethered by continuous belts of tight junctions. However, in our freeze-fracture of enriched endothelial subcultures, the junctions are fascial and form patches of strands separate from neighboring ones. Although some fibrils are anastomotic, many have free ends and are thus discontinuous. The fibrils are closely associated with multiple small gap junctions (Fig. 1A) and in a few developing tight junctions, gap junctions actually interrupt the strands.

Our findings, although tentative, also suggest that while endothelial cells maintained "alone" in vitro are capable of forming tight junctions,[118] their extent and configuration might be brought closer to normal by the presence of astrocytes or some other cell type. In one set of 6-day-old cultures of cerebral endothelial cells that had been cultured alone, the total length of all the tight junctions was about 13 μm (Fig. 1A). When such endothelial cells had been cocultured for 6–17 days with primary cultures of astrocytes from the cerebral cortex of rats, the tight junctions were an order of magnitude longer: 150 μm.[126] The fibrils of the cocultures are continuous, there are more of them, and they form anastomotic loops with few or no free ends (Fig. 1B). Gap junctions, although they occasionally lie between the strands, are far fewer than in the endothelium grown without astrocytes or other cellular contaminants that have been included in the primary cultures (Fig. 1B). It is uncertain whether secondary cultures of astrocytes, which have fewer contaminating cells, are capable of normalizing the junctions. One contaminant, the fibroblast, does not affect the junctions, but we do not know whether pericytes or smooth muscle cells modulate junctional structure or development.

It is unlikely that the greater length and number of fibrils influence permeability of the junctions in the cocultures. Moreover, the additional, long separate fibrils may or may not comprise an uninterrupted seal between the cells. The trauma of removing tissue derived, for example, from the prostate gland and keeping it in various buffers at different temperatures results in the massive formation of long fibrils, where there are normally no tight junctions.[65]

The gap junctions of cerebral endothelial cells in vitro (Fig. 1A) are not a feature of differentiated cerebral capillaries in situ. The gap junctions, intermingled with endothelial tight junctions in vitro and in other organs,[60,73,114] are, as in the aorta, presumably capable of transmitting electrical activity and low-molecular-weight dye from cell to cell.[68] Such gap junctions intrude between the strands of tight junctions in the endo-thelium of cerebral vessels containing periendothelial smooth muscle: cerebral arteries and collecting veins.[88] Cerebral capillaries, which do not have smooth muscle or per-icytes, have only tight junctions. It is, therefore, possible that the endothelial cells having gap junctions in vitro are derived from the large vessels of the adult beef brain. However, the gap junctions are too numerous to be accounted for solely on their origin from larger vessels since such vessels are greatly outnumbered by capillaries that do not have the junctions. It is more likely that the gap junctions reflect a reversion to an earlier stage of development or an adaptation to in vitro conditions. In developing chick brains, cerebral endothelial cells are, in addition to tight junctions, connected by gap junctions that disappear during maturation.[38] The prevalence of gap junctions in developing cerebral endothelium is further exemplified in vitro where recultured endothelial cells from 2-day-old rats are united by gap junctions only.[123]

The diminution in gap junctions of beef brain endothelial cells cocultured with glial cells may thus be due to the close apposition of glial cells. It is conceivable that glial cells, directly abutting endothelium, may inhibit formation of gap junctions. Intervening muscle cells and pericytes could stand in the way of this inhibition. It may be informative to see whether the gap junctions are re-expressed in cultures upon removal of the astrocyte layer and its substitution with smooth muscle cells or pericytes. In summary, glial cells normal-ize the endothelial tight junctions in vitro by enhancing their continuity, length, and complexity while diminishing associated gap junctions. The junctions thus come to re-semble more closely those of mature barrier vessels in situ.

The components of the endothelial cell membrane, which may distinguish it from capillaries elsewhere, are being analyzed by Pardridge et al.[99] Antibodies raised against the plasma membranes of isolated cerebral microvessels selectively immunoprecipitate a $46,000-M_r$ protein. The antibody binds to the lateral portion of the cell membrane and therefore perhaps to the tight junctions. This important analysis makes possible future studies on the regulation of this protein expression.[99] It is possible that glial cells may have an influence on this type of regulation in vitro.

2. PITS, VESICLES, AND CHANNELS

At about the time that the pore hypothesis was published, a morphologic observation was made that had a profound impact on this view of capillary permeability. Palade[96] described a system of pits or Ω-shaped microinvaginations of the endothelial plasma membrane, about 50–100 nm in diameter, in the endothelium of muscle capillaries. Such vesicles, constructed at both luminal and abluminal surfaces of the endothelium, were envisioned as moving into the cytosol of the endothelial cell toward one or the other surface of the cell[96] (Fig. 8A).

2.1. Vesicle Translocation

The vesicles soon came to be regarded as the equivalent of the large pores through which macromolecules traverse such capillaries.[109] Estimates of vesicular turnover sug-gested that vesicular translocation could account for the transfer of molecules greater than $10,000\ M_r$.[109] Opposed to this view was the conclusion[45] that vesicular transport would be too slow and insufficient to account for the rapid entry of large amounts of hydrophilic substances from blood into the interstitium.

In order for a solute to be transferred across a cell, it must first be endocytosed. A morphologic sign of endocytosis is the accumulation of a probe molecule within such depots as lysosomes. Circulating horseradish peroxidase (HRP) is deposited in the lysosomes of cerebral endothelium,[108] but to a much lesser degree when the HRP comes from the CSF or abluminal side.[18] For this reason, a brain–blood barrier has been proposed for macromolecules.[18] However, some abluminal pits could form vesicles that migrate directly to the luminal surface and discharge their HRP. The HRP would not remain in the transferred vesicle, but would, instead, be lost to the blood.

2.2. Most Vesicles May Be Pits

There is now good evidence that the vast majority of what appear to be independent vesicles within the endothelial cytoplasm may actually be members of a system of membrane invaginations that communicate either with the blood or with the perivascular space (Fig. 8F). In the cerebral endothelium of cyclostomes such as the hagfish and sea lamprey, the cytoplasm is riddled with numerous pits, vesicles, and tubules, apparently formed by fusion of successive vesicles. Although the tubules and pits open to one or the other surface of the endothelial cell, no single channel opens to both blood and interstitial spaces simultaneously. Therefore, HRP, circulating in the blood, does not cross the capillary.[23] The vesicles, which can form intercommunicating, grapelike clusters, are regarded as a fixed, immobile system of membrane invaginations.[24] However, the very abundance of the pits suggests that, under the appropriate stimulus, a single vesicle or pit may form and move to an immediate neighbor, fuse with it, and thereby create a momentary transcellular channel through which solutes could traverse the endothelium (Fig. 8F). Tubules, which appear to arise from the coalescence of vesicles, are commonly observed in cerebral endothelium with both low-voltage[14,74] and high-voltage[117] electron microscopy. Nevertheless, an unequivocal opening at both ends of the tubule has yet to be illustrated. A tubule, containing HRP reaction product, when sectioned even the slightest bit obliquely, can give the impression of continuity between blood and perivascular space. The electron density of the grazed membrane belonging to the tubule matches the density of the reaction product and may give the false impression of continuity between luminal and abluminal faces.

A different approach—the cutting of extremely thin, serial, plastic sections—rather than the use of thicker plastic sections viewed with high-voltage electron microscopy, has indicated that even in normally permeable capillaries the endothelium is not traversed by channels. In the mesenteric vessels of the frog, approximately one-half of the spherical profiles in the endothelial cell open to the luminal surface and one-half to the interstitial space.[24] In reconstructions of very thin serial sections of capillaries in striated muscle of the frog, fewer than 1% (4 out of 614 spherical profiles) are actually vesicles. The remainder are pits, opening directly onto either the luminal or abluminal space or vesicles communicating with one of these surface pits.[47,48] No series of pits or vesicles form a transendothelial channel.

2.3. Vesicle Fusion and Fission

An alternative concept, that of transient fusion–fission, holds that a vesicle can form a brief communication with a second vesicle or tubule resulting in the mixing and equilibration of their contents (Fig. 8B). The vesicles then separate, leaving part of the contents with the second vesicle. This concept of solute transfer is based on quantitative considerations. During the steady state, achieved 30–40 sec after the intravascular infusion of ferritin in frogs, the endothelium of mesenteric vessels has luminal pits that are more heavily labeled than cytoplasmic vesicles, which in turn contain more ferritin than do abluminal pits.[32] According to the concept of vesicle translocation, luminal pits, abluminal pits, and vesicles should all have contained about the same amount of ferritin. Instead of vesicles shuttling across the endothelium, it is hypothesized that in this case the momentary fusion and separation of vesicles results in a progression of intravesicular material across the cell via these evanescent intercommunications.[32]

2.4. Cryofixation and Chemical Fixation

These conjectures are based on pictures of aldehyde-fixed brains. The assumption is that such fixed tissue is a reasonable facsimile of the living state. However, chemical fixation may be too slow to capture the transient fusion or fission of vesicles. Brownian and saltatory motion of organelles continue for up to 1 min in cells in vitro after exposure to glutaraldehyde[21] and vesicles may continue to fuse with each other during fixation[13] so that momentary events cannot be captured. Moreover, the number of endothelial pits and vesicles may be artifactually increased as a result of aldehyde fixation.[29,37,74] These aldehyde effects can be avoided by cryofixation: rapid freezing of fresh tissue followed by the substitution of tissue water with acetone at low temperatures and then with osmium tetroxide dissolved in acetone.[58,133] Tissue is well preserved but only for a distance of 15–30 μm from the surface in contact with the freezing block. In the rete mirabile of the leech

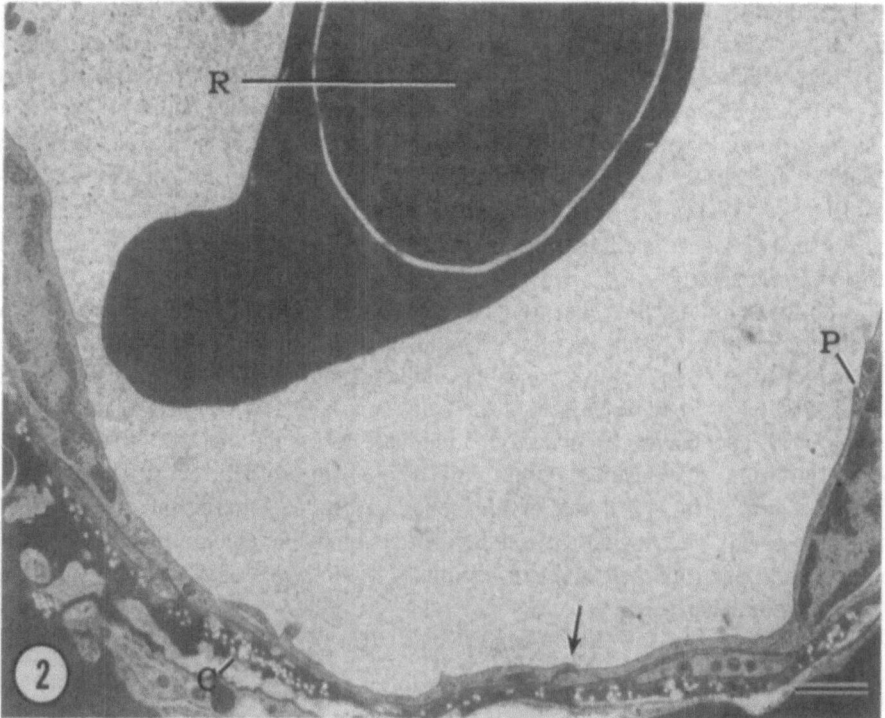

Figure 2. Rapidly frozen pial capillary of frog. Pit (P) indents plasma membrane of endothelial cell tethered to its neighbor by a junction (arrow). Perivascular space includes negatively stained collagen fibrils (C). Lumen contains a nucleated red blood cell (R). Bar = 0.87 nm.

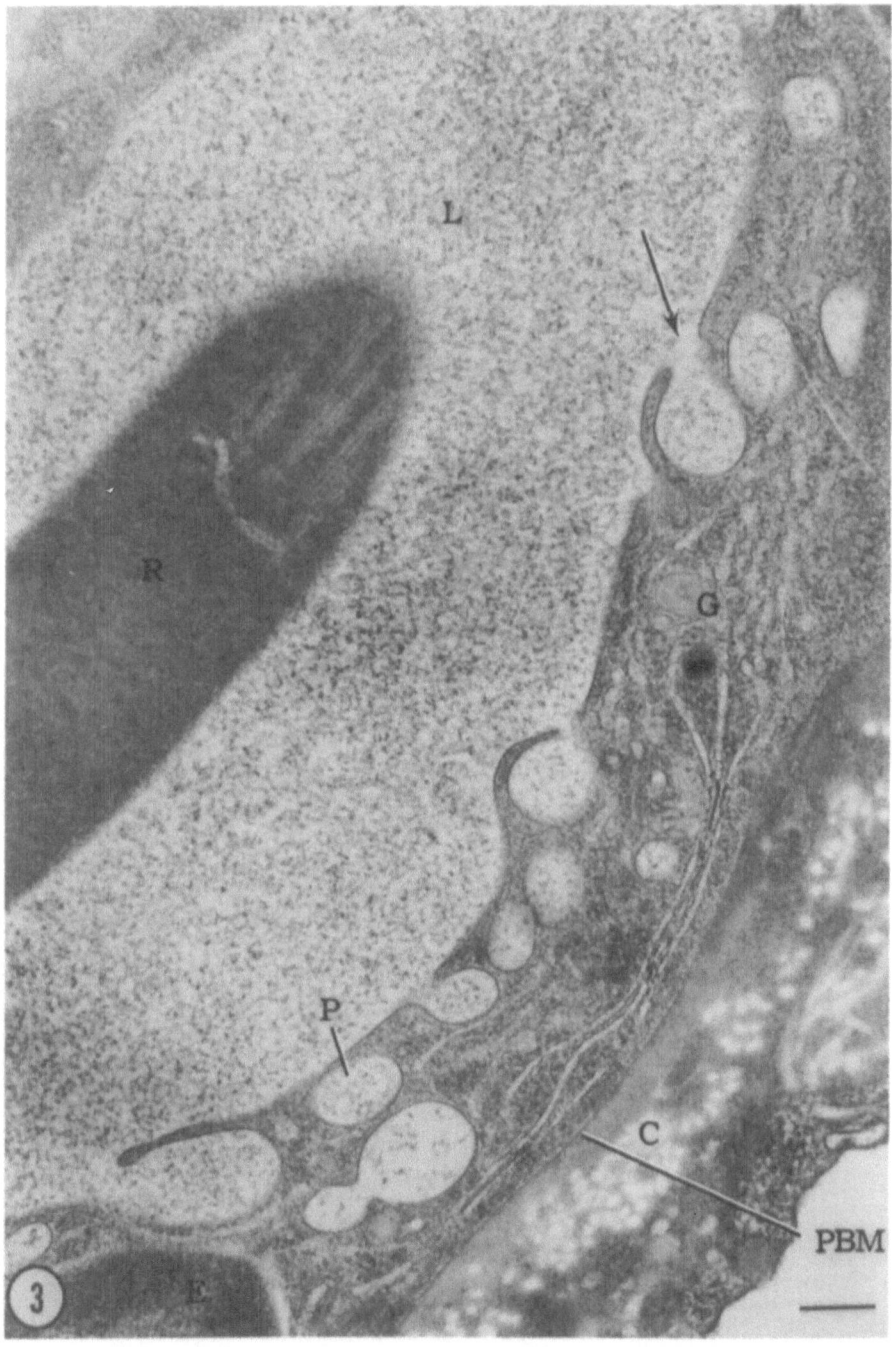

Figure 3. Frog pial capillary rapidly frozen after 30 min exposure to 3 M urea. Endothelial cell (E) emits microvilli (arrow), which may fuse with cell membrane to form large pits (P). Content of the pits is the same as the vessel lumen (L) in which a red blood cell (R) resides. Small vesicles belong to the Golgi complex (G). The perivascular basement membrane (PBM) abuts collagen fibrils (C). Bar = 260 nm.

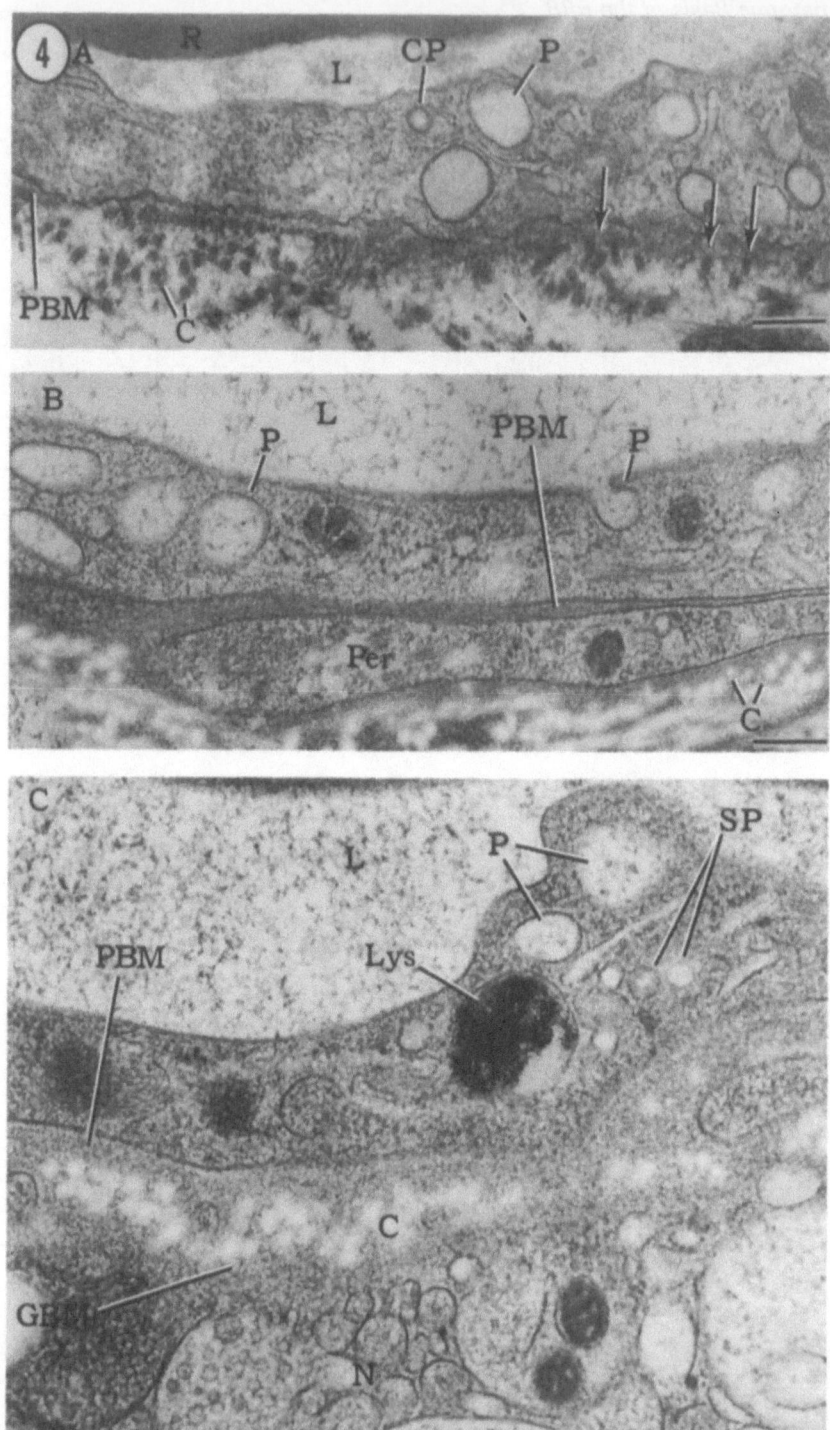

Figure 4. (A) Frog pial capillary, immerse fixed in aldehydes. Endothelial cell contains pits (P), one of them coated (CP), and is subtended by perivascular basement membrane (PBM) which, in places, contacts fibrils (C). Red blood cell (R) in vessel lumen. Bar = 280 nm. (B) Rapidly frozen pial capillary with P shares its PBM with a pericyte (Per). Collagen fibrils (C) abut the basement membrane of the pericyte. Bar = 260 nm. (C) Frog pial capillary rapidly frozen after 20 min exposure to 3 M urea, contains large (P) and small (SP) pits, some of which may be vesicles, near a lysosome (Lys). Collagen fibrils (C) lie between PBM and glial basement membrane (GBM) overlying neuropil (N). Bar = 190 nm.

swim bladder, the aldehyde-fixed endothelium has about three times more vesicles and pits than does the fresh-frozen endothelium.[139] Fewer pits and vesicles in other cryofixed endothelia have also been observed,[29,78,79] although inconsistent preservation of lung vessels renders morphometry questionable in this tissue.[143]

The only known species with pial vessels that are capillaries, and therefore amenable to rapid freezing, is the frog. Pial vessels, like those of the cerebral parenchyma, have a blood–brain barrier (BBB).[22] Our preliminary observations indicate that in cryofixed pial capillaries, the endothelium appears to contain either very few pits and spherical profiles (Figs. 2, 3, and 4C) or large pits, approximately 260 nm in diameter (Figs. 3 and 8G), that communicate with the blood space. In aldehyde fixed pial capillaries, large spherical profiles (Fig. 4A), like the vacuoles in rapidly frozen specimens (Figs. 3, 4B, and 4C), are readily obtained. The vacuoles appear to be far more common in capillaries near a freezing lesion made with a cold probe than they are in nonreactive vessels.

3. RECEPTORS

Is the cerebral endothelium, so refractory to the transcellular conveyance of large hydrophilic solutes, capable of such transfer with respect to certain macromolecules of metabolic importance to neurons and glia? While many macromolecules in the blood are either excluded by cerebral endothelium or endocytosed by it only to be enzymically degraded, there must be some that are transferred, unaltered, across it to the astrocytic face. These molecules would be transferred for utilization by neurons and glial cells. Being lipophobic and large, the solutes would have to be translocated via invaginations or by vesicles moving across the endothelium (Fig. 8D). There is reason to believe that one or more class of luminal pits may initiate such a selective transfer.

Different plasmalemmal pits as well as the vesicles they form bring solutes into the cell in either a selective membrane-associated way or in a nonspecific fluid-associated manner. The ratio of surface area to volume in pits and vesicles is very high. Therefore, a solute bound to the membrane of pits or vesicles is not only internalized selectively, but to a much greater extent than when it is merely dissolved or suspended within the fluid inside these structures.[61] The membrane-associated incorporation is known as adsorptive endocytosis. The bulk uptake, which is not associated with the plasma membrane, is designated as fluid-phase endocytosis and is nonspecific.

Adsorption of a solute to the cell membrane is mediated in at least two ways: receptor binding and coulombic linkage. A variety of biologically active molecules bind selectively to specific receptors on the cell membrane, which then indent as coated pits that give rise to coated vesicles. This process of ligand internalization is termed receptor-mediated endocytosis,[52,100] and is a variant of adsorptive uptake (Figs. 8C and D). Only a small fraction of pits bear a cytoplasmic coat, which consists of a fine network of filamentous protein designated as clathrin.[101] The coated pits are larger than their non-coated neighbors; they internalize and perhaps transport ligand across the cell[51] (Fig. 8D), while other vesicles bearing ligands are shunted to lysosomes, where the ligands are enzymically degraded (Fig. 8C). When intracellular K^+ is lowered, the membrane-associated clathrin coat disappears from Hep_2 cells, and diphtheria toxin can no longer be incorporated by these cells; however, the lectin ricin can.[82] Furthermore, coated pits are not the only type of invagination capable of receptor-mediated endocytosis. Tetanus and cholera toxins are preferentially bound to, and internalized by, small noncoated pits rather than coated ones in cultures of liver cells.[81] Whereas some coated pits bear receptors containing glycoproteins,[83] the receptors on smaller pits consist of glycolipids—sialogangliosides (see Montesano et al.[81] and references cited therein).

The likelihood that cerebral endothelium is capable of transferring certain molecules, unaltered, from blood to brain is considered next. Three ligands—the polypeptide insulin, low-density lipoprotein, and the protein transferrin—serve as examples of selective binding, incorporation, and the probability of transport by cerebral endothelium (Fig. 8D).

3.1. Insulin

In peripheral organs, the transcapillary flux of insulin ($5000\text{-}M_r$) depends, initially, on its association with receptors that have been located in vitro on the cell surface of endothelial cells derived from bovine aorta[67] and in the isolated beating heart.[5] In cultures of endothelial cells from bovine pulmonary arteries, biologically intact insulin rapidly passes through these cells as well.[40] Bovine aortic endothelial cells, grown to confluence in one chamber, transport [^{125}I]insulin to an adjoining chamber.[71] More than 80% of the insulin is transported intact. Since the transport of insulin is temperature dependent, inhibited by unlabeled insulin and by an antibody to insulin receptor, the transport must be receptor mediated.[71]

Insulin may be able to cross the BBB. When infused intravascularly in rats, labeled insulin binds specifically and rapidly to blood vessels throughout the CNS[34] and is receptor mediated.[46,98] However, the entry of circulating exogenous insulin into the CSF of dogs is minor in amount[147] and undetectable in rats.[53] Such largely negative results may be due to the slow rate at which insulin is transported by cerebral endothelium.[53] Isolated capillaries from human brains have a high affinity for insulin and are able to incorporate it and release it at similar rates into the medium, an activity that implies active transfer across the endothelial cell.[98] Most of the insulin is released, metabolically unaltered, by the capillaries. However, as Pardridge et al.[98] are careful to point out, their findings are also consistent with the possibility that the insulin may be released from the luminal front rather than the tissue front of the isolated vessels. Transfer of insulin across the cerebral endothelial cell will be difficult to demonstrate if the transfer rate is so low that only a very small fraction of pits or vesicles is involved at any one time.

The likelihood that insulin passes across cerebral vessels is intimated by the findings that insulin receptors are widely distributed throughout the brain,[56] including the hypothalamus[13] where the hormone has a direct and rapid modulatory effect on spontaneous electrical activity of neurons that regulate food intake.[1,94]

3.2. Low-Density Lipoprotein

A useful strategy for determining receptor number and distribution is to hold cells in culture at 4°C during ligand binding. This temperature does not inhibit binding but it does preclude internalization of the ligand. After maximal binding is achieved, the cells are warmed to 37°C whereupon the receptor membrane with attached ligand is internalized.[2]

The ligand, low-density lipoprotein (LDL), labeled with ^{125}I or ferritin, becomes widely but unevenly distributed over the entire cell surface of human fibroblasts in vitro at 4°C. Most, but not all, of the attached LDL is localized over short segments of cell membrane bearing a clathrin coat and making up only about 2% of the total surface area.[51] At 37°C, the coated segments invaginate as coated pits that form corresponding vesicles that lose their coat and fuse with lysosomes, where the cholesterol is cleaved from its LDL carrier (Fig. 8C). Such ligand-mediated endocytosis in macrophages does not stimulate nonspecific endocytosis.[50]

Endothelial cells also bear LDL receptors on coated patches of the cell membrane.[49,51,70] However, we have found that noncoated pits may also take up LDL, an event that may not be receptor mediated (Fig. 5B). In situ, LDL is not only incorporated by endothelial cells, but it is also transported across the cells. When LDL is perfused through the rat aorta or coronary artery it takes two paths.[136] One route is via high-affinity, receptor-mediated, adsorptive endocytosis with delivery from coated pits to lysosomes for the endothelial cell's metabolic needs. The second nonsaturable route is a nonreceptor one, mediated by noncoated pits and vesicles when large amounts of LDL are perfused. This path leads, within 5 min, to the subendothelial space opposite abluminal pits[135] where the LDL is available to the other cells of the vessel wall.

Our current experiments indicate that a mammalian, cerebral, endothelial cell that in vivo has few pits is induced to form many pits in vitro. In cerebral endothelial cells derived from beef brain and maintained in vitro at 5°C, LDL, labeled with colloidal gold, binds to coated and noncoated pits. When the cells are warmed to 37°C, the LDL-associated pits appear to form vesicles that translocate the LDL to internal regions of the cytoplasm (Fig. 5B). The expectation that normal cerebral endothelium, which does not transport appreciable amounts of proteins such as albumin or HRP, can transfer LDL to its abluminal face for delivery to periendothelial cells awaits experimental verification.

3.3. Transferrin

Another protein that circulates in the blood and that must be transported across capillaries to target tissue is transferrin. The role of this ligand is to sequester iron and carry it to cells, especially those undergoing mitosis.[125,129] Receptors for transferrin occupy the surface of many different tissues, including the microvessels of the mammalian brain.[62] Capillaries, as distinct from dissociated endothelial cells, when isolated from epididymal fat and maintained in vitro, can not only incorporate transferrin but can also release it into the surrounding medium.[140] Fluorochrome-labeled transferrin is endocytosed six to seven times faster than radiolabeled sucrose by these isolated capillaries because of its association with the cell membrane. There also appears to be a large component of fluid-phase endocytosis. When the capillaries are preloaded with tagged transferrin, pelleted, and resuspended in marker-free medium, the ligand is lost from the capillaries, presumably by exocytosis. Vesicular translocation, rather than simple diffusion from a confluent vesicular system,[24] appears to be responsible for transferrin release since the extrusion is temperature dependent.[140] This type of experiment does not indicate, however, whether the ligand is exocytosed from the abluminal face or the luminal face of the endothelium; it does not tell us whether transcellular movement of ligand takes place.

The transferrin receptors on cerebral endothelium[62] presumably initiate the passage of iron from plasma to certain neurons and glial cells. The neurons that accept the iron are widely distributed in the brain,[59] but if there are transferrin receptors on brain cells, they can only be reached after the ligand has traversed the cerebral endothelium, presumably via specialized pits and vesicles.

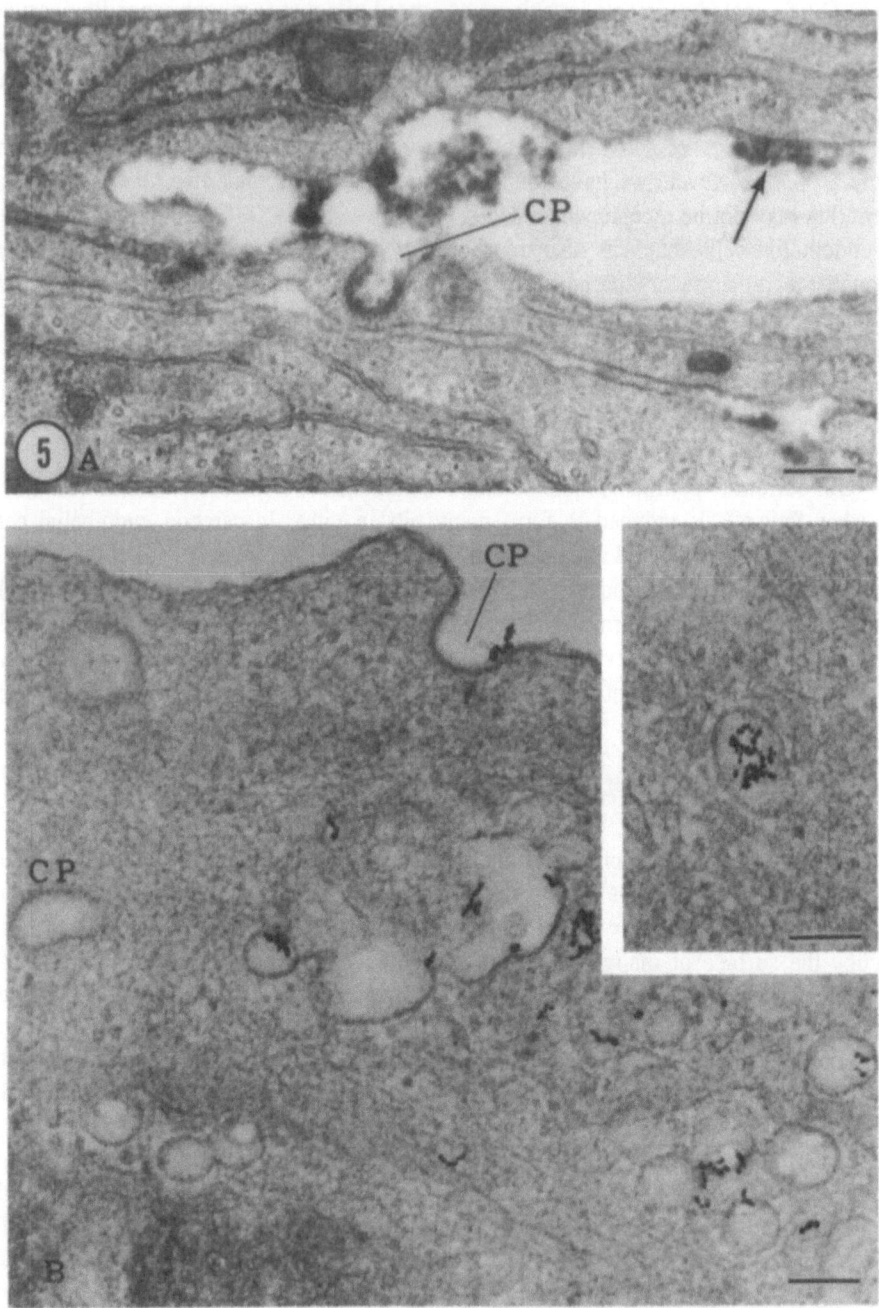

Figure 5. (A) Beef brain endothelial cells 10 days in vitro at 37°C. Secondary culture. Asparagus lectin, conjugated to HRP, binds to outer face of one coated pit (CP) and a segment (arrow) of the plasma membrane of another endothelial cell. Aldehyde fixed. Bar = 160 nm. (B) Beef brain endothelial cells 9 days in vivo at 37°C. Subconfluent culture, sixth passage. Low-density lipoprotein conjugated to colloidal gold (5-nm diameter). Some particles are attached to the membrane of a coated pit (CP), others to noncoated pits or vesicles. Aldehyde fixed. Bar = 160 nm. (Inset) A particularly heavily labeled, noncoated pit or vesicle. Aldehyde fixed. Bar = 240 nm.

4. ELECTRIC CHARGE

Coulombic interaction between the electric charge on a cell membrane and on a molecule that confronts it may, in addition to the size,[3,12] shape, and composition of the molecule, determine whether it enters or crosses the cell. The negative charge on neurons, for example, is attributable, in part, to the sialic acid residues of glycoproteins and glycolipids,[26] the phosphoric groups of phospholipids,[26] and in endothelium, the sulfated groups of heparan.[25,142]

The array of fixed negative charges on its luminal surface enables the endothelial cell to discriminate between otherwise similar molecules. Electrically neutral dextrans ($60,000-70,000-M_r$) traverse renal glomerular capillaries[12] more readily than do negatively charged sulfated dextrans of the same size.

The domains of the anionic sites have been visualized, at the electron microscopic level, by the binding of cationic substances, such as colloidal gold at low pH,[67] ruthenium red,[4,75,116] and cationic ferritin.[122] The amount and distribution of an ionized probe molecule vary with the experimental conditions, such as concentration of the ion, the duration over which the cells are exposed to it,[115] and whether the cells are challenged before or after aldehyde fixation.[26,121] The shortcoming of fixed tissue is that glutaraldehyde may alter the tertiary structure of cell surface proteins, thereby exposing more acidic groups than were present before fixation.[26]

Endothelial cells, isolated from human umbilical cord, bear evenly distributed negative charges on their plasma membrane. These dispersed negative charges are aggregated upon the addition of cationic ferritin (pI 8.8–10.9) before aldehyde fixation. The mosaic of anionic sites reverts to the more diffuse arrangement within a few minutes.[102] Some of the bound cationic ferritin is endocytosed within 30 min.

4.1. Charge and Endocytosis

Cationic ferritin is endocytosed by choroid plexus epithelium and by endothelium to a much greater extent than native ferritin (pI 4.2–4.6[55]) which is anionic at physiologic pH. Cationic ferritin, infused into the rat's cerebral ventricles, is not only endocytosed by choroidal epithelial cells more avidly than native ferritin but is transported to their intercellular clefts.[131] Similarly, when pieces of guinea pig aorta are incubated in cationic ferritin, a few molecules are engulfed by small endocytic pits at the luminal surface of the endothelium and are transferred across the vessel to the elastic lamina.[121]

This type of supravital experiment would be difficult to perform with cerebral vessels, the only accessible ones being the small arteries and veins on the pial surface. Furthermore, the removal of about one-half the sialic acid from polymorphonuclear leukocytes does not affect their phagocytosis of radioactively labeled starch particles.[92] In vivo, the neutralization of negative charges on the luminal surface of the cerebral endothelium results in breaching of the BBB. Vascular infusion of the cation, protamine sulfate, leads to extravasation of circulating HRP.[87] Apparently, the cation opens endothelial junctions, some of which have, according to the illustrations,[87] a gap of approximately 110 nm, through which HRP is exuded. The opening of the junctions is correlated with a diminution in the extent of anionic sites. However, the infusion of the polyanion, heparin, immediately after the protamine results in replenishment of negative charges but fails to prevent the barrier from being opened. Moreover, perfusion of heparin alone opens the barrier by, in some unknown way, affecting the tight junctions.[87]

Although protamine can "fuse" or closely appose the podocytes of the renal glomerulus,[115] it is difficult to see how a polycation can directly affect the endothelial tight junction. The much smaller, trivalent cation, ionic lanthanum does not enter or affect the tight junction in cerebral endothelium.[11,41] Perhaps a critical charge density is required to alter the structure of the tight junction.

The supposition that cerebral endothelium bears anionic sites would be more convincing if electrostatic binding of cations could be demonstrated in vivo. Intravenously administered cationic ferritin (pI 8.4) in rats consistently labels the luminal side of diaphragms in fenestrated vessels, which, as elsewhere, are permeable to large solutes. However, the cationic ferritin consistently fails to label the luminal surface of impermeable, barrier vessels.[113] Surprisingly, luminal anionic sites are also absent in regenerating cerebral endothelium 10 days after a cold lesion is made.[113] It would be expected that the BBB is still disrupted and that protein is exuded at this time; but the results imply that if solute extravasation across these vessels is transcellular, it would be independent of the luminal charge on the endothelium of the damaged area.

In the intact brain of the rat, the fenestrated endothelium of choroid plexus capillaries interacts with intravenously administered ferritin in a charge dependent fashion.[39] Native ferritin (pI 4.5) does not bind to the endothelial surface, but strongly cationic ferritin (pI 9.3) aggregates in luminal pits, abluminal pits, and channels. Significantly, some of this cationic ferritin also accumulates in the perivascular basal lamina and collagen fibrils opposite abluminal pits. This pattern is consistent with a transmural migration of ferritin via a vesiculotubular system[39] across these normally permeable capillaries of circumventricular organs.[14,32]

The distribution of anionic domains on the cell membrane of fenestrated vessels has, regardless of the organ of origin, one common feature: a high density of anionic charges on the fenestral diaphragm[39,119] (Fig. 8H,I). However, the diaphragm is usually impermeable to macromolecules such as ferritin.[104] In the fenestrated endothelium of pancreatic and jejunal capillaries of mice infused intravascularly with cationic ferritin, there is a time-dependent gradient of labeling: fenestral diaphragm > coated pits > plasma membrane (Fig. 8H). The ferritin does not bind to the membrane structures believed to be involved in solute transfer, i.e., noncoated pits, vesicles, or transendothelial channels.[119] This exclusion of cationic ferritin from organelles of transfer differs from the findings that cationic ferritin moves transmurally in aorta[121] and choroidal capillaries. Some factors that supersede coulombic adsorption may determine whether certain molecules can be endocytosed.

4.2. Charge and Lectin Binding

The interplay of surface charge and receptor composition on the binding and internalization of macromolecules is not understood. The saccharide components of glycosoaminoglycans on cell surfaces have been identified by their specific binding to a variety of plant-derived proteins called lectins.[20] Localization of the lectin, hence the specific oligosaccharide, is achieved by labeling the lectin with markers such as fluorochromes, ferritin, or HRP. An example is the binding of HRP-conjugated tetragonolobus purpureas lectin to α-L-fucosyl residues on coated segments and pits of the endothelial cell membrane (Fig. 5A). Simionescu et al.[120] have demonstrated an inverse relationship between the presence of anionic sites and lectin-binding domains. In situ, the fenestral diaphragms in the capillaries of the pancreas and intestine, like those in the choroid plexus,[39] are strongly anionic (Fig. 8I), but the pancreatic and intestinal fenestral diaphragms do not bind lectins or do so only to a slight degree.[120] The coated pits, which have fewer anionic sites[112] bind a number of lectins[120] (Fig. 8H). The adherence of these lectins is greatest to small noncoated pits, transendothelial channels, and their diaphragms, yet none of these structures carries a negative charge.[119] In some unknown way, the lectin receptors are purported to afford a degree of selectivity to the passage of serum-borne proteins from blood to tissue fluid.[119]

The removal of sialic acid residues from carotid artery endothelium by the in situ perfusion of sialidase results in a greatly accelerated uptake of LDL. The endothelial cells do not appear to be damaged, and the heightened internalization is attributed to the removal of anionic sites that would have repelled the LDL, which is negatively charged at physiologic pH.[54] However, as noted by Görög and Born,[54] charge is not the only factor, since macrophages do not internalize LDL but do incorporate negatively charged acetylated LDL.[50]

5. CIRCUMVENTING THE BARRIER

5.1. Retrograde Axonal Transport

When high concentrations of HRP are infused into the femoral vein of mice, the protein leaves the permeable vessels of muscle, skin, and the circumventricular organs of the brain to enter interstitial spaces, including synaptic clefts. From these clefts, presynaptic axon terminals endocytose the HRP, which is transported retrogradely through the axoplasm to the lysosomes of neuronal cell bodies. The protein thus enters the neuronal somata of cranial nerves, neurosecretory neurons, autonomic preganglia, and sensory and lower motor nerves.[17] This axonal route from the periphery into the CNS of blood-borne molecules, such as neurotoxins, and of neurotropic viruses, is a natural means by which the BBB may be circumvented. The transported agents would affect the neuron, provided they were not degraded by the acid hydrolases of the lysosomes, from which they would have to escape into the cytoplasm. Alternatively, the agents might be transported by axoplasmic vesicles and tubules which do not fuse with lysosomes. Here, too, the substances would have to leave these organelles to reach the cytosol.

5.2. Endothelial Alteration

Short of physical damage to blood vessels, several kinds of manipulations can affect the endothelial cell of the brain whereby large hydrophilic molecules cross from blood to interstitial fluid. Since such molecules cannot pass directly across a semipermeable membrane, only two mechanisms can account for the transmural passage: vesiculotubular formation,[14,74,117] and opening of the tight junctions.[16,86] An increase in the number of endothelial pits and vacuoles accompanies diverse insults such as trauma,[6,103,106] induced hypertension[144] and hyperosmotic exposure (reviewed by Brightman,[14] and van Deurs[131]). However, pictures of affected endothelium from single random sections of brains fixed at a given time cannot unequivocally differentiate between endocytosis and transcytosis. If an endothelial cell upstream has been damaged enough to allow proteins to escape from the blood, the HRP can enter abluminal pits retrogradely (Fig. 8E). The resulting simultaneous labeling of both luminal and abluminal pits simulates passage via a vesiculotubular system.[14] A less equivocal means of assessing the role of pits might be the serial sectioning of extremely thin plastic sections.[47]

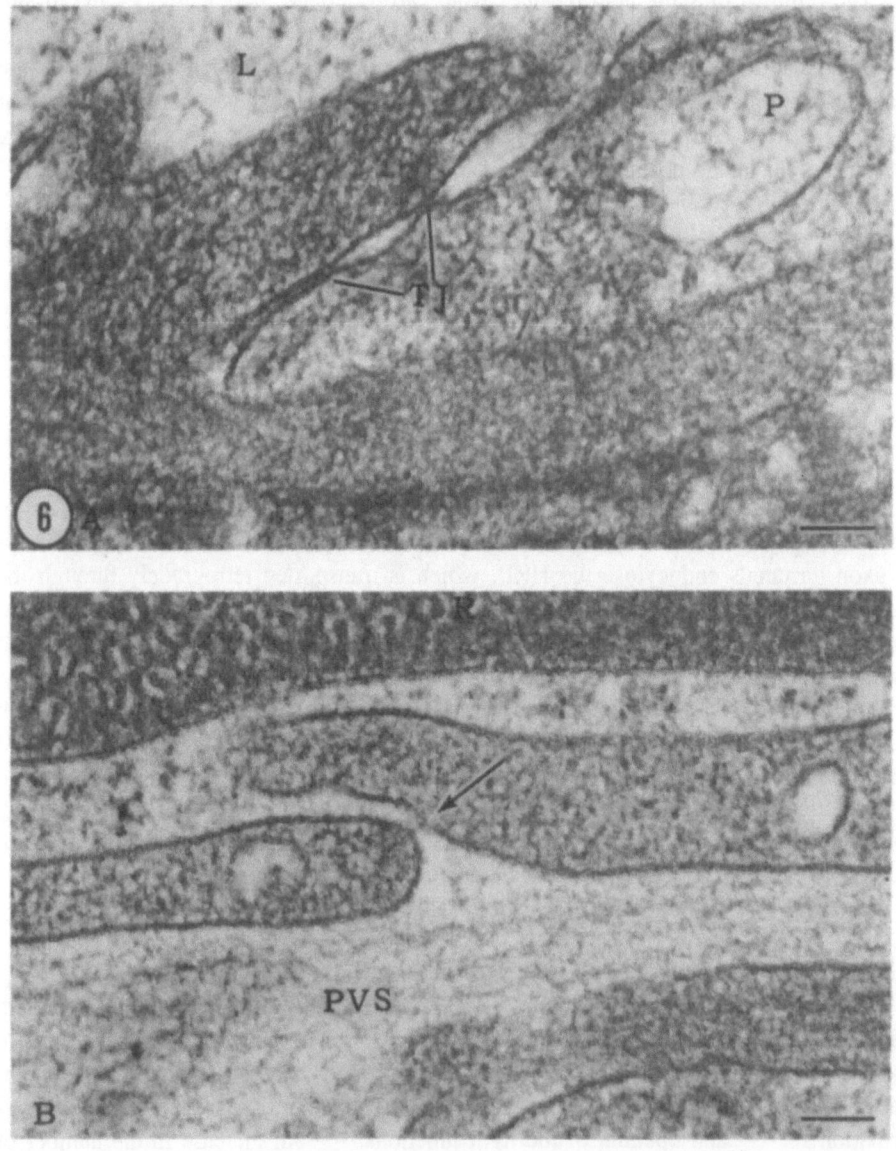

Figure 6. (A) Rapidly frozen pial capillary of frog. Two endothelial cells are connected by tight junctions (TJ) lying near a large pit (P), the contents of which are the same as vessel lumen (L). Bar = 95 nm. (B) Frog pial capillary rapidly frozen 30 min after exposure to 3 M urea. The junction (arrow) between two endothelial cells is now open to the lumen, which contains a red blood cell (R), and to the perivascular space (PVS). Bar = 95 nm.

The intercellular route is a likely one in vessels flushed with hyperosmotic fluid. Osmotic flow of water out of the endothelial cells shrinks the cells sufficiently to deform their junctions. Horseradish peroxidase can then enter successive interjunctional spaces normally inaccessible to protein[16] (Figs. 6A,E). The tight junctions of such capillaries, however, have not been fixed in the open state[16] and their disruption has been denied.[44,117] The likelihood of finding osmotically affected junctions in freeze-fracture replicas[44] is remote not only because the samples are very small but also because the junctional alterations are randomly dispersed. A possible means of capturing the patent mode of the junctions is to prevent the diffusion of water back into the cells during fixation. Cryofixation followed by substitution of tissue water with acetone is a technique whereby this condition can be achieved. Accordingly, a solution of 2.5–3.0 M urea, which opens the barrier to HRP in mammals,[16] has been applied dropwise onto the pial surface of frogs for 3–10 min. As a result, Evans blue escapes from some of the pial blood vessels. If these areas are removed, rapidly frozen, and substituted with acetone-osmium, gaps of about 20 nm can be visualized between some endothelial cells (Figs. 6B, 8G).[85] If this preliminary observation turns out to be the case, it will reinforce the notion that junctional opening during hyperosmotic exposure is the mechanism for transcapillary passage. In mammals, the intracarotid infusion of hyperosmotic saccharides leads consistently to a reversible opening of the barrier.[107] This successful and noninvasive way of introducing chemotherapeutic agents in patients with brain tumors[89] has the disadvantage of randomly disseminating potentially neurotoxic agents throughout the brain.[90]

5.3. Transplants

A highly invasive but localized means of bringing solutes into the cerebral parenchyma can be achieved by placing grafts of peripheral tissue with their permeable vessels onto or within the brain substance. When allografts of superior cervical ganglion are so placed, their extracellular compartment becomes confluent with that of the underlying brain. There is a time-dependent movement of circulating HRP from the permeable vessels of the graft into the interstitial space (IS) of the graft, and then into the IS of the underlying brain.[111] Autogenic grafts or allografts of skeletal muscle or skin inserted into the fourth ventricle become innervated, presumably by cranial nerves. The grafts survive for at least 1 year in rats.[141]

Although ventricular muscle allografts between outbred strains of rats also survive for at least 1 year, allografts between inbred strains are immunologically rejected by 2 months.[141] Intravascular HRP penetrates from autogenic skin and muscle grafts within the ventricle or parenchyma, into surrounding brain parenchyma for a limited distance of about 1 mm (Figs. 7A,B). The intraparenchymal grafts do not become innervated and persist for only a few months. While grafts provide a focal portal for the entry of blood-borne substances into a selected discrete region of the brain, it must be emphasized that, as in hyperosmotic opening, the passage of large hydrophilic solutes across the vessels is passive and nonselective.

Even when the barrier is focally bypassed, there is, in addition to a circumscribed exudation, a simultaneous widespread flow of HRP. This route appears to be along the perivascular basement membrane and is the same one taken by HRP infused into the ventricular or subarachnoid CSF.[110] We may now turn to a more precise consideration of where these substances, once having left the blood, move extracellularly.

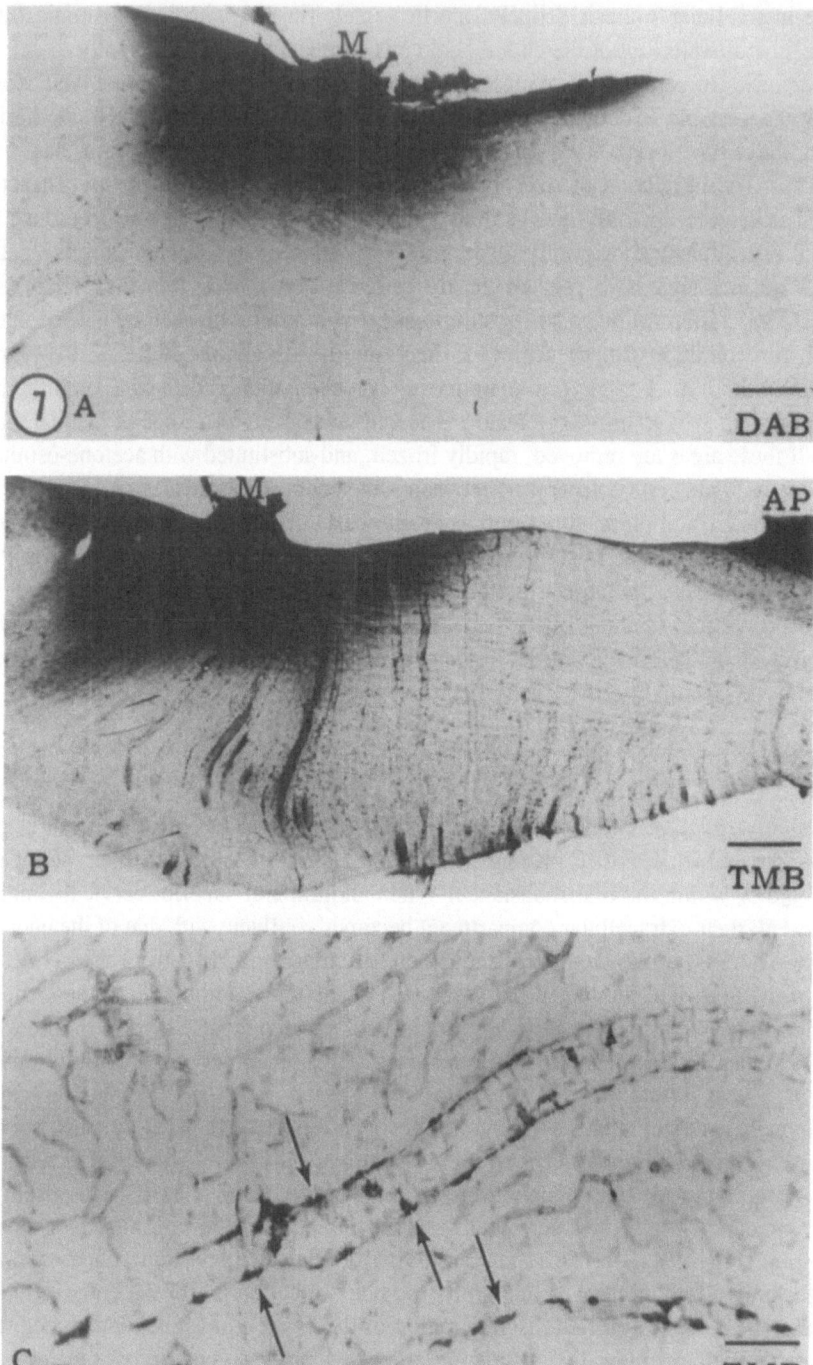

Figure 7. (A) Muscle (M) autografted to pial surface of rat medulla. HRP injected intravenously circulated for 1 hr before fixation. Exudate, about 1 mm deep, demonstrated with diaminobenzidine (DAB). Graft is 1 month old. Bar = 0.8 mm. (B) Same specimen as in (A) but incubated with tetramethyl benzidine (TMB). Walls of large vessels, including pericytes, have been labeled with HRP exuding from graft. Bar = 0.8 mm. (C) TMB incubation. Brain cells lightly stained with cresyl violet. Pericytes (arrows) of large vessels, as well as the outer wall of capillaries, have been labeled with HRP that exuded from the graft. The neural parenchyma between the vessels is free of reaction product. (AP), area postrema. Bar = 100 μm.

6. EXTRAVASCULAR SPREAD OF PEROXIDASE

When the cerebral capillary is rendered permeable or is bypassed, the escaping solutes immediately encounter the perivascular basement membrane (PBM), which is shared by endothelium and astrocyte. The PBM ensheaths all macro- and microvessels of the CNS and is perfectly continuous with the outermost or subpial basement membrane overlying the glia limitans.[91]

The constituents of cerebral basement membranes may influence the movement of molecules through them. Inasmuch as the basement membranes of the CNS have yet to be characterized, however, only general tentative extrapolations can be made from what is known about them in other organs. Structurally, the basement membrane consists of an electron-dense finely filamentous basal lamina situated between two electron-lucent layers. The innermost lucent layer is applied directly to the plasma membrane of the subtended cell and merges with the glycocalyx or glycosaminoglycan surface of the cell. The constituents of the basal lamina are, in general, glycoprotein, laminin, fibronectin, varying types and proportions of collagen, and glycosaminoglycans,[57] including proteoglycans.[4]

6.1. Perivascular Basement Membrane

Like the luminal surface, the PBM around endothelia bears a myriad of negative charges contributed by glycosaminoglycans. Partly because of these fixed negative charges, the PBM acts as an electrostatic filter to anionic solutes that reach it. The effects on solute mobility of altering the composition and organization of the PBM have been examined in the renal glomerulus. Here, the PBM is an effective filter of native ferritin, which is anionic at physiologic pH. Heparinase, perfused through the kidney, removes most of the glycosaminoglycans, so that ferritin is now able to traverse the entire PBM to enter the urinary space.[68] In aminonucleoside-induced nephrosis, heparin sulfate proteoglycan[84] and consequently their anionic charges[30] are lost from the glomerular PBM. Anionic proteins such as catalase[136] can now permeate the charge-depleted basement membrane.

Assuming that cerebral basement membranes are similar in composition to those elsewhere, anionic molecules should be electrostatically repelled by the numerous fixed negative charges within the membrane. When HRP (pI 7.3), electrically neutral at physiologic pH, is infused into the CSF, it readily spreads from the CSF along the outside of vessels.[15,110,138] Likewise, when HRP is infused intravascularly at sufficiently high concentrations and traverses cerebral capillaries that have been rendered permeable, it first enters the basement membrane and quickly spills over into the confluent interstitial spaces (IS).[14]

6.2. Interstitial Compartment

The localization of HRP within the cerebral IS and the measurement of how far it moves through it are determined by the substrate used to develop the reaction product of the enzyme. With diaminobenzidine (DAB), HRP perfused through the cerebral ventricles appears to move slowly, at about 0.5–1.0 mm/hr.[15,43,138] The more sensitive method, using tetramethylbenzidine (TMB), demonstrates a much more rapid and extensive spread from the CSF. When HRP is infused into the CSF of cats and dogs, it is propelled along the abluminal face of all blood vessels in the forebrain and brain stem in about 4–6 min.

The HRP moves from the CSF into the basement membrane network and back out into the CSF.[110] The important experiments conducted by Rennels et al.[110] have also shown that the rapidity and extent of HRP flow depends on pulsations of cerebral arterioles. If the brachiocephalic artery is partially clamped so that blood flow is maintained but pulsations eliminated, the spread of HRP is sharply limited.

The extracellular spread of solutes is convective rather than diffusional. When poly-ethylene glycols of two sizes are injected into the brain substance, they leave the brain via perivascular and periaxonal routes at about the same rate. If diffusion were responsible, the smaller molecule should have been cleared sooner than the larger one.[34]

On the basis of the description noted by Rennels et al. and our use of TMB on sections from graft-bearing brains, the convective spread of HRP from the grafts into the brain appears to have two components: interstitial and perivascular. The interstitial spread of solutes includes the synaptic, periaxonal, peridendritic, perineuronal, and periglial clefts.[10] The interstitial dissemination of HRP is somewhat, but not much, greater when measured after TMB than after DAB incubation and encompasses a 1–1.5-mm wide area (Fig. 7A,B).

6.3. Clearance via Basement Membranes

The perivascular component of spread is far more extensive than the interstitial one. Within minutes after the infusion of HRP into the CSF,[110] the perivascular border of the entire cerebral vasculature is limned by reaction product.[110] It is emphasized, however, that this convective component does not include the IS of the brain. Although the vascula-ture beyond the borders of the ventricle[110] or a graft[141] is extensively outlined by reaction product, the brain tissue between the vessels appears to be free of it (Fig. 7C). The term free is warranted because the TMB method is so sensitive. In normal rats, circulating HRP may be thoroughly flushed from cerebral vessels by perfusion of large volumes (150 ml) of balanced salt solution (BSS), followed by fixative; yet a residual fraction of HRP adsorbed to the luminal surface of the entire vascular tree is still discernible.[141] HRP, coming from the CSF[110] or from the exudate of a graft, enters the ubiquitous basement-membrane network, the surface area of which must be as vast as the vascular bed itself. Once inside the basement membrane, the HRP is rapidly disseminated and diluted. Not only does the basement membrane act as a sink, but the pericytes embedded within it incorporate some of the HRP[14,28,131,138] (Fig. 7B,C). The restriction of the HRP to the perivascular cells and basement membrane and its absence from the intervening IS are counter to the proposal that the basement membrane provides important channels for fluid "exchange with the ECF throughout the neuraxis."[110])

It is suggested here that once a solute such as HRP enters the all-pervasive cerebral basement membrane, it is prevented from exchanging between the CSF and interstitial fluid of the brain. The perivascular network of basement membranes may thus serve as an effluent pathway for the *clearance* of catabolites and other locally produced sub-stances, such as neurotransmitters, from the interstitial fluid rather than their *distribution* through it.

The concept of paravascular flow is partly based upon the inference that the per-ivascular basement membrane is the pathway. However, it is our experience that TMB reaction product, which is extracellular, leaches from the tissue section and that what little remains often straddles the endothelium and basement membrane, rather than being localized within the membrane. Further technical refinements are required for the precise localization of the HRP.

7. ENZYMES

By restricting the passive intercellular exchange of hydrophilic solutes between blood and the interstitial fluid (IF) of the CNS, the zonular tight junctions between cerebral endothelial cells permit the active, *selective* regulation of the composition of the fluid. This energy-dependent selectivity is effected by enzymes associated with the endothelium and its plasma membrane.

As pointed out by Betz et al.,[9] the tight junctions also indirectly play a role in homeostasis by imposing a property upon the cerebral endothelium not shared by the capillaries of other organs: polarity of enzyme distribution and thus of active transport. The zonular tight junctions exert this polarity by preventing the flow of protein between luminal and abluminal parts of the plasma membrane. Thus, the epithelial cells of frog urinary bladder can be dissociated by opening their tight junctions with EDTA. A substance, probably a glycoprotein, normally restricted to the apical surface of adjoined cells, can now move from apical to lateral and thence to basal surfaces of these epithelial cells after their tight junctions have been opened.[105]

The transport of amino acids exemplifies the vectorial distribution of enzymes and directed transport.[8] However, since the same enzyme systems may be shared by neurons and glia, the polarity has been most unequivocally demonstrated in endothelial cells that have been dissociated. Certain large neutral amino acids readily exchange, by a Na^+-independent L system, in either direction between blood and brain, and their carrier enzyme is situated in both luminal and abluminal portions of the endothelial cell membrane.[9] This symmetry was also indicated by localization of α-glutamyltranspeptidase, which may be involved in such transport, in two membrane fractions derived from endothelial cells. By contrast, the transport of certain small, neutral amino acids out of the brain is mediated by a Na^+-dependent A system, confined to the abluminal surface of cerebral endothelium.[8] The polarized distribution of these enzymes permits the preferential transfer of certain small, neutral amino acids out of the brain, so that their amount in the IS can be maintained at a constant level in the face of varying plasma concentrations.[9]

These important inferences have been carried a step further by the demonstration that the polarity of the cerebral endothelium for amino acid transfer may be imposed by its astroglial sheath. Cerebral endothelial cells were grown on one face of a nitrocellulose membrane and C6 glioma cells on the opposite face. The cell-covered membrane was inserted between the two compartments of a bipartite chamber. The small neutral non-metabolizable amino acid, [^{3}H]-α-methylaminoisobutyric acid, was transported from the glial side of the filter across the endothelium to a greater degree than from the endothelial side of the filter across the glial layer.[7]

Other amino acids may be enzymatically degraded and thus prevented from reaching the cerebral interstitium. Thus, L-3,4-dihydroxyphenylalanine is prevented from reaching the brain by enzymatic trapping, i.e., decarboxylation to dopamine and its subsequent metabolism by monoamine oxidase.[95]

The selected energy-requiring flux of sodium and potassium in the brain requires the mediation of Na^+-K^+ activated adenosine triphosphatase (ATPase). Detected cytochemi-

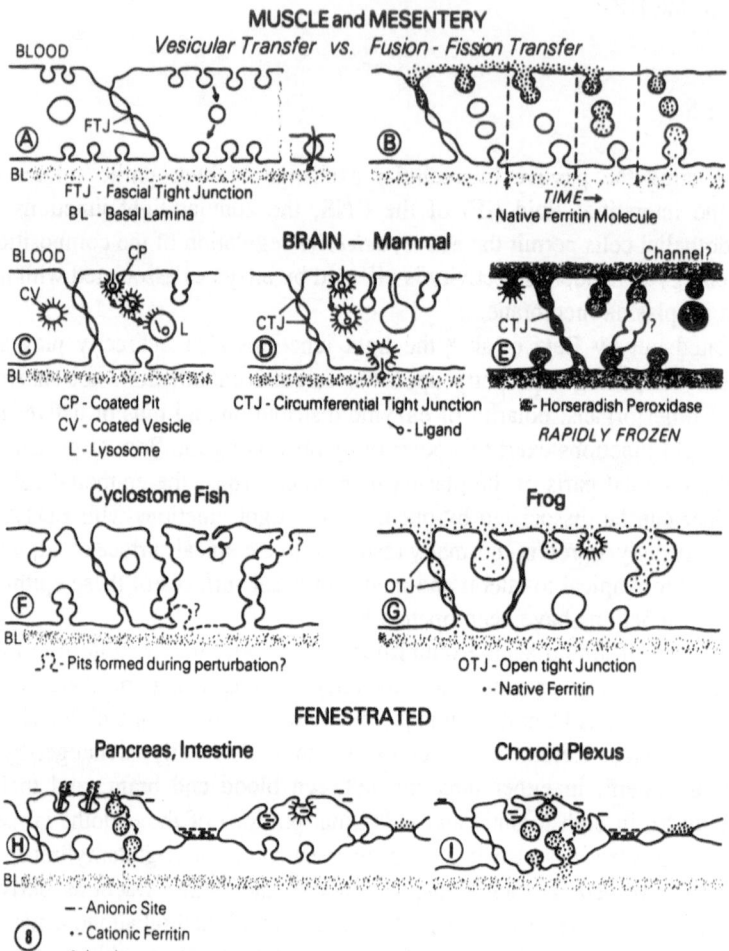

Figure 8. Possible routes for hydrophilic macromolecules across capillary endothelium. (A) Fluid phase endocytosis and transfer. Endothelial cell membrane forms pits that pinch off as vesicles that move through cytoplasm to fuse with opposite side of membrane, disgorging the vesicle contents into the basal lamina. Fascial (interrupted) tight junctions provide a paracellular route for solutes between blood and tissue fluid. (B) A luminal pit containing macromolecules fuses with a vesicle into which some of the molecules enter. The vesicle then separates from the pit (fission) and fuses with another vesicle to which it contributes a fraction of its molecules. (C) Adsorptive endocytosis and transfer. A coated pit, with receptors on its membrane, has a high affinity for a ligand and forms a coated vesicle that fuses with a lysosome where the ligand is enzymically degraded. (D) Alternatively, the ligand is carried by a coated vesicle to the abluminal portion of the cell membrane to be released, unaltered, into the extravascular milieu. (E) Passive paracellular and transcellular passage. Hyperosmotic agents in the blood shrink the endothelial cell sufficiently to open tight junctions. Transcellular channels, another possible route, have not been shown unequivocally. (F) This cerebral endothelium is riddled with confluent pits, some of which are deeply invaginated, but none of which form transcellular channels. However, when perturbed, the endothelium may form new pits which could complete a connection to both surfaces. (G) Hyperosmotic urea opens the zonular tight junctions, as viewed in cryofixed brain, but transcellular channels have not been demonstrated. (H) Coulombic binding and transfer. Cationic ferritin binds to some luminal pits bearing negative charges on fenestrated capillaries and is transported to extracellular fluid. In such extracerebral capillaries, the ferritin does not bind to those pits which bind lectins. Anionic sites are concentrated at fenestrated diaphragms, but cationic ferritin does not cross them. This type of transfer also takes place across nonfenestrated endothelium.

cally with the highly specific substrate, *p*-nitrophenylphosphate, this ATPase has been localized in the luminal and abluminal plasma membrane of cerebral microvessels,[137] but the same cytochemical method may yield reaction product on only the abluminal surface.[9] This one-sided localization has been confirmed biochemically, by restriction of the enzyme to only one of two distinct membrane fractions derived from cerebral endothelium.[9] The abluminal restriction of the ATPase is concordant with a vectorial transport of K^+ out of the brain.[9] Although ATP is involved in ion transport, it does not appear to mediate endocytosis. Metabolic inhibitors, such as 2,4-dinitrophenol, NaN_3, or iodoacetate at concentrations that inhibit ATP, do not quantitatively affect endocytosis by endothelial cells in vitro.[137]

Adenylate cyclase, which produces cyclic adenosine monophosphate (cAMP) from adenosine triphosphate, has been detected in astrocytes and in cerebral capillaries.[64] Since dibutyryl cAMP is purported to enhance endocytosis in cerebral endothelium,[63] Joó surmises that adenylate cyclase could be involved in the regulation of capillary permeability.

Enzymes such as nucleoside diphosphatase (NDPase) involved in the synthesis of polysaccharides and glycoprotein, hence of cell-surface glycosaminoglycans, have been detected in brain cells and blood vessels. Cerebral endothelium,[127] in addition to neurons and glia,[127] has NDPase at the luminal and abluminal surface of capillaries and their basement membrane as well as at the surface of pericytes. NDPase is also present within the cisternae of the endoplasmic reticulum of neurons and glial cells.[137] However, the presence of NDPase on the surface of cerebral endothelial cells has been questioned,[19] because the alkaline phosphatase at this site may cleave the phosphate nonspecifically from the inosine diphosphate substrate used.[137]

Another enzyme, this one having to do with the metabolism of the endothelial cell and subsequently that of brain cells, is glucose 6-phosphatase (G6Pase), which clearly emphasizes a transverse disposition of endoplasmic reticulum peculiar to cerebral endothelium.[112] In the endothelium of cerebral capillaries and arterioles of mice, G6Pase resides within the endoplasmic reticulum and its extension, the perinuclear envelope.[19] The function of this enzyme in endothelia is still a matter of speculation. One interesting proposal is that it may be involved in the synthesis of a readily available cytoplasmic pool of glucose for utilization by the endothelial cell itself. The unused portion that leaves the endothelial cell could then be available to the neural parenchyma beyond it.[19]

8. CONCLUSIONS

There is reason to believe that the normal cerebral endothelium, so refractory to the transcellular transport of large solutes, can transfer certain peptide and protein ligands by way of receptor-bearing portions of its cell membrane. Other solutes are vectorially exchanged between brain and blood by the polarized distribution of enzymes on either the luminal or abluminal parts of the endothelial cell membrane. The active enzymically driven transfer can take place because the passive intercellular route between cerebral endothelial cells is closed by zonular tight junctions.

These tight junctions, which block the intercellular passage of solutes, are formed in close association with gap junctions in vitro by endothelial cells alone. However, the proximity of glial cells is required in vitro for tight junctions to resemble more closely those of cerebral capillaries in situ. The opening of once tight junctions between endothelial cells exposed to hyperosmotic solutions in situ can be captured by rapid freezing of anuran pial capillaries. Alternatively, the BBB can be focally circumvented by grafts to the brain of tissue with intrinsic vessels that are normally permeable.

It is emphasized that these manipulations bypass or open the barrier in a nonselective fashion. Having crossed the endothelium a solute can then enter the interstitial clefts where its distribution may be influenced by its electric charge. The probable receptor-mediated transfer of certain ligands across the BBB and the role of electric charge on the extracellular dissemination of solutes merit further experimental consideration.

ACKNOWLEDGMENT

The author is indebted to Dr. William Pardridge for most helpful discussions.

REFERENCES

1. Anand BK, Chhina GS, Sharma KN, et al: Activity of single neurons in the hypothalamic feeding centers: Effect of glucose. *Am J Physiol* 207:1146–1154, 1964.
2. Anderson RGW, Goldstein JL, Brown MS: Localization of low density lipoprotein receptors on plasma membrane of normal human fibroblasts and their absence in cells from a familial hypercholesterolemia homozygote. *Proc Natl Acad Sci USA* 73:2434–2438, 1986.
3. Arfors KE, Rutili G, Svensjo E: Microvascular transport of macromolecules in normal and inflammatory conditions. *Acta Physiol Scand* 463(suppl):93–103, 1979.
4. Ausprunk DH, Boudreau CL, Nelson DA: Proteoglycans in the microvasculature II. Histochemical localization in proliferating capillaries of the rabbit cornea. *Am J Pathol* 103:367–375, 1981.
5. Bar RS, DeRose A, Sandra A, et al: Insulin binding to microvascular endothelium of intact heart. A kinetic and morphometric analysis. *Am J Physiol* 244:E447–E452, 1983.
6. Beggs JL, Waggener JD: Transendothelial vesicular transport of protein following compression injury to the spinal cord. *Lab Invest* 34:428–439, 1976.
7. Beck DW, Vinters HV, Hart MN, et al: Glial cells influence polarity of the blood–brain barrier. *J Neuropathol Exp Neurol* 43:219–224, 1984.
8. Betz AL, Goldstein GW: Polarity of the blood–brain barrier: Neutral amino acid transport into isolated brain capillaries. *Science* 202:225–227, 1978.
9. Betz AL, Firth JA, Goldstein GW: Polarity of the blood–brain barrier: Distribution of enzymes between the luminal and abluminal membranes of brain capillary endothelial cells. *Brain Res* 192:17–28, 1980.
10. Bodenheimer TS, Brightman MW: A blood–brain barrier to peroxidase in capillaries surrounded by perivascular spaces. *Am J Anat* 122:249–267, 1968.
11. Bouldin TW, Krigman MR: Differential permeability of cerebral capillary and choroid plexus to lanthanum ion. *Brain Res* 99:444–448, 1975.
12. Brenner BVM, Hostetter TH, Humes DH: Molecular basis of proteinuria of glomerular origin. *N Engl J Med* 298:826–833, 1978.
13. Bretscher MS, Whytock S: Membrane-associated vesicles in fibroblasts. *J Ultrastruct Res* 61:215–217, 1977.
14. Brightman MW: Morphology of blood–brain interfaces. In Bito LA, Davson H, Fenstermacher JD (eds): *The Ocular and Cerebrospinal Fluids. Exp Eye Res* 25:1–25, 1977.
15. Brightman MW, Reese TS: Junctions between intimately apposed cell membranes in the vertebrate brain. *J Cell Biol* 40:648–677, 1969.
16. Brightman MW, Hori M, Rapoport SI, et al: Osmotic opening of tight junctions in cerebral endothelium. *J Comp Neurol* 152:317–326, 1973.
17. Broadwell RD, Brightman MW: Entry of peroxidase into neurons of the central and peripheral nervous systems from extracerebral and cerebral blood. *J Comp Neurol* 166:257–283, 1976.
18. Broadwell RD, Balin B, Salcman M: Brain–blood barrier. Yes and no. *Proc Natl Acad Sci USA* 80:7352–7356, 1983.

19. Broadwell RD, Cataldo AM, Salcman M: Cytochemical localization of glucose-6-phosphatase activity in cerebral endothelial cells. *J Histochem Cytochem* 31:818–822, 1983.
20. Brown JC: Lectins. *Int Rev Cytol* 52:277–349, 1978.
21. Buckley IK: Studies in fixation for electron microscopy using cultured cells. *Lab Invest* 29:398–410, 1973.
22. Bundgaard M: Ultrastructure of frog cerebral and pial microvessels and their impermeability to lanthanum ions. *Brain Res* 241:57–65, 1982.
23. Bundgaard M, Cserr H, Murray M: Impermeability of hagfish cerebral capillaries to horseradish peroxidase. *Cell Tissue Res* 198:65–77, 1979.
24. Bundgaard M, Frokjaer-Jensen J, Crone C: Endothelial plasmalemmal vesicles as elements in a system of branching invaginations from the cell surface. *Proc Natl Acad Sci USA* 76:6439–6442, 1979.
25. Buonassisi V, Colburn P: Hormone and surface receptors in vascular endothelium. *Adv Microcirc* 9:76–94, 1980.
26. Burry RW, Wood JG: Contributions of lipids and proteins to the surface charge of membranes. *J Cell Biol* 82:726–741, 1979.
27. Cancilla PA, DeBault LE: Neutral amino acid transport properties of cerebral endothelial cells in vitro. *J Neuropathol Exp Neur* 42:191–199, 1983.
28. Cancilla PA, Baker RN, Pollok PS, et al: The reaction of pericytes of the central nervous system to exogenous protein. *Lab Invest* 26:376–383, 1972.
29. Casley-Smith JR: Freeze-substitution of capillary endothelium: The artefactual nature of trans-endothelial channels and the forms of attached vesicles. *Micron* 11:461–462, 1980.
30. Caulfield JP, Farquhar MG: Loss of anionic sites from the glomerular basement membrane in animonucleoside nephrosis. *Lab Invest* 39:505–512, 1978.
31. Claude P, Goodenough DA: Fracture faces of zonulae occludentes from "tight" and "leaky" epithelia. *J Cell Biol* 58:390–400, 1973.
32. Clough G, Michel CC: The role of vesicles in the transport of ferritin through frog endothelium. *J Physiol (Lond)* 315:127–142, 1981.
33. Crone C, Olesen SP: Electrical resistance of brain microvascular endothelium. *Brain Res* 241:49–55, 1982.
34. Cserr HR: Convection of brain interstitial fluid. In Kovach AGB, Hamar J, Szabb L (eds): *Cardiovascular Physiology Microcirculation and Capillary Exchange.* Proceedings of the 28th Congress of Physiological Sciences. Budapest, 1980 pp. 337–341.
35. Davson H: The blood–brain barrier. *J Physiol (Lond)* 255:1–28, 1976.
36. DeBault LE, Cancilla PA: γ-glutamyl transpeptidase in isolated brain endothelial cells: Induction by glial cells in vitro. *Science* 207:653–655, 1980.
37. De Bruyn PPH, Michelson S, Becker RP: Nonrandom distribution of sialic acid over the cell surface of bristle-coated endocytic vesicles of the sinusoidal endothelium cells. *J Cell Biol* 78:379–389, 1978.
38. Delorme P, Gayet J, Grignon G: Ultrastructural study on transcapillary exchanges in the developing telencephalon of the chicken. *Brain Res* 22:269–283, 1970.
39. Dermietzel R, Th%rauf N, Kalweit P: Surface charges associated with fenestrated brain capillaries. II. In vivo studies on the role of molecular charge in endothelial permeability. *J Ultrastruct Res* 84:111–119, 1983.
40. Dernovsek KD, Bar RS, Ginsberg BH, et al: Rapid transport of biologically intact insulin through cultured endothelial cell. *J Clin Endocrinol Metab* 58:761–763, 1984.
41. Dorovini-Zis K, Sato M, Goping G, et al: Ionic lanthanum passage across cerebral endothelium exposed to hyperosmotic arabinose. *Acta Neuropathol (Berl)* 60:49–60, 1983.
42. Dorovini-Zis K, Bowman P, Betz A, et al: Hyperosmotic arabinose solutions open the tight junctions between brain capillary endothelial cells in tissue culture. *J Brain Res* 302:383–386, 1984.
43. Dunker RO, Harris AB, Jenkins DP: Kinetics of horseradish peroxidase migration through cerebral cortex. *Brain Res* 118:199–217, 1976.
44. Farrell CL, Shivers RR: Capillary junctions of the rat are not affected by osmotic opening of the blood–brain barrier. *Acta Neuropathol (Berl)* 63:179–189, 1984.
45. Fawcett DW: Surface specialization of absorbing cells. *J Histochem Cytochem* 13:75–91, 1965.
46. Frank HJL, Pardridge WM: A direct in vitro demonstration of insulin binding to isolated brain microvessels. *Diabetes* 30:757–761, 1981.
47. Frokjaer-Jensen J: The plasmalemmal vesicular system in capillary endothelium. *Prog Appl Microcirc* 1:17–34, 1983.
48. Frokjaer-Jensen J: The plasmalemmal vesicular system in striated muscle capillaries and in pericytes. *Tissue Cell* 16:31–42, 1984.

49. Goldstein GW, Wolinsky JS, Csejtey J, et al: Isolation of metabolically active capillaries from rat brain. *J Neurochem* 25:715–717, 1975.

50. Goldstein JL, Ho YK, Basu SK, et al: Binding site on macrophages that mediates uptake and degradation of acetylated low density lipoprotein, producing massive cholesterol deposition. *Proc Natl Acad Sci USA* 76:333–337, 1979.

51. Goldstein JL, Richard GW, Anderson RGW, et al: Coated pits, coated vesicles, and receptor–mediated endocytosis. *Nature (Lond)* 279:679–685, 1979.

52. Gonatas J. Steiber A, Olsnes S, et al: Pathways involved in fluid phase and adsorptive endocytosis in neuroblastoma. *J Cell Biol* 87:579–588, 1980.

53. Goodner CJ, Berrie MA: The failure of rat hypothalamic tissues to take up labeled insulin in vivo or to respond to insulin in vitro. *Endocrinology* 1001:605–612, 1977.

54. Görög P, Born GVR: Increased uptake of circulating low-density lipoproteins and fibrinogen by arterial walls after removal of sialic acids from their endothelial surface. *Br J Exp Pathol* 63:447–457, 1982.

55. Graham RC, Karnovsky M: The early stages of absorption of injected horseradish peroxidase in the proximal tubules of mouse kidney; Ultrastructural cytochemistry by a new technique. *J Histochem Cytochem* 14:291–302, 1966.

56. Havrankova J, Roth J, Brownstein M: Insulin receptors are widely distributed in the central nervous system of the rat. *Nature (Lond)* 272:827–829, 1978.

57. Hay ED: Extracellular matrix. *J Cell Biol* 91:205S–223S, 1981.

58. Heuser JE, Reese TS, Landis DMD: Preservation of synaptic structure by rapid freezing. *Cold Spring Harbor Symp Quant Biol* 40:17–24, 1976.

59. Hill JM, Switzer RC: The regional distribution and cellular localization of iron in the rat brain. *Neurology* 11:585–603, 1984.

60. Hüttner I, Boutet M, More RH: Gap junctions in arterial endothelium. *J Cell Biol* 57:247–252, 1973.

61. Jacques PJ: The endocytic uptake of macromolecules. In Trump BF, Arstila AV, (eds): *Pathobiology of Cell Membranes.* Vol. 1. Academic, New York, 1975, pp. 255–282.

62. Jefferies WA, Brandon MR, Hunt SV: Transferrin receptor on endothelium of brain capillaries. *Nature (Lond)* 312:162–163, 1984.

63. Joó F: Effect of N^6O^2-Dibutyryl Cyclic $3',5'$-Adenosine monophosphate on the pinocytosis of brain capillaries of mice. *Experientia* 28:1470–1471, 1972.

64. Joó F, Toth I: Brain adenylate cyclase: Its common occurrence in the capillaries and astrocytes. *Naturwissenschaften* 62:379–398, 1975.

65. Kachar B, Pinto da Silva P: Rapid massive assembly of tight junction strands. *Science* 231:541–544, 1981.

66. Kachar B, Reese TS: Evidence for the lipidic nature of tight junction strands. *Nature (Lond)* 296:464–466, 1982.

67. Kaiser N, Vlodavsky I, Tur-Sinai A, et al: Binding, internalization, and degradation of insulin in vascular endothelial cells. *Diabetes* 31:1077–1083, 1982.

68. Kanwar YS, Linker A, Farquhar MG: Increased permeability of the glomerular basement membrane to ferritin after removal of glycosaminoglycans (Heparin sulfate) by enzyme digestion. *J Cell Biol* 86:688–693, 1980.

69. Karnovsky MJ: The ultrastructural basis of capillary permeability studied with peroxidase as a tracer. *J Cell Biol* 35:213–236, 1967.

70. Kienhuis AM, Krijbolder LH, Van Hinsbergh VWM, et al: Visualization of binding and receptor-mediated uptake of low density lipoproteins by human endothelial cells. *Eur J Cell Biol* 36:201–208, 1985.

71. King GL, Johnson SM: Receptor-mediated transport of insulin across endothelial cells. *Science* 227:1583–1586, 1985.

72. Landis EM, Pappenheimer JR: Exchange of substances through the capillary walls. In Hamilton WF, Dow P, (eds): *Handbook of Physiology.* Section 2: *Circulation,* Vol 2. American Physiology Society, Washington, DC, 1963, pp. 961–1034.

73. Larson DM, Sheridan JD: Intercellular junctions and transfer of small molecules in primary vascular endothelial cultures. *J Cell Biol* 92:183–191, 1982.

74. Lossinsky AS, Vorbrodt AW, Wisniewski HM, et al: Ultracytochemical evidence for endothelial channel–lysosome connections in mouse brain following blood–brain barrier changes. *Acta Neuropathol (Berl)* 53:197–202, 1981.

75. Luft JH: Fine structure of capillary and endocapillary layer as revealed by ruthenium red. *Fed Proc* 25:1773–1783, 1966.

76. Madara JL: Increases in guinea pig small intestinal transepithelial resistance induced by osmotic loads are accompanied by rapid alterations in absorptive-cell tight-junction structure. *J Cell Biol* 97:125–136, 1983.

77. Martinez-Palomo A, Erlij D: Structure of tight junctions in epithelia with different permeability. *Proc Natl Acad Sci USA* 72:4487–4491, 1975.
78. Mazzone RW, Lornblau SM: Pinocytotic vesicles in the endothelium of rapidly frozen rabbit lung. *Microvasc Res* 21:193–211, 1981.
79. McGuire PG, Twietmeyer TA: Morphology of rapidly frozen aortic endothelial cells. *Circ Res* 53:424–429, 1983.
80. Møllgard K, Malinowska DH, Saunders NR: Lack of correlation between tight junction morphology and permeability properties in developing choroid plexus. *Nature (Lond)* 264:293–294, 1976.
81. Montesano R, Roth J, Robert A, et al: Noncoated membrane invaginations are involved in binding and internalization of cholera and tetanus toxins. *Nature (Lond)* 296:651–653, 1982.
82. Moya M, Dautry-Varsat A, Goud B, et al: Inhibition of coated pit formation in Hep cells blocks the cytotoxicity of diphtheria toxin but not that of ricin toxin. *J Cell Biol* 101:548–559, 1985.
83. Murphy TL, Decker G, August JT: Glycoproteins of coated pits, cell junctions, and the entire cell surface revealed by monoclonal antibodies and immunoelectron microscopy. *J Cell Biol* 97:533–541, 1983.
84. Mynderse LA, Hassell JR, Kleinman HK, et al: Loss of heparin sulfate proteoglycan from glomerular basement membrane of nephrotic rats. *Lab Invest* 48:292–302, 1983.
85. Nagy Z, Brightman MW: Cerebral vessels cryofixed after hyperosmosis or cold injury in normothermic and hypothermic frogs. *Brain Res* 440:315–327, 1988.
86. Nagy Z, Pappius HM, Mathieson G, et al: I. Opening of tight junctions in cerebral endothelium. 1. Effect of hyperosmolar mannitol infused through the internal carotid artery. *J Comp Neurol* 185:569–578, 1979.
87. Nagy Z, Peters H, Hüttner I: Charge-related alterations of the cerebral endothelium. *Lab Invest* 49:662–671, 1983.
88. Nagy Z, Peters H, Hüttner I: Fracture faces of cell junctions in cerebral endothelium during normal and hyperosmotic conditions. *Lab Invest* 50:313–322, 1984.
89. Neuwelt EA, Rapoport SI: Modification of the blood–brain barrier in the chemotherapy of malignant brain tumors. *Fed Proc* 43:214–219, 1984.
90. Neuwelt EA, Pagel M, Barnett P, et al: Pharmacology and toxicity of intracarotid adriamycin administration following osmotic blood–brain barrier modification. *Cancer Res* 41:4466–4470, 1981.
91. Nelson E, Blinzinger K, Hager K: Electron microscopic observations on subarachnoid and perivascular spaces of the Syrian hamster brain. *Neurology (NY)* 11:285–295, 1961.
92. Noseworthy J, Korchak H, Karnovsky ML: Phagocytosis of the sialic acid of the surface of polymorphonuclear leukocytes. *J Cell Physiol* 79:91–96, 1972.
93. Oldendorf W, Davson H: Transport in the central nervous system. *Proc R Soc* 60:326–328, 1967.
94. Oomura Y: Significance of glucose, insulin, and free fatty acids on the hypothalamic feeding and satiety neurons. In Novins DW, Wyrwicks L, Bray G, (eds): *Hunger: Basic Mechanisms and Clinical Implications.* Raven, New York, 1976, pp. 145.
95. Owman C, Rosengren E: Dopamine formation in brain capillaries—an enzymic blood–brain barrier mechanism. *J Neurochem* 14:547–550, 1967.
96. Palade GE: Blood capillaries of the heart and other organs. *Circulation* 24:368–384, 1961.
97. Pappenheimer JR: Passage of molecules through capillary walls. *Physiol Rev* 33:387–423, 1953.
98. Pardridge WM, Eisenberg J, Yang J: Human blood–brain barrier insulin receptor. *J Neurochem* 45:1141–1147, 1985.
99. Pardridge WM, Yang J, Eisenberg J, et al: Antibodies to blood–brain barrier bind to brain endothelial lateral membranes and to a 46K protein. *J Cereb Blood Flow Metab* 6:203–211,1986.
100. Pastan IH, Willingham MC: Receptor-mediated endocytosis of hormones in cultured cells. *Annu Rev Physiol* 43:239–250, 1981.
101. Pearse BMF: Clathrin: A unique protein associated with intracellular transfer of membrane by coated vesicles. *Proc Natl Acad Sci USA* 73:1255–1259, 1976.
102. Pelikan P, Grimbrone A, Cotran RS: Distribution and movement of anionic cell surface sites in cultured human vascular endothelial cell. *Atherosclerosis* 32:69–80, 1979.
103. Persson L, Hansson H-A: Reversible blood–brain barrier dysfunction to peroxidase after a small stab wound in the rat cerebral cortex. *Acta Neuropathol (Berl)* 35:333–342, 1976.
104. Pino RM, Essner E: Permeability of rat choriocapillaris to hemoproteins. *J Histochem Cytochem* 29:281–290, 1981.
105. Pisam M, Ripoche P: Redistribution of surface macromolecules in dissociated epithelial cells. *J Cell Biol* 71:907–920, 1976.
106. Povlishock JT, Becker DP, Sullivan HG, et al: Vascular permeability alterations to horseradish peroxidase in experimental brain injury. *Brain Res* 153:222–231, 1978.

107. Rapoport SI, Hori M, Klatzo I: Testing of a hypothesis for osmotic opening of the blood–brain barrier. *Am J Physiol* 223:323, 1972.
108. Reese Ts, Karnovsky MJ: Fine structural localization of a blood–brain barrier to exogenous peroxidase. *J Cell Biol* 34:207–217, 1967.
109. Renkin EM: Transport of large molecules across capillary walls. *Physiologist* 7:13–28, 1964.
110. Rennels ML, Gregory TF, Blaumanis OR, et al: Evidence for a paravascular fluid circulation in the mammalian central nervous system, provided by the rapid distribution of tracer protein throughout the brain from the subarachnoid space. *Brain Res* 326:47–63, 1985.
111. Rosenstein JM, Brightman MW: Circumventing the blood–brain barrier with autonomic ganglion transplants. *Science* 221:879–881, 1983.
112. Schmidley JW, Wissig SL: Abundant uniquely oriented smooth endoplasmic reticulum in CNS capillaries, demonstration using reduced osmium and glucose–6–phosphatase cytochemistry. *Brain Res* 262:9–15, 1983.
113. Schmidley JW, Wissig SL: Anionic sites on the luminal surface of fenestrated and continuous capillaries of the CNS. *Brain Res* 363:265–271, 1986.
114. Schneeberger EE, Karnovsky MJ: Substructure of intercellular junctions in freeze-fractured alveolar-capillary membranes of mouse lung. *Circulation Res* 38:404–411, 1976.
115. Seiler MW, Venkatachalam MA, Cotran RS: Glomerular epithelium: Structural alterations induced by polycations. *Science* 189:390–393, 1975.
116. Shirahama T, Cohen AS: The role of mucopolysaccharides in vesicle architecture and endothelial transport. *J Cell Biol* 52:198–206, 1972.
117. Shivers RR, Harris RJ: Opening of the blood–brain barrier in anolis carolinensis. A high voltage electron microscope protein tracer study. *Neuropathol Appl Neurobiol* 10:343–356, 1984.
118. Shivers RR, Betz AL, Goldstein GW: Isolated rat brain capillaries possess intact, structurally complex, interendothelial tight junctions; freeze–fracture verification of tight junction integrity. *Brain Res* 324:313–322, 1984.
119. Simionescu N, Simionescu M, Palade GE: Differentiated microdomains on the luminal surface of the capillary endothelium. I. Preferential distribution of anionic sites. *J Cell Biol* 90:605–613, 1981.
120. Simionescu M, Simionescu N, Silbert JE, et al: Differentiated microdomains on the luminal surface of capillary endothelium. Distribution of lectin receptors. *J Cell Biol* 94:406–413, 1982.
121. Skutelsky E, Danon D: Redistribution of surface anionic sites on the luminal front of blood vessel endothelium after interactions with polycationic ligand. *J Cell Biol* 71:232–241, 1976.
122. Skutelsky E, Rudich A, Danon D: Surface charge properties of the luminal front of blood vessel walls: An electron microscopical analysis. *Thromb Res* 7:623–634, 1975.
123. Spatz M, Bembry J, Dodson RF, et al: Endothelial cell cultures derived from isolated cerebral microvessels. *Brain Res* 191:577–582, 1980.
124. Stewart PA, Wiley MJ: Developing nervous tissue induces formation of blood–brain barrier characteristics in invading endothelial cells: A study using quail-chick transplantation chimeras. *Dev Biol* 84:183–192, 1981.
125. Sutherland R, Della D, Schneider C, et al: Ubiquitous cell–surface glycoprotein on tumor cells is proliferation–associated receptor for transferrin. *Proc Natl Acad Sci USA* 78:4515–4519, 1981.
126. Tao-Cheng J-H, Nagy Z, Brightman MW: Tight junctions of cerebral endothelium in vitro are enhanced by astroglia. *J. Neurosci* 7:3293–3299, 1988.
127. Torack RM, Barrnett RJ: Nucleoside phosphatase activity in membranous fine structures of neurons and glia. *J Histochem Cytochem* 11:763–772, 1963.
128. Torack RM, Barrnett RJ: The fine structural localization of nucleoside phosphatase activity in the blood–brain barrier. *J Neuropathol Exp Neurol* 23:46–59, 1964.
129. Trowbridge IS, Omary MB: Human cell surface glycoprotein related to cell proliferation is the receptor for transferrin. *Proc Natl Acad Sci USA* 78:3039–3043, 1981.
130. Tsubaki SI, Brightman MW, Nakagawa H, et al: Local blood flow and vascular permeability of autonomic ganglion-transplants in the brain. *Brain Res*, 424:71–83, 1987.
131. van Deurs B: Structural aspects of brain barriers, with special reference to the permeability of the cerebral endothelium and choroidal epithelium. *Intern Rev Cytol* 65:117–119, 1980.
132. van Deurs B, von Bulow F, Moller M: Vesicular transport of cationized ferritin by the epithelium of the rat choroid plexus. *J Cell Biol* 89:131–139, 1981.
133. Van Harreveld A, Crowell J: Electron microscopy after rapid freezing on a metal surface and substitution fixation. *Anat Rec* 149:381–386, 1964.

134. van Houten M, Posner BI: Insulin binds to brain blood vessels in vivo. *Nature (Lond)* 282:623–625, 1979.
135. Vasile E, Simionescu M, Simionescu N: Visualization of the binding, endocytosis, and transcytosis of low-density lipoprotein in the arterial endothelium in situ. *J Cell Biol* 96:1677–1689, 1983.
136. Venkatachalam MA, Cotran RS, Karnovsky MJ: An ultrastructural study of glomerular permeability in aminonucleoside nephrosis using catalase as a tracer protein. *J Exp Med* 132:1168–1180, 1970.
137. Vorbrodt AW, Lossinsky AS, Wisniewski HKM: Cytochemical localization of ouabain-sensitive K$^+$-dependent p-nitro-phenylphosphatase (transport ATPase) in the mouse central and peripheral nervous systems. *Brain Res* 243:225–234, 1982.
138. Wagner HM, Pilgrim CH, Brandl J: Penetration and removal of horseradish peroxidase injected into the cerebrospinal fluid: Role of cerebral perivascular spaces, endothelium and microglia. *Acta Neuropathol* 27:299–315, 1974.
139. Wagner RC, Andrews S: Ultrastructure of the vesicular system in rapidly frozen capillary endothelium of the rete mirabile. *J Ultrastruct Res* 90:172–182, 1985.
140. Wagner RC, Robinson CS, Cross PJ, et al: Endocytosis and exocytosis of transferrin by isolated capillary endothelium. *Microvasc Res* 25:387–396, 1983.
141. Wakai S, Meiselman S, Brightman MW: Focal circumvention of blood–brain barrier with grafts of muscle, skin and autonomic ganglia. *Brain Res* 386:209–222, 1986.
142. Wasteson A, Glimerlius B, Busch C, et al: Effect of a platelet endoglycosidase on cell surface associated heparin sulphate of human cultured endothelial and glial cells. *Thromb Res* 11:309–321, 1977.
143. Weibel ER, Limacher W, Bachofen H: Electron microscopy of rapidly frozen lungs: Evaluation on the basis of standard criteria. *J Appl Physiol* 53:516–527, 1982.
144. Westergaard E, van Deurs B, Bondsted HE: Increased vesicular transfer of horseradish peroxidase across cerebral endothelium, evoked by acute hypertension. *Acta Neuropathol (Berl)* 37:141–152, 1977.
145. Williams SK, Matthews MA, Wagner RC: Metabolic studies on the micropinocytic process in endothelial cells. *Microvasc Res* 18:175–184, 1979.
146. Wissig SL: Identification of the small pore in muscle capillaries. *Acta Physiol Scand Suppl* 463:33–44, 1979.
147. Woods SC, Porte D: Relationship between plasma and cerebrospinal fluid insulin levels of dogs. *Am J Physiol* 233(4):E331–E334, 1977.

4

Quantitation of Blood–Brain Barrier Permeability

Quentin R. Smith

1. INTRODUCTION

The rate of uptake of most hydrophilic solutes into brain is limited by transport across the blood–brain barrier (BBB).[11,97] The barrier between blood and brain extracellular fluid is located at the cerebral capillary endothelium, whereas the barrier between blood and

Symbols and units: B_i, plasma concentration coefficient for elimination of tracer from plasma (dpm/ml); b_i, rate coefficient for elimination of tracer from plasma (sec^{-1}); BUI, brain uptake index; C_A, tracer concentration in arterial blood or plasma (dpm/ml); C_B, tracer concentration in arterial blood (dpm/ml); C_p, tracer concentration in arterial plasma (dpm/ml); $\check{C}$, mean concentration (dpm/ml); C_{pf}, tracer concentration in perfusion fluid (dpm/ml); C', ratio of tracer concentration in venous blood to that in injectate; C_V, tracer concentration in venous blood or plasma (dpm/ml); E, extraction; E_o, unidirectional extraction; f, correction factor for the difference between brain Hct and large-blood-vessel Hct; F, cerebral blood flow (ml/sec per g, also expressed in ml/min per 100g); $h(t)$, transport function that describes the fraction of the indicator injected at the inflow at $t=0$ which arrives at the outflow at time t (sec^{-1}); Hct, hematocrit; IF, integration factor, defined as $\int C_p dt / [C_p(T)T]$ (sec); K_{in}, unidirectional transfer coefficient for influx into brain (ml/sec per g, also expressed as ml/min per 100g); K_{out}, unidirectional transfer coefficient for tracer efflux from brain (ml/sec per g, also expressed as ml/min per 100g); k_{out}, rate coefficient for tracer loss from brain (sec^{-1}) (note: $k_{out} = K_{out}/V_{br}$); K_m, Michaelis constant, defined as that concentration of solute which gives one-half the maximal transport rate (μmol/ml); K_d, constant of nonsaturable diffusion (ml/sec per g); P, cerebrovascular permeability (cm/sec); PS, cerebrovascular permeability–surface area product (ml/sec per g); q_{br}, quantity of tracer in brain parenchyma, i.e. that which has crossed the blood–brain barrier and is present in extravascular brain tissue (dpm/g) (also designated as M, A, Q, or C in the BBB literature); q_{tot}, total quantity of tracer in brain including tracer in brain endothelial cells and tracer in residual blood (dpm/g); r, ratio of tracer concentrations in blood and plasma; S, capillary surface area (cm^2/g); t, time after injection of tracer (sec); T, time of uptake when brain tracer content is determined (sec); T', time when brain tracer content is maximum (sec); $t_{1/2}$, half-time (sec, min, or hr); V_{br}, physical distribution volume of tracer in brain, such as brain extracellular-fluid volume or brain-water volume (ml/g); V'_{br}, effective distribution volume of tracer in brain, defined as K_{in}/k_{out} (ml/g); V_B, volume of blood in brain tissue (ml/g); V_p, volume of plasma in brain tissue (ml/g); V_{max}, maximal rate of saturable transport (μmol/ sec per g); v, fractional distribution volume of tracer in blood (ml/ml).

Quentin R. Smith • Laboratory of Neurosciences, National Institute on Aging, National Institutes of Health, Bethesda, Maryland 20892.

cerebrospinal fluid (CSF) is located at the choroid plexus epithelium and the arachnoid membrane (Fig. 1). Morphologic evidence has established that the barrier at each site is formed by a single layer of cells that are joined by tight junctions (zonulae occludens).[15,102] The cerebral capillaries, which comprise more than 95% of the total surface area of the BBB, represent the principal route for the entry of most solutes into the central nervous system (CNS).

Considerable progress has been made during the past 10 years in methods to quantitate BBB permeability and transport. Advances in the theory of tracer uptake into brain and developments in techniques to measure regional brain radioactivity have resulted in new noninvasive methods that can be used to examine regional BBB transport in humans.[17,48,55,130] These methods are based on the intravenous (IV) administration of short-lived tracers and external detection of regional brain radioactivity using positron emission tomography (PET) or single photon emission tomography (SPECT). They can also be applied to animal studies using quantitative autoradiography. These methods provide accurate values of permeability for both rapidly and slowly penetrating compounds and should prove extremely useful in studies of BBB function in normal and pathologic conditions.

Most studies of BBB permeability use one of five techniques that measure tracer influx into the brain—the IV administration technique,[6,27,100] the brain perfusion technique,[123] the indicator diffusion technique,[23] the brain uptake index (BUI) technique,[75] or the single-injection–external registration technique.[96] Efflux from brain to blood can be measured with either the brain washout technique[13] or the concentration profile technique.[87] In addition, isolated brain capillaries can be used to examine uptake by endothelial cells and to isolate specific capillary transport systems.[42]

This review examines and evaluates each of the five basic methods for measuring unidirectional transport into the brain. Special attention is given to methods that can be used for quantitative determination of BBB permeability in humans. This chapter does not discuss methods for measuring transport at the choroid plexus or at the arachnoid membrane, which are covered in other reviews and in Chapter 9.[31,126,129]

A common set of symbols has not been established in the BBB field. As a result, there are at least five different symbols in use for some commonly used terms, such as the

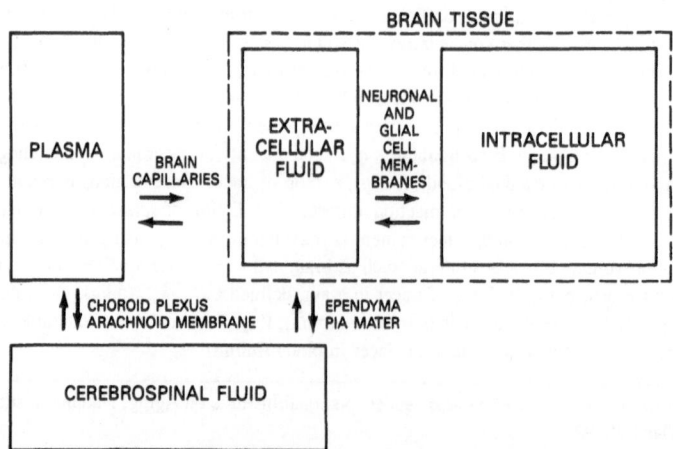

Figure 1. Model for solute exchange among compartments of the central nervous system.

quantity of solute in brain. This review has adopted the set of symbols recently recommended for transport studies in the *American Journal of Physiology*.[4] It is hoped that this effort will minimize confusion and promote the adoption of a single set of symbols by all workers in the BBB field.

2. TWO METHODOLOGIC APPROACHES

All methods to measure influx across the BBB employ one of two different methodologic approaches. In the first approach, a solute is injected as a bolus into the carotid artery; then brain uptake or extraction is determined from a single pass of the bolus through the brain capillaries. This approach was first applied to the brain by Crone[22] in 1963 with the indicator diffusion technique and is the basis of the BUI and single-injection–external registration techniques. It has been extremely popular due to the ease and simplicity of these techniques. However, because permeability is determined from a single pass through the brain, this approach has limited sensitivity and cannot be used to obtain accurate cerebrovascular permeability coefficients for poorly penetrating compounds. Furthermore, this approach cannot be used to examine regional cerebrovascular permeability in humans.

The limitations of the carotid injection techniques have stimulated interest in a second approach in which uptake time is not limited to that of a single pass (5–15 sec). With the IV administration technique, a solute is injected or infused intravenously, and the net amount of solute in brain is determined at any time from 10 sec to several hours thereafter, depending on the cerebrovascular permeability coefficient of the solute. Similarly, with the brain-perfusion technique, solute uptake into brain is determined after perfusion for up to 5 min with blood or saline. Because uptake time can be prolonged, these techniques are 100 times more sensitive than the single-pass methods.[73,117] Both the IV administration and the brain perfusion techniques can be used to examine regional cerebrovascular permeability in animals with either quantitative autoradiography or regional brain sampling. The IV administration technique has also been extended to measure regional cerebrovascular permeability in humans using PET. Thus, the continuous-uptake approach provides two quantitative methods that can be used to obtain accurate measurements of regional transport across the BBB. These methods will enable scientists to localize and explore BBB dysfunction in animals and humans in various pathologic conditions.

3. CONTINUOUS UPTAKE

3.1. Intravenous Administration

The IV administration technique to determine BBB permeability was developed in the pioneering studies of Katzman and Leiderman[59] in 1953 and by Davson[27] in 1955. With this technique, a solute is given parenterally and the plasma concentration monitored until a specific time at which the brain content is determined. In experiments with small animals the subject is killed and the brain removed and analyzed for solute content, whereas in large animals and humans the quantity of solute in the brain can be determined with noninvasive techniques, such as PET or SPECT. Blood–brain barrier permeability is calculated from the brain uptake of solute using a kinetic model that describes solute exchange between plasma and CNS.

In most studies, the solute is given intravenously as a single bolus injection, by constant infusion or by programmed infusion to maintain a constant plasma level.[88] However, the method has been adapted for use after intraperitoneal, intramuscular, or subcutaneous administration.[113,119] Three versions of the method are currently used that differ primarily in the number of time points examined.

3.1.1. Compartmental Analysis

Compartmental analysis is the primary method that has been used to examine regional BBB permeability and transport in humans using PET. With this technique, plasma and brain levels of tracer are determined over a wide range of experimental times after administration. Data are then analyzed with a compartmental model of the brain that describes tracer uptake, distribution, and elimination. The technique, as originally developed by Davson,[27] required a programmed infusion of tracer to maintain a constant plasma concentration. However, since 1978 the method has been extended for use after bolus IV injection and constant IV infusion.[73,98]

3.1.1a. Model. The most frequently used model in BBB studies is the simple, linear two-compartment model (plasma and brain) with blood flow.[12,32,100] This model is usually adequate to describe solute transfer across the BBB. More complex models may be required to describe intracerebral distribution,[7,98] binding,[90,127] metabolism,[38,91,104,120] and brain–CSF exchange.[19,28,98] Because of the prevalence of the two-compartment model in BBB studies, this model is presented in some detail.

The rate of uptake (dq_{br}/dt) of a diffusible radiotracer into brain after IV administration is given as follows by a two-compartment model with blood flow:

$$dq_{br}/dt = K_{in}C_p - k_{out}q_{br} \qquad (1)$$

where q_{br} is the quantity of tracer in parenchymal (extravascular) brain (dpm/g), C_p is the tracer concentration in arterial plasma (dpm/ml), K_{in} is the unidirectional transfer coefficient for influx (ml/sec per g), k_{out} is the rate coefficient for efflux (sec^{-1}), and t is time (sec).[6,27,38,48,55,98] For this model, it is assumed that protein binding and metabolism of tracer are negligible and that tracer enters the brain primarily by crossing the cerebral capillaries.

The rate coefficient for efflux is usually defined as $k_{out} = K_{out}/V_{br}$, where K_{out} is the transfer coefficient for efflux (ml/sec per g) and V_{br} is the physical volume of distribution of tracer in brain (ml/g).[38,41,127] Substitution of K_{out}/V_{br} into Eq. (1) gives $K_{out}(q_{br}/V_{br})$ on the far right-hand side of the equation, where q_{br}/V_{br} represents the effective brain concentration of tracer for backflux into plasma. If transfer across the BBB is symmetric ($K_{in} = K_{out}$), Eq. (1) reduces to,[73,100]

$$dq_{br}/dt = K_{in}(C_p - q_{br}/V_{br}) \qquad (2)$$

In simple terms, Eq. (2) indicates that rate of net uptake of tracer into brain is proportional to the concentration gradient between plasma and brain.

Integration of Eq. (1) from the time of tracer injection ($t = 0$) to the end of the uptake period, T, gives

$$q_{br}(T) = K_{in}e^{-k_{out}T} \int_0^T e^{k_{out}t}C_p(t)dt \qquad (3)$$

which can be fit to the data using nonlinear regression to obtain best estimates of K_{in} and k_{out}.[38,48] This equation does not assume a specific time course of arterial plasma concentration. Alternatively, best-fit estimates of K_{in} and k_{out} can be obtained with the following equations, which are analytic solutions of Eq. (3).[12,72,73,100,101]

For constant plasma concentration ($C_p = B$),

$$q_{br}(T) = C_p (K_{in}/k_{out})(1 - e^{-k_{out}T}) \qquad (4)$$

For IV bolus injection, where $C_p(t) = \sum_{i=1}^{n} B_i e^{-b_i t}$:

$$q_{br}(T) = K_{in} \sum_{i=1}^{n} [B_i/(k_{out} - b_i)](e^{-b_i T} - e^{-k_{out}T}) \qquad (5)$$

For constant IV infusion, where $C_p(t) = \sum_{i=1}^{n} B_i (1 - e^{-b_i t})$:

$$q_{br}(T) = (K_{in}/k_{out}) \sum_{i=1}^{n} B_i \{1 + [b_i/k_{out} - b_i][e^{-k_{out}T} - (k_{out}/b_i)e^{-b_i T}]\} \qquad (6)$$

In addition to these solutions for a two-compartment model, Rapoport et al.[98] derived solutions for a four-compartment model of the CNS that includes plasma, interstitial fluid, intracellular fluid, and CSF.

The recommended method to calculate K_{in} with compartmental analysis is nonlinear regression because it is flexible, unbiased, and simple to perform with a computer.[29] Several good nonlinear regression programs are available, such as the MLAB program for the DEC-10 computer.[61] If the variance of the data is not constant, weighted nonlinear regression should be used. In addition, the goodness of fit and the redundancy of individual parameters should be evaluated statistically.[29] If the investigator does not have access to a computer with a nonlinear regression program, an alternative procedure for data with constant plasma concentration is the graphic analysis procedure of Solomon.[56,68,105,113,121]

Transfer coefficients can be converted to permeability coefficients using the Renkin–Crone model of capillary transfer.[25,32,43] With this model, K_{in} is defined as

$$K_{in} = vF(1 - e^{-PS/vF}) \qquad (7)$$

where F is regional cerebral blood flow (ml/sec per g), P is capillary permeability (cm/sec), S is the surface area of perfused capillaries (cm²/g), and v is the fractional distribution volume of tracer in blood (ml/ml).[31,57] For solutes that do not penetrate red blood cells (RBCs), $v = 1 - Hct$ and vF is cerebral plasma flow. The cerebrovascular permeability–surface area product (*PS*) is obtained from K_{in} by rearrangement of Eq. (7) as

$$PS = -vF\ln(1 - K_{in}/vF) \tag{8}$$

Cerebrovascular permeability can be calculated by dividing *PS* by the surface area (*S*) of perfused capillaries. However, since *S* is difficult to measure, most investigators express their results in terms of a *PS* product.[25] Weiss et al.[125] estimate that approximately 50% of cerebral capillaries are perfused under normal physiologic conditions.

3.1.1b. Brain Tracer Content. Regional measurements of brain tracer content can be obtained in animal experiments using quantitative autoradiography (QAR) or tissue dissection and analysis. The advantages of autoradiography are (1) it provides a visual demonstration of tracer distribution in the brain, and (2) the tissue content of tracer can be determined accurately in small brain regions. For example, the resolution of autoradiography using ^{14}C is 100 μm^2 which represents only 100–200 cells.[6] Figure 2 shows color-transformed autoradiographs of L-[^{14}C]phenylalanine uptake into the brain of a rat.[69] Regional L-[^{14}C]phenylalanine uptake, which was directly proportional to parenchymal brain ^{14}C content, differed by approximately fivefold and correlated significantly with regional cerebral blood flow and capillary surface area.[47] Similar regional differences in brain uptake have been demonstrated for both D-glucose and sucrose using QAR.[39,112] Although tissue dissection does not have the regional resolution of QAR, it does provide a fast, simple method with which to determine brain tracer content. Both autoradiography and tissue dissection are limited in that they can provide values for brain tracer content at only one time point for each animal. Therefore, to describe the entire time course of tracer uptake into brain, many animals must be used.

Rapid sequential measurements of regional brain tracer content can be obtained in large animals and humans with noninvasive techniques, such as PET,[93,94] and SPECT.[64] Both techniques use a large array of external scintillators in order to measure regional brain radioactivity in vivo. However, of the two techniques, PET offers superior resolution because of the physical properties of positrons.[93] When a positron is emitted from a radionuclide, it travels only a short distance in brain tissue before it combines with an electron, resulting in the destruction of both. From the annihilation event, two 0.51-MeV gamma photons are emitted that travel in opposite directions. These photons can be detected with external scintillators positioned on opposite sides of the positron-emitting source and connected by a coincident circuit that records a signal only when two photons arrive almost simultaneously.[93] The use of coincidence detection minimizes the contribution of scattered photons, thereby enhancing the resolution and accuracy of the PET technique.[94] Accurate regional measurements of brain radioactivity are hard to obtain with SPECT because scatter degrades the quality of the image and corrections for absorption are difficult to perform. The resolution of PET scanners has steadily improved during the past few years and is now less than 1 cm. Positron-emitting radionuclides commonly used in studies of brain transport, blood flow, and metabolism are ^{11}C, ^{13}N, ^{15}O, and ^{18}F.[33,41,52,94,103]

Compartmental analysis has been used with PET for quantitative determination of regional cerebrovascular permeability to [^{68}Ga]ethylenediamine tetraacetate (EDTA), which only slowly crosses the intact BBB ($t_{1/2} \simeq 8$ hr).[32] For example, Kessler et al.[60] examined the time-dependent changes in cerebrovascular *PS* to [^{68}Ga]-EDTA in the

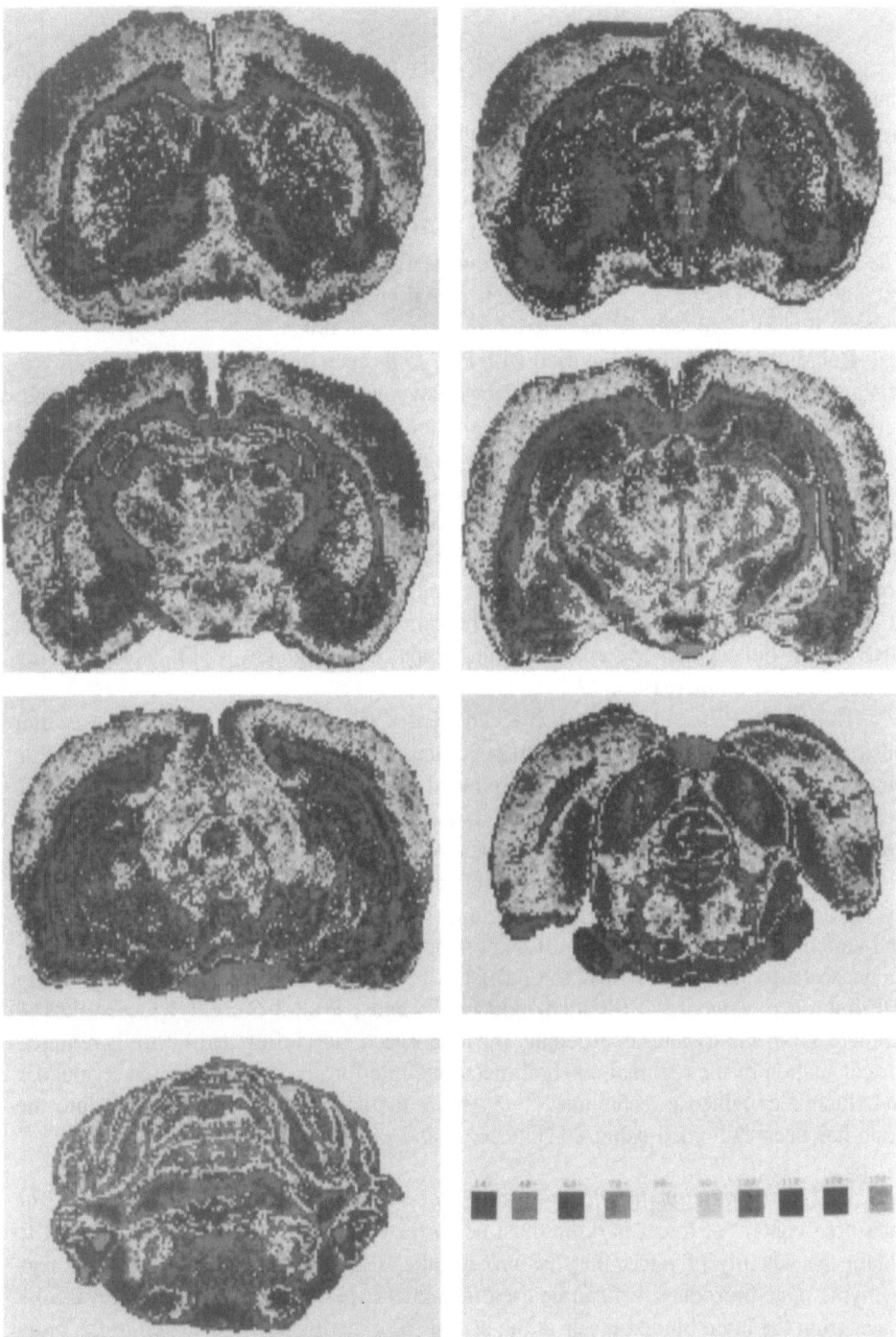

Figure 2. Computerized images of phenylalanine influx into rat brain. L-[^{14}C]phenylalanine was infused into the femoral vein for 90 sec, after which time the rat was killed by decapitation. Regional brain ^{14}C content was determined by a computer-driven densitometer from autoradiographs prepared from coronal sections of the brain. The readings were converted into values of phenylalanine influx and displayed by colors corresponding to discrete levels of transport. Influx was calculated as K_{in} times the plasma phenylalanine concentration. The sections were taken at the level of the following structures: (1) caudate, (2) lateral ventricles, (3) anterior thalamic nuclei, (4) lateral geniculate, (5) medial geniculate, (6) inferior colliculus, and (7) superior olive. The color scale (8) is in units of nanomoles per minute per gram. A color reproduction of this figure appears following p. xxviii. (From Mans et al.[69])

rhesus monkey, following osmotic opening of the BBB. In addition Hawkins et al.[48] and Iannotti et al.[53] examined barrier permeability in humans with brain tumors.

Figure 3 shows successive scans through the brain of a patient following IV injection of 7–10 mCi [^{68}Ga]-EDTA.[48] The patient had a metastatic breast carcinoma in the left cerebellar hemisphere. The round tumor lesion was clearly defined in the PET image and had a *PS* of 4.7×10^{-5} ml/sec per g, which is $\simeq 13$ times greater than normal brain tissue. The [^{68}Ga]-EDTA/PET technique may prove useful in several clinical situations, such as chemotherapy of brain tumors, in which quantitation of BBB permeability will help identify brain lesions and aid in optimal treatment with drugs.[48,55,60]

Rubidium-82 has also been used with PET to measure regional BBB permeability in humans[16,55,63,130] and has certain advantages over [^{68}Ga]-EDTA, such as a lower absorbed dose per administered millicurie, a shorter delay time between repeat scans, and a greater volume of distribution within the brain. However, the cerebrovascular permeability to Rb is tenfold greater than that to EDTA,[32] and therefore Rb will be a less sensitive marker to small changes in BBB permeability.[9] The sensitivity of Rb is further reduced by the significant error ($\simeq 45\%$) associated with determination of K_{in} for ^{82}Rb in normal brain tissue.[55,63] Regardless of these limitations, both Brooks et al.[16] and Jarden et al.[55] have used the ^{82}Rb/PET technique in humans to demonstrate significant increases in BBB *PS* in tumors that are enhanced on computed tomography (CT). Nonenhancing tumors showed no increase in ^{82}Rb uptake.

Thus, both ^{82}Rb and [^{68}Ga]-EDTA can be used to evaluate regional cerebrovascular integrity in humans, with PET. Advantages of these tracers are that they do not need to be obtained from a cyclotron, but can be produced on site using commercial ^{82}Rb and ^{68}Ga generators. In addition, unlike iodinated contrast agents for CT, they are not potentially neurotoxic and do not produce untoward side effects, such as convulsions, if the BBB is altered.[71] However, these tracers provide information only on the passive permeability of the BBB. Information on other BBB functions, such as the facilitated transport of glucose and amino acids into the brain, must be obtained using other radiotracers. For example, monosaccharide transport across the BBB has been studied in humans with PET using [^{18}F]fluorodeoxyglucose, [^{11}C]deoxyglucose, and [^{11}C]-3-*O*-methylglucose.[17,41,103] Knowledge of the transfer coefficients for monosaccharide influx and efflux is required for calculation of the regional cerebral metabolic rate for glucose with the deoxyglucose and fluorodeoxyglucose techniques.[103,104,120] In addition, amino acid transport into the brain has been examined using L-[^{11}C]leucine and L-[^{11}C]methionine.[18,91]

3.1.1c. Correction for Intravascular Tracer. In any study of BBB transport, the measured content of tracer in brain must be corrected for intravascular tracer in order to obtain the quantity of tracer that has crossed the BBB and entered into the brain parenchyma. One procedure that can be used in animal experiments is to wash intravascular tracer from the brain blood vessels at the end of the experiment by perfusion of the brain for 10–60 sec with tracer-free fluid.[2,115] However, this procedure is limited to uptake studies of poorly penetrating solutes for which perfusion time is negligible (<5%) compared with uptake time. A second procedure that can be used for all studies is to calculate q_{br} with the following relation:

$$q_{br}(t) = q_{tot}(t) - V_B C_B(t) \qquad (9)$$

where q_{tot} is the measured brain content (intravascular plus extravascular) of tracer

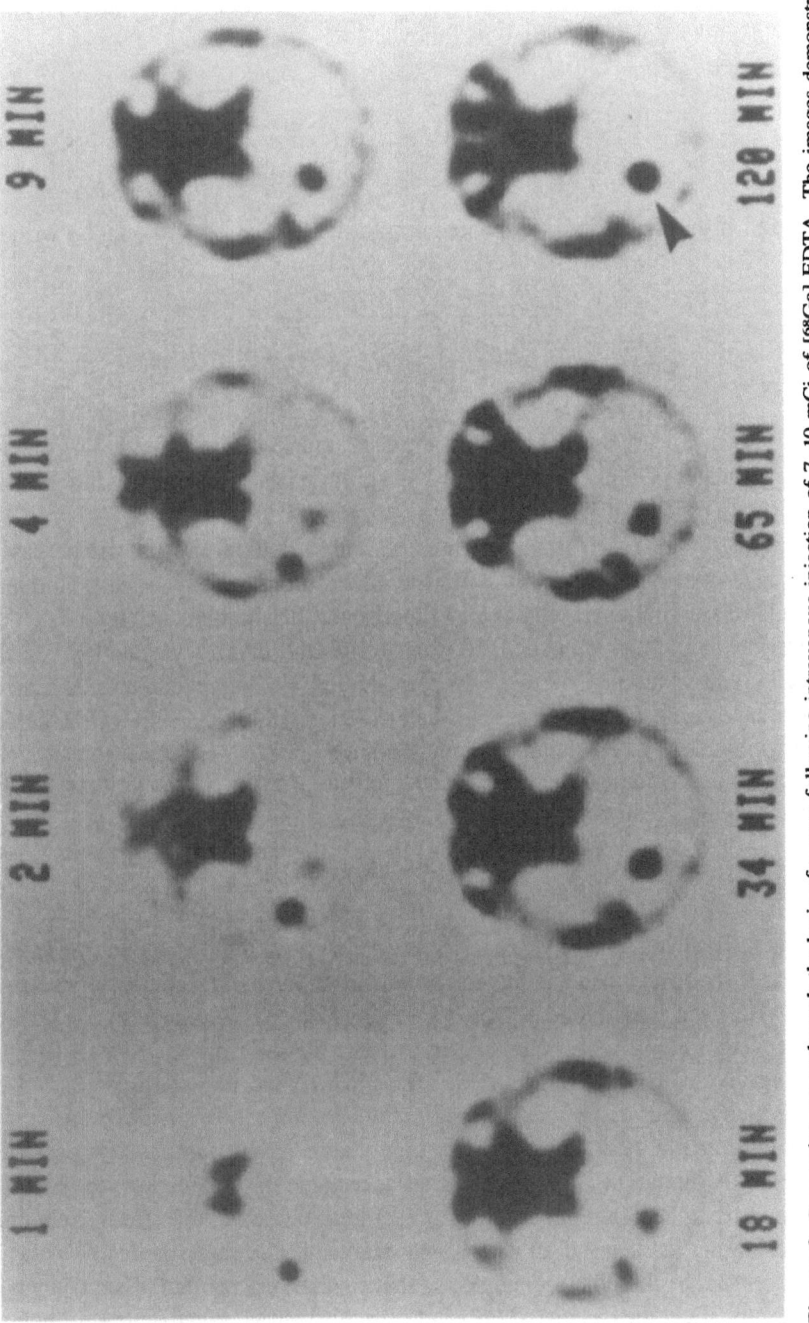

Figure 3. Successive scans through the brain of a man following intravenous injection of 7–10 mCi of [^{68}Ga]-EDTA. The images demonstrate increased uptake of [^{68}Ga]-EDTA by a metastatic breast carcinoma (arrowhead) in the left cerebellar hemisphere. (From Hawkins et al.[48])

(dpm/g), C_B is the tracer concentration in arterial blood (dpm/ml), and V_B is the regional blood or vascular volume of brain (ml/g).

With the compartmental analysis method, V_B can be calculated as a model parameter by combining Eq. (9) with the appropriate equation for the time course of parenchymal brain content.[48] For example, the time course of measured brain content after IV administration is given by Eqs. (3) and (9) as follows:

$$q_{tot}(T) = K_{in}e^{-k_{out}T}\int_0^T e^{k_{out}t}C_p(t)dt + V_BC_B(T) \tag{10}$$

Alternatively, if the tracer does not penetrate into the RBCs, Eq. (10) can be written as follows:

$$q_{tot}(T) = K_{in}e^{-k_{out}T}\int_0^T e^{k_{out}t}C_p(t)dt + V_pC_p(T) \tag{11}$$

Either Eq. (10) or Eq. (11) would be fit to the time course of measured brain content to obtain best estimates of K_{in}, k_{out}, and V_B or V_p. This approach was used by Jarden et al.[55] in their ^{82}Rb/PET study of BBB permeability in humans.

An alternative approach is to determine vascular volume directly with a vascular radiotracer that does not measurably cross the BBB during the experiment. Commonly used vascular tracers in small animal experiments are plasma markers, such as [^{125}I]albumin, [^{3}H]dextran, [^{3}H]inulin, or [^{111}In]transferrin, and erythrocyte markers, such as ^{51}Cr-tagged erythrocytes.[6,21,37,67,111,114] In PET studies, [^{11}C]methylalbumin is used as a plasma marker, whereas erythrocytes are labeled using ^{11}CO or C^{15}O, which combine with hemoglobin (Hb) to form carboxyhemoglobin.[62,92]

Cerebral blood volume, which represents the sum of brain plasma volume and brain erythrocyte volume,[21] can be approximated as

$$V_B \cong q^*_{tot}/C^*_b f \tag{12}$$

where q^*_{tot} is the brain content of vascular tracer (dpm/g), C^*_b is the blood concentration of vascular tracer (dpm/ml), and f is the correction factor for the difference between brain hematocrit (Hct) and large blood vessel Hct.[21,62,92] For an erythrocyte marker, f equals (tissue Hct)/(large blood vessel Hct), whereas for a plasma marker, f equals $(1-$tissue Hct)/($1-$large-vessel Hct). Average values for f in rats and humans are ≈ 0.75 for an erythrocyte marker and ≈ 1.19 for a plasma marker.[21,62,92] However, there are regional differences in cerebral Hct, and brain Hct may change from one physiologic state to another.[21,62,67,107] Equation (12) represents an acceptable approximation for most BBB studies. However, when extreme accuracy is required, blood volume can be determined with both a plasma marker and an erythrocyte marker in the same patient.[21,62] Ideally, BBB transport studies should be conducted so that small errors in brain blood volume will not significantly affect the calculated PS values. In this respect, an advantage of bolus IV injection, as compared with continuous IV infusion, is that the contribution of intravascular tracer to measured brain content progressively declines during the experiment and makes errors in estimating intravascular tracer less important.[73]

If one applies an Hct correction to the measured blood volume (Eq. 12), then one should also apply a Hct correction to the measured concentration of test tracer in large vessel blood. The appropriate test tracer concentration would be calculated as fC_B, where $f = 0.75$ if the test tracer distributes only in erythrocytes, $f = 1.19$ if test tracer distributes only in plasma, and $f = \sim 1.0$ if test tracer distributes evenly between plasma and erythrocytes. If the test tracer and vascular tracer distribute similarly in blood (i.e., both distribute only in plasma or only in erythrocytes), then the f values will cancel out ($V_B C_B = (q_{tot}^*/C_b^* f)(fC_B) = (q_{tot}^*/C_b^*)(C_B))$, and thus no Hct correction is necessary.

For test solutes that do not measurably enter erythrocytes, intravascular volume can be expressed as a plasma volume, V_p, in which case the product of V_p and C_p is subtracted from q_{tot} to obtain q_{br}[36,37]:

$$q_{br}(t) = q_{tot}(t) - V_p C_p(t) \tag{13}$$

This approach does not require a Hct correction.

It is usually advantageous to determine intravascular volume in each individual so that the q_{br} calculation is not based on an average V_B from a second set of experiments. This reduces variation in calculated q_{br}, particularly for poorly penetrating solutes, which may have a significant intravascular fraction of total brain tracer content. However, vascular volume is reported to differ among various solutes, being somewhat greater for small nonelectrolytes than for large polysaccharides and proteins.[111] Furthermore, endothelial uptake or binding may increase the vascular volume for a solute.[6] Therefore, if a vascular marker is used to determine blood volume, the investigator should verify that the volume of the vascular indicator equals that of the test compound, the latter determined either with compartmental analysis [Eqs. (10) or (11)] or with the multiple time point/graphic analysis method.

3.1.1d. Blood Flow. Finally, cerebral blood flow must be known to convert a transfer coefficient to a cerebrovascular PS product.[57] Blood flow is important because the brain uptake of a rapidly penetrating, lipid-soluble compound is likely to be flow limited as well as permeability limited.[25] The relationship between K_{in} and PS is given in Eq. (7). Figure 4 illustrates the relationship between K_{in} and PS for three values of F. K_{in}

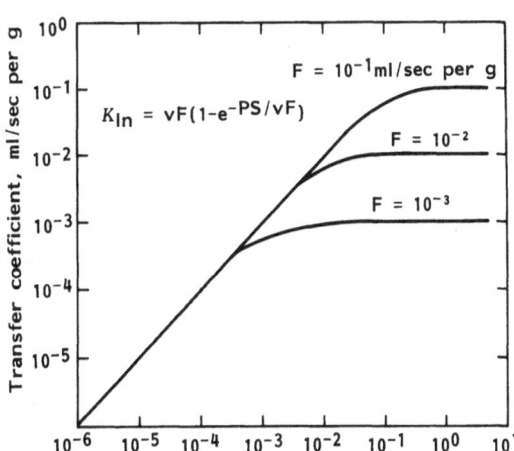

Figure 4. Relation between the cerebrovascular permeability–surface area product, PS, and the transfer coefficient, K_{in}, for three values of cerebral blood flow, F. The units of flow are milliliters per second per gram. PS equals K_{in} (with less than 12% error) when $K_{in} \leq 0.2F$.

equals PS with less than 10% error when $PS \leq 0.2F$, whereas K_{in} approximates F with less than 10% error when $PS \geq 2.3F$. The transfer coefficient depends on both PS and F when $0.2 \leq PS/F \leq 2.3$. Figure 4 shows that accurate estimates of PS are difficult to obtain when $K_{in} \simeq F$, because brain uptake becomes flow limited as the extraction approaches 100%.

Regional cerebral blood flow can be determined in humans with $C^{15}O_2$ or $H_2^{15}O$ using PET.[33,52,93] Similarly, regional cerebral blood flow can be measured in animals with [^{14}C]iodoantipyrine,[108] [^{14}C]butanol,[110] [^3H]nicotine,[74] [^{125}I]isopropyliodoamphetamine,[66] or labeled microspheres.[70] Furthermore, several methods have been developed for the simultaneous measurement of F and PS.[5,39,106]

Thus, the compartmental analysis version of the IV administration technique has several advantages for studies of BBB transport. Quantitative estimates of both the cerebrovascular K_{in} for influx and the rate constant for efflux can be obtained using a simple two-compartment model. Alternatively, the method can be expanded to examine intracerebral distribution of tracer with models that incorporate factors such as brain cell uptake, binding, metabolism, and brain–CSF exchange. Unlike the single time-point and multiple time-point methods, compartmental analysis does not require unidirectional uptake kinetics. Furthermore, intravascular volume can be determined directly with the test tracer or, if desired, with a separate intravascular marker. The method can be used to measure cerebrovascular PS accurately over a 10^5-fold range from as low as 2×10^{-7} ml/sec per g (i.e., the PS to inulin) to as high as 2×10^{-2} ml/sec per g (i.e., the PS to antipyrine).[32,73] Compartmental analysis can be used on conscious or anesthetized animals to examine cerebrovascular transport in specific brain regions using QAR or tissue dissection. In addition, the method is ideally suited to study BBB transport in humans using PET or SPECT, which can provide rapid sequential scans of brain radioactivity in the same patient.

Although the compartmental analysis method has many advantages, the method also has several disadvantages that apply primarily to small animal studies. For example, many animals are usually required to describe the entire time course of brain tracer uptake in small animal experiments. This can be costly and time consuming for the investigator. The series of animals that are used must have similar values of K_{in}, k_{out} and of F and V_B for the model to be valid. As a result, the method cannot be used in small animal experiments to determine PS in brain tumors or in strokes because of the large interindividual variability in the transport parameters.[8,46] In addition, the method cannot be used when PS changes with time, such as occurs after osmotic opening of the BBB.[99] These limitations can be overcome in experiments with large animals or humans by measurement of the entire time course of brain tracer content in the same patient, using PET or SPECT.[60] The compartmental analysis method is subject to errors in the measurement of solute uptake due to radiotracer metabolism either in the brain or in the periphery. For example, K_{in} may be significantly overestimated if there is substantial peripheral metabolism of the labeled solute to rapidly penetrating compounds, such as $^{14}CO_2$ or ^3H_2O.[1,23] Alternatively, K_{in} may be underestimated if there is significant peripheral metabolism of the labeled solute to less permeant compounds or if there is substantial brain metabolism to compounds that leave the CNS rapidly.[38,39] As a result, chromatography should be performed in all studies to determine the radiochemical identity of tracer in plasma and, when possible, in brain. The IV administration technique provides no simple means of controlling plasma solute concentrations for studies of facilitated transport into the brain. This limitation applies not only to the solute under study but also to any competitors or modifiers of transport that may be present in the plasma as well.[26,106,117] Compartmental analysis is open to criticism concerning the appropriate model to describe tracer uptake and distribution within the brain. Results for BBB transfer coefficients can vary depending on whether a two-compartment, three-compartment, or four-compartment (series or parallel) model is used for analysis.[68] Furthermore, most models do not take into account intracerebral diffusion of tracer. Diffusion of tracer from regions of high permeability to regions of low permeability can lead to significant errors in measured transfer coefficients, as has been shown for the ions, ^{22}Na and ^{36}Cl in normal brain tissue[114,115] and for [^{14}C]aminoisobutyric acid in brain tumors.[7,58]

3.1.2. Multiple Time-Point/Graphic Analysis

The IV administration technique, as originally developed by Davson,[27] required a constant plasma concentration of tracer and measurements of brain tracer content over a wide range of experimental times until the brain level reached a steady state. Transfer coefficients were calculated, as described in the previous section, by analysis of the data with a simple two-compartment model. This approach, which was used routinely by BBB researchers for 20 years, was adequate, but it was also costly and time consuming, requiring many animals to obtain a single PS value. Furthermore, doubts concerning the kinetics of tracer distribution within the CNS often clouded conclusions concerning tracer exchange across the BBB.[68,113]

Bradbury was the first to point out that accurate estimates of transfer across the BBB could be obtained more readily with the IV administration technique from analysis of the initial linear portion of the brain uptake curve.[10–12,109] Bradbury recognized that during the initial phase of uptake, when $q_{br} \approx 0$, Eq. (1) simplifies to

$$dq_{br}/dt = K_{in}C_p \qquad (14)$$

When C_p is constant, Eq. (14) can be integrated and combined with Eq. (13) to give

$$q_{tot}(T) = K_{in}C_pT + V_pC_p \qquad (15)$$

Division of Eq. (15) by C_p yields

$$q_{tot}(T)/C_p = K_{in}T + V_p \qquad (16)$$

Thus, Bradbury realized that the transfer coefficient for influx could be obtained as the slope of the linear portion of the uptake curve when the data are plotted as q_{tot}/C_p versus t. This analysis method does not require assumptions concerning intracerebral distribution of tracer or vascular volume[85,86] and necessitates fewer animals than does compartmental analysis to obtain an accurate estimate of K_{in}. The method was extended and generalized to all IV administration studies by Gjedde[37] in 1981 and by Patlak et al.[86] in 1983. The method is well suited for the determination of PS in large animals and humans using PET, which can provide sequential measurements of brain radioactivity in the same subject.

The experimental procedures for the multiple time-point method are essentially the same as those described for compartmental analysis. However, instead of determining the entire time course of tracer uptake into brain, measurements are concentrated on the initial linear phase of the uptake curve. A constant arterial concentration of tracer is not required.

In fact, no specific time course of arterial concentration is assumed. Thus, the tracer may be given by IV bolus injection or constant infusion. Three measurements are required for each time point: $q_{tot}(T)$, $\int C_p dt$, and $C_B(T)$ or $C_p(T)$. With unidirectional uptake, measured brain content of tracer is given as

$$q_{tot}(T) = K_{in}\int C_p dt + V_p C_p(T) \tag{17}$$

Division by $C_p(T)$ gives an equation for a straight line,[37,86] with slope K_{in} and intercept V_p:

$$q_{tot}(T)/C_p(T) = K_{in}[\int C_p dt/C_p(T)] + V_p \tag{18}$$

If tracer is taken up into erythrocytes, Eq. (18) can be expressed as[34]

$$q_{tot}(T)/C_B(T) = K_{in}[\int C_p dt/C_B(T)] + V_B \tag{19}$$

In Eqs. (18) and (19), $[\int C dt/C_p(T)]$ and $[\int C_p dt/C_B(T)]$ have the units of time and represent the equivalent length of time that would be required to obtain the same plasma integral with a constant concentration equal to $C_p(T)$ or $C_B(T)$. When C_p is maintained constant with a programmed infusion, $[\int C_p dt/C_p(T)] = T$ and Eq. (18) reduces to Eq. (16). With IV bolus injection, $[\int C_p dt/C_p(T)] > T$, whereas with constant IV infusion, $[\int C_p dt/C_p(T)] < T$.

To apply the multiple time-point/graphic analysis procedure, the data are plotted using the format of either Eqs. (16), (18) or (19); the graph is then inspected to determine the linear unidirectional phase of uptake.[6,11,86] An example is shown in Fig. 5, which illustrates data for the uptake of [^{14}C]mannitol and [^{14}C]inulin into rat brain during constant plasma concentration experiments.[1] Significant backflux is indicated when the data points fall below that predicted by the initial straight line. Finally, K_{in} can be converted to PS using Eq. (8) if F is known.

Hawkins et al.[48] used both the multiple time-point method and compartmental analysis to calculate cerebrovascular K_{in} in their [^{68}Ga]-EDTA/PET study of BBB permeability in human brain tumors. Values obtained with graphic analysis were comparable to, but slightly lower than, those obtained with compartmental analysis using Eq. (3). Iannotti et al.[53] have also employed the graphic analysis procedure to determine BBB integrity to [^{68}Ga]-EDTA in humans with PET.

It should be noted that V_B or V_p as obtained with the multiple time-point method may differ from the true blood volume or plasma volume as measured with the classic markers. This is shown in Fig. 5, where V_p values for [^{14}C]mannitol and [^{14}C]inulin are approximately threefold greater than values obtained in the same animals with [^{113m}In]transferrin. Possible reasons for these differences may include binding or accumulation of tracer by the capillary endothelium, tracer uptake by arteries, veins, or any meningeal tissue that has been included with the brain sample, tracer exchange between plasma and erythrocytes, and differences in Hct between brain capillaries and systemic arteries.[6,21,111]

Multiple time-point/graphic analysis shares many of the advantages of compartmental analysis. It can be used for accurate determination of cerebrovascular PS for rapidly, moderately, and slowly penetrating compounds.[37] Furthermore, it can be used to measure regional transport with either QAR or PET. As demonstrated by Hawkins et al.,[48] IV administration data can be analyzed with both compartmental analysis and multiple time-

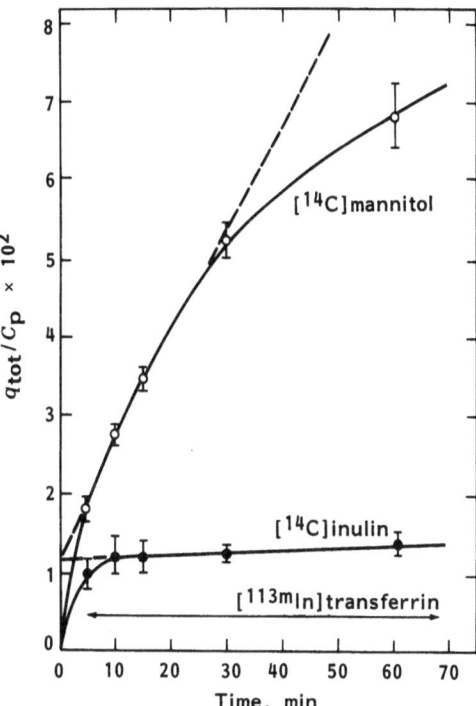

Figure 5. Time course of q_{tot}/C_p for [^{14}C]mannitol and [^{14}C]inulin in rat cerebral cortex. Each point represents a mean $\pm$SE for 5–7 rats. The average q_{tot}/C_p for the plasma marker, [^{113m}In]transferrin, is shown as well. Plasma tracer concentration was maintained constant in each rat by a programmed IV infusion. The blood-–brain transfer coefficient, K_{in}, and V_p were calculated by fitting Eq. (16) to the linear portion of the uptake curve. For [^{14}C]mannitol, K_{in} = 2.2 $\pm$ 0.1 $\times$ 10^{-5} ml/sec per g and V_p = 0.014 $\pm$ 0.001 ml/g, whereas for [^{14}C]inulin, K_{in} equals 4.9 $\pm$ 0.3 $\times$ 10^{-7} ml/sec per g and V_p = 0.013 $\pm$ 0.001 ml/g. The mean value of q_{tot}/C_p for [^{113m}In]transferrin, 0.0044 ml/g, was significantly less than V_p to [^{14}C]mannitol and [^{14}C]inulin. (Modified from Amtorp.[1])

point analysis, for comparison. The multiple time-point method produces a relatively model-independent transfer coefficient for influx.[85,86] This represents a significant advantage over compartmental analysis for tracers that have complex intracerebral distribution or metabolism. Calculated transfer coefficients with compartmental analysis are often specific to the particular model chosen. The multiple time-point method requires fewer data points than does compartmental analysis for accurate determination of K_{in}. Furthermore, the method does not require the use of a separate vascular indicator because the vascular volume is determined directly with this technique. Finally, the assumption of unidirectional uptake kinetics can be verified by inspection of the graph.

The limitations of the multiple time-point method include many of those of the compartmental analysis method. For example, the method is subject to errors due to tracer metabolism, does not provide a simple means to control plasma solute concentrations for studies of saturable transport, requires a series of animals with similar values of K_{in}, V_p, and F, and cannot be used when K_{in} changes with time. Unlike compartmental analysis, the multiple time-point method may tend to underestimate K_{in} and overestimate V_p due to difficulty in determination of the linear portion of the graph. Inclusion of data points with modest yet significant backflux will decrease the slope and increase the intercept (Fig. 5). The method cannot be used to measure regional cerebrovascular *PS* in each animal unless two different isotopic forms of the same solute are employed, such as [^{14}C]sucrose and [^{3}H]sucrose. In addition, the method may provide ambiguous results if $k_{out} \gg K_{in}$ or if V_{br} is very small. Last, although the method is less dependent on model selection than is compartmental analysis, results can be affected by intracerebral diffusion or metabolism,[7,109,115] especially in experiments with long uptake times.

3.1.3. Single Time-Point Analysis

Single time-point analysis represents the simplest form of the IV administration technique. The method has not been used to determine cerebrovascular *PS* in humans because both PET and SPECT provide sequential measurements of brain radioactivity in the same patient. As a result, the data are best analyzed using the multiple time-point or compartmental analysis methods. However, in small animal experiments, data can be obtained on brain uptake at only one time-point for each animal. This is because the animal must be killed if one is to measure brain tracer content with either QAR or tissue dissection. Therefore, to obtain a PS value in each animal, the single time-point analysis is the only version of the IV administration technique that can be employed. This property, along with the simplicity and flexibility of the technique, have made it the principal method for the study of cerebrovascular transport in small animals.

The experimental procedures for the single time-point method are identical to those of the multiple time-point method, except that uptake is determined with only one time-period. Three measurements are obtained from each animal: $q_{tot}(T)$, $\int C_p dt$, and $C_p(T)$ or $C_B(T)$.

With the single time-point method, uptake is limited so that only a small quantity of tracer accumulates in the brain ($q_{br}/V_{br} << C_p$), and thus backdiffusion of tracer from brain can be assumed to be negligible. With this condition, uptake is unidirectional,[10,73] and the time course of brain tracer content is given by Eq. (17) as

$$q_{tot}(T) = K_{in}\int C_p dt + V_p C_p(T)$$

This equation can be rearranged and solved for K_{in} to give

$$K_{in} = [q_{tot}(T) - V_p C_p(T)]/\int C_p dt \qquad (20)$$

Alternatively, if tracer is taken up into RBCs, K_{in} is given as

$$K_{in} = [q_{tot}(T) - V_B C_B(T)]/\int C_p dt \qquad (21)$$

PS can be calculated from K_{in} using Eq. (8) if F is known.

The single time-point method was developed by Ohno et al.[73] in 1978 and by Rapoport et al.[100] in 1979. These investigators were the first to derive the equation to calculate cerebrovascular PS from measurements of brain uptake at a single time point. In addition, they made the important observation that plasma tracer concentration does not have to be maintained constant to determine PS in unidirectional uptake experiments. Prior to that time, most experiments with the IV administration technique had used a programmed infusion to maintain a constant plasma concentration of tracer, as originally suggested by Davson.[27] However, a constant plasma concentration is not required for Eqs. (20) or (21). In fact, no specific time course is assumed. Ohno et al.[73] used an IV bolus injection procedure that is simple to perform and that permits plasma tracer concentration to decline with time, thus reducing the contribution of intravascular tracer to measured brain content.

Figure 6 illustrates a single time-point experiment in which a rat was killed 15 min after bolus injection of [14C]mannitol into the femoral vein. Blood samples were collected at various times after injection from a cannula in the femoral artery. $\int C_p dt$ was obtained by using a nonlinear least-squares procedure to fit the equation, $C_p(t) = \Sigma\, B_i e^{-b_i t}$, to the data and then by computer integrating the equation from $t = 0\text{–}15$ min.[73,100,115] Alternatively, $\int C_p dt$ can be determined by continuous arterial blood withdrawal at a constant rate with a pump to obtain an average concentration, $\bar{C}_p$, where $\int C_p dt = \bar{C}_p T$.[36,37] Transfer coefficients were calculated using Eq. (20) where V_p was determined in the same animal using [3H]inulin.[114,115]

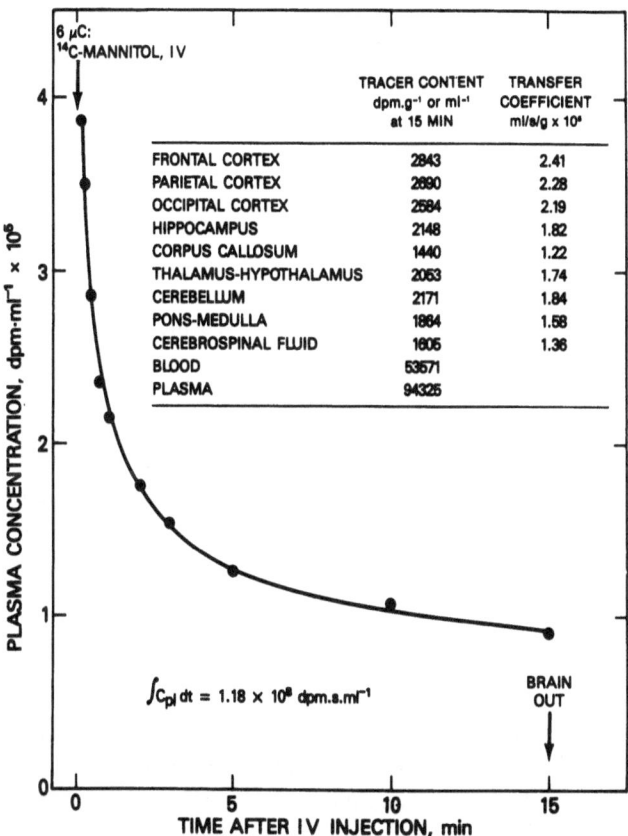

Figure 6. Arterial plasma concentrations of [^{14}C]mannitol following IV injection of 6 μC [^{14}C]mannitol into a rat. The curve is the computer fit of the equation $C_p(t) = \Sigma B_i e^{-b_i t}$, to the data. Regional brain content, q_{br} at 15 min, represents parenchymal radioactivity (total minus intravascular).

The single time-point method requires knowledge of intravascular volume in order to correct the measured brain content (q_{tot}) for residual intravascular tracer. Intravascular volume is usually measured using a vascular radiotracer that does not measurably cross the BBB during the experiment. Care should be taken, however, to ensure that the vascular volume of the impermeant tracer is comparable to that of the test compound (see Section 3.1.1). Significant differences can occur, as shown in Fig. 5 where V_p for [^{14}C]mannitol is approximately threefold greater than that for [^{113m}In]transferrin. An alternative procedure is to wash intravascular tracer from the brain blood vessels by perfusion of the brain with tracer-free fluid. With this procedure $V_B C_B(T) \approx 0$ and Eq. (21) reduces to

$$K_{in} \cong q_{tot}(T)/\int C_p dt \qquad (22)$$

The uptake time of each experiment should be chosen to ensure accurate determination of $q_{br}(T)$. As a general rule, each experiment should be run until $q_{tot}(T) \geq 3V_B C_B(T)$. If uptake time is too short and $V_B C_B(T)$ comprises more than over one-third of $q_{tot}(T)$, small errors in the determination of intravascular tracer could markedly affect the calculated value for $q_{br}(T)$ and thus K_{in}. When $q_{tot}(T) > 3V_B C_B(T)$, a 20% error in the measurement of intravascular tracer will produce less than a 10% error in the determination of K_{in}.

A key assumption of the single time-point method is that tracer uptake is unidirectional during the measurement of K_{in}.[73] For a two-compartment model, influx is unidirectional when $k_{out} \int q_{br} dt << K_{in} \int C_p dt$. There are three ways to evaluate this assumption. In the first, the calculated brain concentration, $q_{br}(T)/V_{br}$, is compared with the average plasma concentration, $\bar{C}_p = \int C_p dt/T$. If $q_{br}(T)/V_{br} < 0.2\bar{C}_p$, then net tracer uptake can be considered unidirectional with less than 10% error. This calculation requires knowledge of V_{br} and is valid only when $K_{in} = K_{out}$. In the second, K_{in} is determined at several early uptake times and the period of unidirectional influx is obtained from the period over which K_{in} is constant.[115] Significant backflux is indicated by a decrease in calculated K_{in}. In the third, the period of unidirectional uptake is determined using either the compartmental or multiple time-point analysis techniques. The assumption of unidirectional influx should be verified in all studies using the single time-point method.[11,37,86]

Thus, the single time-point version of the IV administration method is well suited to determine regional cerebrovascular PS in individual small animals using QAR or tissue dissection. The method is simple to perform and permits accurate measurements of PS over a 10^5-fold range (Fig. 7). Both K_{in} and F can be measured simultaneously for accurate determination of PS for rapidly penetrating compounds. The method is less subject to errors due to intracerebral metabolism or diffusion than the compartmental or multiple time-point techniques, as uptake time can be minimized.[37,115] Furthermore, the method does not require a series of animals with similar values for K_{in}, V_B, and F, and can therefore be used to obtain quantitative information on BBB permeability and transport in various pathologic conditions, such as brain tumors, where variability in K_{in}, V_B, and F values prevent use of the compartmental and multiple time-point techniques.[8,46] Last, knowledge of the specific time course of plasma radioactivity is not required. The investigator need only determine $\int C_p dt$ and $q_{br}(T)$ to calculate K_{in}.

The limitations of the single time-point method include possible error in determination of residual intravascular tracer $[V_B C_B(T)]$ and in the assumption of negligible backflux. Appropriate choice of uptake time is critical to minimize the influence of these factors on the calculation of K_{in}.[9] Ideally, uptake should be long enough to permit sufficient tracer to enter the brain to obtain accurate estimates of parenchymal radioactivity with minimal correction for intravascular tracer $[q_{tot}(T) > 3V_B C_B(T)]$. Yet, uptake

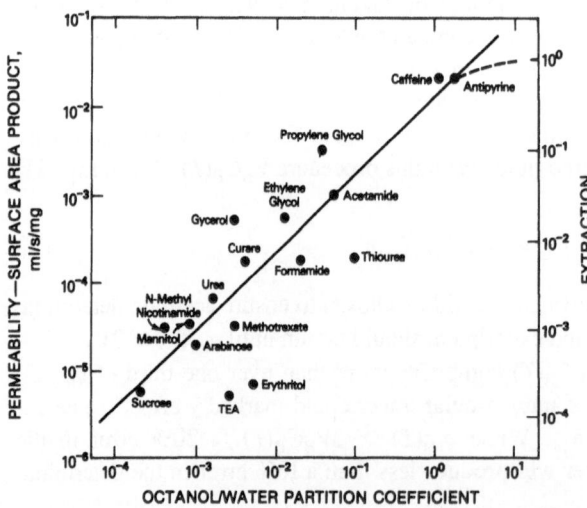

Figure 7. Relationship between cerebrovascular permeability–surface area product, PS, and octanol/water partition coefficient for organic electrolytes and nonelectrolytes. The calculated unidirectional extraction, E_o, is shown as well for $F = 3 \times 10^{-2}$ ml/sec per g. The solid line is the least-squares fit to the data and is given as $\log_{10} PS = -1.92 \ (\pm 0.34) + 0.866 \ (\pm 0.139)\log_{10}$ partition coefficient. The dotted line shows the calculated extraction, which approaches a maximum of $E_o = 1$. (Modified from Rapoport et al.[100])

time should not be so long that tracer accumulates within the brain and backflux becomes significant as compared with influx [$k_{out} \int q_{br}dt \ll K_{in} \int C_p dt$]. If good estimates of K_{in} and V_{br} are available for a compound, the appropriate time interval for unidirectional uptake can be calculated as

$$(2rV_B/K_{in}IF) < T < (0.2V_{br}/K_{in}) \tag{23}$$

where $IF = \int C_p dt/TC_p(T)$, $r = C_B(T)/C_p(T)$ and K_{in} is assumed to equal K_{out}. During that time interval, backflux will be negligible (<10%) and intravascular tracer will contribute less than 33% to measured brain content. If $K_{in} \neq K_{out}$, Eq. (23) can still be used by substitution of V'_{br} for V_{br} where $V'_b = K_{in}/k_{out}$ (V'_{br} = the effective volume of distribution).[38,41] If K_{in} and V_{br} are not known for a compound, the appropriate uptake time may be difficult to predict, although an initial estimate of K_{in} can be obtained from the octanol/water partition coefficient of the compound and the empirical relationship between the partition coefficient and cerebrovascular PS.[100,123] In the end, the investigator may be required to try several different times and analyze the data using the multiple time-point or compartmental analysis techniques. If regional K_{in} differs by >20-fold, it may be impossible to obtain a single time point that fulfills both requirements for minimal intravascular tracer and negligible backflux in all brain regions,[115] unless the tracer has a large intracerebral volume of distribution.[6,12,55] Finally, the method is subject to the same limitations as the multiple time-point and compartmental analysis techniques concerning tracer metabolism and control of plasma solute concentration.

3.2. Brain-Perfusion Technique

Although brain-perfusion techniques are not applicable to the study of BBB transport in humans, they do provide many possibilities for the study of cerebrovascular transport in experimental animals. Their principal advantage, as compared with the IV administration method, is that they provide virtually absolute control over perfusate composition and thus simplify experiments that are difficult or impossible under normal in vivo conditions. Whereas various perfusion methods have been developed during the past 20 years, their use has been limited by the difficulty of the surgical procedures and uncertainty concerning the physiologic status of the perfused brain.[35,128,132] In addition, the accuracy and range of the transport measurements were limited with these techniques, because single-pass methods were used to measure uptake into the brain. These limitations, however, have been overcome recently by the development of simple brain-perfusion methods that permit accurate measurements of BBB transport. The rat brain-perfusion technique of Takasato et al.[123] was the first method to use this new approach and has been followed by several alternative procedures for use in other small animals.[44,133]

With the brain-perfusion technique of Takasato et al.,[123] the right cerebral hemisphere of an anesthetized rat is perfused by retrograde infusion of fluid into the right external carotid artery. Cerebrovascular PS is calculated from the brain parenchymal uptake of tracer during perfusion. Because perfusion time can be extended beyond that of a single pass (5–15 sec), this method is more sensitive than the indicator diffusion, BUI, or single-injection–external-registration techniques. The principal advantage of this method, as compared with the IV administration technique, is that virtually absolute control is permitted over perfusate composition.

A diagram of the perfusion system is shown in Fig. 8. Prior to perfusion, the right external carotid artery of an anesthetized rat is catheterized, and the right pterygopalatine

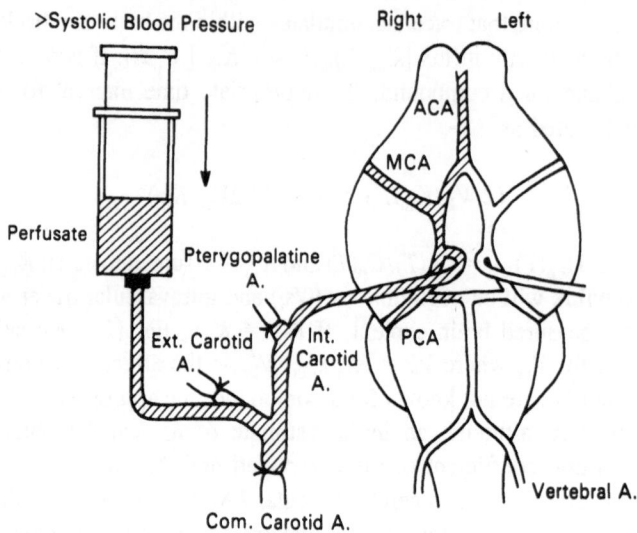

Figure 8. Diagram of technique for perfusing the right cerebral hemisphere of a rat. ACA, anterior cerebral artery; MCA, middle cerebral artery; PCA, posterior cerebral artery. (From Takasato et al.[123])

artery is ligated. The cannula to the right external carotid artery is connected to a syringe containing a test tracer and an impermeant intravascular tracer dissolved in physiologic saline, plasma, or blood. The right common carotid artery is ligated 1 sec before perfusion. Perfusion fluid is infused into the external carotid artery at a rate that minimizes (<0.3%) systemic blood flow to the right cerebral hemisphere and maintains a perfusion pressure similar to systemic blood pressure.[116,122,123] During perfusion with blood, cerebral blood flow and blood volume are comparable to respective values in the conscious rat, whereas perfusion with saline or plasma increases F by three- to fourfold due to the low viscosity of these fluids.[123] Perfusion with blood for 300 sec or with saline for 60 sec does not alter the permeability of the BBB.[116] The perfusion is terminated by decapitation of the rat, after which samples from six brain regions and perfusion fluid are analyzed for radiotracer content.

The calculation of cerebrovascular PS is equivalent to that of the IV administration technique, except that perfusion fluid concentration, C_{pf}, is inserted for C_p in the equations. If perfusion time is limited to restrict tracer accumulation in the brain and thus minimize backdiffusion, cerebrovascular PS can be calculated as,[100,123]

$$PS = -vF\ln[1 - (q_{br}(T)/vFTC_{pf})] \qquad (24)$$

When backflux cannot be ignored ($q_{br}/V_{br} > 0.2\ C_{pf}$), PS can be calculated from a single time-point with Eqs. (4) and (8), if V_{br} is known. Similarly, both the multiple time-point and compartmental analysis methods can be used with brain-perfusion data when uptake is measured over two or more time intervals.

Regional cerebral perfusion fluid flow can be measured in a separate set of animals with [14C]iodoantipyrine during blood perfusion or [14C]diazepam during saline perfusion.[123] Alternatively, both PS and F can be measured simultaneously in short (5–20 sec) perfusions by substituting a flow indicator for the intravascular marker.

The in situ brain-perfusion technique has several advantages in common with the IV

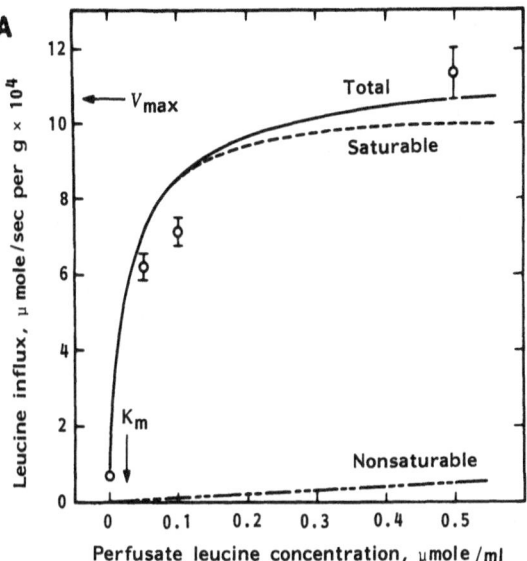

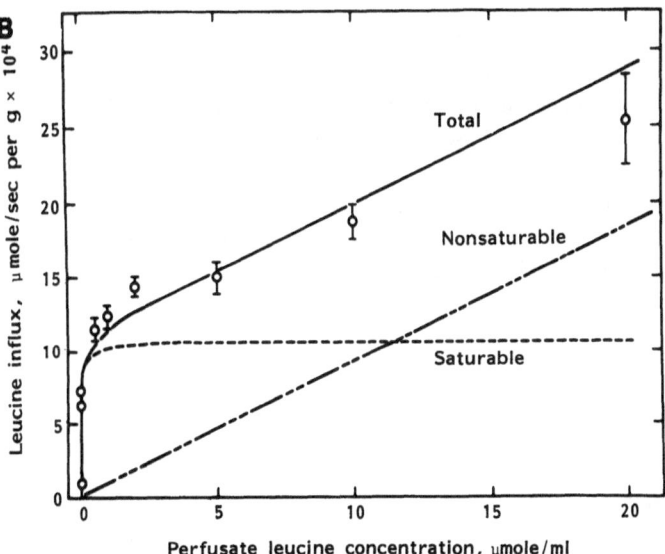

Figure 9. Relationship of unidirectional L-leucine influx into rat parietal cortex to L-leucine concentration of physiologic saline perfusion fluid. The perfusate did not contain any other amino acids. (A) Perfusate concentration range: 0–0.5 µmoles/ml. (B) Perfusate concentration range: 0–20 µmoles/ml. Each point represents a mean ±SE for three animals. The curve is the best fit of the equation

$$\text{Influx} = V_{max}C_{pf}/(K_m + C_{pf}) + K_d C_{pf}$$

where V_{max} and K_m are the Michaelis–Menten constants of the saturable component and K_d is the constant of nonsaturable diffusion. The best-fit values are $V_{max} = 1.07 \times 10^{-3}$ µmoles/sec per g, $K_m = 0.026 \pm 0.001$ µmoles/ml, and $K_d = 6.8 \pm 0.7 \times 10^{-5}$ ml/sec per g. (From Smith et al.[117])

administration technique. Cerebrovascular *PS* products can be measured over a 10^5-fold range from a minimum of $\sim 5 \times 10^{-6}$ ml/sec per g. Both K_{in} and F can be determined simultaneously in order to obtain accurate *PS* values for rapidly penetrating compounds. In addition, a *PS* value can be obtained for each experimental animal.

A few advantages of the brain-perfusion technique are not shared by the IV administration technique. First, the perfusion method permits total control of perfusate composition. Specific solute concentrations in the perfusate can be manipulated to examine saturation, competition, and inhibition of carrier-mediated transport (Fig. 9).[116–118,122] Furthermore, pH, osmolality, ionic content, and protein concentrations can be varied over a greater range than would be tolerated systemically. Second, the brain-perfusion technique avoids errors due to radiotracer metabolism by tissues other than the brain.[123] Last, the three- to fourfold greater F with saline perfusion minimizes flow-related errors in the determination of *PS* for rapidly penetrating substances,[102] because as $K_{in}/F \rightarrow 0$, $K_{in} \rightarrow PS$.

The advantages of the brain-perfusion technique, as compared with the indicator diffusion, BUI, and single-injection–external-registration techniques, are that regional cerebral perfusion fluid flow is known and constant during perfusion, and that *PS* products can be measured accurately for poorly penetrating compounds, such as urea, mannitol, and sucrose. Furthermore, unlike the BUI method, there is negligible mixing (<0.3%) of perfusion fluid with blood before the perfusate reaches the brain capillaries.[116] Mixing with blood, when significant, can lead to large errors in the determination of brain influx for compounds that cross the BBB by facilitated diffusion.[116–118,122]

Whereas several isolated brain-perfusion methods have been reported,[35,128,132] the in situ brain-perfusion technique of Takasato et al.[123] requires far less surgery and has a shorter perfusion time than the isolated brain methods.

The limitations of the in situ brain-perfusion technique are that, as of now, only one cerebral hemisphere is perfused and that the experiments use anesthetized, as opposed to conscious, rats. Furthermore, the brain may become hypoxic after perfusion for 30 sec or longer with physiologic saline, because of insufficient delivery of oxygen to the brain.

4. SINGLE-PASS UPTAKE

4.1. Indicator Diffusion

The cerebrovascular permeability to moderately and rapidly penetrating compounds can be determined with the indicator-diffusion technique, which was first applied to the brain by Crone[22] in 1963. The method was extended for use in humans by Lassen et al.[65] in 1971 but is rarely used in clinical research because it requires direct injection into the carotid artery and measures permeability only for the brain as a whole with no regional resolution.[89]

With the indicator-diffusion technique, a buffered saline solution containing the test tracer and an impermeant reference tracer is injected as a bolus into the carotid artery. Commonly used reference tracers, such as [^{99m}Tc]albumin, ^{24}Na, ^{36}Cl, and [^{113m}In]-DTPA, do not measurably cross the BBB in a single pass.[23,50,65,131] Immediately following the injection, serial blood samples are collected for 15–30 sec from a cannula in the superior sagittal sinus or the internal jugular vein (Fig. 10).

For each sample, a fractional extraction can be determined as follows from the difference in the arterial (C_A) and venous (C_V) concentrations of test tracer.

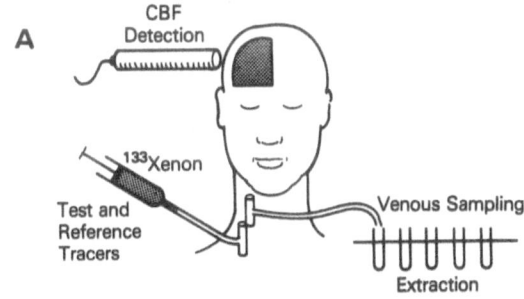

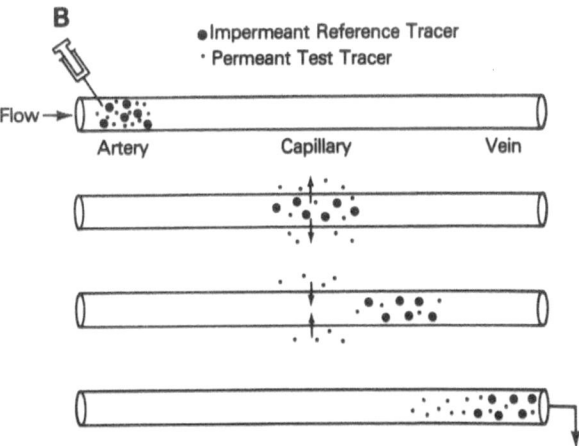

Figure 10. (A) Schematic illustration of the indicator-diffusion technique in humans, combined with measurement of cerebral blood flow with ^{133}Xe. (Modified from Paulson and Hertz.[89]) (B) Schematic representation of the extraction of a permeant test tracer relative to an impermeant reference tracer as the bolus passes through the cerebral vasculature. (Modified from Granger and Perry.[43])

$$E = (C_A - C_V)/C_A \qquad (25)$$

With the indicator-diffusion technique, however, C_A is not measured directly but is obtained indirectly from the venous concentration of reference tracer. The impermeant reference tracer corrects for dilution with blood as the bolus passes through the cerebral vasculature, thereby indicating what the concentration of test tracer would have been if no brain uptake had occurred (Fig. 11). Then, Eq. (25) can be expressed as[22–25]

$$E = [(C_V)_{ref} - (C_V)_{test}]/(C_V)_{ref} \qquad (26)$$

where ref is the impermeant reference tracer and test is the test tracer. For Eq. (26), it is assumed that the injectate solution contains equal concentrations of test and reference tracers. In most experiments, however, different concentrations are employed. In that case, the venous concentration of each tracer is normalized for the injectate concentration and E is calculated as

$$E = 1 - C'_{test}/C'_{ref} \qquad (27)$$

where $C' = (C_V/C_{injectate})$.

　　During the initial phase of tracer uptake, extraction is unidirectional, E_o, and can be related to PS with the Crone equation[22]:

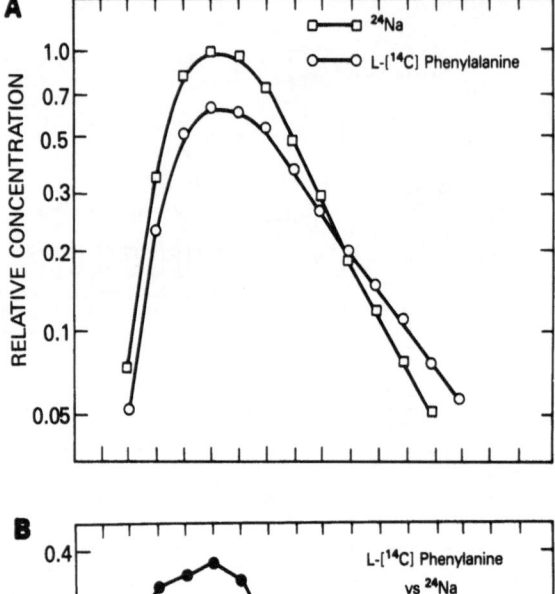

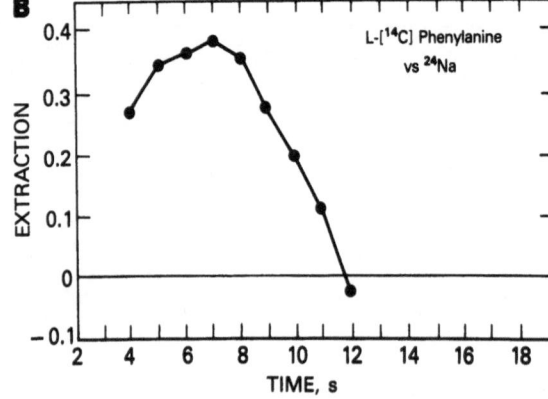

Figure 11. (A) Relative concentrations of L-[^{14}C]phenylalanine and ^{24}Na in samples of blood from the internal jugular vein after simultaneous injection of both tracers into the internal carotid artery of a man. Relative concentration equals $C_v/C_{injectate}$. (B) Calculated extraction E, as given by Eq. (27). (Modified from Hertz and Paulson.[51])

$$PS = -vF\ln(1 - E_o) \qquad (28)$$

where F equals the flow of the bolus through the brain.

　　The advantages of the indicator-diffusion technique are that several measurements of E_o can be obtained on the same animal and that the technique can be used on human subjects[50,51,65,89] as well as on large[22] and small animals.[49] Furthermore, cerebral blood flow and extraction can be measured in the same patient to ensure accurate determination of PS for rapidly penetrating compounds.

　　However, the indicator-diffusion technique has several limitations. E_o can be accurately measured only from 0.05 to 0.9, which limits the permeability range of the technique. For example, with $E_o = 0.05$ and $F = 1 \times 10^{-2}$ ml/sec per g, the lower limit of PS for the technique is 5×10^{-4} ml/sec per g, which is 100 times the PS to sucrose (Fig. 5). Thus, the method provides PS values for poorly penetrating solutes, such as urea, mannitol, and sucrose, that are either insignificant or spuriously high.[23]

　　An additional problem is that E is not constant at different time-points of the venous outflow curve (Fig. 11). E will increase in the rapidly rising part of the outflow curve due to capillary heterogeneity.[50,51] Then, in the falling part of the outflow curve, E will decrease with time due to backdiffusion.[24,43,65] Furthermore, E can vary throughout the curve because of separation of test and reference tracers in the blood vessels as a result of

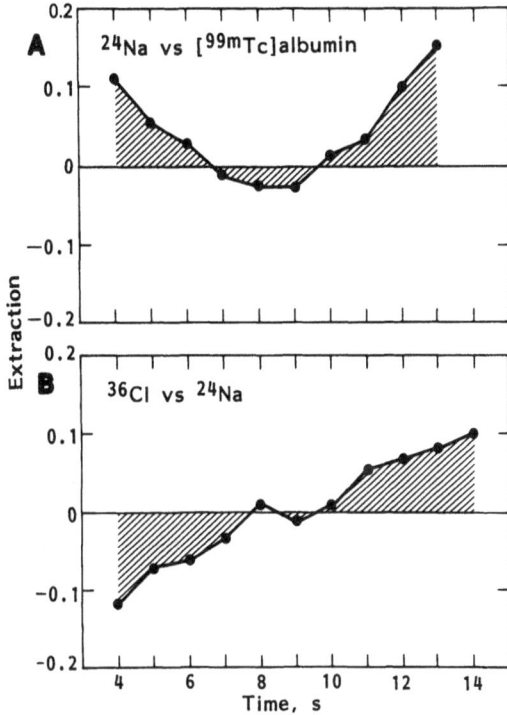

Figure 12. Artifactual extraction of impermeant reference tracers in man as a result of intravascular separation. The true unidirectional extractions of ^{24}Na, ^{36}Cl, and [^{99m}Tc]albumin are less than 1%.[65] (A) Artifactual extraction of ^{24}Na relative to [^{99m}Tc]albumin due to intralaminar diffusion (Taylor's phenomenon[103]). Because ^{24}Na has a greater diffusibility than [^{99m}Tc]albumin, there is a greater movement of ^{24}Na from the faster-moving central core of blood to the slower-moving peripheral layers in the blood vessels. As a result, the ^{24}Na concentration is less than the [^{99m}Tc]albumin concentration in the first samples of venous blood, thus producing a positive calculated extraction for ^{24}Na. After the peak ^{24}Na concentration has passed, ^{24}Na extraction becomes negative because diffusion of ^{24}Na is reversed, i.e., from the slower-moving peripheral layers of blood to the faster-moving central core.[65,89,124] (B) Artifactual extraction of ^{36}Cl relative to ^{24}Na due to erythrocyte carriage. ^{36}Cl, which penetrates erythrocytes, has a shorter transit time than does ^{24}Na, which remains in plasma, because erythrocytes move more rapidly than plasma through the cerebral vasculature. As a result, the concentration of ^{36}Cl is greater than that of ^{24}Na in the first samples of venous blood, thus producing a negative calculated extraction for ^{36}Cl.[89] (Modified from Paulson and Hertz.[89])

intralaminar or Taylor diffusion[24,65,124] and of erythrocyte carriage[50,65] (Fig. 12). However, tracer separation can be reduced by using a reference tracer with similar diffusion and erythrocyte-penetration properties to those of the test tracer[65,89] (Fig. 12).

Because E is not constant and is therefore ambiguous (Fig. 11), several papers have dealt with the issue of which part of the outflow curve should be used to calculate unidirectional extraction and thus PS.[24,43,65,89] For example, Bass and Robinson[3] have proposed a method to calculate a secure lower bound to cerebrovascular PS from indicator diffusion data, using the following equation

$$PS = -vF \int h(t)\ln[1 - E(t)]dt \qquad (29)$$

where $h(t)$ is the frequency function of the transit times of the intravascular reference tracer, $E(t)$ is the time-dependent extraction fraction of the indicator, and the integration is extended over all times for which $E(t)$ is positive. Eq. (29) gives a lower bound for PS because no explicit correction is made for backdiffusion.

Extracerbral contamination of sinus blood can be a significant problem with the indicator-diffusion technique when blood samples are obtained rapidly, and may require the use of a second reference tracer.[49] The rapid intracarotid injection may also transiently elevate cerebral blood flow, which should be known and constant during the measurement period. Last, the indicator-diffusion technique measures an average cerebrovascular permeability for the brain and cannot be used to examine regional permeability.

4.2. Brain Uptake Index

The BUI technique was introduced by Oldendorf[75] in 1970 as an intracarotid injection–single-pass method to measure cerebrovascular transport and permeability. In contrast to the indicator-diffusion technique, which measures E from the concentration of tracer in venous blood, the BUI method determines E from the quantity of tracer in brain after a single pass. Although the BUI technique is not used on humans, it has been widely used in small animal experiments due to the incredible speed, simplicity, and flexibility of the method.[76,80,84] In Oldendorf's laboratory alone, the method has been used on more than 10,000 animals to examine the transport of at least 200 different compounds.[78]

The customary animal in most BUI studies is the adult rat. A 200-μl bolus of buffered saline containing a test tracer and a permeant reference tracer is injected rapidly (<1 sec) into the carotid artery (Fig. 13). In some studies, an impermeant vascular tracer, such as [113mIn]-EDTA, is included in the bolus to correct for residual intravascular test tracer in the brain.[79] After 5–15 sec the rat is decapitated, and the brain is removed and analyzed for tracer contents.

Because only a small, variable fraction of injectate goes to the brain, the brain uptake of test tracer is normalized by the use of a permeant reference tracer of known uptake.[77] The classic reference for most BUI studies is [3H]-H2O.[75,77] However, other reference tracers have been proposed and tested recently, such as [14C]butanol, [14C]thiourea, [3H]tryptamine, [125I]N-isopropyl-p-iodoamphetamine, and labeled microspheres.[54,77,82]

The BUI for test tracer was defined by Oldendorf as[75]

$$\text{BUI} = [(q_{\text{tot}}/C_{\text{inj}})_{\text{test}}/(q_{\text{tot}}/C_{\text{inj}})_{\text{ref}}] \tag{30}$$

where ref is the permeant reference tracer. If the injectate contains equal concentrations of test and reference tracer, the BUI reduces to

$$\text{BUI} = (q_{\text{tot}})_{\text{test}}/(q_{\text{tot}})_{\text{ref}} \tag{31}$$

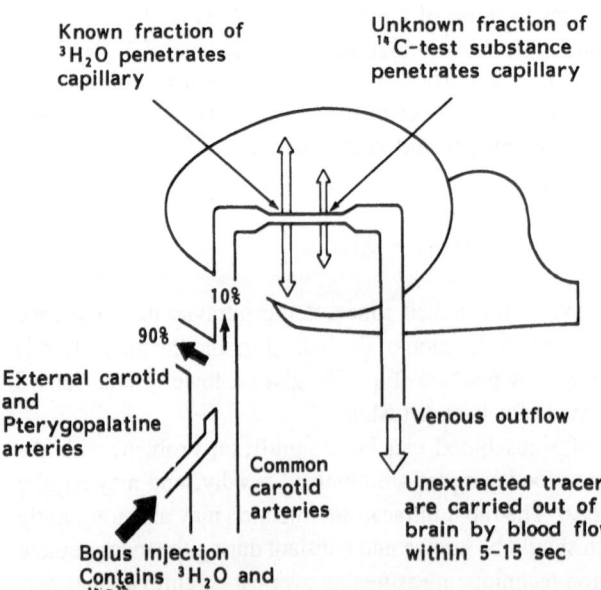

Figure 13. Diagram of the brain uptake index technique. A buffered solution containing a [14C]-labeled test solute and 3H2O is injected as a bolus into the carotid artery of a rat, and then the rat is decapitated 15 sec later. (Modified from Oldendorf.[77])

The BUI can be converted into an E_o if a correction is made for residual intravascular test tracer in the brain tissue, if the brain extraction of reference tracer is known, and if the washout of both test and reference tracers from brain is known for the 5–15 sec period before decapitation.[40,77] Assuming negligible backflux of test tracer, Oldendorf related BUI to E_o as

$$(E_o)_{test} \cong E_{ref}[(BUI_{test} - BUI_{vas})/(1 - BUI_{vas})] \tag{32}$$

where BUI_{vas} = BUI of the impermeant reference tracer and E_{ref} is the net extraction of permeant reference tracer.[77] E_{ref} equals ~0.42 for ^3H_2O and ~0.77 for [^{14}C]butanol in the cerebral hemisphere of a conscious rat at 15 sec after intracarotid injection.[14,40,82] Cerebrovascular PS can be obtained from E_o with Eq. (28) where F equals the flow of the bolus through the brain.

The advantages of the BUI technique are that the experiments are fast, inexpensive, and simple to perform, that either conscious or anesthetized animals can be used,[14,81] and that absolute control is allowed of injectate composition. Because solute concentration in the injection fluid can be manipulated easily, this method has been used frequently to examine saturable transport systems at the cerebrovascular endothelium.[80,84] The major limitation of the BUI technique is the difficulty in relating the measured BUI to cerebrovascular PS. E_{ref}, which relates BUI of the test tracer to cerebrovascular E_o, depends on several variables, such as blood flow, BBB permeability, brain region, and decapitation time; these variables may change under different experimental conditions.[13,32,40,45,82] In addition, the flow of the bolus, which has never been measured, must be known to calculate PS [Eq. (28)].

Cerebral blood flow should not be used for F in Eq. (26), because the intracarotid injection may transiently elevate flow.[45] The fact that a significant fraction of injectate travels down the common carotid into the aorta[77] indicates that the injection elevates carotid pressure above systolic blood pressure. The rise in carotid pressure would increase flow to the brain, which should be constant and known during the experimental procedure. In addition to the injection artifact, F should increase transiently due to the low viscosity of the saline injectate.[123]

There are additional limitations of this method. The BUI technique measures transport only in one cerebral hemisphere. The single-pass uptake limits the sensitivity of the method and prevents the method from obtaining accurate permeability coefficients for poorly penetrating compounds.[20] Last, there is substantial mixing ($\approx 5\%$) of the carotid bolus with systemic blood before the injectate reaches the brain capillaries.[83] Mixing, which probably occurs at the injection site and at the circle of Willis, can alter solute concentrations in the intracarotid bolus and lead to erroneous values for transport into brain, especially for the neutral and basic amino acids.[116–118,122]

4.3. Single Injection–External Registration

The single-injection–external-registration technique of Raichle et al.[95] is a single-pass method that uses external detection to measure brain extraction of a radiotracer. A test solute labeled with a γ- or positron-emitting radionuclide is injected into the carotid artery, and the time course of total brain radioactivity is measured for 30–60 sec with an external NaI detector. Because external detection is used, the method requires the use of a large animal, such as a monkey or dog, and has potential use for the study of cererbrovascular transport in humans.

Total brain radioactivity, q_{tot}, increases rapidly after injection and reaches a peak within 1–3 sec, when all the injected tracer is in either brain blood vessels or brain parenchyma (Fig. 14). Thereafter, total brain radioactivity decreases with time. The decline can be resolved into two components: a rapid phase ($t_{1/2} \leq 1$ sec), which primarily reflects washout of intravascular brain tracer, and a slow phase ($t_{1/2} \leq 10$ sec), which reflects tracer efflux from parenchymal brain tissue into blood[30,95,96] (Fig. 14).

Unidirectional brain extraction is calculated as

$$E_o = q_{br}(T')/q_{tot}(T') \tag{33}$$

where $T' \rightarrow$ is time of peak total brain radioactivity. $q_{br}(T')$ is obtained by graphicly extrapolating the slow component of the total brain radioactivity curve back to time T'.[30,96] Cerebrovascular *PS* is calculated from E_o with Eq. (28).

One advantage of the single-injection–external registration technique is that, unlike the indicator diffusion and BUI methods, a reference tracer is not required. In addition, several measurements of E_o can be obtained on the same animal. Finally, external detection permits possible use for the study of BBB transport in man.

However, this method is limited to the determination of E_o for moderately to rapidly

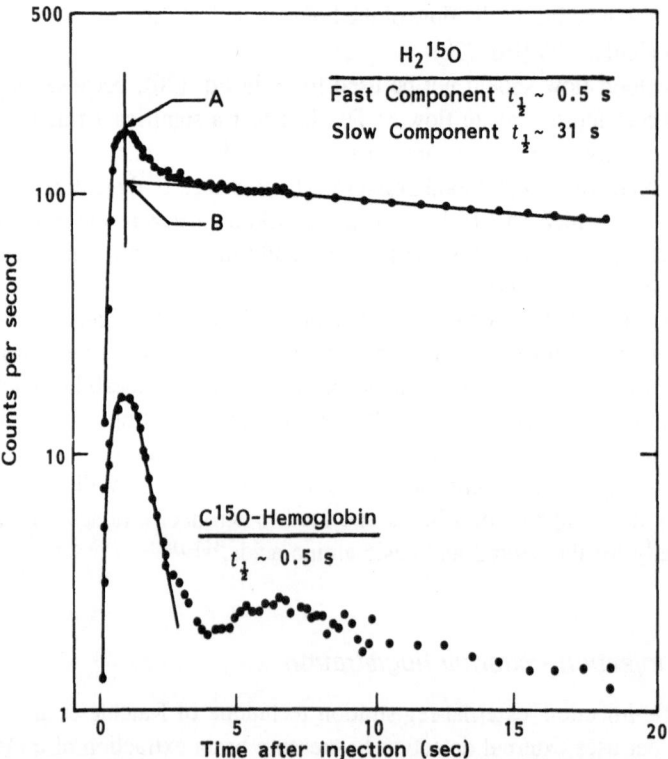

Figure 14. Time courses of $H_2{}^{15}O$ and of $[C^{15}O]$-hemoglobin in brain after intracarotid injection. Unidirectional extraction is calculated as the ratio of parenchymal brain radioactivity (B) divided by the total brain radioactivity (A) at the peak of the curve. (Modified from Eichling et al.[30])

penetrating compounds ($E_o \geq 0.10$). Furthermore, the method measures only an average E_o for whole brain and cannot be used to determine regional cerebrovascular PS.

5. SUMMARY

During the past two decades there has been considerable progress in the development of methods to examine BBB permeability quantitatively in animals and humans. Currently, the IV administration and in situ brain-perfusion techniques are the most versatile and sensitive methods to measure transport into the brain. These techniques are 100 times more sensitive than the single-pass methods and can accurately measure PS over a 10^5-fold range from a minimum of $\simeq 1 \times 10^{-6}$ ml/sec per g. Furthermore, cerebral blood flow and brain uptake can be measured independently in order to obtain accurate permeability coefficients for rapidly penetrating compounds. The IV administration technique can be used to measure regional cerebrovascular PS in humans with PET. In addition, the method can be used in conscious or anesthetized animals with either quantitative autoradiography or tissue dissection. The brain-perfusion technique, which complements the IV administration method, permits total control of perfusate composition to examine saturable transport at the BBB.

REFERENCES

1. Amtorp I: Estimation of capillary permeability of inulin, sucrose and mannitol in the rat brain cortex. *Acta Physiol Scand* 110:337–342, 1980.
2. Bachelard HS, Daniel PM, Love ER, et al: The transport of glucose into the brain of the rat in vivo. *Proc R Soc Lond B* 183:71–82, 1973.
3. Bass L, Robinson PJ: Capillary permeability of heterogeneous organs: A parsimonious interpretation of indicator diffusion data. *Clin Exp Pharmacol Physiol* 9:363–388, 1982.
4. Bassingthwaighte JB, Chinard FP, Crone C, et al: Terminology for mass transport and exchange. *Am J Physiol* 250:H539–H545, 1986.
5. Betz AL, Iannotti F: Simultaneous determination of regional cerebral blood flow and blood–brain glucose transport kinetics in the gerbil. *J Cereb Blood Flow Metab* 3:193–199, 1983.
6. Blasberg RG, Fenstermacher JD, Patlak CS: Transport of α-aminoisobutyric acid across brain capillary and cellular membranes. *J Cereb Blood Flow Metab* 3:8–32, 1983.
7. Blasberg RG, Hiraga S, Nakagawa H, et al: Diffusion of AIB into brain from the CSF–brain interface: Influence on the calculated influx constants across brain capillaries. *Acta Neurol Scand* 72:91, 1985.
8. Blasberg RG, Kobayashi T, Patlak CS, et al: Regional blood flow, capillary permeability, and glucose utilization in two brain tumor models: Preliminary observations and pharmacokinetic implications. *Cancer Treatm Rep* 65:3–12, 1981.
9. Blasberg RG, Patlak CS, Fenstermacher JD, et al: Selection of experimental conditions for the accurate determination of blood–brain transfer constants from single-time experiments: A theoretical analysis. *J Cereb Blood Flow Metab* 3:215–225, 1983.
10. Bradbury MWB: Ontogeny of mammalian brain barrier systems. In Cserr H, Fenstermacher JD, Fencl V (eds): *Fluid Environment of the Brain*. Academic, New York, 1975, pp. 81–103.
11. Bradbury M: *The Concept of a Blood–Brain Barrier*. Wiley, Chirchester, 1979.
12. Bradbury MWB, Kleeman CR: Stability of the potassium content of cerebrospinal fluid and brain. *Am J Physiol* 213:519–528, 1967.
13. Bradbury MWB, Patlak CS, Oldendorf WH: Analysis of brain uptake and loss of radiotracers after intracarotid injection. *Am J Physiol* 229:1110–1115, 1975.
14. Braun LD, Miller LP, Pardridge WM, et al: Kinetics of regional blood–brain barrier glucose transport and cerebral blood flow determined with the carotid injection technique in conscious rats. *J Neurochem* 44:911–915, 1985.

15. Brightman MW, Reese TS: Junctions between intimately apposed cell membranes in the vertebrate brain. *J Cell Biol* 40:648–677, 1969.

16. Brooks DJ, Beaney RP, Lammertsma AA, et al: Quantitative measurement of blood–brain barrier permeability using rubidium-82 and positron emission tomography. *J Cereb Blood Flow Metab* 4:535–545, 1984.

17. Brooks DJ, Beaney RP, Lammertsma AA, et al: Glucose transport across the blood–brain barrier in normal human subjects and patients with cerebral tumours studied using [^{11}C]-3-0-methyl-D-glucose and positron emission tomography. *J Cereb Blood Flow Metab* 6:230–239, 1986.

18. Bustany P, Henry JF, de Rotrou J, et al: Correlations between clinical state and positron emission tomography measurement of local brain protein synthesis in Alzheimer's dementia, Parkinson's disease, schizophrenia and gliomas, In Greitz T, Ingvar DH, Widen L (eds): *The Metabolism of the Human Brain Studied with Positron Emission Tomography*. Raven, New York, 1985, pp. 241–249.

19. Collins JM, Dedrick RL: Distributed model for drug delivery to CSF and brain tissue. *Am J Physiol* 245:R303–R310, 1983.

20. Cornford EM, Braun LD, Oldendorf WH, et al: Comparison of lipid-mediated blood–brain barrier penetrability in neonates and adults. *Am J Physiol* 243:C161–C168, 1982.

21. Cremer JE, Seville MP: Regional brain blood flow, blood volume and haematocrit values in the adult rat. *J Cereb Blood Flow Metab* 3:254–256, 1983.

22. Crone C: The permeability of capillaries in various organs as determined by use of the indicator diffusion method. *Acta Physiol Scand* 58:292–305, 1963.

23. Crone C: The permeability of brain capillaries to non-electrolytes. *Acta Physiol Scand* 64:407–417, 1965.

24. Crone C: Capillary permeability—techniques and problems, In Crone C, Lassen NA (eds): *Capillary Permeability*, Academic, New York, 1970, pp. 15–31.

25. Crone C, Levitt DG: Capillary permeability to small solutes, In Renkin EM, Michel CC, (eds): *Handbook of Physiology*. Section 2: *The Cardiovascular System*. Vol. 4: Microcirculation. Part 1. American Physiological Society, Bethesda, 1984, pp. 411–466, 1984.

26. Daniel PM, Pratt OE, Wilson PA: The transport of L-leucine into the brain of the rat in vivo: Saturable and non-saturable components of influx. *Proc R Soc Lond B* 196:333–346, 1977.

27. Davson H: A comparative study of the aqueous humor and cerebrospinal fluid in the rabbit. *J Physiol (Lond)* 129:111–133, 1955.

28. Davson H, Welch K: The permeation of several materials into the fluids of the rabbit's brain. *J Physiol (Lond)* 218:337–351, 1971.

29. Draper N, Smith H: *Applied Regression Analysis*, 2nd Ed. Wiley, New York, 1981, pp. 458–517.

30. Eichling JOM, Raichle ME, Grubb RL, et al: Evidence of the limitations of water as a freely diffusible tracer in brain of the rhesus monkey. *Circ Res* 35:358–364, 1974.

31. Fenstermacher JD, Blasberg RG, Patlak CS: Methods for quantifying the transport of drugs across brain barrier systems. *Pharmacol Ther* 14:217–248, 1981.

32. Fenstermacher JD, Rapoport SI: Blood–brain barrier. In Renkin EM, Michel CC (eds): *Handbook of Physiology*. Section 2: *The Cardiovascular System*. Vol. 4: *Microcirculation*. Part 2. American Physiological Society, Bethesda, 1984, pp. 969–1000.

33. Frackowiak RSJ, Lenzi GL, Jones T, et al: Quantitative measurement of regional blood flow and oxygen metabolism in man using ^{15}O and positron emission tomography: Theory, procedure, and normal values. *J Comput Assist Tomogr* 4:727–736, 1980.

34. Fuglsang A, Lomholt M, Gjedde A: Blood–brain transfer of glucose and glucose analogs in newborn rats. *J Neurochem* 46:1417–1428, 1986.

35. Gilboe DD: Perfusion of the isolated brain, In Lajtha A (ed): *Handbook of Neurochemistry*. Vol. 2. Plenum, New York, 1982, pp. 301–330.

36. Gjedde A: Rapid steady state analysis of blood–brain glucose transfer in rat. *Acta Physiol Scand* 108:331–339, 1980.

37. Gjedde A: High and low-affinity transport of D-glucose from blood to brain. *J Neurochem* 36:1463–1471, 1981.

38. Gjedde A: Calculation of cerebral glucose phosphorylation from brain uptake of glucose analogs in vivo: A re-examination. *Brain Res Rev* 4:237–274, 1982.

39. Gjedde A, Diemer NH: Double-tracer study of the fine regional blood–glucose transfer in the rat by computer assisted autoradiography. *J Cereb Blood Flow Metab* 5:282–289, 1985.

40. Gjedde A, Rasmussen M: Blood–brain glucose transport in the conscious rat: Comparison of the intravenous and intracarotid injection methods. *J Neurochem* 35:1375–1381, 1980.

41. Gjedde A, Weinhard K, Heiss WD, et al: Comparative regional analysis of 2-fluorodeoxyglucose and methylglucose uptake in brain of four stroke patients. With special reference to the regional estimation of lumped constant. *J Cereb Blood Flow Metab* 5:163–178, 1985.

42. Goldstein GW, Betz AL, Bowman PD: Use of isolated brain capillaries and cultured endothelial cells to study the blood–brain barrier. *Fed Proc* 43:191–195, 1984.

43. Granger DN, Perry MA: Permeability characteristics of the microcirculation, In Mortillaro NA (ed): *The Physiology and Pharmacology of the Microcirculation,* Vol. 1. Academic, New York, 1983, pp. 157–208.

44. Greenwood J, Luthert PJ, Pratt OE, et al: Maintenance of the integrity of the blood–brain barrier in the rat during an in situ saline-based perfusion. *Neurosci Lett* 56:223–227, 1985.

45. Hardebo JE, Nilsson B: Estimation of cerebral extraction of circulating compounds by the brain uptake index method: Influence of circulation time, volume injection, and cerebral blood flow. *Acta Physiol Scand* 107:153–159, 1979.

46. Hasegawa H, Yuritaka U, Hayakawa T, et al: Changes of the blood–brain barrier in experimental metastatic brain tumors. *J Neurosurg* 59:304–310, 1983.

47. Hawkins RA, Mans AM, Biebuyck JF: Amino acid supply to individual cerebral structures in awake and anesthetized rats. *Am J Physiol* 242:E1–E11, 1982.

48. Hawkins RA, Phelps ME, Huang SC, et al: A kinetic evaluation of blood–brain barrier permeability in human brain tumors with ^{68}Ga-EDTA and positron computed tomography. *J Cereb Blood Flow Metab* 4:507–515, 1984.

49. Hertz MM, Bolwig TG: Blood–brain barrier studies in the rat: An indicator dilution technique with tracer sodium as an internal standard for estimation of extracerebral contamination. *Brain Res* 107:333–343, 1976.

50. Hertz MM, Paulson OB: Heterogeneity of cerebral capillary flow in man and its consequences for estimation of blood–brain barrier permeability. *J Clin Invest* 65:1145–1151, 1980.

51. Hertz MM, Paulson OB: Transfer across human blood–brain barrier: Evidence for capillary recruitment and for paradox glucose permeability increase in hypocapnia. *Microvasc Res* 24:364–376, 1982.

52. Huang SC, Carson RE, Hoffman EJ, et al: Quantitative measurement of local cerebral blood flow in humans by positron computed tomography and ^{15}O–water. *J Cereb Blood Flow Metab* 3:141–153, 1983.

53. Iannotti F, Alfano B, Pozzilli C, et al: Quantitative assessment of blood–brain barrier permeability to ^{68}Ga-EDTA by positron emission tomography in human brain tumors. *Acta Neurol Scand,* 72:104, 1985.

54. Irwin GH, Preskorn SH: A dual label radiotracer technique for the simultaneous measurement of cerebral blood flow and the single-transit cerebral extraction of diffusion limited compounds in rats. *Brain Res* 249:23–30, 1982.

55. Jarden JD, Dhawan V, Poltorak A, et al: Positron emission tomographic measurement of blood-to-brain and blood-to-tumor transport of ^{82}Rb: The effect of dexamethasone and whole brain radiation therapy. *Ann Neurol* 18:636–646, 1985.

56. Johanson CE, Woodbury DM: Uptake of [^{14}C]urea by the in vivo choroid plexus–cerebrospinal fluid–brain system: Identification of sites of molecular sieving. *J Physiol* 275:167–176, 1978.

57. Johnson JA, Wilson TA: A model for capillary exchange. *Am J Physiol* 210:1299–1303, 1966.

58. Juhler M, Blasberg RG, Fenstermacher JD, et al: A spatial analysis of the blood–brain barrier in experimental allergic encephalomyelitis. *J Cereb Blood Flow Metab* 5:545–553, 1985.

59. Katzman R, Leiderman PH: Brain potassium exchange in normal adult and immature rats. *Am J Physiol* 175:263–270, 1953.

60. Kessler RM, Goble JC, Bird JH, et al: Measurement of blood–brain barrier permeability with positron emission tomography and ^{68}Ga-EDTA. *J Cereb Blood Flow Metab* 4:323–328, 1984.

61. Knott GD: M Lab—A mathematical modeling tool. *Comput Programs Biomed* 10:271–280, 1976.

62. Lammertsma AA, Brooks DJ, Beaney RP, et al: In vivo measurement of regional cerebral hematocrit using positron emission tomography. *J Cereb Blood Flow Metab* 4:317–332, 1984.

63. Lammertsma AA, Brooks DJ, Frackowiak RSJ, et al: A method to quantitate fractional extraction of rubidium-82 across blood–brain barrier using positron emission tomography. *J Cereb Blood Flow Metab* 4:523–534, 1984.

64. Lassen NA: Regional cerebral blood flow in cerebrovascular disease by SPECT (single photon emission computed tomography). *J Neuroradiol* 10:181–184, 1983.

65. Lassen NA, Trap-Jensen J, Alexander SC, et al: Blood–brain barrier studies in man using the double-indicator method. *Am J Physiol* 220:1627–1633, 1971.

66. Lear JL, Ackermann RF, Kameyama M, et al: Evaluation of [^{123}I]isopropyliodoamphetamine as a tracer

for cerebral blood flow using direct autoradiographic comparison. *J Cereb Blood Flow Metab* 2: 179–185, 1982.

67. Levin VA, Ausman JI: Relationship of peripheral venous hematocrit to brain hematocrit. *J Appl Physiol* 26:433–437, 1969.

68. Levin V, Patlak CS: A compartmental analysis of ^{24}Na kinetics in rat cerebrum, sciatic nerve and cerebrospinal fluid. *J Physiol* 224:559–581, 1972.

69. Mans AM, Biebuyck JF, Shelly K, et al: Regional blood–brain barrier permeability to amino acids after portacaval anastomosis. *J Neurochem* 38:705–717, 1982.

70. Marcus ML, Heistad DD, Ehrhardt JC, et al: Total and regional cerebral blood flow measurement with 7-, 10-, 25-, and 50 μm microspheres. *J Appl Physiol* 40:501–507, 1976.

71. Neuwelt EA, Rapoport SI: Modification of the blood–brain barrier in the chemotherapy of malignant brain tumors. *Fed Proc* 43:214–219, 1984.

72. Ohno K, Chiueh CC, Burns EM, et al: Cerebrovascular integrity in protein-deprived rats. *Brain Res Bull* 5:251–255, 1980.

73. Ohno K, Pettigrew KD, Rapoport SI: Lower limits of cerebrovascular permeability to nonelectrolytes in the conscious rat. *Am J Physiol* 235:H299–H307, 1978.

74. Ohno K, Pettigrew KD, Rapoport SI: Local cerebral blood flow in the conscious rat as measured with ^{14}C-antipyrine, ^{14}C-iodoantipyrine and ^{3}H-nicotine. *Stroke* 10:62–67, 1979.

75. Oldendorf WH: Measurement of brain uptake of radiolabeled substances using a tritiated water internal standard. *Brain Res* 24:372–376, 1970.

76. Oldendorf WH: The blood–brain barrier. In Bito LZ, Davson H, Fenstermacher JD (eds): *The Occular and Cerebrospinal Fluids*. Academic, New York, 1977, pp. 177–190.

77. Oldendorf WH: Clearance of radiolabeled substances by brain after arterial injection using a diffusible internal standard. In Marks N, Rodnight R (eds): *Research Methods in Neurochemistry*, Vol. 5. Plenum, New York, 1981, pp. 91–112.

78. Oldendorf WH: Speculations on functions of the blood–brain barrier. *Adv Physiol Sci* 7:349–353, 1982.

79. Oldendorf WH, Braun LD: [^{3}H]Tryptamine and ^{3}H-water as diffusible internal standards for measuring brain extraction of radio-labeled substances following carotid injection. *Brain Res* 113:219–224, 1976.

80. Pardridge WM: Brain metabolism: A perspective from the blood–brain barrier. *Physiol Rev* 63:1481–1535, 1983.

81. Pardridge WM, Crane PD, Mietus LJ, et al: Kinetics of regional blood–brain barrier transport and brain phosphorylation of glucose and 2-deoxyglucose in the barbiturate-anesthetized rat. *J Neurochem* 38:560–568, 1982.

82. Pardridge WM, Fierer G: Blood–brain barrier transport of butanol and water relative to *N*-isopropyl-*p*-iodoamphetamine as the internal reference. *J Cereb Blood Flow Metab* 5:275–281, 1985.

83. Pardridge WM, Landaw EM, Miller LP, et al: Carotid artery injection technique: Bounds for bolus mixing by plasma and by brain. *J Cereb Blood Flow Metab* 5:576–583, 1985.

84. Pardridge WM, Oldendorf WH: Transport of metabolic substrates through the blood–brain barrier. *J Neurochem* 28:5–12, 1977.

85. Patlak CS, Blasberg RG: Graphical evaluation of blood-to-brain transfer constants from multiple-time uptake data: Generalizations. *J Cereb Blood Flow Metab* 5:584–590, 1985.

86. Patlak CS, Blasberg RG, Fenstermacher JD: Graphic evaluation of blood-to-brain transfer constants from multiple-time uptake data. *J Cereb Blood Flow Metab* 3:1–7, 1983.

87. Patlak CS, Fenstermacher JD: Measurements of dog blood–brain transfer constants by ventriculocisternal perfusion. *Am J Physiol* 229:877–884, 1975.

88. Patlak CS, Pettigrew KD: A method to obtain infusion schedules for prescribed blood concentration time courses. *J Appl Physiol* 40:458–463, 1976.

89. Paulson OB, Hertz MM: The indicator dilution method: Assumptions and applications to brain uptake. In Lambrecht RM, Rescigno A (eds): *Tracer Kinetics and Physiologic Modeling*. Springer-Verlag, Berlin, 1983, pp. 429–444.

90. Perlmutter JS, Larson KB, Raichle ME, et al: Strategies for in vivo measurement of receptor binding using positron emission tomography. *J Cereb Blood Flow Metab* 6:154–169, 1986.

91. Phelps ME, Barrio JR, Huang SC, et al: Measurement of cerebral protein synthesis in man with positron computerized tomography: Model, assumptions and preliminary results, In Greitz T, Ingvar DH, Widen L, (eds): *The Metabolism of the Human Brain Studied with Positron Emission Tomography*. Raven, New York, 1985, pp. 215–232.

92. Phelps ME, Huang SC, Hoffman EJ, et al: Tomographic measurement of cerebral blood volume with ^{11}C-labeled carboxyhemoglobin. *J Nucl Med* 20:328–334, 1979.

93. Powers WJ, Raichle ME: Positron emission tomography and its application to the study of cerebrovascular disease in man. *Stroke* 16:361–376, 1985.

94. Raichle ME: Positron emission tomography. *Ann Rev Neurosci* 6:249–267, 1983.

95. Raichle ME, Eichling JO, Grubb RL: Brain permeability of water. *Arch Neurol* 30:319–321, 1974.

96. Raichle ME, Eichling JO, Straatmann MG, et al: Blood–brain barrier permeability of ^{11}C-labeled alcohols and ^{15}O-labeled water. *Am J Physiol* 230:543–552, 1976.

97. Rapoport SI: *Blood–brain barrier in Physiology and Medicine*. Raven, New York, 1976, pp. 1–206.

98. Rapoport SI, Fitzhugh R, Pettigrew KD, et al: Drug entry into and distribution within brain and cerebrospinal fluid: [14C]urea pharmacokinetics. *Am J Physiol* 242:R339–R348, 1982.

99. Rapoport SI, Fredericks WR, Ohno K, et al: Quantitative aspects of reversible osmotic opening of the blood–brain barrier. *Am J Physiol* 238:R421–R431, 1980.

100. Rapoport SI, Ohno K, Pettigrew KD: Drug entry into the brain. *Brain Res* 172:354–359, 1979.

101. Rapoport SI, Ohno K, Pettigrew KD: Blood–brain barrier permeability in senescent rats. *J Gerontol* 34:162–169, 1979.

102. Reese TS, Karnovsky MJ: Fine structural localization of a blood brain barrier to exogenous peroxidase. *J Cell Biol* 34:207–217, 1967.

103. Reivich M, Alavi A, Wolf A, et al: Glucose metabolic rate kinetic model parameter determination in humans: The lumped constant and rate constants for [18F]fluorodeoxyglucose and [11C]deoxyglucose. *J Cereb Blood Flow Metab* 5:170–192, 1985.

104. Reivich M, Kuhl D, Wolf A, et al: The [18F]fluorodeoxyglucose method for the measurment of local cerebral glucose utilization in man. *Circ Res* 44:127–137, 1979.

105. Riggs DS: *The Mathematical Approach to Physiological Problems*. MIT Press, Cambridge, Massachusetts, 1963, pp. 120–167.

106. Sage JI, Van Uitert RL, Duffy TE: Simultaneous measurement of cerebral blood flow and unidirectional movement of substances across the blood–brain barrier: Theory, method, and application to leucine. *J Neurochem* 36:1731–1738, 1981.

107. Sakai F, Nakazawa K, Tazaki Y, et al: Regional cerebral blood volume and hematocrit measured in normal human volunteers by single photon emission computed tomography. *J Cereb Blood Flow Metab* 5:207–213, 1985.

108. Sakurada O, Kennedy C, Jehle J, et al: Measurement of local cerebral blood flow with iodo[14C]antipyrine. *Am J Physiol* 234:H59–H66, 1978.

109. Sarna GS, Bradbury MWB, Cavanagh J: Permeability of the blood–brain barrier after portocaval anastomosis in the rat. *Brain Res* 138:550–554, 1977.

110. Schafer JA, Gjedde A, Plum F: Regional cerebral blood flow in rat using n-[14C]butanol. *Neurology (NY)* 26:394, 1977.

111. Sisson WB, Oldendorf WH: Brain distribution spaces of mannitol[3]-H, inulin-^{14}C, and dextran-^{14}C in the rat. *Am J Physiol* 221:214–217, 1971.

112. Smith QR, Gnaedinger JM: Comparison of [14C]urea and [14C]sucrose as indicators of regional blood–brain barrier permeability using quantitative autoradiography. *Soc Neurosci Abstr*, 12:1258, 1986.

113. Smith QR, Johanson CE, Woodbury DM: Uptake of [36]Cl and [22]Na by the brain–cerebrospinal fluid system: Comparison of the permeability of the blood–brain and blood–cerebrospinal fluid barriers. *J Neurochem* 37:117–124, 1981.

114. Smith QR, Rapoport SI: Carrier-mediated transport of chloride across the blood–brain barrier. *J Neurochem* 42:754–763, 1984.

115. Smith QR, Rapoport SI: Cerebrovascular permeability coefficients to sodium, potassium and chloride. *J Neurochem* 46:1732–1742, 1986.

116. Smith QR, Takasato Y. Kinetics of amino acid transport at the blood–brain barrier studied using an in situ brain perfusion technique. *Ann NY Acad Sci*, 481:186–201, 1986.

117. Smith QR, Takasato Y, Rapoport SI: Kinetic analysis of L-leucine transport across the blood–brain barrier. *Brain Res* 311:167–170, 1984.

118. Smith QR, Takasato Y, Sweeney DJ, et al: Regional cerebrovascular leucine transport as measured by the in situ brainperfusion technique. *J Cereb Blood Flow Metab* 5:300–311, 1985.

119. Smith QR, Woodbury DM, Johanson CE: Kinetic analysis of [36Cl]-, [22Na]- and [3H]mannitol uptake into the in vivo choroid plexus–cerebrospinal fluid–brain system: Ontogeny of the blood–brain and blood–CSF barriers. *Dev Brain Res* 3:181–198, 1982.

120. Sokoloff L: The radioactive deoxyglucose method: Theory, procedure, and applications for the measurement of local glucose utilization in the CNS. *Adv Neurochem* 4:46–59, 1982.
121. Solomon AK: Compartmental methods of kinetic analysis, In Comar CL, Bronner F (eds): *Mineral Metabolism.* Vol. 1. Academic, New York, 1960, pp. 119–167.
122. Takasato Y, Momma S, Smith QR: Kinetic analysis of cerebrovascular isoleucine transport from saline and plasma. *J Neurochem* 45:1013–1020, 1985.
123. Takasato Y, Rapoport SI, Smith QR: An in situ brain perfusion technique to study cerebrovascular port in the rat. *Am J Physiol* 247:H484–H493, 1984.
124. Taylor G: The dispersion of soluble matter in solvent flowing slowly through a tube. *Proc R Soc Lond A* 219:186–203, 1953.
125. Weiss HR, Buchweitz E, Murtha TJ, et al: Quantitative regional determination of morphometric indices of the total and perfused capillary network in the rat brain. *Circ Res* 51:494–503, 1982.
126. Welch K: The principles of physiology of the cerebrospinal fluid in relation to hydrocephalus including normal pressure hydrocephalus. In Friedlander WJ (ed): *Advances in Neurology.* North-Holland, Amsterdam, 1975, pp. 247–332.
127. Wong DF, Gjedde A, Wagner HN: Quantification of neuroreceptors in the living human brain. I. Irreversible binding of ligands. *J Cereb Blood Flow Metab* 6:137–146, 1986.
128. Woods HF, Youdim MBH. The isolated perfused rat brain preparation—A critical assesment. *Essays Neurochem* 3:49–69, 1978.
129. Wright EM: Transport processes in the formation of the cerebrospinal fluid. *Rev Physiol Biochem Pharmacol* 83:1–34, 1978.
130. Yen CK, Budinger TF, Friedland RP, et al: Brain tumor evaluation using Rb-82 and positron emission tomography. *J Nucl Med* 23:532–537, 1982.
131. Yudilevich DL, DeRose N: Blood–brain transfer of glucose and other molecules measured by rapid indicator dilution. *Am J Physiol* 220:841–846, 1971.
132. Zivin JA, Snarr JF: A stable preparation for rat brain-perfusion: Effect of flow rate on glucose uptake. *J Appl Physiol* 32:658–663, 1972.
133. Zlokovic BV, Begley DJ, Djuricic YY, et al: Measurement of solute transport across the blood–brain barrier in the perfused Guinea pig brain: Method and application to N-methyl-α-aminoisobutyric acid. *J Neurochem* 46:1444–1451, 1986.

5

Transport across the Blood–Brain Barrier

Michael W. Bradbury

1. PHYSICOCHEMICAL FACTORS INFLUENCING PERMEABILITY

1.1. Solute Transport

In Chapter 2, Hugh Davson clearly tells the story of how the idea of blood–brain barrier (BBB) and blood–cerebrospinal fluid (CSF) barriers arose at the turn of the century. Parallel to the early barrier studies, the experiments of Hans Meyer and of Overton indicated that both the ability of compounds to cause narcosis in animals and the permeability of the cell membrane depended on the solute's oil/water partition coefficient. Later, Barlund and Collander demonstrated that molecular size also influences the rate of entry of compounds into certain plant cells. Small molecules such as formamide, methanol, and water itself permeate the plasma membrane more rapidly than do large molecules of similar lipid solubility, leading to the concept that the lipid cell membrane might also contain small water-filled pores of molecular dimensions.

In 1946, Krogh,[40] in a review of ionic exchange across living membranes, drew attention to a few observations on movements of solutes between blood and brain. These observations led him to the conclusion that the properties of the cerebral capillaries are those of the cell membrane and, in the search for drugs that act on the central nervous system (CNS) one should be guided by their solubility in lipid rather than by other properties, such as their electric charge. Subsequent experimentation has fully confirmed the prime role of lipid solubility in determining rate of entry of a solute into brain and CSF. Notable studies, establishing this fact during the period prior to 1965, were those of Davson,[19] Rall et al.,[62] Brodie et al.,[6] and Crone.[11] More recent studies, which have again confirmed the rule, are those of Oldendorf[49] and of Rapoport et al.[64] A full discussion of the relationship appears in Fenstermacher and Rapoport.[25] A good method of comparison is to plot permeability against oil/water partition coefficient divided by the square root of the molecular weight.[25] As the latter investigators point out, linearity for this plot covers a change of at least four orders of magnitude in both the ordinate and the abscissa (Fig. 1).

Michael W. Bradbury • Department of Physiology, King's College, London WC2R 2LS, England.

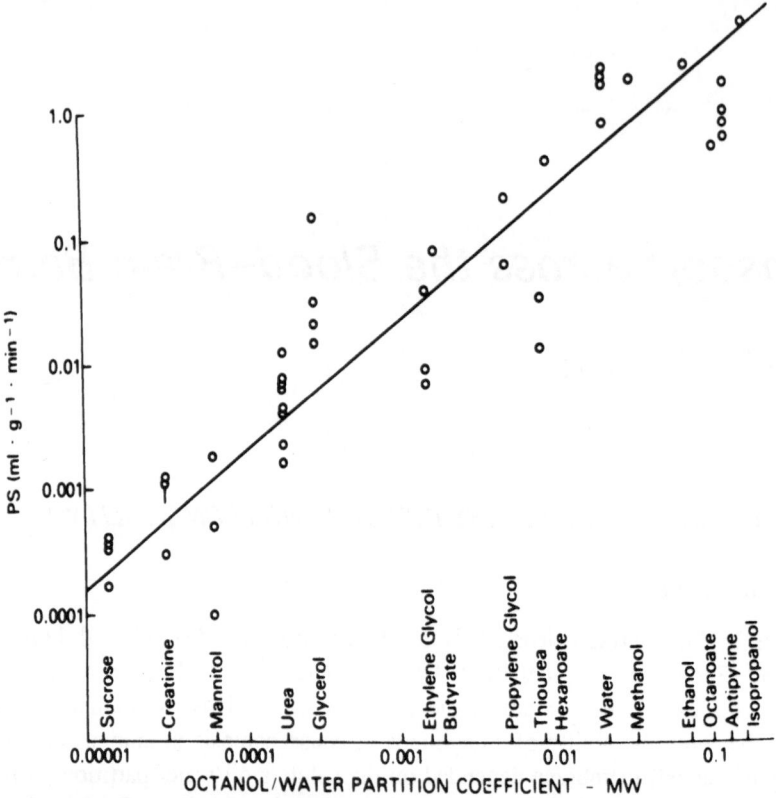

Figure 1. The permeability–surface area product of the blood–brain barrier (BBB) to 16 different compounds, plotted against their respective octanol/water partition coefficient divided by the square root of molecular weight. The line has a slope of one and is fitted to the points by the method of least squares. (From Fenstermacher and Rapoport.[25])

Further relationships follow from the primary correlationship with lipid solubility. First, since it is the un-ionized fraction of an acidic or basic compound that is more lipid soluble, penetration into brain and CSF will fall with increasing dissociation into ions. Rall et al.[62] studied antipyrine and a series of six sulfonamides. Entry into CSF, as estimated from the rate constant for uptake correlated reasonably well with both the percentage of drug un-ionized at pH 7.4 and with its chloroform/water partition coefficient. It was made clear that percentage binding to protein is also important. For a diffusion-limited solute, it may be anticipated that the bound fraction of a substance in plasma will not be available immediately for transport into brain. If the unbound concentration is reduced by transport into brain, however, it may be replenished by dissociation of the protein bound substance during a single capillary pass. The influence of protein binding on blood–brain transport is fully discussed in this chapter.

Generally, the rate of uptake into brain will be primarily determined by the lipid solubility of the un-ionized and unbound fraction of the compound. Attempts have been made to make this relationship more quantitative by defining a partition parameter that contains the product of partition coefficient and fraction of un-ionized unbound compound.[63]

Table I. Electrical Resistance, Potassium Permeability, and Filtration Coefficient of Frog Capillaries[a]

	Resistance ($\Omega \cdot cm^2$)	Permeability (10^{-5} cm/sec)	Filtration coefficient (10^{-9} cm/sec per cm H_2O)
Brain endothelium	1900	$(0.1-0.15)$[b]	1.15[c]
Skeletal muscle endothelium	20–30	5–15	74
Mesenteric endothelium	1–3	70	779

[a]Values for resistance and permeability compiled from Crone and Levitt[14] and those for filtration coefficient by Dallas and Fraser.[18]
[b]Mammalian value from Bradbury.[4]
[c]Value from Dallas and Fraser[18] is for water-only pathway.

The relationship of entry to lipid solubility suggests that the two lipid plasma membranes of the capillary endothelium provide the primary limitation to solute movement. Might there be an additional route that permits movement of polar solutes through water-filled channels either in the tight interendothelial junctions or through the two plasma membranes? Certainly, all the points for water and for glycerol are above the line of Fig. 1. The water permeability of thin lipid membranes, both natural and artificial, has been found to be higher than the partition coefficient of water would suggest. This is probably because of the ability of water molecules to squeeze between the hydrocarbon chains of lipid bilayers. Otherwise, there is no obvious evidence from this type of plot for a significant penetration of solutes through water-filled channels, as might occur in the interendothelial region.

The presence of water-filled channels is also suggested, if the osmotic permeability of a membrane is greater than its diffusional permeability to tracer water. Experimental observations suggest that this may be so for the BBB. The observed greater osmotic permeability could be explained, however, by the presence of unstirred layers. Recently, Crone[13] has estimated the selectivity of single capillaries in the frog brain to small ions. Dilution potentials were measured during the microperfusion of salt solutions. The results could best be explained by a small fraction, perhaps 0.1%, of the length of the tight endothelial junctions being open and permitting nonselective passage of small ions. This was considered not incompatible with the high electrical resistance, about 2000 $\Omega \cdot cm^2$ of the capillary wall measured previously[15] (Table I). Thus while there may be a small number of water-filled channels in the barrier, their total area is probably not sufficient to influence the magnitude of solute transport markedly.

Last, at high values of the partition coefficient, entry of the compound into brain is so fast that the capillary blood is depleted of the solute during a single pass. As the partition coefficient, hence permeability of the capillary increase, the entry of a solute into brain will be progressively more dependent on blood flow until, at the highest lipid solubility, uptake is linear with flow. This is illustrated in Fig. 2, in which uptake into brain of a solute (extraction × blood flow) has been calculated for different flows, abscissa, and for different PS products, noted on each line. The calculation was made from the equation derived by Renkin[66] and by Crone[10]

$$PS = Q \cdot \ln (1 - E)$$

where PS is the permeability–surface area product, Q the blood flow, and E the extraction. The equation assumes a linear uniform capillary with no backdiffusion from the tissue compartment. As shown in Fig. 2, uptake is almost unaffected by flow, provided that the latter is above about 0.5 ml/g per min and the PS product is below about 0.2 ml/g per min Similarly, uptake is dominated by flow when the PS product is approximately 5 ml/g per min or above.

Flow limitation has long been recognized either implicitly or explicitly as applying to oxygen and carbon dioxide transport between blood and all tissues including the brain. It is approached by many anesthetic gases and drugs, such as thiopental. Extraction from the cerebral circulation of the fat-soluble alcohols, ethanol, propanol, and butanol, is greater than 90%.[11,61] The property of flow limitation in a solute provides a means of measuring cerebral blood flow from either its rate of uptake into brain or from its rate of washout. Compounds of high fat solubility, used for this purpose, include N_2, krypton, xenon, hydrogen, ethanol, butanol, and iodoantipyrine.

1.2. Water

Water movement across membranes can be described in two ways. First, net flux or bulk flow of water occurs in response to either a hydrostatic or an electrochemical gradient, or both. Since the activity of water varies inversely with increasing solute concentration, this net flow will occur toward the side having the greater concentration of osmotically active particles. Volume flow, J_v, can be related to both hydrostatic and osmotic pressure gradients by

$$J_v = K(\Delta P - \Sigma \sigma \cdot \Delta \pi)$$

where K is the hydraulic conductivity or filtration coefficient of the membrane, ΔP is the difference in hydrostatic pressure and $\Sigma \sigma \cdot \Delta \pi$ is the sum of the products of reflection coefficient, σ, with total possible osmotic pressure for each solute.

Second, self-diffusion of water occurs without bulk flow and can be estimated from the movement of isotopically labeled water, just as self-diffusion of a solute can be measured in absence of a concentration gradient. If there are no water-filled channels or pores in the membrane, the osmotic permeability of the membrane, P_f, i.e., the filtration coefficient, will equal the diffusional permeability to water, P_d, provided that both are expressed in the same units. If such channels exist, P_f may be anticipated to be greater than P_d.

The filtration coefficient of the BBB has been measured by several methods in several species. Fenstermacher and Johnson[24] estimated brain shrinkage in the rabbit by connecting the intracranial cavity to a saline-filled tube. Blood osmolality was raised by intravenous infusion of different compounds of which sucrose and raffinose were judged to have a reflection of unity, i.e., to be impermeable. Paulson et al.[58] measured net water movement into the human brain in response to the intracarotid injection of a hypertonic bolus, while Dallas and Fraser[18] used a modification of the Landis occlusion technique on single pial capillaries in frog brain, water movement in response to raised external osmolality being estimated from the increasing concentration of a fluorescent impermeable marker in the capillary. Values for the filtration coefficient for multiple capillaries are close at about 0.8 and $0.74 \times 10^{-9} \mathrm{cm/sec \cdot cmH_2O^{-1}}$, whereas that for single capillaries in the frog was somewhat higher (Table I). It may be noted that in the frog, muscle

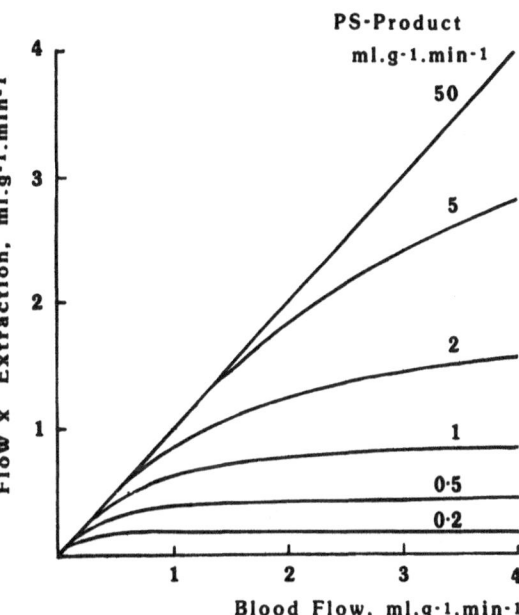

Figure 2. Theoretical curves based on the Crone–Renkin equation for flow extraction (uptake) plotted against blood flow. Note that the flow × extraction product represents volume of arterial blood cleared of solute per min, is identical to the influx rate constant, and is directly proportional to net flux or mass of solute entering brain per min. Since the equation assumes no backdiffusion, this product is also proportional to unidirectional flux of solute from blood into brain. (From Bradbury.[4])

capillaries are about 60 times more permeable to water than are those of the brain, while the ratio for mesenteric capillaries to those of the BBB is about 700.

The *PS* product of the BBB to labeled water is difficult to measure directly, since its size is of the same order as that of cerebral blood flow, i.e., diffusion of water into or out of brain is normally partly dependent on flow and at low blood flows is completely flow limited.[23] Nevertheless, estimates have been made[23] and lie between 0.65 and 1.75 ml/g per min. Paulson et al.[58] measured both P_f and P_d in humans and found the ratio of the first to the second to be about 4.3. This ratio of greater than unity has been interpreted as due to the presence of an unstirred layer within the endothelium and not necessarily to the presence of water-filled channels across the capillary wall.[57] If there are such channels, their influence on the function of the barrier is small.

In the applied context, it should be noted that since the normal BBB has a very low permeability to ions and to polar compounds of even low molecular weight, σ for all such solutes is close to unity. This means that potential gradients of osmotic pressure across the BBB are much greater than any likely gradient of hydrostatic pressure. Any initial filtration, due to hydrostatic pressure, will immediately be limited by an osmotic gradient in the opposite direction. If, however, the BBB is pathologically opened, filtration of fluid from plasma with or without protein becomes possible. Such a mechanism is probably responsible for so-called vasogenic cerebral edema.

1.3. Influence of Protein Binding

Although the state of a solute in blood plasma must obviously influence uptake into brain, many recent studies have neglected this factor. It was generally assumed in the past, either explicitly[62] or implicitly, that flux into brain of a compound that is partially protein bound would be proportional to the concentration. It follows from this that the unbound concentration should be used in calculations of the *PS* product. However, development of the intracarotid injection technique[46] has permitted direct measurement of the brain uptake of labelled compounds in the absence or in the presence of different concentrations of protein, usually albumin. This direct view appears to be an over simplification.

Tryptophan has attracted interest, as its concentration in blood plasma is an important determinant of the content of the neurotransmitter serotonin in brain despite the fact that this amino acid is about 90% bound to plasma albumin in blood. Yuwiler et al.[80] confirmed the marked binding in vitro but demonstrated that the presence of albumin had surprisingly little effect on brain uptake of [^{14}C]tryptophan from an intracarotid bolus. It appeared that bound tryptophan was stripped from the albumin during the single capillary pass. Suggested mechanisms were rapid dissociation due to uptake into brain, local physiologic pH changes, or metabolite displacements that would alter the association constant of the complex. Similarly, most steroid hormones are heavily bound in plasma to either albumin or to a specific binding globulin, leaving only a few percentiles free. Uptake, though somewhat reduced by the presence of serum containing binding protein, was little affected by the presence of even large amounts of albumin,[55] [^{3}H]progesterone, and [^{3}H]testosterone, having brain uptake indices of 80–100%, either with or without this protein. Dissociation of the steroid albumin complex must again be taking place during the capillary pass. Unlike the steroid hormones that are generally highly lipid soluble, thyroid hormones are polar and enter brain by a specific transport system.[51] However, stripping from albumin again occurs in the brain capillary.

Pardridge and Landaw[54] mathematically analyzed the dissociation and transport of protein bound hormones and drugs during a single capillary pass, transport data being fitted to a modification of the Renkin–Crone equation of capillary physiology. The results were best fitted by an equation in which the apparent in vivo dissociation constant of the ligand protein complex was 5- to 50-fold greater than the dissociation constant measured in vitro. This is an interesting possibility but how dissociation of the ligand from plasma protein is brought about within the capillary remains unknown.

1.4. Convection in Cerebral Interstitial Fluid

In extracerebral tissues, large-molecular-weight compounds or particles in suspension, which cannot readily cross the capillary wall from interstitial fluid to blood, are removed by drainage with the lymph. Clearance from brain of even small polar solutes presents a similar problem, since these cannot easily pass across the BBB from interstitial fluid to blood, unless there is a specific transport mechanisms in the endothelium. The presence of a nonspecific clearance mechanism is indicated by the fact that many slowly penetrating polar solutes, e.g., [^{35}S]sulfate and [^{14}C]inulin, have a low CSF–plasma ratio in the steady state and have brain spaces much lower than the 20% extracellular space of cerebral cortex. Davson[20] attributed such findings to the sink action of CSF. As a polar solute slowly enters cerebral interstitial fluid from blood, there will be a steady slow removal because of its diffusion into adjacent CSF in the ventricular or subarachnoid spaces. Since CSF, deficient in the solute, is continually being renewed by secretion of bulk fluid, the concentrations of the solute in interstitial fluid and CSF are maintained below that in blood plasma—the smaller the permeability of the solute at the blood–brain and blood–CSF barrier, the lower the steady-state concentrations in cerebral extracellular fluids. The process may be anticipated to keep concentrations much lower in CNS tissue close to CSF than in that more distant. In general, profiles of solutes, such as [14C]sucrose, within brain substance conform to this idea.[60] The efficiency of the sink action as a mechanism for influencing the composition of interstitial fluid deep within a large brain, such as that of humans, may be doubted.

A further consideration is that there may be convective or bulk flow of the actual interstitial fluid of brain. Cserr and colleagues[16] have perfected a technique for making a microinjection of 0.5 µl fluid containing a tracer deep into the brain substance in vivo. Movements of interstitial fluid may be studied by observing the distribution of large molecular weight tracers in brain sections at set times after microinjection. Tracers such as colored dextrans, Evans blue–albumin, and horseradish peroxidase (HRP), initially appear to diffuse concentrically out from the injection site but soon become concentrated in the perivascular spaces, seeming to move centrifugally in these channels. Even more striking is the conclusion derived from measuring the rate of disappearance of three radiolabeled compounds from the whole brain: polyethylene glycol (PEG) of 900 M_r, PEG of 4000, and albumin of 69,000.[17] All three left brain according to the same exponential rate after microinjection into the caudate nucleus of the rat (Fig. 3). Since the diffusion coefficient of PEG-900 in water is five times that of albumin, diffusion into CSF can be excluded as a major factor responsible for exit of these large molecules from brain. Cserr attributed the phenomenon to secretion of interstitial fluid by the capillary endothelium, followed by an outward movement of this fluid, occurring largely in the perivascular spaces. The rate of interstitial fluid secretion was estimated from the rate of turnover and varied between 0.2 and 0.3 µl/g per min depending on the region of microinjection.[74]

The single rate constant could be explained by convection due to causes other than secretion. Using an especially sensitive reaction, Rennels et al.[67] have recently demon-

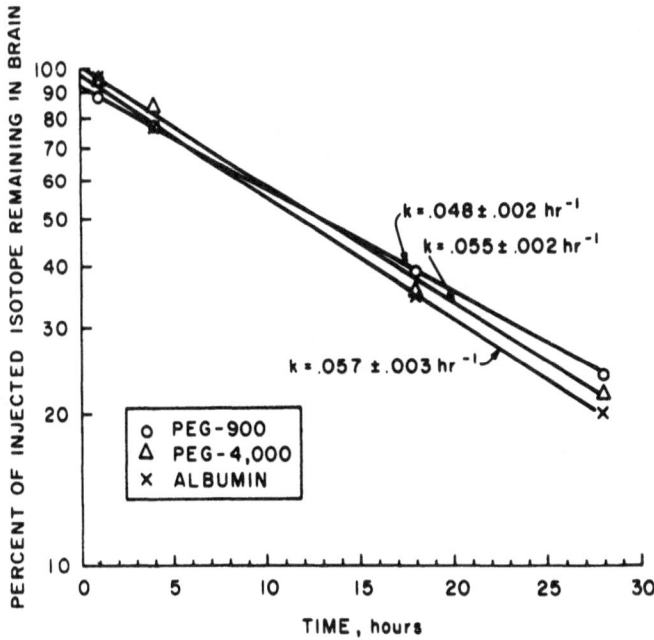

Figure 3. Disappearance of radiolabeled polyethylene glycols (PEGs), following microinjection into rat caudate nucleus. *K* values are first-order rate constants for total efflux from brain, and points are the means for five to seven animals. (From Cserr et al.[17])

strated that the introduction of HRP into the ventricular CSF rapidly penetrates the arterial perivascular spaces, even reaching the perivascular basal laminae of capillaries. Rennels and co-workers attribute this to a pulse-driven flow of CSF, entering brain through the arterial perivascular spaces and leaving through the venous spaces. An alternative explanation is a to-and-fro movement of CSF, occurring principally in the arterial perivascular spaces—going out during systole (arterial expansion) and in during diastole (arterial reduction in diameter). The three mechanisms (i.e., secretion of interstitial fluid, circulation through perivascular spaces, and tidal movements particularly in arterial spaces) are not mutually exclusive. Further experiments are needed to determine which of these three mechanisms is the main determinant of the observed equality of rate constants for drainage of different sized molecules from brain.

2. SPECIFIC TRANSPORT

If solubility in lipid were the sole determinant of the rate of entry into brain of a compound, this organ would be deprived of adequate amounts of several metabolic substrates on which it depends. In particular, the partition coefficient of glucose is so low that unfacilitated diffusion would be quite inadequate to supply the amounts of this compound required for cellular respiration. Crone[12] used the indicator diffusion technique to make the first demonstration of specific transport of glucose at the BBB. The introduction of the intracarotid tissue-sampling technique by Oldendorf[46] provided a simple, rapid, reproducible method for characterizing specific transport at the endothelium. Use of the technique has led to the recognition of specific mechanisms facilitating the transport of certain compounds within several groups: monosaccharides, monocarboxylic acids, neutral amino acids, dicarboxylic amino acids, and certain amines, of which choline is the main representative (Table II). In each case, the transport is stereospecific, saturable, and subject to competitive inhibition. Since access to the abluminal membrane of the capillary endothelium is difficult, it has not generally been possible to determine whether a given mechanism is energy dependent and transports preferentially in one direction across the endothelium or whether it is a passive equilibrating system. There is, however, a strong supposition that the monosaccharide transporter solely facilitates diffusion, whereas those

Table II. Various Transport Systems at the Blood–Brain Barrier for Physiologically Important Nonelectrolytes[a]

Class of compounds and representative	Maximum transport capacity (V_{max}) (μmoles/g per min)	Apparent Michaelis constant (K_m) (mM)
Monosaccharides, D-glucose[29,32,52]	2–4	7–11
Monocarboxylic acids, L-lactate[52]	90	2
Neutral amino acids, L-leucine[52,69]	30–60	0.025–0.1
Basic amino acids, L-arginine[52]	8	0.09
Dicarboxylic amino acids, L-glutamate[50]	Low	—
Amine, choline[52]	11	0.34
Nucleosides, adenosine[52]	0.75	0.025
Purines, adenine[52]	0.05	0.01

[a]All values, except those for glucose and leucine, were estimated by the Oldendorf technique; see text for explanation.

for the neutral and dicarboxylic amino acids are involved in pumping at least some of their substrates from interstitial fluid to blood.

2.1. Monosaccharides

The transporter is highly stereospecific, the D-sugar being selected. It has a high affinity to D-glucose, as is to be expected, whereas the uptake of L-glucose is barely measurable.[47] The rank order of affinities of the transporter for D-hexoses, based on either the Michaelis constant for uptake of the sugar or its inhibition of glucose uptake is 6-chloro-6-deoxyglucose > 2-fluoro-2-deoxyglucose > 2-deoxyglucose > glucose > 3-O-methyl-glucose > mannose > galactose.[47,52] The transport is sodium independent and is not influenced by insulin. Phloretin is a more powerful inhibitor than phlorridzin and cytochalasin B inhibits both specifically and potently. The mechanism is thus very similar in its characteristics to the equilibrating system of the red blood cell (RBC) membrane and to that of the guinea pig placenta.[79]

A good deal of effort, interest, and argument have been expended on measurement of the kinetic constants for glucose in the system, i.e., the Michaelis constant, K_m, and the maximum transport capacity V_{max}. These are important both in relationship to the fact that brain metabolism depends on glucose and that glucose and deoxyglucose transport at the BBB contribute to the lumped constant used in estimations of glucose utilization by the deoxyglucose method.[70] Estimations of K_m can at best be apparent—although the composition of the fluid in cerebral capillaries can be temporarily fixed by intracarotid injection or infusion, there can be little control of the composition of cerebral interstitial fluid, hence of net flux of glucose across the BBB during the course of the measure-ment.[42] The maximum transport capacity is even more subject to variation in that it is sensitive to anesthesia and to other factors influencing glucose utilization in brain, proba-bly acting via redistribution of blood flow (see discussion below). Recent measurements of the apparent K_m have given values varying between 7 and 11 mM.[29,31,52] The V_{max} determined by the Oldendorf technique has generally been lower than those estimated after single intravenous injection of tracer glucose, by the method developed by Gjedde and colleagues.[28,34] This discrepancy may be related to a lack of RBCs in the Oldendorf bolus and to the large correction necessary for a diffusional component in this method. Values for the V_{max} after IV injection are close to 4 μmole/g per min in the awake to lightly anesthetized rat,[29,32] and after intracarotid injection, vary from 1.4 to 3.5 μmoles/g per min.[33,52] There is considerable interregional variation in V_{max}, and this parameter obtained in one study may not always be directly comparable with that obtained in another in this respect.[9] There does, however, seem to be more agreement on the important generalization that unidirectional influx of glucose is usually about twice net glucose influx, i.e., utilization, under given conditions.[9,52] Recently, photoaffinity labeling with cytochalasin B has been used to isolate the glucose transporter obtained from isolated cerebral capillaries.[1,22] It appears to be an integral membrane protein with a molecular weight of 53,000. Interestingly it cross-reacts with antiserum raised against the erythrocyte glucose transporter.[22]

2.2. Monocarboxylic Acids

Following the observation of Nemoto and Severinghaus[45] that L-lactate is preferen-tially transported into brain compared with D-lactate, Oldendorf[48] demonstrated the presence of a transporter at the BBB for a number of monocarboxylic acids. The rank order of affinities appears to be α-ketomethiobutyrate (the ketoacid of methionine) > α-ketoiso-caproate (the ketoacid of leucine) > butyrate > pyruvate > lactate > β-hydroxybuty-rate.[52] This transporter is of interest in that it permits escape of lactate from brain and allows entry of the ketone bodies. Its capacity is subject to a high degree of modulation in relationship to the availability of certain metabolic substrates.

2.3. Amino Acids

Three independent systems at the BBB have been characterized, again largely due to the work of Oldendorf and colleagues. Rapid exchange occurs between blood and brain of a group of large neutral amino acids, many of which are essential and that correspond to those of Christensen's L system.[47] The small neutral amino acids have much smaller brain uptake indices (BUI),[47] but experiments with isolated cerebral capillaries suggest that A group amino acids, as represented by the specific substrate α-methylaminoisobutyrate, may be actively moved from interstitial fluid to blood by a sodium-dependent mechanism.[2] Affinity constants K_m for the large neutral group have been estimated to vary from 0.1 to 0.5 mM by the Oldendorf technique.[52] However, development of a rapid brain perfusion technique[75] has enabled the amino acid composition of the intracapillary fluid to be well maintained, without mixing with the animal's blood, during the perfusion period of 30 sec. Through the use of this improved method, the K_m of leucine has been estimated to be about 0.024 mM, i.e., about 25% of the Oldendorf estimate of 0.1 mM.[69] This means that at normal plasma amino acid levels, the carrier will be 96% saturated with large neutral amino acids. As Smith has pointed out, a general increase or decrease in the level of all these amino acids in plasma will lead to little change in the influx of each amino acid into brain. If one amino acid is increased in isolation, its influx will change linearly with its concentration, the influx of all the other competitors being reduced in compensation.

Distinct and quite separate from the transport of neutral amino acids is a system that transports the three basic amino acids: arginine, lysine, and ornithine. These basic amino acids have K_m values that are of similar magnitude of the K_m of the L-group amino acids for their specific transporter. However, the estimated V_{max} values of the basic amino acids are about a fifth of those for the large neutral amino acids.[52]

Finally, a third system, yielding low uptakes, has been described for the dicarboxylic amino acids, aspartate and glutamate.[50]

2.4. Vitamins, Cofactors, and Nucleic Acid Precursors

A number of polar compounds other than the metabolic substrates and neurotransmitter precursors themselves are needed by the brain. These include the precursors of DNA and RNA, as well as cofactors derived from vitamins. The presence of specific saturable transport at the BBB has been demonstrated for certain purines, including adenine, guanine, and hypoxanthine,[7] for certain nucleosides, including adenosine, inosine, guanosine, and uridine,[7] and for thiamine.[35]

In contrast to the proven presence of certain transporters in the cerebral capillaries is the interesting suggestion that certain essential polar compounds reach brain via the CSF after prior transport through the choroid plexuses. This hypothesis, largely developed by Spector and his colleagues, provides an attractive mechanism for regulating and distributing certain substances only required in small amounts by the brain. Undoubtedly the choroid plexus in vitro is able to accumulate actively and specifically ascorbic acid, folates, thymidine, and probably certain other deoxynucleosides (reviewed by Spector and Eells[73]), riboflavin[72] and vitamin B$_6$.[71] Evidence that the choroid plexuses and CSF together provide an important route for entry into brain is most comprehensive in relationship to ascorbic acid. In addition to the marked accumulation by the plexuses in vitro, the evidence for transport via CSF includes concentrations of ascorbic acid in CSF higher than in plasma, an apparent low permeability of the brain capillaries to ascorbic acid, and autoradiographic demonstration of prior entry of [^{14}C]ascorbic acid into the choroid plexuses, followed by penetration into CSF and those parts of brain adjacent to CSF.

2.5. Peptides

Increasing awareness of the wide distribution of peptide neurotransmitters in brain, coupled with observations that certain peptides administered peripherally may influence behavior, has stimulated interest in, and experimentation on, the possible presence of transport mechanisms for these compounds at the BBB. The polarity and size of most peptide molecules would preclude significant transport across the barrier, unless a special mechanism(s) were present. Since the subject has spawned both extravagant claims and skepticism, it is appropriate to list some ways in which behavior (i.e., function of the CNS) might be influenced by a systemically administered peptide without its passing directly across the cerebral endothelium into the general interstitial fluid of the brain:

1. The peptide might influence the internal environment of the body in a way that would also affect brain function. Thus vasopressin delays the extinction of active shock-avoidance behavior in rats, from which it has been argued that the peptide directly enhances memory consolidation.[21] It is obviously important to exclude the possibility that the effect is mediated via the state of hydration of the animal or more reasonably by its pressor action raising blood pressure and thus leading to increased arousal.[41]

2. The peptide might specifically or generally stimulate afferent peripheral somatic or visceral nerves and thus secondarily affect the CNS. Such a sequence of events is plausible in light of the realization that there is a complex nervous system associated with the gut and that this system contains numerous peptidergic neurons.

3. The agent might specifically bind to the cerebral endothelium and hence generate a signal that might be conveyed to neurons by intracellular and extracellular messenger systems. Undoubtedly, insulin binds specifically to receptors on the luminal surface of the cerebral endothelium[76] so that a stimulus could be received from this hormone in blood, but further links in the chain are speculative.[53]

4. Any peptide can be anticipated to penetrate locally into the cerebral interstitial fluid at the regions of high vessel permeability. These include the area postrema, subfornical region, median eminence of the hypothalamus, infundibulum, neurohypophysis, and the pineal gland.[4] Dendrites or cell bodies sensitive to the peptide might then activate neuronal systems, mediating certain behavioral responses. A good example is the causation of drinking behavior (thirst?) via angiotensin in the blood acting at the subfornical region.[26]

5. Finally the choroid plexuses are known to be rather more leaky than the cerebral blood vessels; also, the BBB is circumvented by a route due to the direct physical

continuity of plexus interstitial fluid (accessible to blood-borne solutes and the true interstitial fluid of brain).[65] If a solute were to leak or be specifically transported at the choroid plexuses, the circulation of CSF might lead to its distribution to appropriate regions of the brain adjacent to this fluid. Despite its attraction, this hypothesis has generated few or no experimental results in its favor.

These possible ways in which peptides may influence behavior from outside the brain are illustrated diagrammatically in Fig. 4. Since possible effects and movements of peptides, other than those involving transport across the BBB, have been dealt with, it is now pertinent to list the general types the mechanism which might be implicated in transport of whole peptides or fragments of peptides across the cerebral endothelium itself:

1. It has already been suggested that the lipid solubility and molecular size of most peptides are not such as to permit significant passage through the lipid membranes of the endothelium. This statement begs the question of what is meant by significant, but certainly, when extraction during a single capillary pass has been measured and hydrolysis before barrier transport has been excluded, the uptake of enkephalins, thyrotropin-releasing hormone, angiotensin II, and insulin is close to

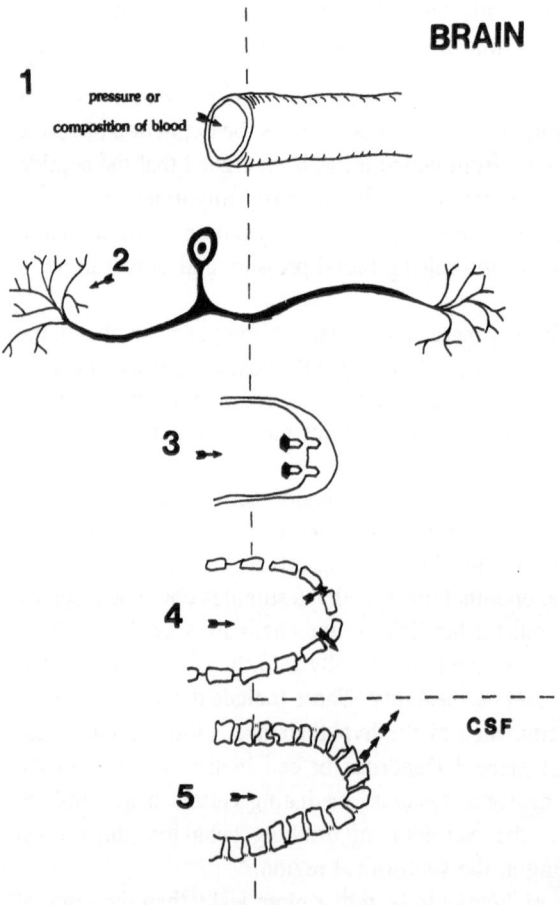

BRAIN

CSF

Figure 4. Possible ways by which an active peptide may influence CNS function without traversing the tight endothelium of cerebral capillaries: (1) by inducing a chemical or physical change in the blood perfusing the brain; (2) by influencing the activity of afferent neurons; (3) by binding to a receptor on the luminal membrane of the endothelium and thus activating a hypothetical communication system; (4) by crossing the fenestrated endothelium of a region, lacing a blood–brain barrier, thereby modulating local neurons; and (5) by entering CSF via the choroid plexuses, thereby being distributed by CSF to a site of sensitive neurons.

the value of sucrose, i.e., the blood background.[53,82,83] The cyclic dipeptide (Leu–Gly) enters CSF in appreciable amount from blood,[37] uptake into brain not having been measured. This may well be related to an anticipated high lipid solubility of this ring structure which has no free polar carboxyl or amino groups.[53] Paradoxically, cyclosporin, a cyclic decapeptide with marked lipid solubility, is excluded from CSF and has a low brain uptake over 2 hr (D. J. Begley and L. Squires, personal communication). This may be due to a combination of very low water solubility and very high protein binding for this compound.

2. Movement of peptides of low molecular weight might be facilitated by combination with membrane transporters of the general type involved in transport of metabolic substrates. Either these might be specific carriers for certain di-, tri-, or larger peptides, or it is a possibility that a small peptide, having an L-system amino acid in a terminal position might hitch a ride on the L-amino acid transporter. Zlokovic et al.[81] have examined this possibility in some detail. Two ^{14}C-labeled dipeptides, glycyl-L-leucine and glycyl-L-phenylalanine, had BUIs by the Oldendorf method equivalent to that of sucrose, indicating no passage across the barrier. Also, the large BUIs of the L-system amino acids leucine and phenylalanine were quite unaffected by either of the dipeptides at 10 mM concentration. By contrast, the BUIs of ^{3}H-labeled leucine-enkephalin and D-alanine-D-leucine-enkephalin were some 4% higher than the sucrose background.[82] The small uptake of leucine-enkephalin was abolished by either 5 mM tyrosine or by the aminopeptidase inhibitor bacitracin (Table III). Leucine-enkephalin at 5 mM had no effect on the BUI of tyrosine. The D-alanine enkephalin analogue had a residual BUI of about 2% in the presence of bacitracin, suggesting slight transport of the intact molecule during a single pass. These results indicate hydrolysis of the enkephalins during a single capillary pass but provide little evidence of transport of intact small peptides and no evidence of small peptides, containing L-system amino acids, themselves interacting with the L-system transporter.

3. This leaves the possibility of endocytosis/exocytosis in the endothelium as a system for transferring smaller amounts of larger peptides or proteins from blood to brain. It has already been mentioned that insulin binds specifically to the

Table III. Brain Uptake Indices for [³H]-Leucine-Enkephalin in the Presence of Potential Inhibitors[a]

Injectate	BUI for Leu-Enk, corrected for blood background, BUI-sucrose (%)	Inhibition (%)
Injectate medium only	3.8 ± 0.6	—
+5 mM Leu-Enk	3.5 ± 0.5	7
+5 mM tyrosine	0.2 ± 0.1	96
+10 mM BCH[c]	0.8 ± 0.4	80
+2 mM bacitracin[b]	−0.8 ± 0.2	120

[a]From Zlokovic et al.[82]
[b]An aminopeptidase inhibitor.
[c]2-Aminobicyclo(2,2,1)heptane-2-carboxylic acid, a specific L-system substrate.

luminal membrane of the cerebral endothelium.[27,76] Pardridge et al.[56] prepared microvessels from human brain. Not only was insulin avidly bound, with a dissociation constant 1.2 nM, but 75% was internalized. This internalized and largely intact insulin was re-exported into a radiotracer-free medium with a half-time of about 70 min. The findings are consistent with, but do not prove, the hypothesis that insulin may be translocated from blood into interstitial fluid by specific binding to the luminal membrane of the endothelium, followed by endocytosis and finally exocytosis through the abluminal membrane. It is not clear whether this postulated mechanism might contribute to the relatively large transport of endogenous and exogenous insulin from blood to CSF, observed by Woods and Porte.[78]

3. FACTORS MODIFYING PERMEABILITY

Chapter 4 describes the experimental and mathematical approach to estimation of the permeability of cerebral capillaries. Since the precise area of endothelium per unit weight of brain is not generally known, a *PS* product, rather than a permeability coefficient, is often used. At a given solute concentration in plasma, this apparent permeability might vary in two ways. First, changes in the distribution of cerebral perfusion might alter the effective area of endothelium available for transport or, in the case of specific transport, alter the access to transport sites. Second, again in the case of specific transport, there might be a modulation of the transporter itself with a change in either V_{max} or K_m, or both. An increased V_{max} suggests the presence of more transport sites. It is not always easy to distinguish between the two primary mechanisms leading to an altered *PS* product. If the transport of several independent solutes is changed by the same amount, an alteration in endothelial area is suggested, whereas if the change in transport is restricted to a single solute or to a group of chemical analogues, modulation of the specific transporter is indicated.

3.1. Blood Flow and Permeability

If *PS* product has been calculated by the use of an equation that takes account of the fall in solute concentration along the length of the capillary, e.g., Eq. (6) in Chapter 4, it should be independent of the magnitude of flow in a perfused capillary. If the apparent permeability increases with increasing blood flow, it is suggestive of the recruitment of more capillaries from a pool of unperfused microvessels, thereby providing a larger endothelial surface area. This has long been recognized as occurring in skeletal muscle, but as has generally been supposed, there are normally no unperfused capillaries in brain, hence no scope for recruitment. Certainly, the apparent permeability of several solutes not subject to specific transport increases when blood flow is raised by hypercapnia or electrically induced convulsions[3,36,59] and decreases when flow is reduced by hypocapnia.[36,59] The topic has been reviewed by Bradbury.[5] It seems likely that, if recruitment does not occur and it is by no means certain that it does not,[77] redistribution of blood flow takes place in such a way that at higher total flows, a greater fraction of the blood must be exposed to higher surface areas of endothelium or to more transport sites.

Pentobarbital anesthesia reduces both maximal transport V_{max} into, and utilization of glucose by brain to about 50%.[33] Similarly, Cremer et al.[9] showed that treatment of rats with the tremor-inducing stimulant of cerebral metabolism, cismethrin, not only stimulated regional glucose utilization and blood flow, but also markedly increased unidirectional influx of glucose into brain; i.e., there was an increase in the apparent permeability. Interestingly, the parameter that correlated best with the increased transport capacity of the capillaries was the residual blood content of the tissue, suggesting capillary recruitment or at least better distribution of perfusing blood. Since unidirectional glucose influx is normally only about twice that of glucose utilization, the increased influx with raised activity helps prevent metabolism being limited by glucose transport. The increased *PS* product thus appears to be due to changes in vascular factors, but a rapid modulation of the transporter itself has not yet been excluded. There is no reason why vascular redistribution and a change in the transport mechanism itself could not occur together.

3.2. Modulation of Transport Mechanisms

A true adaptation of a transporter was first demonstrated by Gjedde and Crone.[30] The greatly increased utilization of ketone bodies by brain in starvation is supported by a large increase in the V_{max} and to some extent in the K_m for the system that transports β-hydroxy-butyrate and other monocarboxylic acids into brain. The same transporter also has a greatly increased capacity during suckling in the baby rat whose diet is rich in fat and lactose.[8] The activity of the transporter is specifically suppressed after portacaval anastomosis[68] when the ability of the liver to produce ketone bodies is severely impaired. Portacaval anastomosis produces other changes in transport at the BBB presumably in response to some aspect of the metabolic disturbance. Influx of the large neutral amino acids phenylalanine, tryptophan, and leucine into brain is markedly increased in a way that indicated greater capacity of the transporter.[38,43] This enhancement of transport can also be induced by prolonged infusion of ammonium salts[39,44] and seems to depend on the ability of the brain to synthesize glutamine.[39]

4. CONCLUSIONS

Since the blood–brain barrier (BBB) is the cerebral endothelium, the functional characteristics of the barrier are those of the single cylindrical layer of endothelial cells, the main component of the brain capillary. The transport properties of this endothelium are very similar to those of a tight epithelium. It has a high electrical resistance, i.e., a low permeability to ions. In general, passage across it into brain is highly dependent on the solubility of the diffusing compound in lipid. Again like an epithelium, it probably slowly secretes fluid by means of a sodium–potassium pump from blood to interstitial fluid. The endothelium contains a number of specific transporters that permit movement of metabolic substrates, transmitter precursors, and so forth. In general, the permeability of the endothelium to peptides and small proteins is very low, but there are a number of possible mechanisms by which active peptides in the blood may influence cerebral function and hence behavior. The apparent permeability of the BBB to most solutes may be influenced by the magnitude and distribution of cerebral blood flow. In addition, the capacity and affinity of certain transporters may be modulated in relationship to the energy needs of the brain.

REFERENCES

1. Baldwin SA, Brester F, Cairns MT, et al: Identification of a D-glucose-sensitive cytocholasin B-binding component of isolated ovine cerebral microvessels. *J Physiol (Lond)* 357:75 (abstr), 1984.
2. Betz AL, Goldstein GW: Polarity of the blood–brain barrier: Neutral amino acid transport into isolated brain capillaries. *Science* 202:225–227, 1978.
3. Bolwig TG, Hertz MM, Paulson OB, et al: The permeability of the blood–brain barrier during electrically induced seizures in man. *Eur J Clin Invest* 7:87–93, 1977.
4. Bradbury MWB: *The Concept of a Blood–Brain Barrier.* Wiley, New York, 1979.
5. Bradbury MWB: The blood–brain barrier: Transport across the cerebral endothelium. *Circ Res* 57:213–222, 1985.
6. Brodie BB, Kurz H, Schanker LS: The importance of dissociation constant and lipid solubility in influencing the passage of drugs into cerebrospinal fluid. *J Pharmacol* 130:20–25, 1960.
7. Cornford EM, Oldendorf WH: Independent blood–brain barrier transport systems for nucleic acid precursors. *Biochim Biophys Acta* 394:211–219, 1975.
8. Cremer JE, Braun L, Oldendorf WH: Changes during development in transport processes of the blood–brain barrier. *Biochim Biophys Acta* 448:633–637, 1976.
9. Cremer JE, Cunningham VJ, Seville MP: Relationshipships between extraction and metabolism of glucose, blood flow and tissue blood volume in regions of rat brain. *J Cereb Blood Flow Metab* 3:291–302, 1983.
10. Crone C: The permeability of capillaries in various organs as determined by use of the indicator diffusion method. *Acta Physiol Scand* 58:292–305, 1963.
11. Crone C: The permeability of brain capillaries to nonelectrolytes. *Acta Physiol Scand* 64:407–417, 1965.
12. Crone C: Facilitated transfer of glucose from blood into brain. *J Physiol (Lond)* 181:103–113, 1965.
13. Crone C: Lack of selectivity to small ions in paracellular pathways in cerebral and muscle capillaries of the frog. *J Physiol (Lond)* 353:317–337, 1984.
14. Crone C, Levitt DG: Capillary permeability to small solutes. In Renkin EM, Michel CC (eds): *Handbook of Physiology.* Section 2: *Microcirculation.* American Physiological Society, Washington, DC, 1984, pp. 411–466.
15. Crone C, Olesen SP: Electrical resistance of brain microvascular endothelium. *Brain Res* 241:49–55, 1982.
16. Cserr HF, Cooper DN, Milhorat TH: Flow of cerebral interstitial fluid as indicated by the removal of extracellular markers from rat caudate nucleus. *Exp Eye Res* (Suppl) 25:461–473, 1977.
17. Cserr HF, Cooper DN, Suri PK, et al: Efflux of radiolabelled polyethylene glycols and albumin from rat brain. *Am J Physiol* 240:F319–F328, 1981.
18. Dallas AD, Fraser PA: Filtration coefficient of blood–brain barrier (BBB) capillaries in the frog. *J Physiol (Lond)* 358:115 (abstr), 1985.
19. Davson H: A comparative study of the aqueous humour and cerebrospinal fluid in the rabbit. *J Physiol (Lond)* 129:111–133, 1955.
20. Davson H: The cerebrospinal fluid. *Erg Physiol* 52:21–75, 1963.
21. De Wied D, Versteeg DHG: Neurohypophyseal principles and memory. *Fed Proc* 38:2348–2354, 1979.
22. Dick APK, Harik SI, Klip A, et al: Identification and characterization of the glucose transporter of the blood–brain barrier by cytocholasin B binding and immunological reactivity. *Proc Natl Acad Sci USA* 81:7233–7237, 1984.
23. Eichling JOM, Raichle ME, Grubb RL, et al: Evidence of the limitations of water as a freely diffusible tracer in brain of the rhesus monkey. *Circ Res* 35:358–364, 1974.
24. Fenstermacher JD, Johnson JA: Filtration and reflection coefficients of the blood–brain barrier. *Am J Physiol* 211:341–346, 1966.
25. Fenstermacher JD, Rapoport SI: Blood–brain barrier. In Renkin EM, Michel CC (eds): *Handbook of Physiology. Microcirculation.* American Physiological Society, Washington, DC, 1984, pp. 969–1000.
26. Fitzsimons JT: *The Physiology of Thirst and Sodium Appetite.* Cambridge University Press, Cambridge, 1979.
27. Frank HJ, Pardridge WM: A direct in vitro demonstration of insulin binding to isolated brain microvessels. *Diabetes* 30:757–761, 1981.
28. Gjedde A: Rapid steady-state analysis of blood–brain glucose transfer in the rat. *Acta Physiol Scand* 108:331–339, 1980.
29. Gjedde A: Blood–brain transfer of galactose in experimental galactosemia, with special reference to the competitive interaction between galactose and glucose. *J Neurochem* 43:1654–1662, 1984.

30. Gjedde A, Crone C: Induction processes in blood–brain transfer of ketone bodies during starvation. *Am J Physiol* 229:1165–1169, 1975.
31. Gjedde A, Diemer NH: Autoradiographic determination of regional brain glucose content. *J Cereb Blood Flow Metab* 3:303–310, 1983.
32. Gjedde A, Rasmussen M: Blood–brain glucose transport in the conscious rat: comparison of the intravenous and intracarotid injection methods. *J Neurochem* 35:1375–1381, 1980.
33. Gjedde A, Rasmussen M: Pentobarbital anesthesia reduces blood–brain glucose transport in the rat. *J Neurochem* 35:1382–1387, 1980.
34. Gjedde A, Hansen AJ, Siemkowicz E: Rapid simultaneous determinations of regional blood flow and blood–brain glucose transfer in brain of rat. *Acta Physiol Scand* 108:321–330, 1980.
35. Greenwood J, Love ER, Pratt OE: Kinetics of thiamine transport across the blood–brain barrier in the rat. *J Physiol (Lond)* 327:95–103, 1982.
36. Hertz MM, Paulson OB: Transfer across the human blood–brain barrier. Evidence for capillary recruitment and for a paradox glucose permeability increase in hypocapnia. *Microvasc Res* 24:364–376, 1982.
37. Hoffman PL, Walter R, Bulat M: An enzymatically stable peptide with activity in the central nervous system: Its penetration through the blood–CSF barrier. *Brain Res* 122:87–94, 1977.
38. James JH, Escourrou J, Fischer JE: Blood–brain neutral amino acid transport activity is increased after portacaval anastomosis. *Science* 200:1395–1397, 1978.
39. Jonung T, Rigotti P, James JH, et al: Effect of hyperammonaemia and methionine sulfoxime on the kinetic parameters of blood–brain transport of leucine and phenylalanine. *J Neurochem* 45:308–318, 1985.
40. Krogh A: The active and passive exchanges of inorganic ions through the surfaces of living cells and through living membranes generally. *Proc R Soc Lond B* 133:140–200, 1946.
41. Le Moal M, Koob JF, Koda LY, et al: Vasopressor receptor antagonist prevents behavioral effects of vasopressin. *Nature (Lond)* 291:491–493, 1981.
42. Lund-Anderson H: Transport of glucose from blood to brain. *Physiol Rev* 59:305–352, 1979.
43. Mans AM, Biebuyck JF, Shelly K, et al: Regional blood–brain barrier permeability to amino acids after portacaval anastomosis. *J Neurochem* 38:705–717, 1982.
44. Mans AM, Biebuyck JF, Hawkins RA: Ammonia selectively stimulates neutral amino acid transport across blood–brain barrier. *Am J Physiol* 245:C74–C77, 1983.
45. Nemoto EM, Severinghaus JW: Stereospecific permeability of rat blood–brain barrier to lactic acid. *Stroke* 5:81–84, 1974.
46. Oldendorf WH: Measurement of brain uptake of radiolabeled substances using a tritiated water internal standard. *Brain Res* 24:372–376, 1970.
47. Oldendorf WH: Brain uptake of radiolabelled amino acids, amines and hexoses after arterial injection. *Am J Physiol* 221:1629–1639, 1971.
48. Oldendorf WH: Carrier-mediated blood–brain barrier transport of short chain monocarboxylic organic acids. *Am J Physiol* 224:1450–1453, 1973.
49. Oldendorf WH: Lipid solubility and drug penetration of the blood–brain barrier. *Proc Soc Exp Biol Med* 147:813–816, 1974.
50. Oldendorf WH, Szabo J: Amino acid assignment to one of three blood–brain barrier amino acid carriers. *Am J Physiol* 230:94–98, 1976.
51. Pardridge WM: Carrier-mediated transport of thyroid hormones through the rat blood–brain barrier: Primary role of albumin-bound hormone. *Endocrinology* 105:605–612, 1979.
52. Pardridge WM: Brain metabolism: A perspective from the blood–brain barrier. *Physiol Rev* 63:1481–1535, 1983.
53. Pardridge WM: Neuropeptides and the blood–brain barrier. *Ann Rev Physiol* 45:73–82, 1983.
54. Pardridge WM, Landaw EM: Tracer kinetic model of blood–brain barrier transport of plasma protein bound ligands. Empiric testing of the free hormone hypothesis. *J Clin Invest* 74:745–752, 1984.
55. Pardridge WM, Mietus LJ: Transport of steroid hormones through the rat blood–brain barrier. Primary role of albumin bound hormone. *J Clin Invest* 64:145–154, 1979.
56. Pardridge WM, Eisenberg J, Yang J: Human blood–brain barrier insulin receptor. *J Neurochem* 44:1771–1778, 1985.
57. Patlak CS, Paulson OB: The role of unstirred layers for water exchange across the blood–brain barrier. *Microvasc Res* 21:117–127, 1981.
58. Paulson OB, Hertz MM, Bolwig TG, et al: Filtration and diffusion of water across the blood–brain barrier in man. *Microvasc Res* 13:113–124, 1977.

59. Phelps ME, Huang S-C, Hoffman EJ, et al: Cerebral extraction of N-13 ammonia: Its dependence on cerebral blood flow and capillary permeability–surface area product. *Stroke* 12:607–619, 1981.
60. Pollay M, Kaplan RJ: Effect of the cerebrospinal fluid sink on sucrose diffusion gradients in brain. *Exp Neurol* 30:54–56, 1971.
61. Raichle ME, Eichling JO, Straatman MG, et al: Blood–brain barrier permeability of ^{11}C-labeled alcohols and ^{15}O-labeled water. *Am J Physiol* 230:543–552, 1976.
62. Rall DP, Stabenau JR, Zubrod C: Distribution of drugs between blood and cerebrospinal fluid: General methodology and effect of pH gradients. *J Pharmacol* 125:185–193, 1959.
63. Rapoport SI: *Blood–Brain Barrier in Physiology and Medicine.* Raven, New York, 1976.
64. Rapoport SI, Ohno K, Pettigrew KD: Drug entry into the brain. *Brain Res* 172:354–359, 1979.
65. Reese T: Blood–Brain Barriers to Microperoxidase. In *Proceedings the Wates Foundation. Symposium on the Blood–Brain Barrier.* Truex, Oxford, 1970, pp. 181–188.
66. Renkin EM: Transport of potassium-42 from blood to tissue in isolated mammalian skeletal muscle. *Am J Physiol* 197:1205–1210, 1959.
67. Rennels ML, Gregory TF, Blaumanis OR, et al: Evidence for a paravascular fluid circulation in the mammalian central nervous system, provided by the rapid distribution of tracer protein throughout the brain from the subarachnoid space. *Brain Res* 326:47–63, 1985.
68. Sarna GS, Bradbury MWB, Cremer JE, et al: Brain metabolism and specific transport at the blood–brain barrier after portacaval anastomosis in the rat. *Brain Res* 160:69–83, 1979.
69. Smith QR, Takasato Y, Sweeney DJ, et al: Regional cerebrovascular transport of leucine as measured by the in situ brain perfusion technique. *J Cereb Blood Flow Metab* 5:300–311, 1985.
70. Sokoloff L, Reivich M, Kennedy C: The ^{14}C-deoxyglucose method for the measurement of local cerebral glucose utilization: Theory, procedure and normal values in the conscious and anesthetized rat. *J Neurochem* 28:897–916, 1977.
71. Spector R: Vitamin B_6 transport in the central nervous system: In vitro studies. *J Neurochem* 30:889–897, 1978.
72. Spector R: Riboflavin transport in the central nervous system. Characterization and effect of drugs. *J Clin Invest* 66:821–831, 1980.
73. Spector R, Eells J: Deoxynucleoside and vitamin transport into the central nervous system. *Fed Proc* 43:196–200, 1984.
74. Szentistvanyi I, Patlak CS, Ellis RA, et al: Drainage of interstitial fluid from different regions of rat brain. *Am J Physiol* 246:F835–F844, 1984.
75. Takasato Y, Rapoport SI, Smith QR: An in situ brain perfusion technique to study cerebrovascular transport in the rat. *Am J Physiol* 247:H484–H493, 1984.
76. Van Houten M, Posner BI: Insulin binds to brain vessels in vivo. *Nature (Lond)* 282:623–625, 1979.
77. Weiss HR, Buchweitz E, Murtha TJ, et al: Quantitative determinations of morphometric indices of the total and perfused capillary network in rat brain. *Circ Res* 51:494–503, 1982.
78. Woods SC, Porte D: Relationship between plasma and cerebrospinal fluid insulin in dogs. *Am J Physiol* 233:E331–334, 1977.
79. Yudilevich DL, Eaton BM, Short AH, et al: Glucose carriers at maternal and fetal sides of the trophoblast in guinea pig placenta. *Am J Physiol* 237:C205–C212, 1979.
80. Yuwiler A, Oldendorf WH, Geller E, et al: Effect of albumin binding and amino acid competition on tryptophan uptake into brain. *J Neurochem* 28:1015–1023, 1977.
81. Zlokovic BV, Begley DJ, Chain DG: Blood–brain barrier permeability to dipeptides and their constituent amino acids. *Brain Res* 271:65–71, 1983.
82. Zlokovic BV, Begley DJ, Chain–Eliash DG: Blood–brain barrier permeability to leucine-enkephalin, D-alanine²-D-leucine⁵-enkephalin and their N-terminal amino acid (tyrosine). *Brain Res* 336:125–132, 1985.
83. Zlokovic BV, Segal MB, Begley DJ: Permeability of the blood–cerebrospinal fluid and blood–brain barriers to thyrotropin releasing hormone. *Brain Res* 358:191–199, 1985.

6

Pharmacology of the Blood–Brain Barrier

Joseph D. Fenstermacher

1. INTRODUCTION

All so-called permeability studies of the blood–brain barrier (BBB) actually measure one of several transfer constants (e.g., the extraction fraction) and not a true permeability coefficient. The permeability coefficient or, more simply, the permeability is classically defined as the flux (i.e., the rate of unidirectional solute flow) per unit membrane area divided by the driving forces for the flux, which are the concentration, pressure, and electrical gradients, in most cases. Experimental measurements of permeability involve an assessment of the flux across the membrane of predetermined surface area that separates two solutions of nearly identical composition. For the BBB, which is generally believed to be formed by the capillary endothelium, the classic definition of the permeability coefficient should be retained because it can be used to understand BBB function more clearly. By contrast, the standard methods of measuring the permeability coefficient cannot be employed for the cerebral capillaries, since neither the capillary surface area nor the composition of the blood within the capillaries and the interstitial fluid surrounding them can be tightly controlled or precisely known.

These experimental complications for assessing BBB permeability arise for several somewhat obscure reasons. First, the number of perfused capillaries and thus the effective capillary surface area S is uncertain for the system. Accordingly, the measured flux is for some unknown surface area of capillary membrane. Second, although the composition of arterial blood can be altered and measured, the actual composition of the fluid within the capillary varies as it passes along the capillary because of exchange across the BBB: the extent of this change in intracapillary fluid composition depends on the BBB permeability and effective distribution volume of each solute within the intracapillary fluid as well as the linear velocity of blood flow along the capillary. This means that the driving force for the flux across the BBB can only be estimated. For these and other reasons, the capillary

Joseph D. Fenstermacher • Departments of Neurological Surgery, and Physiology and Biophysics, State University of New York, Stony Brook, New York 11794.

permeability coefficient within the brain cannot be directly measured, and the constant obtained is referred to as a transfer constant.

Although strict adherence to the principle that brain capillary permeability cannot be directly measured could be viewed as the "hangup" of an ideologue or purist, it has definite advantages, since it forces the investigator to consider other factors such as variations in capillary surface area and blood flow as causes of changes or differences in blood–brain exchange and to design experiments to evaluate the appropriate variables of the system. With this purpose in mind, a brief treatment of methods of measuring transfer constants and converting them into expressions of capillary permeability is presented first. Following this, mechanisms of transfer across the BBB, regional and local differences in these transfer processes, and finally drug distribution between blood and brain are discussed.

2. METHODS OF MEASURING DRUG TRANSFER

The parameters that can be measured when investigating blood–brain transport are the amounts of the test material in plasma and brain samples and the sampling times. One of three transfer constants—the extraction fraction (E), the influx constant (K_i), or the efflux constant k_o—is then calculated from these data. None of these constants is a membrane permeability coefficient per se; however a functional expression of capillary permeability that combines the two unknown variables that set the flux rate across the BBB, the permeability coefficient P and the effective capillary surface area S, can be derived from them.[26,27] This expression is known as the permeability–surface area or PS product.

The extraction fraction is determined by injecting a bolus containing both a test substance plus a reference material into one carotid artery and measuring brain uptake during the next several seconds. It is defined as

$$E = B/A \qquad (1)$$

where B is the amount of test substance taken up by brain tissue and A is the total amount of extractable test material that has flowed into brain capillaries during the experimental period, as indicated by the reference material. The indicator diffusion,[18,19] the brain-uptake index (BUI),[53,54] and the external registration[70] techniques are used to measure E. Useful values of E are obtained by these three techniques when the capillary is moderately to highly permeable to the test material, the experimental duration is very short (usually less than 5 sec), and the backflux of test material from brain to blood is minute compared with the influx.[26,27] The extraction fraction is the transfer constant which should be measured in most instances when dealing with a rapidly transformed drug or substrate.

The influx constant is measured in experiments in which the test solute is intravenously administered by either bolus injection[32,33,51] or continuous infusion,[3,9] and samples of plasma and central nervous system (CNS) tissue are obtained at various times thereafter. This constant is calculated from these data by the relationship

$$K_i = B/\int_0^T C_a(t)dt \qquad (2)$$

where B is the amount which has passed across the BBB into the brain parenchyma, as in Eq. (1), T is the experimental duration, and $C_a(t)$ is the activity or concentration of free and exchangeable test material in arterial plasma during the experiment (i.e., from zero time to T).[25,27] The denominator in Eq. (2) is analogous to A in Eq. (1) and represents the amount of test material available for transfer across the BBB during the experiment, i.e., the exposure. For the resulting K_i to be meaningful, B must indicate the amount in the brain parenchyma and not the entire tissue sample, which often includes blood entrapped in the sample plus vascular and meningeal tissue, and the experimental period must be brief enough that backflux from parenchyma to blood is minuscule relative to influx. With respect to the latter point, acceptable experimental durations can range from about 10 sec for moderately permeable substances such as glucose[32] to 1 hr or more for relatively impermeable solutes such as sucrose.[51] This approach can be used to assess the transfer of moderately permeable to nearly impermeable drugs as long as metabolism, in the case of a radioactively labeled drug, is negligible during the experimental period.

Two of the methods for measuring the efflux constant involve the determination of the tissue washout or clearance of test material following brain loading by intracarotid injection[10] or ventriculocisternal perfusion.[66] The efflux constant is determined from the tissue concentration which changes as a function of time when the brain is loaded by intracarotid injection and as a function of distance from the ventricular surface when the brain is loaded by ventriculocisternal perfusion.[26,27] Under the appropriate conditions (i.e., the best of all possible worlds), the following relationship holds between the influx and efflux constants:

$$K_i = k_o V_b \qquad (3)$$

where V_b is the space in the brain in which the test material distributes during loading and from which it washes out as it clears from the tissue. The nuances of Eq. (3) and V_b have been dealt with in several reviews.[26,27] Suffice it to say that these kinds of k_o estimates require, among other things, fairly rapid uptake and clearance of the test substance, functionally single-compartment distribution in the tissue, and trivial CNS metabolism; hence these techniques will yield useful k_o values for only a few drugs.

The most commonly used procedure for measuring blood–brain transfer involves the intravenous administration of the test material and sampling of blood and tissue as described for the influx constant but the transfer constants are subsequently calculated by compartmental analysis rather than Eq. (2). To produce sufficient data for compartmental analysis, a large number of experiments which vary widely in duration (e.g., from several min to several hr) must be performed. The transfer constants are then obtained by assuming a blood–tissue distribution model of two or more compartments and of specific configuration (e.g., serially arranged), combining all the experimental data and processing the grouped data according to the appropriate format. The simplest system is the two-compartment model (one blood plus one tissue compartment), which yields estimates of k_o, V_b, and K_i. More complicated models that produce an array of transfer constant and space estimates have been used for the brain (see, e.g., Hironaka et al.[40] or Levin and Patlak[45]) and are appealing when working with complex molecules such as drugs. Nonetheless, the results of compartmental analysis are always specific for the model chosen, are susceptible to sizeable errors when metabolism is significant, and are often misleading or ambiguous.

3. CAPILLARY TRANSPORT MODELS AND PS PRODUCTS

The purpose of capillary transport models is the integration of the processes that determine blood–tissue exchange and their relative importance in material distribution. Among the various transport processes considered in such models are the flow of fluid and entrained solutes through the capillary (blood flow) and the movement of solutes and water across the capillary wall (permeability). The following discussion considers simple models that deal with blood flow and permeability in order to indicate first how BBB permeability can be evaluated from experimental data and second how blood–tissue distribution is affected by blood flow and permeability and by variation in these two processes.

The capillary PS product can be calculated from E, K_i, or k_o by assuming a particular capillary–tissue exchange model and knowing or approximating the rate of intracapillary fluid flow F and the effective distribution volume of the exchangeable test material in intracapillary fluid V_f, which is a complex function of the rates of various intracapillary distributional processes such as plasma protein and red blood cell (RBC) flows and ligand–protein dissociation. The single capillary model is the simplest, most commonly used model and assumes the capillary to be a tube of fixed diameter and uniform permeability surrounded by a cylinder of tissue. The velocity of intracapillary fluid flow is assumed to be constant during the time of the experiment and changes in F are produced by changes in flow velocity. Provided that E, K_i, or k_o truly indicates unidirectional transfer or flux, the following relationships hold:

$$E = 1 - \exp(-PS/FV_f) \tag{4}$$

$$K_i = FV_f[1 - \exp(-PS/FV_f)] \tag{5}$$

$$k_o = (FV_f/V_b)[1 - \exp(-PS/FV_f)] \tag{6}$$

where exp symbolizes exponentiation of the term $(-PS/FV_f)$, and all other terms are as defined earlier; incidentally, the PS and FV_f products must be expressed in the same units, (e.g., milliliters per gram per minute).

The exponential term is common to Eqs. (4)–(6), accounts for the change in concentration that occurs as the test material passes along the capillary and exchanges with the tissue, and yields an approximation of the mean capillary concentration, which is the driving force for blood-to-tissue flux. If permeability is much greater than flow (i.e., $PS > 3 \times FV_f$), then (1) the exponential term approaches zero, (2) the mean intracapillary concentration is very low relative to the arterial concentration or $C_a(t)$, (3) the extraction is virtually complete ($E = 1.0$), (4) the value of K_i approaches that of FV_f, and (5) tissue uptake is mainly limited by flow or delivery to the capillary. If permeability is low relative to flow ($PS < 0.1 \times FV_f$), then (1) the exponential term approaches 1, (2) $C_a(t)$ approximates the mean capillary concentration, (3) the extraction is immeasurably low ($E < 0.10$), (4) the term within the brackets of Eqs. (5) and (6) approximates PS/FV_f, (5) the value of K_i approaches that of PS, and (6) tissue uptake is mostly limited by the permeability or PS product of the capillary.

The brain and spinal cord contain a myriad of capillaries; hence a multiple capillary treatment of blood–tissue exchange is more fitting. Two commonly used multicapillary models differ in the way variations or changes in blood flow are viewed. In the variable velocity model, all capillaries (or a fixed number) are perfused at all times, and blood flow differences are the result of variations in linear velocity of blood flow through the perfused capillaries. In the capillary recruitment model, linear velocity of flow through the perfused capillaries is fixed or constant, and blood flow differences are the result of variations in the number of perfused capillaries per unit weight of tissue.

Eqs. (4)–(6) describe the interaction between drug extraction or flux and blood flow for the variable velocity multicapillary model as well as for the single capillary model. These equations indicate that differences in capillary blood flow (which, for example, might occur between awake and anesthetized subjects) lead to differences in drug extraction and flux but no differences in the *PS* product. For this case, drug extraction will be less at high flow (i.e., high FV_f) than at low flows, whereas drug influx (as indicated by K_i) will be greater at high flow than at low flow.

A set of equations that are similar but not identical to Eqs. (4)–(6) describe the relationship between drug extraction or flux and blood flow for the capillary recruitment model. The capillary recruitment equations, which are not presented here, indicate that differences in blood flow lead to differences in drug flux and *PS* product but no differences in drug extraction. To be specific, drug flux and *PS* product will be greater at high flows than at low flows; intuitively, this is easy to comprehend for the capillary recruitment model because more capillaries and capillary surface area are available for exchange at high flows than at low flows.

Good experimental support for the variable-velocity model was obtained by Raichle et al.,[70] who altered cerebral blood flow by changing arterial CO_2 pressure, measured both blood flow and extractions of several alcohols plus water, and observed that E decreased as flow increased in the way predicted by Eq. (4).

On the other hand, experimental evidence for the recruitment model has been produced. The morphometric studies of Weiss et al.[87] indicated that about 50% of cerebral capillaries are normally perfused at one time and that the percentage of perfused capillaries significantly increases immediately after asphyxia. The extraction fraction observations by Phelps et al.[67] with ammonia, by Hertz and Paulson[39] with propanolol and thiourea, and by Pardridge and Fierer[63] with water and butanol indicated that some capillary recruitment occurs when blood flow is altered as in these studies.

Since blood flow rates vary among CNS regions under normal conditions, between awake and anesthetized states, and among various physiologic and pathologic conditions, the causes of these variations need to be known not only to evaluate *PS* products from experimental data but also to understand the consequences of these flow differences in the exchange of drugs, nutrients, and other materials at the local level. Confusingly, neither of the two multicapillary models presented is likely to hold for all circumstances; even more complex models such as the saturation-recruitment model described by Phelps et al.,[67] which combines or blends the features of the two simple models, may have to be used to interpret the data in some cases. In summary, the use of capillary models may seem esoteric and overpowering but does force an appreciation of the complexity of the variables of the blood–brain transfer system.

4. MECHANISMS OF TRANSPORT TO AND THROUGH THE BBB

As material is brought to the capillaries by the blood, it may be associated with binding sites on plasma protein and RBCs and be in free solution in plasma and RBC water. These intravascular distribution possibilities are mathematically covered by the V_f term in Eqs. (4)–(6), which represents a dynamic distribution volume in blood. In all cases, the material in plasma water is available for exchange and interaction with the capillary wall in passing between blood and CNS parenchyma; accordingly, V_f for all substances equals at least the plasma water space in whole blood, which is approximately 0.51 in mammals. For materials that very rapidly exchange across RBC membranes and distribute in RBC water (e.g., urea), V_f will include both RBC and plasma water, namely whole blood water, which is about 0.81. Variations in V_f are not only found among compounds but also among species. For instance, glucose very rapidly moves in and out of human RBCs but only slowly enters rat RBCs; consequently, the glucose V_f is approximately 0.8 for humans but 0.5 for rats.

Many substances, including drugs,[15] tryptophan,[58] and hormones,[59] bind to plasma proteins and are potentially available for blood–brain exchange. If the rate of ligand dissociation from plasma proteins is small relative to the rate of protein passage through the capillaries (the mean capillary transit time of protein is about 1 sec for most brain areas), the V_f need only account for the free ligand in plasma and RBC water. If, on the other hand, the rate of protein–ligand dissociation is large relative to the velocity of flow within the capillary, then V_f must reflect both the free and the protein dissociable ligand. Plasma protein–ligand interaction in blood–brain exchange have been both experimentally (Cornford et al.[15] and Pardridge[58]) and theoretically (Pardridge[58,59]) analyzed; the findings clearly indicate that simply considering plasma protein binding on the basis of equilibrium dialysis data can be very deceiving and that V_f must reflect the dynamics of protein–ligand binding.

For most noncerebral capillaries, aqueous extracellular channels or pores, which seemingly involve less than 0.1% of the surface area, are considered the major route of transcapillary exchange for water-soluble substances;[20] however few, if any, solutes penetrate the tight junctions, (i.e., the extracellular channels) of the BBB. Two different types of experimental evidence support this statement. First, electron dense markers ranging from horseradish peroxidase (HRP) (40,000 M_r, diameter 5–6 nm) to ionic lanthanum (139 M_r, diameter 0.25 nm) do not pass through the series of tight junctions that join adjacent endothelial cells of brain capillaries.[11,12,22,75,82] Second, brain capillary PS products for a series of compounds extending from sucrose, a relatively impermeable molecule, to isopropyl alcohol, a highly permeable molecule, are directly proportional to lipid solubility and diffusivity.[24,26] This relationship agrees with Overton's rule of solute permeability for nonporous membranes and experimental findings with planar lipid bilayer membranes[52] and indicates that the extracellular channels of the BBB are essentially closed to solute transfer.

If the intercellular junctions of the BBB are tight as just suggested, then virtually all solutes must move between blood and CNS parenchyma by passing through the endothelial cell. Experimental findings have shown two transcellular mechanisms of passage across the BBB: (1) simply dissolving in and diffusing through the endothelial cell, and (2) carrier-mediated transfer. Another often suggested mechanism of material transfer through cerebral endothelial cells is vesicular transport.[82] Since actual solute transfer across the BBB via vesicles has not been rigorously demonstrated for either control or pathologic conditions, this mechanism of transendothelial movement should be considered possible but not proven.

All materials pass to some extent across the BBB by dissolving in and diffusing through the luminal and abluminal membranes plus the cytoplasm of the endothelial cell. This is the only way that many solutes enter the CNS parenchyma from the blood, and the PS products of such solutes for cerebral capillaries are directly proportional to (1) their hydrocarbon or lipid solubility, usually expressed as the octanol : water partition (*PC*); and (2) their membrane diffusivity (D_m), customarily approximated by the diffusion coefficient in water or the reciprocal of the square root of the molecular weight ($1/MW^{1/2}$). In recent reviews, this relationship was graphically tested by plotting *PS* versus *PC* divided by $MW^{1/2}$ for 16 substances[24,27]; *PS* was found to be a linear function of *PC* divided by $MW^{1/2}$ over a 10^4 range (Fig. 1 is a modified version of these graphs). For drugs the likely range of partition coefficients is very broad (10^6)[43] but of diffusion coefficients is relatively narrow (about tenfold); consequently, differences in drug influx or *PS* product are mostly the result of differences in lipid solubility and not of molecular size or diffusivity.

The transport of some solutes through the cerebral endothelial cell is facilitated by specific carrier systems. Such BBB carrier systems, like carrier system in other membranes, are stereospecific, saturable, and competitively inhibited by structurally similar compounds. Moreover, the *PS* products of substrates transported by BBB carrier are larger than would be expected with respect to their lipid solubility and diffusivity.[24] On the basis of stereospecificity, saturability, and competitive inhibitability, highly selective BBB carrier systems have been demonstrated for several classes of compounds (e.g., substrates are indicated in parentheses): hexoses (D-glucose), monocarboxylic acids (lactate), large neutral amino acids (L-phenylalanine), acidic amino acids (L-glutamate), basic amino acids (L-arginine), some purine bases (adenine), and some nucleosides (adenosine).[26,60,61]

The blood-to-tissue transfer of some neuroactive compounds such as dopamine, acetylcholine (ACh), and encephalins apparently is partly blocked by endothelial cell enzyme systems, which convert them to inactive and/or less permeable metabolites; thus, the BBB can function not only as a physical barrier but as an enzymic barrier as well. For instance, brain capillaries contain monoamine oxidase for monoamine (e.g., dopamine and serotonin) inactivation,[4,35,36,37] cholinesterase for ACh metabolism,[76] and aminopeptidase for encephalin degradation.[64] In addition, brain endothelial cells have dopadecarboxylase[4,5] which converts monoamine precursors such as L-3,4-dihydroxyphenylalanine (L-dopa) and L-5-hydroxytryptophan, which are apparently inadvertently transported into and through the BBB by the large neutral amino acid carrier,[83] into dopamine and serotonin, respectively. These monoamines would then be further metabolized by endothelial monoamine oxidase into their respective end products. In order for the enzymic BBB to function effectively, the neuroactive compound must pass through the endothelial cell and not appreciably through the tight junctions between cells; furthermore, the metabolites must be cleared or excreted in some manner. Experimental evidence of such clearance from endothelial cells is lacking.

Cytochemical findings have shown that the transport enzyme Na–K ATPase is located in cerebral endothelial membranes.[6] The presence of this and the previously described enzyme systems plus the observation that brain capillaries contain many more mitochondria than other capillaries[56] suggest that the BBB is quite metabolically active. In view of this, it is distinctly possible that drug and other substrate metabolizing systems are also part of the enzymic BBB.

The capillaries within certain structures that are adjacent to the ventricles of the brain are markedly more permeable than typical BBB capillaries. These structures include the choroid plexus (the transport barrier of the plexus is located at the choroidal epithelium), the median eminence, the organum vasculosum of the lamina terminalis, the pineal gland, the subfornical organ, the neural lobe of the pituitary, and the area postrema. Because they are positioned around the ventricles, these structures are collectively referred to as the circumventricular organs (CVOs). The capillaries in these structures are permeable to vital dyes and electron dense markers such as HRP as well as to N-acetyl-4-aminoantipyrine and sulfaguanidine.[11,85,86,88] Most of the CVOs are involved in neuroendocrine activities, and their permeable capillaries permit the ready blood–tissue uptake and release of humoral agents. For example, the neural lobe stores and releases vasopressin and oxytocin, whereas the subfornical organ has receptors for angiotensin II and is apparently involved in the central action of this peptide. Because of the leakiness of their capillaries, CVOs may also serve as points of entry into adjacent brain areas for other blood-borne materials, such as drugs, antibodies, and neurotoxins. Price et al.[68] present some excellent findings that indicate that aspartate moves from blood into CVO structures such as the median eminence and subfornical organ and subsequently spreads into adjacent, non-CVO brain tissue; however, more quantitative evidence of such CVO-to-surrounding brain transfer is needed before it can be concluded that significant penetration of solutes into non-CVO brain tissue occurs along this route.

Material can enter the CNS not only directly from the blood across the BBB but also by transport from the blood to CSF, and then to brain. The major site of blood–CSF exchange is the choroid plexus; other sites could be microvessels in the pia–arachnoid and dura–arachnoid membranes and the circumventricular organs. Once a particular substance reaches the ventricular or subarachnoidal CSF, it would be carried by bulk flow through the CSF system and readily enter adjacent CNS tissue, since both the ependymal[66] and pial–glial[46] membranes are highly permeable. Evidence for entry into brain via the CSF has been produced for folates, ascorbic acid, and deoxyribonucleosides in rabbits,[80] chloride in rats,[79] and urea in the dogfish shark.[25] Because the normal mechanism of entry into the brain from the CSF is diffusion, the depth of penetration will be limited by distance (diffusion is relatively inefficient beyond 1–2 mm) and fate within the tissue (e.g., cellular uptake and metabolism). As discussed by Spector and Eells,[80] delivery of solutes by the CSF route would only be physiologically or pharmacologically significant if the sites of action are near the CSF–tissue interface and the amount needed is very small, as might be the case for vitamins and CNS-active hormones and peptides.

For many years, the BBB was considered a simple, single-layered membrane that functions as a highly restrictive, nearly inert filter between blood and brain. Over the past decade, however, the view of the BBB has broadened. Although the tightness of the intercellular junctions is still the cornerstone of BBB function, we have come to recognize that these capillaries consist of luminal and abluminal membranes that contain several carrier transfer systems, of enzyme systems that breakdown various neuroactive compounds and related substances, and of organelles such as mitochondria which suggest significant metabolic activity.

5. LOCAL DIFFERENCES IN INFLUX

Blood flow varies widely throughout the CNS. Under normal conditions, blood flow is several-fold larger to gray matter structures than to white matter areas; in the conscious rat, for example, flow rates are greater than 1.0 ml/g per min in most gray matter structures but around 0.4–0.5 ml/g per min in white matter areas.[77] Among gray matter structures, blood flow also varies fairly broadly; the highest flow areas ($F > 2.0$ ml/g per min) include the auditory cortex, inferior colliculus (the highest), and medial geniculate.[77] In normal animals, these variations in cerebral blood flow correlated well with differences in capillary density within the brain[28] and are probably the result of this cerebrovascular feature, but this may not be the case under abnormal conditions.

Regardless of the cause of local and regional differences in blood flow (i.e., differences in flow velocity or differences in numbers of perfused capillaries) they indicate that highly permeable drugs such as thiopental[50] and nicotine[55] will distribute during the first-pass period (< 10 sec) mainly to high-flow areas of the CNS. Subsequently, the drug will rapidly pass back out of the tissue and redistribute if not metabolized or otherwise trapped. With respect to the latter possibility, high flow areas will have significantly greater exposure to the drug (i.e., a larger concentration–time product) than will low-flow areas if tissue metabolism and/or trapping greatly reduces redistribution.

Capillary surface area (the S in the PS product) seems to vary widely throughout the CNS. Such variation was first suggested by Craigie,[17] who reported that capillary density is much greater in gray matter than in white matter and that capillary density varies considerably both among gray matter structures and among white matter areas. Current estimates of capillary surface area indicate values (in mm^2/mm^3) of 15 for inferior colliculus (a high-flow gray matter structure), 10 for parietal cortex, 6.6 for the ventromedial nucleus (hypothalamus), and 4.1 for corpus callosum (white matter).[28] Since there are regional and local differences in S throughout the CNS, similar differences in PS products would be expected and have been reported for phenylalanine,[38] two synthetic small neutral amino acids (AIB and methyl-AIB) plus diethylenediaminetetracetic acid,[8] glucose,[31] and chloride.[79] In turn, regional and local dissimilarities in drug influx and efflux across the BBB are likely and should be considered when employing chemotherapy.

The preceding reports infer that capillary permeability, i.e., the P part of the PS product, is the same for BBB capillaries; however, the capillaries of the choroid plexus, median eminence, and other circumventricular organs have patent intercellular clefts and fenestrae and are readily permeated by electron-dense markers and vital dyes,[11,85,86] aspartate,[68] N-acetyl-4-aminoantipyrine,[88] and sulfaguanidine.[88] These observations imply that most solutes will readily permeate CVO capillaries and that most drugs will quickly enter these areas and produce effects, if locally active. Moreover, drugs will spread to adjacent non-CVO tissue if connecting diffusional pathways exist and if appreciable concentrations of free untransformed drug persist in the interstitium for more than several min. The study conducted by Price et al.[68] indicates that these two conditions hold for aspartate, which can be neurotoxic, and implies that drugs that are neither vigorously trapped nor metabolized in CVOs and that have moderate to long plasma half-lives will penetrate into surrounding neural tissue.

Drugs and other molecules readily move between brain and CSF across the pial–glial and ependymal membranes.[7,11,46,66,71] In most respects, solute transfer across the BBB and across the epithelium of the choroid plexus, the principal site of direct blood–CSF transport, is fairly similar. Consequently, differences between CSF and brain interstitial fluid (ISF) concentration and CSF-to-brain or brain-to-CSF flows are not likely for most substances once a steady state among blood, brain, and CSF is reached.

There are, however, several cases in which the concentration deviates significantly between ISF and CSF and uptake in tissue adjacent to the CSF may be modified by ISF–CSF exchange. First, materials such as ascorbic acid, folates, and some pyrimidine deoxynucleosides (primarily thymidine and deoxycytidine) are preferentially transferred across the choroidal epithelium vis-à-vis the BBB and subsequently flow from CSF into brain,[80] but only limited brain penetration of these materials is likely because of relatively large diffusion distances (> 2 mm). Second, the choroid plexus may be able to take up and accumulate solutes such as thiocyanate[69] and penicillin[29] from CSF and thereby keep their concentrations low in at least some CSF compartments and the adjacent CNS tissue. This hypothesized flow from CSF to choroid plexus and ultimately to blood is referred to as the sink action of the CSF. A similar efflux system or pump has been postulated for the BBB, but unequivocal demonstration of either of these two systems has not been forthcoming. Third, because of spatial and flow differences between the vascular and CSF systems, blood–CSF transfer appears to be less dynamic for most solutes than does blood–ISF exchange. Brain capillaries are fairly evenly and closely distributed throughout the tissue (the maximum distance any tissue site is from its nearest capillary is about 50 μm); much greater distances separate the choroid plexus from most of the CSF. Moreover, CSF flows fairly slowly through the ventricles and subarachnoid space (the half-time of turnover is more than 60 min) whereas blood flows very briskly through cerebral capillaries (mean transit time 1.0 sec). Accordingly, following a change in the blood level of a solute, its CSF concentration will lag behind its ISF concentration for short periods of time even if the permeabilities of the BBB and choroidal epithelium are similar, and a steady state (or near steady state) among plasma, brain ISF, and CSF may never be reached with a drug or other material that is rapidly cleared from the blood. This complicates many pharmacologic studies in which CSF samples are taken and assumed to represent brain levels.[13]

6. DRUG PERMEABILITY AND DISTRIBUTION

In the previous sections of this chapter, a physiologic framework for investigating and understanding the transfer of any compound between blood and CNS tissue was developed. Three parts of this development will shape the following discussion. First, since there is little or no movement of drugs between blood and brain through aqueous channels or pores (except in the circumventricular organs), drugs enter most areas of the CNS by passing through the endothelial cells. Second, there are two established mechanisms of transfer across typical brain capillaries. All drugs will move to some extent from blood into brain by simply dissolving in and diffusing through the BBB; in addition some drugs may also permeate CNS capillaries by specific carrier systems. Third, in many cases, drug uptake by brain will be partially or nearly completely limited by components of the distribution system other than the BBB.

Fenstermacher[24] and Fenstermacher and Rapoport[26] determined transfer constants from the literature for 16 different substances, which are relatively stable and nonreactive in biologic systems and are commonly used in membrane transport studies and calculated PS products from these transfer constants (this yielded more than 50 estimates of PS since two or more measurements of transfer constants were obtained for many of these materials). In turn, these PS products were plotted against their respective products of lipid solubility (approximated by their partition coefficients or PC) times their membrane diffusivity (approximated by their diffusion coefficients in water), i.e., $PC \times D_m$. The graphs published in these two reviews indicated for these substances that (1) PS is linearly related to $PC \times D_m$ and (2) the BBB acts like a continuous homogeneous (i.e., nonporous) membrane between blood and CNS tissue. These observations on BBB permeability also suggested several further applications of these kinds of graphs. First, if PC

and D_m are known, a hypothetical estimate of the PS product for simple diffusional penetration of the BBB can be obtained for a drug or other substance by using the regression line on one of of the published graphs (hereafter referred to as the correlation line). Second, when PS, PC, and D_m have been measured for a drug, the mechanism of blood–tissue exchange can be examined by checking the fit of this point to the correlation line. For example, if the point lies above the correlation line, then blood–brain transfer was more rapid than that attributable to simple diffusional entry, and a second transport mechanism, e.g., a carrier system, was probably involved as well in the movement of this substance across the BBB.

This usage of PS versus $PC \times D_m$ plots to analyze for transport mechanism was suggested by some of the data in the graphs published by Fenstermacher[24] and Fenstermacher and Rapoport.[26] Although the array of points on these graphs fairly closely fits the correlation line, the PS products of three substances were conspicuously grouped either above or below the correlation line (a similar deviation from the correlation line of 2 out of 10 substances was also found by Ohrbach and Finkelstein[52] in their permeability study of lipid bilayers). The thiourea and antipyrine points all fell below the correlation line. Since both compounds bind appreciably to plasma proteins, the arterial concentrations of free exchangeable thiourea and antipyrine were probably overestimated; thus the derived transfer constants and PS products were underestimated [Eqs. (1)–(5)]. The points for water all lay above the correlation line. For water, passage through exceedingly small aqueous channels in cellular[49] and artificial lipid membranes[52] has been suggested and may occur in cerebral endothelial membranes; therefore, water may penetrate the BBB by diffusing both through lipoidal pathways and through aqueous channels, which are too small and restrictive for other materials. Further support for using these graphs to ascertain mechanisms of blood–brain transport was produced by plotting PS products for substrates which can move across the BBB by carrier systems such as D-glucose and L-leucine and finding that all points lay considerably above the correlation line; for example, the experimental PS products of D-glucose are 50–200 greater than the hypothetical PS product of D-glucose.[24]

An updated plot of PS versus $PC \times D_m$ is given in Fig. 1; with respect to the original graphs of Fenstermacher[24] and Fenstermacher and Rapoport,[26] all thiourea, antipyrine, and water data have been eliminated because of the complications listed in the previous paragraph. All points for the following compounds were eliminated (with reasons for elimination indicated in parentheses): glycerol (rapid metabolism), butyrate (carrier-mediated transfer),[54] and octanoic acid (backflux). One mannitol estimate[44] was eliminated because it was derived from results with galactitol, and four new mannitol values[1,21,30,81] were added. In addition, new points were included for sucrose,[1,30,81] ethylene glycol,[81] urea,[41,81] hexanoic acid,[14] caffeine,[14] ethanol,[14,55] and isopropanol.[14,55] The PS products in Fig. 1 are mostly based on either whole brain or forebrain samples; accordingly, Fig. 1 should be viewed as a general brain or forebrain plot and may only modestly approximate the situation for any particular brain structure such as the olfactory cortex or corpus callosum. The equation of the correlation line is

$$\log (PS) = 5.236 + 1.0 \log(PC \times D_m) \tag{7}$$

and was determined by regression analysis with the slope fixed at 1.0, which is a constraint of the simple solution–diffusion permeability model.[52]

The search of the drug uptake literature yielded acceptable transfer constant estimates for 33 drugs. The results on eight of these drugs were not further analyzed, however,

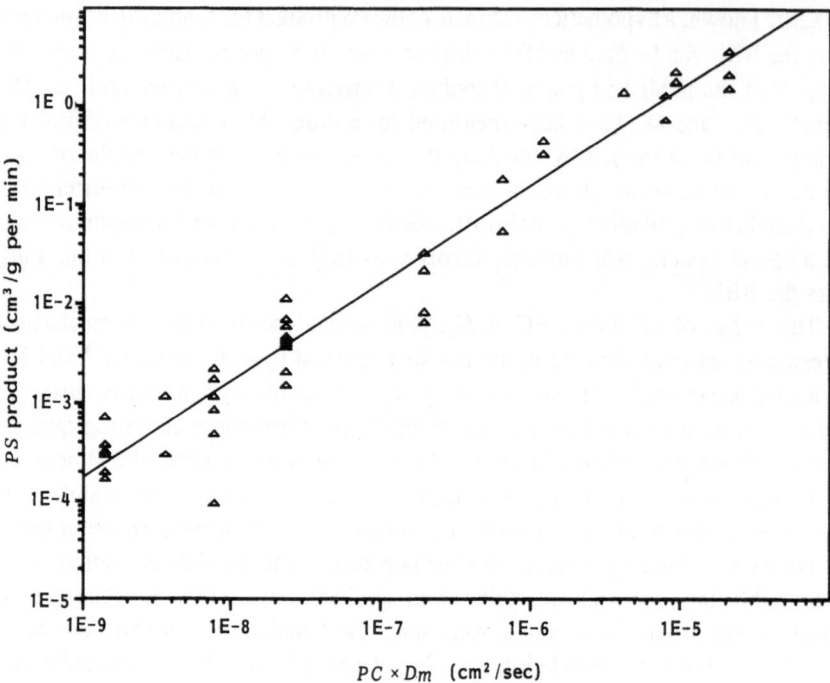

Figure 1. Log–log plot of *PS* product versus octanol/water partition coefficient times diffusion coefficient ($PC \times D_m$) for 11 compounds. The *PS* products are from Fenstermacher and Rapoport[27] plus some additional sources listed in the text. Two or more *PS* products are plotted for 10 of these 11 compounds. Starting from the left, the compounds (with number of *PS* estimates indicated in parentheses) are sucrose (7), creatinine (2), mannitol (6), urea (10), ethylene glycol (4), hexanoic acid (2), methanol (1), caffeine (2), ethanol (3), and referred to as the correlation line. The notation on both axes (1E-X) indicates powers to the base 10, e.g., 1E-5 denotes 1×10^{-5}.

because their octanol:water partition coefficients were unknown or uncertain. The *PS* products of the remaining 25 drugs are given in Table I, where they are listed by increasing order of $PC \times D_m$ product.

Examination of Table I indicates reasonable agreement between the hypothetical and experimental *PS* products for only three agents, Baker's antifol (Triazinate), caffeine, and ethanol. For all the other drugs listed, each hypothetical *PS* is four or more times larger than its respective experimental *PS*; in some cases (e.g., phenobarbital and BCNU), the hypothetic *PS*: experimental *PS* ratio is > 100. This general discrepancy in *PS* products suggests that either the theory is wrong, the $PC \times D_m$ values of these drugs were improperly assessed, or the transfer constants were incorrectly measured.

As far as the theory and the proper assessment of $PC \times D_m$, Fig. 1 shows good agreement between hypothetical and experimental *PS* products over a $PC \times D_m$ range extending from 1×10^{-9} cm²/sec (lower than the $PC \times D_m$ of Baker's antifol) to 3×10^{-5} cm²/sec (slightly less than the $PC \times D_m$ of morphine). Moreover, the work of Ohrbach and Finkelstein[52] indicates agreement between theory and results for lipid bilayer membranes over a slightly higher range ($5 \times 10^{-8} - 1.2 \times 10^{-4}$ cm²/sec). The hypothesized relationship between *PS* and $PC \times D_m$ therefore seems valid and established for values of $PC \times D_m < 2 \times 10^{-4}$ cm²/sec. There is no a priori reason for this relationship to break down for drugs with significantly higher $PC \times D_m$ values; thus, the general lack

Table I. Hypothetical and Experimental PS Products

Drug	$PC \times D_m$ (cm²/sec)	Hypothetical PS (ml/g per min)	Experimental PS (ml/g per min)	Reference
Baker's antifol	2×10^{-8}	0.004	0.002	44
Methotrexate (MTX)	8×10^{-8}	0.014	0.002	74
Dianhydrogalactitol (DAG)	5×10^{-7}	0.09	0.02	44
5-Fluorouracil (5-FU)	1.3×10^{-6}	0.2	0.015	44
Ftorafur	3×10^{-6}	0.5	0.06	44
Dibromodulcitol	4×10^{-6}	0.7	0.16	44
Adriamycin	4×10^{-6}	0.7	<0.001	44
Caffeine	8×10^{-6}	1.4	1.0–1.3	14,55,74
Ethanol	9×10^{-6}	1.6	1.8	70
Procarbazine	1×10^{-5}	1.8	0.2	44
Valproic acid	1.4×10^{-5}	2.5	0.5	16
Morphine[a]	4×10^{-5}	7.1	<0.03	55
Codeine[a]	1.2×10^{-4}	21	0.2	55
Nicotine[a]	1.5×10^{-4}	27	>3.0	55
Propranolol[a]	1.5×10^{-4}	27	0.5–1.2	39,65
Phenobarbital[a]	2×10^{-4}	36	0.2	16,55
BCNU[a]	3×10^{-4}	53	1.2–2.5	7,44
Lidocaine[a]	4×10^{-4}	71	>3.0	65
D-Amphetamine[a]	5×10^{-4}	89	0.9	62
Phenytoin[a]	3×10^{-3}	462	0.3	15
Vincristine[a]	3×10^{-3}	498	0.006	44
Pyrimethamine[a]	4×10^{-3}	711	2.0	44
Diazepam[a]	5×10^{-3}	1067	>3.0	50
CCNU[a]	6×10^{-3}	1067	1.4	44

[a]The rate of uptake of these drugs should be set by the rate of blood flow or delivery because they are highly lipid-soluble and should readily cross the blood–brain barrier.

of agreement between the hypothetical and experimental *PS* values in Table I is probably the result of poorly determined transfer constants.

The latter remark, although it may be taken as pejorative, is intended to point out the difficulties in accurately measuring transfer constants when dealing with biologically active materials such as drugs and to indicate that those processes which complicate accurate assessments of *PS* products also complicate CNS uptake of drugs. Accordingly, not only the BBB but also other distribution variables (e.g., plasma protein-binding) must be considered when seeking to optimize drug delivery to brain and spinal tissue. To emphasize further that the BBB is not the only problem in CNS drug delivery, just four of the drugs listed in Table I have $PC \times D_m$ and hypothetical *PS* products in the range in which BBB permeability virtually limits uptake (i.e., $PS < 0.2$ ml/g per min); they are Baker's antifol, methotrexate (MTX), dianhydrogalactitol (DAG), and 5-fluorouracil (5-FU).

Other than the BBB, the most commonly recognized complicating factor in CNS uptake of pharmacologic agents is plasma-protein binding, which is certainly a process worthy of drug distributional notoriety. More than one-half the drugs listed in Table I bind significantly to plasma-proteins (equilibrium dialysis indicating greater than 10% binding). Baker's antifol[48] and MTX[74] are about 50% plasma-protein bound, and slowly penetrate the BBB; the differences in their hypothetical and experimental *PS* products probably are to a significant extent the result of such binding. Many of the more lipid-soluble and more permeable drugs also bind to plasma-proteins. Thus, valproic acid,[16] propranolol,[23] and phenobarbital[16] are 60% or more bound to plasma-proteins, and their uptake is probably affected by this binding. As suggested, the experimental *PS* is much less than the hypothetical *PS* for all three drugs.

Rapoport[73] indicated that drug binding to albumin depends on the interaction of lipoidal regions of drug and protein and that the same kind of interaction occurs between drug and cellular membranes. Accordingly, as lipid solubility of drug increases, both drug–albumin binding and the ability of free drug to permeate the BBB would rise. The combination of effects may lead to relatively less blood–brain uptake of lipid-soluble drugs than expected because extensive plasma-protein binding will reduce the amount of free or exchangeable drug in plasma. As a result, attempts to increase brain uptake of drug by producing more lipid soluble analogues may be thwarted by more extensive plasma-protein binding.

In connection with this, Cornford and co-workers[15,16] presciently suggested that not only the equilibrium state of protein–drug binding but also the rate of dissociation are important in setting drug transfer across the BBB; they interpreted their results as indicating that (1) for phenobarbital[15] the rate of dissociation is slow relative to the capillary transit time of plasma proteins (about 1 sec) and thus the phenobarbital which is bound to plasma proteins is unavailable for exchange, whereas (2) for valproic acid[15] the rate of dissociation is much more rapid and a significant fraction of the protein-bound drug entering the microvessels subsequently becomes free and available for exchange during passage through the capillaries. To carry this a bit further, the rates of both drug dissociation and association with plasma-proteins should be considered when measuring and analyzing drug uptake and when selecting drugs for administration to patients with either normal or abnormal plasma-protein profiles.

The influx of a highly permeable drug ($PS/FV_f > 3.0$) is almost completely limited by the rate of blood flow. Since a reasonable approximation of mean forebrain blood flow is around 1.0 ml/g per min, the influx of drugs with PS products greater than 3.0 ml/g per min should be virtually flow limited. In theory, the influx of all drugs that rapidly distribute in blood water and have $PC \times D_m$ products $> 2 \times 10^{-5} cm^2/sec$ (i.e., from morphine to CCNU in Table I) should be limited mainly by blood flow or, more precisely, FV_f. Of the 13 drugs in Table I that fit into this category, only nicotine, lidocaine, diazepam, and thiopental seem to have experimentally determined transfer constants (i.e., extraction fractions and/or influx rate constants) that yield PS products > 3.0 ml/g per min.

There are three prominent reasons for the low PS products of the remaining 10 drugs. First, significant differences in PS products yield virtually undetectable differences in E and K_i when $PS/FV_f > 2.0$ (i.e., when $E > 0.9$ or $K_i > 0.9 \ FV_f$); hence PS cannot be determined precisely (and will almost always be underestimated) for such permeable drugs. Second, protein binding will retard the uptake of many drugs in this group, e.g., propranolol,[23] phenobarbital,[15] *d*-amphetamine,[2] and phenytoin.[15] Interestingly, diazepam[34] and thiopental[50] strongly bind to plasma-proteins but, like valproic acid, their rates of dissociation from these proteins may be rather high and much of the protein-bound drug may be available for transcapillary exchange. Third, rapid transformation of the drug to a less lipid-soluble less permeable form by simple chemical breakdown (as happens with BCNU[47] and CCNU[57]) or by metabolism (as occurs with *d*-amphetamine[78]) would lead to an underestimation of influx in experiments of duration similar to or longer than the half-time of the transformation process. Despite the uptake-restricting effects of the latter two processes, highly permeable drugs will preferentially distribute to the high-flow areas of the CNS during the first moments after entering the bloodstream and may never reach therapeutically effective levels in areas in which blood flow is pathologically or physiologically low if there is little drug recirculation after the first passage through the microvessels. The preceding statement points to another restrictor of tissue uptake; that is, rapid clearance of the drug from the circulation. The driving force for drug transfer from blood to brain is the concentration in plasma over time ($C \times t$), commonly referred to as the exposure. Among the drugs listed in Table I that are rapidly cleared from the plasma and that thus have relatively small $C \times t$ products are phenobarbital,[72] d-amphetamine,[78] phenytoin,[72] and CCNU.[57]

The CNS uptake of highly lipid-soluble drugs may also be limited by sequestration and accumulation in RBCs, the larger blood vessels proximal to the capillaries (e.g., the middle cerebral and penetrating arteries), and the endothelial cells of all vessels including the microvessels. For instance, Levin[44] favors endothelial cell trapping as the cause of the very limited penetration of vincristine into the brain over several hours, and the auto-radiographic images presented by Lear et al.[42] suggest significant arterial and arteriolar trapping of isopropyliodoamphetamine, a highly lipid-soluble material. Endothelial cell sequestration and subsequent metabolism could, however, serve a highly protective function by limiting the uptake of neurotoxic drugs.

Some of the moderately to highly lipid-soluble drugs listed in Table I are putative antineoplastic agents and apparently permeate the BBB at reasonable rates; nonetheless, these drugs are seemingly ineffective for the treatment of CNS tumors. Drugs that rapidly enter brain tissue from the blood will also readily pass back into the blood as plasma concentration falls if they remain free and exchangeable in brain fluids. Therefore, in pharmacology as in romance, it is "easy come, easy go" unless there are significant attachments between the two parties—the drug and the tissue. The BBB cannot be blamed for the lack of effectiveness for such unattached drugs, and agents that are active against the particular disease within the environment of the brain must be sought.

The ready washout or backflux of BBB-permeable materials mentioned may also explain some of the discrepancies between the hypothetical and experimental PS products in Table I. Any drug that enters the tissue and remains free and exchangeable can also leave the tissue. Provided that there is negligible trapping in the parenchyma, this backflux of drug commences immediately after influx begins. Unless experiments are carefully designed and executed and the resulting data carefully scrutinized for evidence of back-flux, significant loss of drug (and its metabolites) from the tissue will occur during the period of observation, and the measured transfer constant will not accurately indicate unidirectional extraction or influx. In this situation, E and K_i and, in turn, PS will be underestimated. This problem is especially troubling when dealing with highly permeable materials and may be almost insurmountable.

A number of substances are partly transported across the BBB by so-called carrier systems. Notably, one drug that principally moves between blood and brain by carrier-mediated transport is L-dopa.[84] Some evidence exists that the transfer of propranolol and lidocaine[65] as well as d-amphetamine[62] is partly carrier mediated; however, these findings must be viewed with some restraint because (1) simple transfer by dissolving and diffusing through the endothelial cell should be immense relative to carrier flow; (2) exceedingly high concentrations of drug (10–100 mM) were used; and (3) the distribution of these drugs is very complex and difficult to interpret because it depends appreciably on such variables as blood flow and protein binding. Finally, the experimental PS of a drug listed in Table I that moves across the BBB by a carrier system as well as by simple transcellular diffusion should be greater than its matching hypothetical *PS*. The *PS* products in Table I indicate the opposite situation and suggest no appreciable carrier-mediated transfer of any of these drugs across the BBB.

7. SUMMARY

There are five major points in this review of BBB pharmacology. First, all experimental methods for measuring the permeability of the BBB to a drug yield a blood–brain transfer constant and not a permeability coefficient. Second, in many cases, the transfer constant of a drug can be used to estimate its permeability if blood flow has been measured and a suitable capillary tissue exchange model is assumed; in connection with such estimations, several of these models are discussed. Third, drugs are apparently unable to penetrate the tight junctions that join adjacent endothelial cells and close the interstitium of the capillary wall; thus, all drugs permeate the BBB by simply dissolving in and diffusing through the endothelial cell. In addition, some drugs may also cross the BBB to an appreciable extent by carrier-mediated transport. Fourth, the rate of drug uptake varies throughout the CNS because of local differences in blood flow, capillary surface area, capillary permeability (circumventricular organ capillaries being notably more permeable than those of the rest of the brain), and proximity to the CSF. Fifth, drug influx is always affected to some extent by blood flow, plasma-protein binding, plasma half-life or disappearance rate, and endothelial and RBC sequestration as well as by capillary permeability and surface area; thus, the ready blood–brain transfer of many drugs is restricted by distribution variables other than the BBB.

ACKNOWLEDGMENTS

The assistance of Beverly Gabriel, Kurt Gruber, and Virgil Acuff in the preparation of this review is gratefully acknowledged. The material presented is the product of many discussions held with various National Institutes of Health colleagues including Dr. R. G. Blasberg, J. M. Collins, R. L. Dedrick, A. M. Guarino, C. S. Patlak, and D. P. Rall. I thank them for sharing their thoughts about drug distribution. Finally, critical review of this manuscript by Dr. R. L. Dedrick is gratefully acknowledged.

REFERENCES

1. Amtorp O: Estimation of capillary permeability of inulin, sucrose and mannitol in rat brain cortex. *Acta Physiol Scand* 110:337–342, 1980.
2. Baggot JD, Davis LE, Neff CA: Extent of plasma-protein binding of amphetamine in different species. *Biochem Pharmacol* 21:1813–1816, 1972.
3. Bänos G, Daniel PM, Moorhouse SK, et al: The influx of amino acids into the brain of the rat in vivo: The essential compared with some non-essential amino acids. *Proc R Soc (Lond) B* 183:59–70, 1973.
4. Bertler A, Falck B, Rosengren E: The direct demonstration of a barrier mechanism in brain capillaries. *Acta Pharmacol Toxicol* 20:317–321, 1963.
5. Bertler A, Falck B, Owman C, et al: The localization of mono-aminergic blood–brain barrier mechanisms. *Pharmacol Rev* 18:369–385, 1966.

6. Betz LA, Firth JA, Goldstein GW: Polarity of the blood–brain barrier: Distribution of enzymes between luminal and antiluminal membranes. *Brain Res* 192:17–28, 1980.

7. Blasberg R, Patlak C, Fenstermacher J: Intrathecal chemotherapy: Brain tissue profiles after ventriculocisternal perfusion. *J Pharmacol Exp Ther* 195:73–83, 1975.

8. Blasberg R, Fenstermacher J, Patlak C: Transport of alpha-aminoisobutyric acid across brain capillary and cellular membranes. *J Cereb Blood Flow Metab* 3:8–32, 1983.

9. Bradbury MWB, Kleeman CR: Stability of the potassium content of cerebrospinal fluid and brain. *Am J Physiol* 213:519–528, 1967.

10. Bradbury MWB, Patlak CS, Oldendorf WH: Analysis of brain uptake and loss of radiotracers after intracarotid injection. *Am J Physiol* 229:1110–1115, 1975.

11. Brightman MW, Reese TS: Junctions between intimately apposed cell membranes in the vertebrate brain. *J Cell Biol* 40:648–677, 1969.

12. Bundgaard M: Ultrastructure of frog cerebral and pial microvessels and their impermeability to lanthanum ions. *Brain Res* 241:57–65, 1982.

13. Collins JM, Dedrick RL: Distributed model for drug delivery to CSF and brain tissue. *Am J Physiol* 245:R303–310, 1983.

14. Cornford EM, Braun LD, Oldendorf WH, et al: Comparison of lipid-mediated blood–brain-barrier penetrability in neonates and adults. *Am J Physiol* 243:C161–C168, 1982.

15. Cornford EM, Pardridge WM, Braun LD, et al: Increased blood–brain barrier transport of protein–bound anticonvulsant drugs in the newborn. *J Cereb Blood Flow Metab* 3:280–286, 1983.

16. Cornford EM, Diep CP, Pardridge WM: Blood–brain barrier transport of valproic acid. *J Neurochem* 44:1541–1550, 1985.

17. Craigie EH: On the relative vascularity of various parts of the central nervous system of the albino rat. *J Comp Neurol* 31:429–464, 1920.

18. Crone C: Permeability of capillaries in various organs as determined by use of the indicator diffusion method. *Acta Physiol Scand* 58:292–305, 1963.

19. Crone C: The permeability of brain capillaries to non-electrolytes. *Acta Physiol Scand* 64:407–417, 1965.

20. Crone C, Levitt DG: Capillary permeability to small solutes. In Renkin EM, Michel CC (eds): *Handbook of Physiology*. Section 2: *The Cardiovascular System*, Vol. IV: *Microcirculation*. American Physiology Society, Bethesda, 1984, pp. 411–466.

21. Daniel PM, Lam DKC, Pratt OE: Comparison of the vascular permeability of the brain and the spinal cord to mannitol and inulin. *J Neurochem* 45:647–649, 1985.

22. Dorvini-Zis K, Sato M, Goping G, et al: Ionic lanthanum passage across cerebral endothelium exposed to hyperosmotic arabinose. *Acta Neuropathol (Berl)* 60:49–60, 1983.

23. Evans GH, Nies AH, Shand DG: The disposition of propranolol. III. Decreased half-life and volume of distribution as a result of plasma binding in man, monkey, dog, and rat. *J Pharmacol Exp Ther* 186:114–122, 1973.

24. Fenstermacher JD: Drug transfer across the blood–brain barrier. In Breimer DD, Speiser P (eds): *Topics in Pharmaceutical Sciences*. Elsevier, Amsterdam, 1983, pp. 143–154.

25. Fenstermacher JD, Patlak CS: CNS, CSF, and extradural fluid uptake of various hydrophilic materials in the dogfish. *Am J Physiol* 232:R45–53, 1977.

26. Fenstermacher JD, Rapoport SI: Blood–brain barrier. In Renkin EM, Michel CC (eds): *Handbook of Physiology*.Section 2: *The Cardiovascular System*. Vol. IV: *Microcirculation*. American Physiological Society, Bethesda, 1984, pp. 969–1000.

27. Fenstermacher JD, Blasberg RG, Patlak CS: Methods for quantifying the transport of drugs across brain barrier systems. *Pharmacol Ther* 14:217–248, 1981.

28. Fenstermacher JD, Sposito NM, Nornes SE, et al: Relationship of capillary density to glucose utilization and blood flow in white and gray matter of the rat brain. *Microvasc Res* 29:219–220, 1985.

29. Fishman RA: Blood–brain and CSF barriers to penicillin and related organic acids. *Arch Neurol Chi* 15:113–124, 1966.

30. Gjedde A: High- and low-affinity transport of D-glucose from blood to brain. *J Neurochem* 36:1463–1474, 1981.

31. Gjedde A, Diemer NH: Double tracer study of the fine regional blood–brain glucose transfer in the rat by computer-assisted autoradiography. *J Cereb Blood Flow Metab* 5:282–289, 1985.

32. Gjedde A, Hansen AJ, Siemkowicz E: Rapid simultaneous determination of regional blood flow and blood–brain glucose transfer in brain of rat. *Acta Physiol Scand* 108:321–330, 1980.

33. Go KG, Pratt JJ: The dependence of blood to brain passage of radioactive sodium on blood pressure and temperature. *Brain Res* 93:329–336, 1975.

34. Greenblatt DJ, Ochs HR, Lloyd BL: Entry of diazepam and its major metabolite into cerebrospinal fluid. *Psychopharmacology* 70:89–93, 1980.
35. Hardebo JE, Owman C: Characterization of the in vivo uptake of monoamines in brain microvessels. *Acta Physiol Scand* 108:223–229, 1980.
36. Hardebo JE, Falck B, Owman C, et al: Studies on the enzymic blood–brain barrier: Quantitative measurements of DOPA decarboxylase in the wall of microvessels as related to the parenchyma in various CNS regions. *Acta Physiol Scand* 105:453–460, 1979.
37. Hardebo JE, Emson PC, Falck B, et al: Enzymes related to monoamine metabolism in brain microvessels. *J Neurochem* 35:1388–1393, 1980.
38. Hawkins RA, Mans AM, Biebuyck JF: Amino acid supply to individual cerebral structures in awake and anesthetized rats. *Am J Physiol* 242:E1–E11, 1982.
39. Hertz MM, Paulson OB: Transfer across the human blood–brain barrier: Evidence for capillary recruitment and for a paradoxical glucose permeability in increase in hypocapnia. *Microvasc Res* 24:364–376, 1982.
40. Hironaka T, Fuchino K, Fuji T: Absorption of diazepam and its transfer through the blood–brain barrier after intraperitoneal administration in the rat. *J Pharmacol Exp Ther* 229:809–815, 1984.
41. Johanson CE, Woodbury DM: Uptake of [^{14}C]urea by the in vivo choroid plexus–cerebrospinal fluid–brain system: Identification of sites of molecular sieving. *J Physiol (Lond)* 275:167–176, 1978.
42. Lear JL, Ackermann RF, Kameyama M, et al: Evaluation of [^{123}I]isopropyliodoamphetamine as a tracer for local cerebral blood flow using direct autoradiographic comparison. *J Cereb Blood Flow Metab* 2:179–185, 1982.
43. Leo A, Hansch C, Elkins D: Partition coefficients and their uses. *Chem Rev* 71:525–616, 1971.
44. Levin VA: Relationship of octanol/water partition coefficient and molecular weight to rat brain capillary permeability. *J Med Chem* 23:682–684, 1980.
45. Levin VA, Patlak CS: A compartmental analysis of ^{24}Na kinetics in rat cerebrum, sciatic nerve, and cerebrospinal fluid. *J Physiol (Lond)* 224:559–581, 1972.
46. Levin VA, Fenstermacher JD, Patlak CS: Sucrose and inulin space measurements of the cerebral cortex in four mammalian species. *Am J Physiol* 291:1528–1553, 1970.
47. Loo TL, Dion RL, Dixon RL, et al: The antitumor agent, 1,3-Bis-(2-chloroethyl)-1-nitrosourea. *J Pharm Sci* 55:492–497, 1966.
48. Loo TL, Benjamin RS, Lu K, et al: Metabolism and disposition of Baker's antifolate (NSC-129104), ftorafur (NSC-148958), an dichlorallyl lawsone (NSC-126771) in man. *Drug Metab Rev* 8:137–150, 1978.
49. Macey RI: Transport of water and urea in red blood cells. *Am J Physiol* 246:C195–C203, 1984.
50. Mayer S, Maickel RP, Brodie BB: Kinetics of penetration of drugs and other foreign compounds into cerebrospinal fluid and brain. *J Pharmacol Exp Ther* 127:205–211, 1959.
51. Ohno K, Pettigrew KD, Rapoport SI: Lower limits of cerebrovascular permeability to nonelectrolytes in the conscious rat. *Am J Physiol* 235:H299–H307, 1978.
52. Ohrbach E, Finkelstein A: The nonelectrolyte permeability of planar lipid bilayer membranes. *J Gen Physiol* 75:427–436, 1980.
53. Oldendorf WH: Brain uptake of radiolabeled amino acids, amines, and hexoses after arterial injection. *Am J Physiol* 221:1629–1639, 1971.
54. Oldendorf WH: Carrier-mediated blood–brain barrier transport of short-chain monocarboxylic organic acids. *Am J Physiol* 224:1450–1453, 1973.
55. Oldendorf WH: Lipid solubility and drug penetration of the blood–brain barrier. *Proc Soc Exp Biol Med* 147:813–816, 1974.
56. Oldendorf WH, Cornford ME, Braun WJ: The large apparent work capability of the blood–brain barrier. A study of the mitochondrial content of capillary endothelial cells in brain and other tissue of the rat. *Ann Neurol* 1:409–417, 1977.
57. Oliverio VT, Vietzke WM, Williams MK, et al: The absorption, distribution, excretion, and biotransformation of the carcinostatic 1-(2-chloroethyl)-3-cyclohexyl-1-nitrosourea in animals. *Cancer Res* 30:1330–1337, 1970.
58. Pardridge WM: Tryptophan transport through the blood–brain barrier: In vivo measurement of free and albumin-bound amino acid. *Life Sci* 25:1519–1528, 1979.
59. Pardridge WM: Transport of protein-bound hormones into tissues in vivo. *Endocrine Rev* 2:103–123, 1981.
60. Pardridge WM: Neuropeptides and the blood–brain barrier. *Annu Rev Physiol* 45:73–82, 1983.
61. Pardridge WM: Brain metabolism: A perspective from the blood–brain barrier. *Physiol Rev* 63:1481–1535, 1983.
62. Pardridge WM, Connor JD: Saturable transport of amphetamine across the blood–brain barrier. *Experientia* 29:302–304, 1973.

63. Pardridge WM, Fierer G: Blood–brain barrier transport of butanol and water relative to *N*-isopropyl-*p*-iodoamphetamine as the internal reference. *J Cereb Blood Flow Metab* 5:275–281, 1985.

64. Pardridge WM, Mietus LJ: Enkephalin and blood–brain barrier: Studies of binding and degradation in isolated brain microvessels. *Endocrinology* 109:1138–1143, 1981.

65. Pardridge WM, Sakiyama R, Fierer G: Blood–brain barrier transport and brain sequestration of propranolol and lidocaine. *Am J Physiol* 247:R582–R588, 1984.

66. Patlak CS, Fenstermacher JD: Measurements of dog blood–brain transfer constants by ventriculocisternal perfusion. *Am J Physiol* 229:877–884, 1975.

67. Phelps ME, Huang SC, Hoffman EJ, et al: Cerebral extraction of N-13 ammonia: Its dependence on cerebral blood flow and capillary permeability–surface area product. *Stroke* 12:607–619, 1981.

68. Price MT, Pusateri ME, Crow SE, et al: Uptake of exogenous aspartate into circumventricular organs but not other regions of adult mouse brain. *J Neurochem* 42:740–744, 1984.

69. Pollay M: Cerebrospinal fluid transport and the thiocyanate space of the brain. *Am J Physiol* 210:275–279, 1966.

70. Raichle ME, Eichling JO, Straatman MG, et al: Blood–brain barrier permeability of ^{14}C-labeled alcohols and ^{15}O-labeled water. *Am J Physiol* 230:543–552, 1976.

71. Rall DP, Oppelt WW, Patlak CS: Extracellular space of brain as determined by diffusion of inulin from the ventricular system. *Life Sci* 1:43–48, 1962.

72. Ramsay RE, Hammond EJ, Perchalski RJ, et al: Brain uptake of phenytoin, phenobarbital, and diazepam. *Arch Neurol* 36:535–539, 1979.

73. Rapoport SI: *Blood–Brain Barrier in Physiology and Medicine*. Raven, New York, 1976.

74. Rapoport SI, Ohno K, Pettigrew KD: Drug entry into brain. *Brain Res* 172:354–359, 1979.

75. Reese TS, Karnovsky MJ: Fine structural localization of a blood–brain barrier to exogenous peroxidase. *J Cell Biol* 34:207–217, 1967.

76. Renkawek K, Murray MR, Spatz M, et al: Distinctive histochemical characteristics of brain capillaries in organotype culture. *Exp Neurol* 50:194–206, 1976.

77. Sakurada O, Kennedy C, Jehle J, et al: Measurement of local cerebral blood flow with iodo[^{14}C]antipyrine. *Am J Physiol* 234:H59–H66, 1978.

78. Simpson LL, Barkai A: Kinetic studies on the entry of *d*-amphetamine into the central nervous system: I. Cerebrospinal fluid. *J Pharmacol Exp Ther* 212:541–545, 1980.

79. Smith QR, Rapoport SI: Carrier-mediated transport of chloride across the blood–brain barrier. *J Neurochem* 42:754–763, 1984.

80. Spector R, Eells J: Deoxynucleoside and vitamin transport into the central nervous system. *Fed Proc* 43:196–200, 1984.

81. Takasato Y, Rapoport SI, Smith QR: An *in situ* brain perfusion technique to study cerebrovascular transport in the rat. *Am J Physiol* 247:H484–H493, 1984.

82. Van Deurs B: Structural aspects of brain barriers, with special reference to the permeability of the cerebral endothelium and choroidal epithelium. *Int Rev Cytol* 65:117–191, 1980.

83. Wade LA, Katzman R: Rat brain regional uptake and decarboxylation of L-DOPA following carotid injection. *Am J Physiol* 228:352–359, 1975.

84. Wade LA, Katzman R: Synthetic amino acids and the nature of L-DOPA transport at the blood–brain barrier. *J Neurochem* 25:837–842, 1975.

85. Weindl A: The blood–brain barrier and its role in the control of circulating hormone effects on the brain. In Ganten D, Pfaff D (eds): *Central Cardiovascular Control*. Springer-Verlag, Berlin, 1983, pp. 152–186.

86. Weindl A, Joynt RJ: The median eminence as a circumventricular organ. In Knigge KM, Scott DE, Weindl A (eds): *Brain–Endocrine Interaction: Median Eminence, Structure and Function*. Karger, Basel, 1972, pp. 16–22.

87. Weiss HR, Buchweitz E, Murtha TJ, et al: Quantitative regional determination of morphometric indices of the total and perfused capillary network in the rat brain. *Circ Res* 51:494–503, 1982.

88. Wilson CWM, Brodie BB: The absence of blood–brain barrier from certain areas of the central nervous system. *J Pharmacol Exp Ther* 133:332–334, 1961.

7

Ontogeny and Phylogeny of the Blood–Brain Barrier

Conrad E. Johanson

1. INTRODUCTION

Penetration of agents into the immature central nervous system (CNS) is a topic of fundamental significance, basically to neurobiologists as well as therapeutically and toxicologically to physicians. The orderly maturation of neuronal and glial systems demands a carefully regulated flow of nutrients and minerals across the blood–brain barrier (BBB) and blood–cerebrospinal fluid (CSF) barrier. Certain drugs and toxins that gain access to developing neural tissue can induce severe malfunctions. Since the rapidly growing CNS is vulnerable to alterations in the composition of brain interstitial fluid (ISF), it is pertinent to elucidate the respective roles played by numerous physiologic factors affecting rate and extent to which solutes penetrate neural tissue.

The concept of effective cerebrovascular permeability entails more than actual permeability of endothelial elements. In an integrated view considering morphologic and physiologic features, Levin[72] convincingly advanced the argument for lack of synonymity between the BBB and capillary permeability. Thus, the global view of the BBB encompasses such diverse factors as CSF sink action, barrier cell enzymic activities, and endothelial and glial foot active transport. In adults, the blood–CSF barrier, i.e., the choroid plexus (CP), works in concert with the BBB. In neonates, the primary dependence of cerebral function on CP transport is suggested by Russian physiologists' findings of generalized atrophy of the brain following extirpation of the four CPs in young dogs.[63]

Thus, discussion of the developing BBB takes on more perspective if it is related to simultaneous events occurring in the blood–CSF barrier. The purpose of this chapter is to review recent observations, in particular those findings with mammals over the last half-dozen years or so, and to relate them to earlier reports in the field. The ontogenetic evidence that has been marshalled expands on established concepts; in some cases,

Conrad E. Johanson • Department of Clinical Neurosciences, Brown University and Rhode Island Hospital, Providence, Rhode Island 02902.

however, the newer information prompts reconsideration of ideas. Finally, the need for research in certain developmental areas is discussed.

2. ONTOGENETIC AND PHYLOGENETIC ANIMAL MODELS OF BARRIERS

It is a principle of development that ontogenetic processes are repetitive of phylogenetic ones, i.e., the development of the individual organism bears many similarities to the evolution of the phyla. The available information on the progressive development of the CNS barrier systems in various phyla can furnish expectations, qualitatively, for corresponding maturational phenomena in the mammal. Thus, the leakier neural vasculature generally associated with primitive forms (e.g., some invertebrates) is predictive of the readily permeable BBB in the mammalian fetus. The period of rapid closure of the BBB varies greatly from one mammal to another,[11] and is approximately coincidental with the phase of rapid growth of cerebral cortex. The period of critical cortical proliferation occurs either prenatally (sheep, guinea pig) or postnatally (rat, cat); therefore, mammalian BBB tightening takes place either before or after birth (Fig. 1).

Many invertebrates do not have an endothelial–glial BBB, exceptions being insects and cephalopods. The insect BBB is impermeable to water-soluble solutes, and this tightness protects nervous tissue against plasma high in [K] and [Mg^{+2}]. All vertebrate classes investigated, including primitive cyclostomes, have a BBB greatly restricting inulin passage.[27] Thus, the vertebrate BBB is uniformly tight whether polar non-electrolyte diffusion is thwarted by endothelial membranes (mainly the case) or by capillary-enveloping glial processes (rarely, as in elasmobranch sharks). Less uniform among the vertebrates is the presence of certain ion-transport and secretory systems, perhaps having developed as secondary features to the essential tightness of the barrier.[27] The lack of CP in the hagfish, a primitive vertebrate, suggests this species as a model to evaluate deductively how the BBB in higher forms affects and is affected by blood–CSF barrier function.

Systematic studies of the immature mammalian BBB have their origin in the seminal work of Behnsen who, in 1926–1927, injected Trypan blue into postnatal mice at various

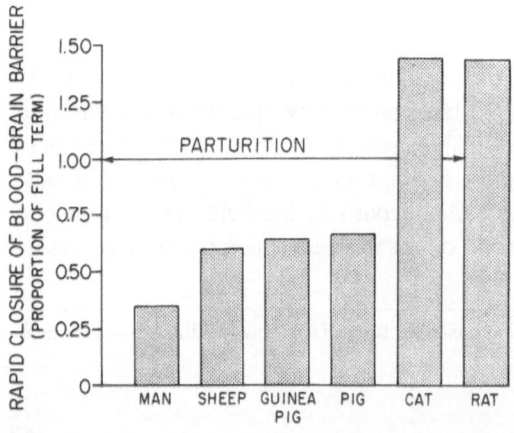

Figure 1. Approximate developmental stage at which there is rapid reduction in the permeability of the cerebral capillaries, expressed as the proportion of the time of the full gestation period for each of several mammals. Normal delivery time is indicated by the horizontal line running through 1.0. Rapid closure of the blood–brain barrier, which coincides approximately with the growth spurt of cerebral cortex, occurs by midterm in some species and a few weeks after birth in others. Information was obtained from several laboratories and summarized by Bradbury.[11]

stages of development. Rodents have proved to be expeditious models in BBB research because the wide spectrum of processes associated with closure of the BBB can be examined postnatally; this precludes many difficulties inherent in analyzing fetal animals in utero or by exteriorization procedures.

Ontogenetically, there is a close temporal relationship between maturation of the BBB and blood–CSF barrier. It is hard to imagine efficient functioning of one mature barrier concurrently with the other barrier at a grossly immature stage. The approximate coincidental development of barrier systems in the individual makes it difficult to sort out physiologic factors determining the permeability of a particular barrier. Early developmental alterations in BBB permeability may be masked by kinetic factors associated with a larger volume of ventricular CSF or by a greater surface area of CP, relative to brain size.[101]

Thus, delineation of maturational characteristics unique to cerebral capillaries and to the CP, demands an examination of the functional interrelationships among CNS compartments. Much of the recent work reviewed in this chapter represents that carried out in mammals from which plasma and CSF as well as brain and CP were obtained from different regions. The functional interrelation between distributive phenomena (diffusion, active transport) associated with cerebral capillaries and the CP–CSF system is addressed in this review within the context of extracellular fluid (ECF) homeostasis.

3. STRUCTURAL CONFIGURATIONS OF BARRIER SYSTEMS IN INFANT MAMMALS

Cross-sectional schematics of anatomic elements comprising the adult mammalian BBB and blood–CSF barrier are presented in Fig. 2. In cerebral capillaries, the tight junctions (zonulae occludens) together with the luminal membrane of endothelial cells render the microvasculature relatively impermeable to most water-soluble solutes. In CP the apical tight junctions together with the basolateral (plasma-facing) membrane of the epithelial cells constitute the rate-limiting barrier to hydrophilic materials. However, in both barrier systems of adults, there are other structures that impede solute permeation from blood ultimately to brain parenchyma and CSF. The following anatomic discussion on development refers to rats, mice, and rabbits, all of which are born grossly immature with poor vascularization and visual development.

3.1. Blood–Brain Barrier

To gain access to neurons, a solute in plasma must penetrate the cerebral capillary wall either paracellularly (through tight junctions) or transcellularly (via luminal and abluminal membranes, sequentially), or both. As revealed by freeze-fracture studies, the tight junctions between endothelial cells are not significantly different in appearance in fetal versus adult brains.[79] From earliest development, the tight junctions thwart the diffusion of large proteins. Thus, the morphologic substrate for the continually attenuating permeability of the postnatal BBB to many molecules may be found mainly in altered structure of membranes surrounding or within endothelial cells (e.g., endoplasmic reticulum). Greater permeability of the BBB in immature mice (1–3 weeks postnatal) has been attributed to vesiculocanalicular transport from luminal to abluminal membranes;[75]

Figure 2. Cross-sectional diagrams of structural elements composing the adult blood–brain barrier (BBB) (i.e., cerebral capillary) (a) and the adult blood–CSF barrier (i.e., choroid plexus) (b). Disc-shaped elements at the core of each diagram are residual erythrocytes. The interstitial fluid (ISF) compartments are the darkened areas outside of the capillary lumina. ISF volume is about 15% and 5% of brain and choroid plexus weights, respectively. Arrows point to tight junctions. In the idealized diagram (a), the astrocyte foot processes are knoblike structures abutting on nearly all the outside circumferential surface area of the capillary wall. N, neuron. The choroid plexus (b) is drawn approximately to scale, and is adapted from Murphy and Johanson.[84] The choroid plexus ISF separates capillaries from the basolateral membrane–tight junction complex of the epithelium, which constitutes the blood–CSF barrier.

after 3 weeks, such transendothelial transport disappears. Postnatal tightening of the BBB may also be related to the continual increase in glial sheath envelopment of capillaries. In the rat, the envelopment increases to 66–84%, 84–94%, and 94–97% during the first, second, and third postnatal weeks, respectively.[52] Another factor affecting solute movement in the developing brain is volume of ISF. As the ECF is gradually displaced by the expanding neuropil and glial processes, there is consequently more hindrance to diffusing molecules. Overall, the diverse structural elements of the rat BBB at 3 weeks resemble the mature barrier in adults more closely than the incompletely developed barrier at 1 week.

3.2. Blood–CSF Barrier: Choroid Plexus

The infant rabbit (and rat), at 1 week after birth displays many differences, compared with adults, in CP structure.[124] In general, there is less overall impediment to diffusion from choroidal plasma to CSF; there are also less-well-developed external limiting membranes (apical and basolateral poles of the epithelium), which are sites for carrier-transport systems. In comparison with adult tissue, the infant animal CP is characterized by similar structure of the capillary wall, thinner basement membranes, less collagen and connective tissue in the greater volume of ISF, comparable tight junction structure, less extensive infoldings of the basolateral membrane, a smaller number of mitochondria (the energy source for carriers), and fewer microvilli on the cells' apices. By 3 weeks postpartum, the choroidal epithelium and subjacent extracellular matrix have developed progressively to a configuration resembling the adult form.

4. QUANTITATION OF SOLUTE PERMEATION INTO CNS

Measurement of in vivo permeability and transport parameters poses challenging problems in work with diminutive-size and nonreadily accessible animals like infant rats and fetal kittens. Microtechniques for sampling blood (0.1–0.2 ml) and CSF (10–20 μl) in tiny perinatal rats have permitted kinetic determination of distribution volumes and half-times for several solutes.[113] Moreover, exteriorization of fetal lambs has made feasible the carrying out of kinetic analyses of tracers.[35]

Over the past decade the Oldendorf technique, i.e., the brain-uptake index (BUI), and the arterial integral method have been used widely to quantify transport characteristics (K_m and V_{max}) and cerebrovascular permeability PS in adults of several species. Most commonly, the rat has been employed. Infant rats (< 1 week postpartum; 10–20 g) do not lend themselves readily to carotid injection or femoral artery cannulation. Therefore, there is a paucity of Michaelis–Menten and PS products for murine species less than 1 week old. Ten years ago Sershen and Lajtha[105] adapted the Oldendorf method to newborn rats (5–6 g), using a 27-gauge needle to inject a double-label mixture into the left cardiac ventricle; while providing useful qualitative information about solute transport in the BBB, this modification is not quantitatively rigorous because bolus mixing with blood is considerable. In independent studies, Cremer et al.[26] and Sarna et al.[99] applied the BUI carotid injection procedure to 19-day-old rats. More recently, Lefauconnier and Trouve[70] adapted the BUI method to 5-, 12-, and 19-day-old rats; recognizing that the method is valid only if a free arterial flow past the needle persists throughout the procedure, these workers injected retrogradely into the brachial artery with a resultant clear bolus in the carotid. On the other hand, the larger size of the neonatal rabbit has permitted carotid injections.[89]

There is a conspicuous lack of PS coefficients for infant rats, as obtained by the classical arterial integral method. However, Banos et al.[6] used a 3-min brain uptake experiment, in which solute fluxes expressed in nanomoles per minute per gram (nmole/min/g) were calculated from steady-state concentration data for amino acids in plasma.

Alternatively, PS parameters (e.g., rate constant, k, and volume of distribution, V_d) can be obtained for perinatal rats by analyzing time course of uptake of tracer, with each animal as a point on the uptake curve.[113] With this approach, the inconvenience of using a large number of animals is balanced by the ready availability and reasonable cost of each specimen. Another strategy has been to use infant rats to analyze a particular time point, e.g., 1-hr uptake by the CNS. Thus, a comparison of the ratio of the 1-hr values for V_d of slowly penetrating nonelectrolytes, e.g., mannitol to inulin, is comparable to the determination of the ratio of their respective PS coefficients obtained by curve stripping.[108] To maintain steady levels of radioisotope in plasma, it is straightforward to ligate both renal arteries; this procedure presumably has a negligible effect on plasma chemistry, since renal function in infant rats, to begin with, is normally at low capacity.

The PS coefficient is conventionally expressed in milliliters per hour per gram (ml/hr/g) (or equivalent units). Thus, the clearance unit expressed in milliliters per hour (ml/hr) is normalized for weight of brain tissue (grams). Since brain water content decreases with advancing age more rapidly than does plasma water (Fig. 3), it is expedient for each stage of maturation to correct a given PS value,[113] thereby adjusting for the age-variant ratio of tissue solids to H_2O. This ensures comparability of data in ontogenetic

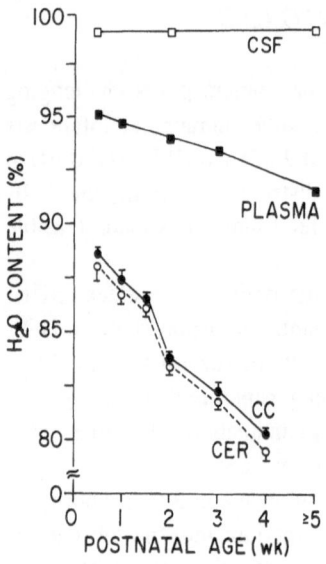

Figure 3. Ontogenetic time course of alteration in the water content of various regions in the developing CNS of Sprague–Dawley rats. Limits are SEM for cerebral cortex (CC) and cerebellum (CER), and are for four animals.[87] SEM values for plasma and cerebrospinal fluid (CSF) ($n = 6$) are less than the width of the respective symbols. Plasma was from abdominal aorta,[58] and CSF from cisterna magna. (From Burton and Johanson, unpublished data.)

studies (and in regional analyses of adults), a significant correction when H_2O content values range from 70–90% of tissue weight.

5. PHYSIOLOGIC FACTORS AFFECTING SOLUTE PERMEATION INTO CNS

5.1. Cerebrovascular Variables

Cerebral blood flow (CBF), a parameter significantly affecting distribution of lipid-soluble materials, increases substantially after birth in species born with their eyes closed, such as the rat. The caudate nucleus in neonatal rats (1–3 days) is perfused at a rate one-third that in adults,[93] as determined by rate of H_2 saturation or elimination. Similarly, analysis of tritiated water permeation revealed that blood flow to cerebrum in infant rats was 0.3 ml/min per g compared with 1.1 in adults (Fig. 4). Newborn rabbits also have a relatively low CBF of 0.25 ml/min per g.[89]

Dogs and sheep have considerably higher rates of perfusion at birth, consistent with more advanced stages of CNS development.[45] Also, blood flow to the cerebrum, cerebellum and brain stem of neonatal piglets is relatively high, i.e., 1.1–1.2 ml/min per g. Perfusion of piglet brain is not altered by aminophylline, an agent frequently used to treat perinatal breathing disorders in humans.[119] However, blood flow to CP is reduced 50% by therapeutic doses of aminophylline (6 mg/kg).

An important factor controlling velocity of CBF in adults is P_{CO_2}. Vascular reactivity to P_{CO_2} is found in animals that can see at birth (dogs, sheep), but not in rats, which are born with eyes shut. Thus, inhalation of 10–15% CO_2 by newborn rats did not alter regional CBF or blood volume.[93] Reactivity of cerebral vessels to increasing arterial pressure, i.e., autoregulation of CBF, is present in dogs and sheep[94] at parturition. The autoregulatory plateau in puppies extends over a smaller blood pressure range than in adult dogs,[48] a reflection of normally lower blood pressure in neonates. Possible lack of autoregulation in infant rats should be considered when interpreting transport/permeability phenomena in experiments in which MABP is substantially elevated. Although human infants have autoregulation of CBF, it can be impaired in distressed situations.[76] Hypercarbia compromises autoregulation of brain blood flow in newborn piglets, particularly in the brain stem.[46]

Circulation time, or the interval required for an intraarterial bolus to complete the initial circulatory pass through the brain, is a critical consideration for decapitation time when Oldendorf-type techniques are used. After carotid injection in rabbits, 10–15 sec are necessary for the bolus to move through the brain in neonates and adults, respectively.[23,24] Following retrograde injection into the brachial artery in rats, 20, 15, and 10 sec are required for the bolus to enter and leave the brain in 5-, 12-, and 19-day-old animals.[70] Thus, circulation time is inversely related to developmental age.

For polar nonelectrolytes that diffuse into the brain from plasma, their rate of penetration is proportional to the permeability P as well as surface area S of the capillary bed in a particular region. Transendothelial and perivascular permeabilities are undoubtedly altered between 1 and 3 weeks (see Section 3.1). In respect to estimates of S by light microscopy, the number of blood vessel profiles per square millimeter in the cuneate nucleus of normal rats doubled between 1 and 3 weeks, and plateaued thereafter.[29] Morphometric analyses of rat cortical vasculature by Bar[7] revealed a trebling in surface area (per unit volume of tissue) of the capillaries between the first and third postnatal week (Fig. 5a), likely attributable to capillary elongation (Fig. 5b). Functional studies also indicate that blood volume in rat cortex increases between two- and threefold between 1 and 3 weeks (Fig. 6). Overall, there is substantial evidence that proliferation culminating in maximal vascular density (i.e., surface area) occurs in rat CNS primarily during the second and third week.

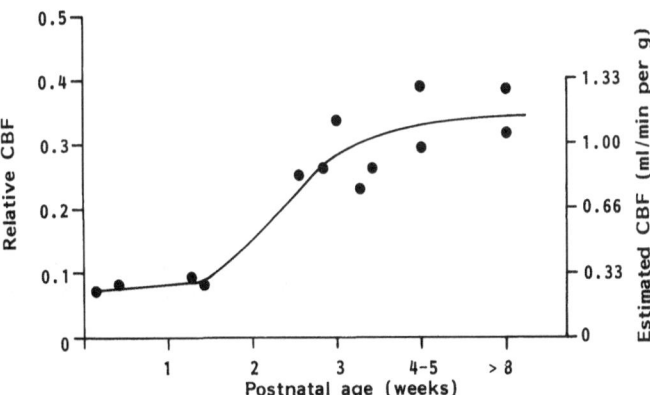

Figure 4. Cerebral blood flow (CBF) in Sprague–Dawley rats, as a function of postnatal age. Each point represents a measurement in a single animal. Each data point is a 40-sec brain/serum tritiated water ratio (ml/g) after IP injection and is a relative value of CBF (left ordinate). The tritiated water volume of distribution is proportional to CBF; thus, the estimated absolute value of CBF is given on the right ordinate. Data and assumptions are taken from Moore et al.[82,83]

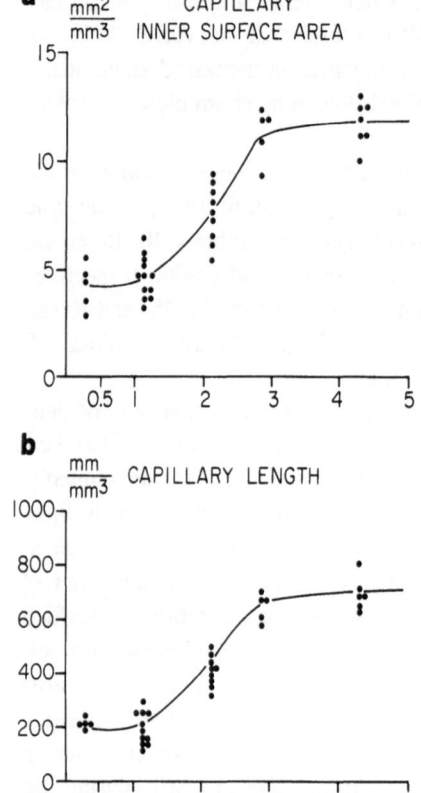

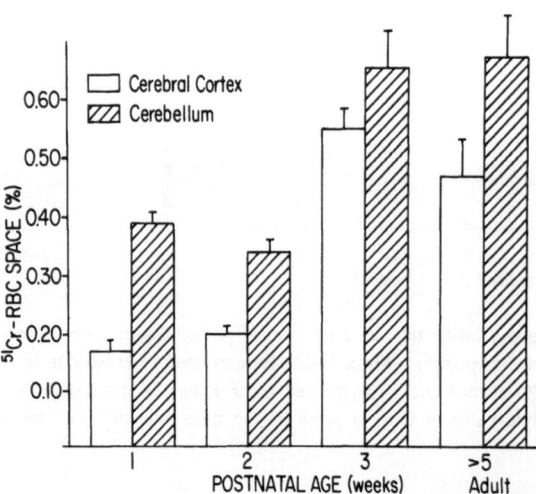

Figure 5. Morphometric evaluation in Sprague–Dawley rats of capillaries in different layers of the cerebral cortex, by automated image analysis. Each data point represents the mean value of 40 measuring fields per section of visual cortex. In each section, four vertical strips through the cortex were measured, while each strip was subdivided horizontally into 10 layers of equal thickness. Data points are replotted from a 1980 publication by Bar.[7] Reconstructed curves were drawn by eye.

Figure 6. Residual erythrocyte volume in cerebral cortex and cerebellum, as a function of postnatal age of Sprague–Dawley rats. Erythrocytes (RBCs) tagged with ^{51}Cr were injected intravenously and allowed to circulate for 10 min.[52] Brain tissue was then analyzed for radioactivity (dpm/g). Residual RBC volume (%) = 100 × dpm/g of CC or CER, divided by dpm/g RBC. Residual blood volume in cortex, like cortical capillary surface area (Fig. 5), increases about threefold between 1 and 3 weeks. Means ± SEM for $n = 5$. Values at 1 and 2 weeks are generally significantly different from corresponding values in adults, $P < 0.05$, by multiple range test.

5.2. Extracellular Fluid Volume

Electron microscopic studies,[20] kinetic analyses of tracer distribution,[113] and stable electrolyte concentration profiles (Fig. 15) have provided compelling evidence that rat cortical extracellular fluid volume (ECFV) decreases steadily with age. Table I summarizes radiolabeled ion and nonelectrolyte techniques to estimate ECFV in postnatal rat brain. The findings from the most seemingly reliable quantifications indicate that ECFV in cerebral cortex decreases from about 22–25% to 12–15% during the period from 1 week to adulthood. Postnatally, the ECFV in the cerebellum is similar to that in the cortex.[113] However, the quantitative relationship between infant and adult values for ECFV in brain is distinctly different from that in non-neural tissues such as lungs and liver (Fig. 7).

The fraction of cortical capillary surface in contact with free ECF decreases from 0.26 to 0.13 in the first postnatal week.[52] As the ECF surrounding parenchymal elements is reduced progressively, consequent to proliferation of neuropil and glia, the net uptake of water-soluble solutes from blood is proportionally decreased due, in part, to enhanced reflection of molecules in the denser interstitium. The greater impediment to molecular movement in older animals may promote backdiffusion into capillaries. The greater PS products (i.e., V_d times k) for ^{22}Na and ^{36}Cl in immature cortical and cerebral tissues are consistently associated with larger (compared with adult) V_d values (Table II). Although it is risky to ascribe kinetically determined V_d values to anatomic compartments, it is likely that the largest V_d values for tracer ions found in the youngest rats are a manifestation of the greatest ECFV (Table III).

5.3. Cerebrospinal Fluid Sink Action

Since higher steady-state concentration of solute in brain could be due to lack of a CSF sink rather than to absence of a barrier,[101] it is pertinent to distinguish alternatives by

Table I. Estimates of the Extracellular Fluid Volume in the Cerebral Cortex of Immature and Mature Sprague–Dawley Rats

Indicator	Method	ECFV (%) Infant (1 wk)	ECFV (%) Adult (>4 wk)
[^{14}C]inulin	Steady-state space of tracer after IP injection	22%[127]	5%[a127]
[^{14}C]urea	Volume of distribution of fast component of uptake	25%[b87]	15%[56]
^{36}Cl		24%[113]	13%[113]
^{22}Na	Initial slope of curve for brain space vs. CSF space	25%[113]	15%[113]
[^{14}C]inulin	6–8 hr of perfusion of ventricles and plasma with tracer	Not measured	14%[131]
[^{14}C]sucrose	Kinetic analysis of tracer uptake by brain	Not measured	14%[97]
[^{14}C]inulin	2–3 hr of perfusion of ventricular system with inulin	11%[c55]	5%[c55]
[^{14}C]sucrose		7%[d36]	54%[d36]
[^{14}C]inulin	Ratio of brain to CSF space 24 hr after IP injection	39%[d36]	41%[d36]

[a]Because of appreciable CSF sink action on inulin in the plasma of adult (but not infant) rats, this value substantially underestimates ECFV.
[b]Calculated as the product of the fractional volume of distribution of the fast component times brain tissue water content.
[c]Since the relatively short time of perfusion probably does not permit equilibration of tracer between perfusion fluid and brain interstitial fluid, these values are undoubtedly underestimates of ECFV.
[d]Pathophysiologic phenomena associated with 24-hr nephrectomy may seriously alter the measurement of ECFV.

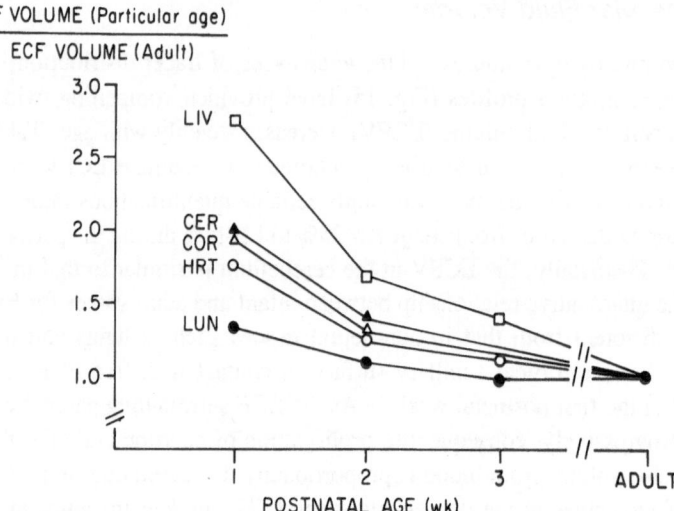

Figure 7. Relationship between extracellular fluid (ECF) volume in immature and adult Sprague–Dawley rats, for five different tissues. Two neural tissues (COR, cerebral cortex; CER, cerebellum) are compared with three non-neural tissues (LIV, liver; HRT, heart; LUN, lungs). ECF volume measurements were determined from distribution of tracers.[51,113] ECF volume decreases with age in all tissues investigated.[51,58,113] The developmental relationship between infant and adult values is similar for cortical and cerebellar tissues, but it varies significantly for non-neural tissues.

Table II. Volumes of Distribution and Rate Constants for ^{22}Na and ^{36}Cl in Developing Rat Brain Regions[a]

Volume of distribution[b] (%)	Sodium		Chloride	
	Cortex	Cerebellum	Cortex	Cerebellum
1 wk	49	42	41	34
2 wk	35	29	31	25
5 wk	31	31	27	24
Rate constant k (hr^{-1})				
1 wk	0.46	0.58	0.43	0.58
2 wk	0.63	0.87	0.46	0.63
5 wk	0.46	0.63	0.41	0.58

[a]Volume of distribution V_d times rate constant k equals the permeability–surface area product PS. By curve stripping,[108] the V_d is the y intercept of the regression slope $-k$. Even in 1-week-old rats the cerebral capillary wall is the rate-limiting step for ion penetration into the brain (Fig. 11).
[b]The 95% confidence limits are given in table III of Smith et al.[113]

Table III. Relationship between Tracer Uptake (V_d) and the Extracellular Fluid Volume in Brain Tissue of Developing Sprague–Dawley Rats[a,b]

	Cerebral cortex			Cerebellum		
	1 wk	2 wk	Adult	1 wk	2 wk	Adult
ECFV (%)	24.1	16.1	12.5	24	17.5	12.5
1-hr V_d of [³H]mannitol	5.1	5.0	3.0	7.8	4.6	2.6
$\dfrac{V_d \text{ mannitol}}{\text{ECFV}}$	0.21	0.31	0.24	0.33	0.26	0.21
1-hr V_d of [³H]inulin	1.9	1.8	0.6	2.8	1.5	0.6
$\dfrac{V_d \text{ inulin}}{\text{ECFV}}$	0.08	0.11	0.05	0.12	0.09	0.05

[a]V_d = volume of distribution = 100 × dpm ³H/g tissue, divided by dpm/g plasma H_2O. V_d is corrected for activity in residual blood.
[b]Extracellular fluid volume (ECFV) data are taken from Smith et al.,[113] and the tracer uptake data from Johanson.[52]

conducting kinetic analyses.[87] Sink action is the extent to which the concentration of a solute in brain is held below that in plasma, by net transependymal diffusion from brain ISF into CSF. Molecular sieving of solute in cerebral capillary walls and in the basolateral membrane of CP, together with continuous bulk flow of CSF from ventricles to venous sinuses, are factors that permit establishment of a solute concentration gradient from plasma to brain ISF to CSF. Net diffusion is a consequence of the gradient. Sink action is a function of the size of the penetrating molecules, hence a relative phenomenon. Thus, in neonatal rats there is sink action on albumin[2] but not on urea.[87] Theoretically, the sink effect on tracer distribution is less problematic in short-term experiments (30–60 min) than in longer experiments (several hr).

An effective way to quantify sink action is to compare the steady-state CSF to plasma ratio for a particular solute, R_{CSF}, at various stages of development. R_{CSF} for urea is unity in fetal pigs[37] and infant rats.[87] After 1 week postpartum the R_{CSF} for urea steadily decreases to 0.7 in adult rats, due to gradual development of CSF secretion and BBB tightening.[87] Clearly, the degree of sieving in barrier systems affects the extent as well as the rate of molecular penetration into CSF and brain (Fig. 8). In the immature rat, molecular sieving of urea in the blood–CSF barrier commences between 1 and 2 weeks, perhaps slightly preceding that in the BBB of cortex (see fig. 5 in ref. 87).

5.4. Brain Cell Metabolism

What is the rate-limiting step in solute (e.g., glucose) uptake by brain parenchyma in fetal and neonatal animals? Dobbing[32] proposed that the generally more permeable BBB in immature versus mature organisms is more apparent than real, due to a relatively greater rate of uptake of nutrients by rapidly synthesizing cells, early in development, rather than to leakier elements in the cerebral capillary wall. Although it is true that amino acid and glucose penetration is matched to metabolic needs of the immature brain, the kinetic analyses of inert tracer (e.g., inulin, mannitol) distribution did not lend support to the Dobbing model[32] of unchanging BBB permeability throughout development (see Section 6.3). There is intimate coupling among CBF, capillary transport and parenchymal

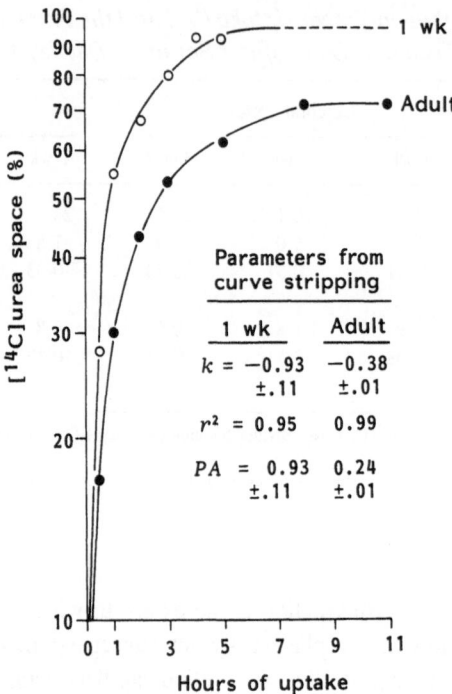

Figure 8. Penetration of [^{14}C]urea into cerebrospinal fluid (CSF) of immature and mature Sprague–Dawley rats. CSF was sampled from cisterna magna and blood from abdominal aorta. [^{14}C]urea space = 100× dpm/g CSF, divided by dpm/g plasma H$_2$O. Each data point represents the mean for four to six animals. Both the rate of uptake (i.e., *PS* coefficient) and extent of uptake (steady-state plateau on uptake curve) are significantly greater in infant rats. Parameters for CSF uptake represent the slow component of tracer flux across CNS barriers. The level of ^{14}C activity in plasma was held constant throughout the experiment.[56,87] k, rate constant (hr^{-1}). $PS = k \times V_d$, where V_d is the *y* intercept of the least-squares fit of individual uptake points subtracted from steady-state value. r = correlation coefficient for least-squares regression.

cell uptake in adults; moreover, there well may be different set points for the linkage of substrate permeation to metabolism as the BBB undergoes development.

5.5. Macromolecular Composition of CSF, Cells, and Plasma

Quantitatively and qualitatively, the CSF composition of proteins undergoes substantial changes throughout perinatal development. Active secretion of proteins into CSF makes it risky to attribute elevated protein content in prenatal CSF solely to a more permeable BBB.[33,86,100] The steady-state volume of distribution of protein-binding drugs in incompletely developed CNS regions can be substantially affected by age variations in the tissue solids/water ratio, the CSF protein level, and the neuronal/glial cell volume ratios. Characterization of parenchymal cell macromolecular composition is a significant ontogenetic problem awaiting elucidation. Already, there is evidence that the protein composition of barrier cells, in the CP, undergoes changes with advancing age.[103]

The amount and type of proteins in plasma are altered postnatally in mammals. Such developmental alterations affect the penetration into CNS of drugs and endogenous solutes that bind to plasma proteins. The permeability of the BBB to [^{125}I]albumin ($PS = 13 \times 10^{-7}$ ml/sec) in the infant piglet is the same at 2 days and 2 weeks postpartum[66]; however, over the same ontogenetic period there is a three- to fourfold reduction in *PS* bilirubin[67] in several regions of the brain. These recent findings furnish additional insight on the longstanding controversy about the pathogenesis of kernicterus. Although the reduced serum bilirubin-binding capacity observed in preterm infants may largely account for their increased risk of kernicterus as compared with full-term infants, it is now evident that another facet to the problem is that the perinatal BBB is more permeable than the adult to unbound bilirubin.

5.6. Acid–Base Effects

Arterial pH in mammalian infants is more acidic than in adults (Fig. 9), due largely to respiratory acidosis associated with Pco_2.[61] The precipitous drop in arterial Pco_2 ($Paco_2$) and H ion in rats between 1 and 2 weeks may relate to maturation of chemoreceptors or increasing levels of carbonic anhydrase in blood and tissues. It is of mechanistic interest that the progressive increase in CBF in infants occur in spite of declining $Paco_2$. Greater acidity of blood in early development can alter plasma protein binding to drugs and the degree of ionization of endogenous and xenobiotic compounds. CSF pH changes less than arterial pH throughout postnatal development (Fig. 10).

The magnitude of pH gradients among various compartments[50] affects partitioning of organic electrolytes among plasma, CSF, and brain. Compared with the adult, the pH gradient in the infant is significantly smaller across both the BBB and blood–CSF barrier (Fig. 10). The more acidic plasma in neonates promotes ionization of organic bases and suppresses ionization of acids. The rate of penetration of a base ($R-NH_2$), e.g., into the brain parenchyma, would probably be slowed, since in the younger animals a higher proportion of base would be as the charged moiety ($R-NH_3^+$), which permeates cerebral capillaries more slowly than the nonionized form. The extent of base uptake by brain cells might also be less in infants since the mean pH gradient, i.e., intracellular pH minus extracellular (CSF) pH, is less on the average in infants (-0.15) than in adults (-0.35) (Fig. 10).

Acid–base variables other than pH gradients can modify solute permeation and

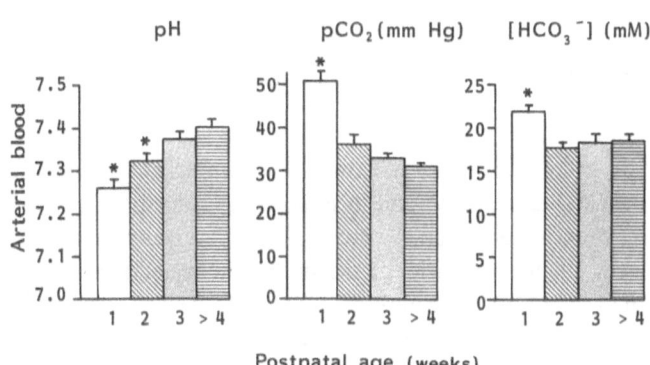

Figure 9. Acid–base composition of arterial blood in Sprague–Dawley rats, as a function of postnatal development. Means ± SEM for 10 or more animals anesthetized with ketamine. pH and Pco_2 were measured directly by electrodes in an Instrumentation Laboratory blood gas machine at 37°C.[61] Bicarbonate concentration was calculated by the Henderson–Hasselbalch equation, using experimentally determined values for pH and Pco_2 and 6.1 for pKa'. $P < 0.05$, 1, 2, or 3 weeks versus > 4 weeks, by multiple range test.

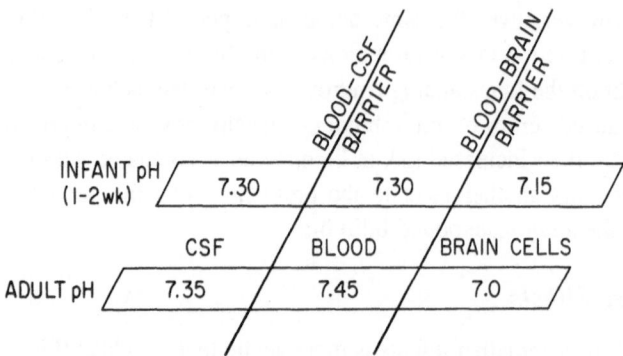

Figure 10. Ontogeny of pH in rat CNS compartments. Mean values of pH for blood, cerebrospinal fluid (CSF), and brain, in incompletely and completely developed Sprague–Dawley rats. The values are averages obtained from several publications.[53,59,61,91] Blood was sampled from abdominal aorta, CSF from cisterna magna, and neural tissue from cerebral cortex. Animals were anesthetized with ether, ketamine, or pentobarbital, all in acute experiments. Blood pH was measured directly by electrode with an Instrumentation Laboratories blood-gas machine. CSF and brain pH values were calculated from the steady-state distribution of [¹⁴C]dimethadione (DMO) or of HCO_3/CO_2. Diagonal lines represent barriers, separating major regions of the CNS (depicted as trapezoidal boxes). Standard errors are generally ±0.02 pH unit.

distribution. When analyzing the rate of uptake of organic compounds by immature versus mature brain, one must also consider that the infant BBB may be somewhat more permeable to ionized moieties. Indeed, the 1-week rat barrier systems are readily permeable to bicarbonate.[59] Moreover, since the extent of parenchymal uptake of drugs is also determined by cell macromolecular binding (itself pH dependent), it is important to ascertain developmental differences in tissue sequestering abilities. According to the pH partition hypothesis, release of organic acids (e.g., phenobarbital) from tissues should occur in systemic alkalosis. However, the movement of phenobarbital out of newborn piglet brain exposed to respiratory alkalosis or to seizures did not conform to prediction.[81] Partitioning of drugs according to pH gradients deserves further analysis in the developing organism.

6. PERINATAL DEVELOPMENT OF BARRIERS REGULATING SOLUTE MOVEMENT

A variety of carrier mechanisms exists for the transport of ions and metabolic substrates into the CNS. Such active and facilitating mechanisms undergo developmental alterations in their transporting characteristics (affinity, capacity). On the other hand, to evaluate ontogenetic changes in barrier permeability, a series of nonelectrolyte probes has been used to quantify the passive permeation of inert substances. Overall, recent perinatal findings for several mammals indicate that there are substantial developmental modifications in barrier properties. The following discussion relates earlier work to current observations.

6.1. Anions

One of the first radioisotopic probes used to analyze BBB permeability was ³²P. During the late 1940s and 1950s, several investigators noted that radioactive PO_4^{-3}

uptake from blood by the brains of rats, cats, and rabbits was significantly greater in neonates than in adults (Table IV). Bakay recognized the difficulty in expressing BBB permeability in absolute figures, but his analysis of time concentration (i.e., specific activity) curves for perinatal rabbits convincingly demonstrated that cerebrovascular permeability to ^{32}P was enhanced at birth,[4] declining gradually over 7 weeks postpartum to the adult level. He observed hardly any difference in infants between the ^{32}P content of the CP and that of the brain, a striking contrast to the situation in adults. However, Bakay[4,5] appropriately advanced the argument that there was still a barrier to ^{32}P penetration in the immature brain (including human), although the cerebral capillary wall in the fetus was considerably less restrictive than in the adult. Unfortunately, the choice of barrier probes by early investigators was complicated by cerebral metabolic sink factors, and this undoubtedly contributed to the controversy in the early 1960s about the locus of the BBB and its true permeability in neonates.[32] Nonetheless, his pioneering work concerning ^{32}P distribution pointed to developmental differences in effective permeability, and it considered the possibility of a more leaky cerebral capillary wall in immature mammals.

The movements of $^{36}ClO_4$ and ^{131}I into the developing cerebrum were analyzed by Woodbury,[128] who demonstrated in acute experiments that these anions attain concentrations in the infant brain that are 100–150% higher than in the adult. By analyzing CSF as well as brain isotopic activity, he showed that brain accumulation of tracer $^{36}ClO_4$ and ^{131}I from blood was a function of transport at both the BBB and blood–CSF barrier. Brain isotope concentration was intimately related to that in CSF, in both the incompletely and completely developed CNS. The use of stable perchlorate in tracer experiments made it

Table IV. Summary of Ontogenetic Studies of Anion and Cation Penetration of Mammalian CNS Barrier Systems

Isotopic probes	Subject	Developmental stage [a]	Regions analyzed [b]	Investigators
		Anions		
$^{32}PO_4$	Rat	N, I, J, A	B	Fries and Chaikoff[39]
$^{32}PO_4$	Cat	F, A	B	Stern and Marshall[117]
$^{32}PO_4$	Rabbit	F, N, I, J, A	B, P	Bakay[4]
$^{32}PO_4$	Human	F, A	B	Bakay[4]
^{36}Cl	Rat	I, A	B	Vernadakis and Woodbury[127]
$^{36}ClO_4$	Rat	I, A	B	Woodbury[128]
^{131}I	Rat	I, A	B	Woodbury[128]
$H^{12}CO_3$	Rat	I, A	C	Johanson et al.[59]
^{36}Cl	Rat	I, J, A	B, C, P	Smith et al.[113]
		Cations		
^{24}Na	Cat	F, A	B	Bakay[5]
^{22}Na	Rat	N, A	B, P	Luciano[77]
^{22}Na	Rat	I, J, A	B, C, P	Smith et al.[113]
^{42}K	Rat	N, I, J, A	B, C	Katzman and Leiderman[62]
^{45}Ca	Rat	I, A	B	Graziani and Escriva[44]
^{65}Zn	Rat	I, A	B, C, P	Burton and Johanson[18]
^{24}Mg	Rat	J, A	C	Olsen and Sorensen[85]

[a] F, fetal; I, infant; N, neonatal; J, juvenile; A, adult.
[b] B, brain; C, CSF; P, choroid plexus.

possible to demonstrate self-saturation and competitive inhibition, i.e., evidence for anion distribution by a carrier mechanism in barrier systems. The anion carrier was postulated to have less reabsorbtive capacity (i.e., transport from CNS to blood) in infants. Developmental differences in the anion excretory abilities of the CP (and perhaps cerebral capillaries) represent a blood sink for ions such as iodide, which increases with age.[8] Thus, in the neonate, the smaller clearance to blood in combination with greater permeability of the BBB promotes maximal net uptake of certain anions.

Whereas in adults there is avid clearance of iodide from CNS to blood, there is concurrently a large-capacity secretion of Cl^- from blood into CSF and brain. The substantial net movement of Cl across the CP is integral to the CSF secretory process,[107,112] while carrier translocation of Cl^- across cerebral capillaries[110] contributes to anion metabolism carried out by neurons and glia. In 1965, Vernadakis and Woodbury[127] published the initial systematic investigation of ^{36}Cl penetration into the immature CNS, and they concluded from half-time analysis that Cl^- permeates the infant rat brain more readily than the adult brain.[127]

Because Cl transport is part of numerous physiologic processes (e.g., cell pH and volume regulation, fluid formation), it is hardly surprising that the movement of this anion across CNS barriers continues to be investigated in ontogenetic models. In 1-week, 2-week, and adult rats, the time course of uptake of ^{36}Cl by cerebral cortex was compared with that in skeletal muscle and salivary gland (Fig. 11). In immature as well as mature animals, the rate of buildup in ^{36}Cl concentration in cortical (and cerebellar) tissue was always slower than that in the two non-neural tissues. One interpretation is that capillaries even in the youngest rats impede Cl^- movement into brain; in other words, at each developmental stage the capillary permeability P is less in brain than in muscle or salivary gland. An alternative explanation, not mutually exclusive but less likely, is that the

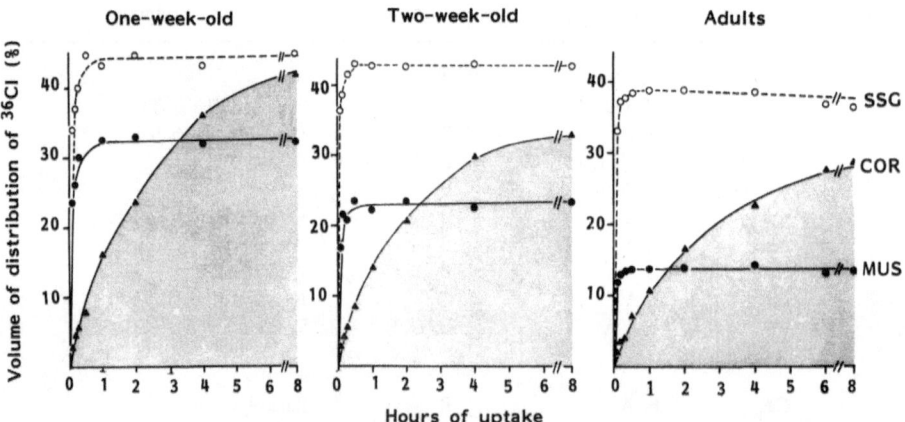

Figure 11. Comparative analysis of the movement of ^{36}Cl into cerebral cortex (COR) versus non-neural tissue, at three different stages of postnatal development in Sprague–Dawley rats. SSG, submaxillary salivary gland. MUS, skeletal muscle. Each symbol represents the mean for three to five animals. (From Smith et al.[113] and unpublished data from author's laboratory.) Five to 10 min after IP injection (time zero), the level of ^{36}Cl in plasma was stable. Volume of distribution (V_d) = 100 × dpm/g tissue, divided by dpm/g plasma H_2O. Stippled area under curve for cortex represents integrated (total) uptake of ^{36}Cl by brain tissue from steady level of radioisotope in plasma.

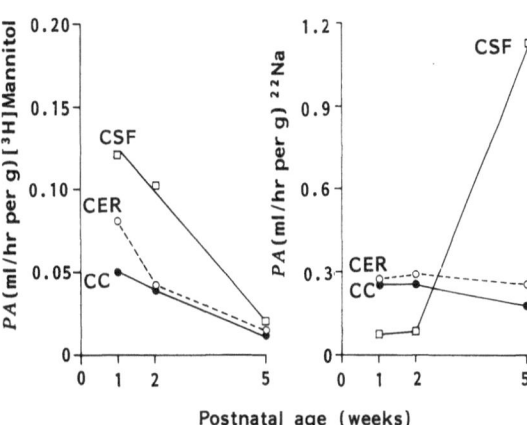

Figure 12. Permeability–surface area products (*PS*) for [³H]mannitol and ²²Na, in CNS barrier systems of developing Sprague–Dawley rats.[113] Each symbol is the mean for at least four animals. CC, cerebral cortex; CER, cerebellum. *PS* values for CSF (from cisterna magna) represent fast-component uptake, i.e., primarily across the blood–CSF barrier (choroid plexus).

capillary surface area A is less in neural than in non-neural tissue.[51] Still, BBB permeability is a relative phenomenon, ontogenetically. Thus, the rate of ³⁶Cl uptake (e.g., compare V_d of ³⁶Cl at 1 hr) was 16%, 13%, and 11% in 1-week, 2-week, and adult rats, respectively (Fig. 11).

There are also developmental differences in the extent of uptake of ³⁶Cl (Fig. 11). In all tissues, the steady-state V_d of ³⁶Cl was inversely related to age. Thus, in the least mature rats, ³⁶Cl distributed into the largest volume of fluid. The simplest explanation is that Cl⁻ distribution is mainly extracellular, and so the alterations with age in steady-state ³⁶Cl V_d run more-or-less parallel to those in ECFV (see Section 5.2).

6.2. Cations

Bakay noted 25 years ago that the brains of fetal kittens contained twice the concentration of radioactive Na⁺ as the maternal brain, 5 hr after the intravenous (IV) injection of ²⁴Na into a pregnant cat; however, even at the midtime gestation ²⁴Na exchange between plasma and brain was slowed, although not as extensively as in the adult.[5] To explain greater uptake by the fetus, the presumed larger ECFV of fetal brain was considered a plausible explanation in addition to more permeable cerebral capillaries. A later autoradiographic and specific activity analysis of ²²Na entry into rat brain furnished penetration half-times of 40 and 75 min for neonates and adults, respectively.[77] Thus, Luciano postulated that Na⁺ entered the brain more rapidly at birth due mainly to the incompletely developed pericapillary basement membrane and glial sheath.

More recently, the kinetics of distribution of ²²Na into several regions of rat CNS has been analyzed comparatively according to age and other tracers.[113] At all stages of postnatal development, the kinetics of distribution of ²²Na was similar to that of ³⁶Cl. An analysis of *PS* coefficients for ²²Na in cerebral cortex and cerebellum revealed that adult values were only slightly less than infant (Fig. 12b). However, the *PS* values for [³H]mannitol (180 M_r) decreased markedly with age in regions analyzed (Fig. 12a). Thus, the ability of the isotopic probes to penetrate the developing BBB in two regions is apparently a sensitive function of solute size (and perhaps charge). In other words, throughout postnatal development the cerebral capillaries progressively impeded the passage of mannitol (more than Na⁺) from blood to brain. Another significant observation was that the *PS* for ²²Na transport into CSF substantially increased after 2 weeks, as the fluid secretory mechanism in the CP becomes operational (Fig. 12). No such phenomenon was detected in the BBB of cortex or cerebellum,[113] even though brain capillaries in adults may slowly secrete fluid into the interstitium.[12]

Other cations have been investigated, but not as extensively as radio-Na^+. ^{42}K influx into cerebral hemispheres of one-half and 2-week-old rats was 3.9 mEq/kg per hr, a flux 35% greater than that in adults.[62] Studies of multivalent ions have been particularly sparse. ^{65}Zn uptake by CNS regions, determined by a modified Oldendorf technique, was similar in 1 week and adult rats.[18] However, in in vivo uptake experiments, the specific activity of brain Ca^{2+} equaled that in blood within 3 and 9 hr, respectively, for suckling and adult rats,[44] even though stable $[Ca^{2+}]$ in brain of immature rats is about 50% greater than in mature animals.[19] Clearly, there is need for investigations of the response of the developing brain–CSF system to plasma loaded with stable cation, similar to the experimental design of Olsen and colleagues, who analyzed CNS Mg^{+2} homeostasis in young animals.[85]

6.3. Nonelectrolytes

Water-soluble nonelectrolytes with graded molecular weight, such as inulin, sucrose, mannitol, and urea, are ideal permeability probes, since they are not bound to plasma proteins, metabolized, or actively transported (Table V). Systematic investigations by Ferguson and Woodbury[36] in the late 1960s unequivocally demonstrated that radioinulin (5500 M_r) readily penetrated the fetal and neonatal rat cerebral hemispheres; e.g., brain/plasma ratios of 50–60% were obtained in perinatal animals when several hours were allowed for tracer distribution. By contrast, Evans et al.[35] in 1974 showed that the BBB to sucrose (342 M_r) was tight in lambs at birth; indeed, the cerebral capillaries started to impede this hydrophilic tracer even 2 months before term. This was a significant finding because it revealed that not all mammals are born with a leaky BBB. However, a follow-up study of midgestation and near-term lambs by Dziegielewska et al.,[34] to include other lipid-insoluble nonelectrolytes like erythritol (122 M_r), provided evidence for unrestricted diffusion of molecules as large as inulin across the CP and cerebral capillaries. With respect to the inverse relationship between the size of the tracer molecule and its steady-state level in CSF (and brain), these investigators attributed differences with advancing age to progressively increasing CSF sink action and to decreasingly effective (but not total) surface area for diffusion at the transport interfaces.

More recent ontogenetic studies by Johanson and colleagues[52,113] have quantified the permeation of mannitol and inulin into CSF, choroid plexus, and various regions of the brain in rats. The 1-hr uptake of tritiated mannitol and inulin in cerebral cortex (CC) and cerebellum (CER) was 200% greater in neonates than in adults (Fig. 13). The radioactive nonelectrolyte penetration into ECFV is summarized in Table III. Mannitol (180 M_r) and inulin (5500 M_r) have to distribute into a larger ECFV in younger animals. Yet, the short-term experiments demonstrate that 1 hr after intraperitoneal (IP) administration, both tracers have filled a larger fraction of the greater total ECFV in infant animals, compared with adults. Such findings deductively indicate that cerebrovascular permeability is highest at 1 to 2 weeks, because at this stage the smaller vascular surface area, CBF, and CSF sink action cannot explain the greater uptake of nonelectrolyte tracers.

Another quantification, i.e., by PS products, showed that the PS for mannitol was threefold higher (both in the BBB and blood–CSF barrier) in 1-week compared with 5-week animals.[113] Urea is the smallest nonelectrolyte tested in immature animals,[87] and

Table V. Summary of Ontogenetic Studies of Nonelectrolyte and Sugar Penetration of Mammalian CNS Barriers

Isotopic probes	Subject	Developmental stage[a]	Regions analyzed[b]	Investigators
		Nonelectrolytes		
[^{14}C]inulin	Rat	I, A	B, C, P	Vernadakis and Woodbury[127]
[^{14}C]sucrose	Rat	F, N, I, J, A	B, C	Ferguson and Woodbury[36]
[^{14}C]inulin	Rat	F, N, I, J, A	B, C	Ferguson and Woodbury[36]
[^{3}H]sucrose	Sheep	F, N, A	B, C	Evans et al.[35]
[^{14}C]erythritol	Sheep	F	B, C, P	Dziegielewska et al.[34] Johanson[52]
[^{3}H]mannitol	Rat	N, I, J, A	B, C, P	Johanson[52]
[^{3}H]inulin	Rat	N, I, J, A	B, C, P	Smith et al.[113]
[^{3}H]mannitol	Rat	I, J, A	B, C, P	Parandoosh and Johanson[87]
[^{14}C]urea	Rat	I, J, A	B, C, P	
		Sugars		
D-Glucose,[c] 3-MG	Rat	N, I, J, A	B	Moore et al.[82]
D-Glucose[d]	Rat	J, A	B	Daniel et al.[28]
D-Glucose[e]	Rat	I, J, A	B	Cremer et al.[26]
D-Glucose[e]	Rabbit	N, A	B	Braun et al.[14]
3-MG[f]	Rat	I	B	Betz and Goldstein[9] Cornford et al.[21]
D-Glucose[e]	Rabbit	N, I, J, A	B	Gjedde and Fuglsang[41]
D-Glu, 2-DG, 3-MG, mann, gal[c]	Rat	N	B	

[a],[b]See abbreviations in Table IV.
[c]Tissue uptake curve.
[d]1-min influx.
[e]BUI technique.
[f]Isolated capillaries.

even its uptake across barriers into CSF differs with the developmental stage of rats (Fig. 8). Urea undergoes molecular sieving in the CNS barriers of adult animals.[56] However, the lack of sieving of urea in the cerebral capillaries and CP of infant rats constitutes strong evidence for incompletely developed barrier systems in the immature brain.[87]

6.4. Sugars

Early investigators inappropriately deduced that D-glucose rapidly penetrates neonatal brain capillaries by unrestricted diffusion, a phenomenon they linked to the enhanced ability of the newborn to survive energy-deprived states (Table V). However, Moore and colleagues[82] in 1971 demonstrated that sugar entry into rat cerebrum occurs by the facilitating action of a carrier identified in the infant, as in the adult, by transport competition, selectivity, and counterflow. The initial observation of selectivity has been upheld by the recent finding of Gjedde and Fuglsang,[41] who used mannose, galactose, and glucose as probes. Interestingly, although the flux of glucose (V_{max}) into the brain of the newborn is only one-fifth of the adult rate, the concentration of glucose is considerably higher in immature than in mature brain. This is due to the much slower rate of metabolism in

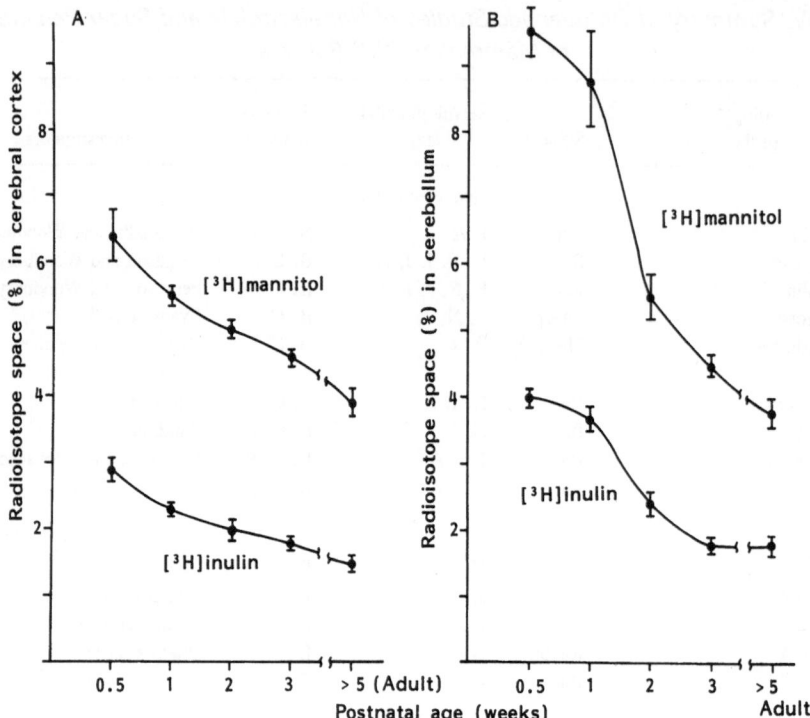

Figure 13. One-hr volume of distribution (space) of [³H]inulin and [³H]mannitol, in (A) cerebral cortex and (B) cerebellum, of developing Sprague–Dawley rats.[52] % Space = 100× dpm/g tissue, divided by dpm/g plasma H_2O. Plasma level of ³H activity was held constant throughout the 1-hr experiment. Means ± SEM for *n* = five animals at each age.

young animals. Using rats aged from 2–116 weeks, Daniel et al.[28] later corroborated carrier-mediated transport in young animals, concluding that the slow influx of glucose into the brain of the suckling animal is due to lower V_{max} rather than to higher K_m. Their developmental data conformed to simple Michaelis–Menten kinetics, indicating that passive diffusion (K_d) of glucose from capillary to brain interstitium played a minor role in sugar translocation at the maturing barrier. However, Cremer et al.[26] found a significant nonsaturable diffusion component (K_d) in immature rats, especially significant at high concentrations of glucose in plasma (8 mM); thus, in hyperglycemia, diffusion accounted for 35% of infant brain uptake.

BUI values of about 20 and 30% for glucose were reported by Cremer et al.[25,26] for 2-week and adult rats, respectively. The same laboratory quantitated that the K_m for glucose was age invariant, although V_{max} in 2-week rats was half that in adults. On the other hand, the BUI for glucose is quite uniform in rabbits over six ontogenetic stages between birth and adulthood.[21]

The above mentioned studies indicate that the transendothelial transport of glucose displays characteristics that are age variant (V_{max}) and age invariant (K_m). The uptake of glucose into rat cerebral capillaries has been analyzed by Betz and Goldstein,[9] who marshaled evidence for similarities in in vitro transport by endothelia from 10- and 30-day-old animals. Thus, at both ages, the hexose uptake was rapid (not rate-limiting for capillary metabolism), inhibited by cytochalasin B, but not reduced by ouabain or 2,4-dinitrophenol.

6.5. Amino Acids

The earliest findings, 25 years ago, indicated greater uptake by neonatal brain of glutamic acid,[49] α-aminoisobutyric acid,[64] and lysine.[65] Subsequent investigations are summarized in Table VI. Cerebral uptake of several amino acids in vivo by newborn mice was analyzed by Seta et al.[106] Their general conclusion was that for a given increase of an amino acid in blood, there was a greater corresponding elevation of that amino acid in newborn than in adult brain. Subsequently, capillary transport of amino acids in the developing rat brain was described by Sershen and Lajtha.[105] With BUI methodology, they found that, following injection of a large dose of amino acids, the essential acids were elevated to a greater degree in brain than the nonessential ones, both in infants and adults. Furthermore, in cerebral capillaries in immature as well as mature animals, the transport of essential neutral amino acids was more rapid than that of the basic ones, whereas the transport of Gly, Pro, Glu, and γ-aminobutyric acid (GABA) was below detectable limits. Therefore, Lajtha and colleagues established that although the capillary permeability barriers in young brain are weaker, nevertheless the general pattern of carrier transport of amino acids in neonates is qualitatively similar to that in adults.

Penetration of amino acids has also been quantified in acute experiments (3–12 min) by relating brain accumulation of tracer to specific activity of plasma.[6] Influx rates (nanomoles per minute per gram whole brain) for most amino acids were found to be several-fold greater in suckling (1–2-week-old) than in adult (8–10-week-old) rats. Thus, Banos and colleagues found a continual decline with age in penetration rate for most

Table VI. Developmental Studies of Amino and Monocarboxylic Acid Transport across the Mammalian Blood–Brain Barrier

	Subject	Developmental stage[a]	Experimental technique[b]	Investigators
Amino acids				
Various (10)[c]	Mouse	N, A	Tissue uptake curve	Seta et al.[106]
Various (12)[c]	Rat	N, A	BUI	Sershen and Lajtha[105]
Various (11)[c]	Rat	I, J, A	3-min influx	Banos et al.[6]
Arg	Rabbit	N, A	BUI	Braun et al.[14]
Tyr, Lys	Rat	J	Tissue uptake	Freedman et al.[38]
Trp	Rabbit	N, I, J, A	BUI	Cornford et al.[21]
Trp	Rat	I, J, A	BUI	Sarna et al.[99]
Various (13)[c]	Rat	N, I, J	BUI	Lefauconnier and Trouve[70]
Ala, Phe	Rat	I, J, A	Arterial integral	Lefauconnier and Bernard[68]
Monocarboxylic acids				
BHB[d]	Rat	N, I, J, A	Tissue uptake curve	Moore et al.[83]
Lact, Pyr	Rat	I, J, A	BUI	Cremer et al.[26]
BHB	Rabbit	N, A	BUI	Braun et al.[14]
Lact	Rabbit	N, I, J, A	BUI	Cornford et al.[21]
Lact	Dog	N	Arteriovenous diff.	Hellmann et al.[47]

[a]N, neonate; I, infant; J, juvenile; A, adult.
[b]Experimental techniques are described in corresponding references.
[c]The number of different amino acids tested is given in parentheses.
[d]BHB, β-OH butyrate; Lact, lactate; Pyr, pyruvate.

amino acids across cerebral capillaries during the second to seventh postnatal weeks. The rapid uptake in infants was attributed primarily to carrier-mediated transport, not to an incomplete BBB. Regional analyses of amino acid uptake have been infrequently done. However, Freedman et al.[38] observed that the rate of uptake of Tyr was similar in forebrain, brainstem, and CER in weanling rats (3 weeks).

As determined by BUI analysis in rabbits, the penetration of Arg into cerebrum is sixfold greater in neonates than in adults.[14] This age-related alteration in Arg uptake is more substantial than that estimated for 1-day-old versus adult rats.[105] The BUI of Arg in rabbits is halved during the initial week postpartum, and it continues to decrease to the adult like value obtained at 3 weeks.[21] With respect to Trp transport, Braun et al.[14] reported BUI values of 56 and 36% for newborns and adults, respectively; and Cornford and colleagues found a fourfold reduction over the first month postnatally.[21] However, comparable BUI values in immature versus mature rabbits have been determined for Phe (55%) and Glu (4%).

Because of the diminutive size of the carotid artery, a paucity of BUI data have been generated for infant rats. The study by Sarna et al.[99] described a greater influx of Trp in 19-day-old animals than in 6-month rats, reflecting higher values for both V_{max} and K_d (diffusion). Thus, BUI values for Trp in immature rats are similar to infant rabbits, i.e., about 50%. A recent modification of Oldendorf's method, involving retrograde injection of bolus into the brachial artery, was accomplished by Lefauconnier and Trouve to estimate BUI in rats 5, 12, and 19 days after birth.[70] At each age, Glu and Gly had low BUI values (5–10%) close to sucrose, in agreement with previous findings for immature rodents.[14,105] Patterns of neutral amino acid uptake differed with development. Especially prominent in older rats (near weaning) were two types of transporting mechanisms: an ASC-like system for Ala, Ser, Cys, and Thr (with BUI values of 10–20) and an L system for Met, Leu, Phe, and Trp (with BUI values of 40–60). Evidence was also obtained for a β-amino acid transporter, because the BUI of taurine was lowered by the addition of cold amino acid. The facilitated entry of taurine by a carrier system in immature rats agrees with Sturman's finding that taurine passes much more rapidly through the perinatal than adult BBB.[120]

6.6. Monocarboxylic Acids

The lack of behavioral symptoms in immature rats rendered severely hypoglycemic led early investigators to conclude that substances other than glucose could be transported into and used by the developing brain (Table VI). An important contribution in this regard was Moore's finding that the permeability to the ketone body, β-hydroxybutyrate (BHB), rose sevenfold throughout the rat suckling period and then declined after weaning to low newborn values.[83] Significantly, the age dependence of BHB distribution differed from those of CBF, nonelectrolytes, and dimethadione (DMO) (cell pH-sensitive distribution). A carrier for BHB, modulable by diet, was therefore postulated. Later, Braun's BUI analysis of newborn rabbit BBB furnished a value of 14% for BHB uptake[14]; this can be compared with rat BUI values of 26% at 2 weeks[25] and 4–7% in adults.[25,40]

Lactate and pyruvate are also transported into adult brain by the monocarboxylic carrier (Table VI). Single injections into the carotid artery of suckling rats (15–21 days) have yielded V_{max} values of 2 μM/g per min for both L-lactate and pyruvate; by adulthood, V_{max} had decreased tenfold.[26] Significantly, the K_m for these two acids was 5–10 times greater in immature rats versus adults; and K_d (diffusion) for lactate was elevated (by fourfold) in the youngest animals. A later study by Cornford et al.[21] using rabbits, provided BUI values of about 35% for infants (birth to 2 weeks) and values that progressively declined throughout postnatal development. The adult BUI was 7% for L-lactate. The arteriovenous (A–V) difference method was used in newborn dogs by Hellmann et al.,[47] who demonstrated that the rate of uptake of lactate by brain was proportional to its arterial concentration. They were not able to distinguish between translocation by diffusion or by a carrier with saturability greater than 8 mmoles/liter, i.e., hyperlactemia.

6.7. Monoamine Neurotransmitters

It is generally thought that circulating monoamines do not readily enter the adult brain due to the existence of a dual BBB. A physical barrier is present in the luminal membrane of the cerebral endothelium; in addition a chemical barrier is represented enzymically as monoamine oxidase (MAO) degradative activity in endothelial cytoplasm. However, the BBB to monoamines (MA) may not be fully developed in infant rats. A fluorescence histochemical study by Loizou[73] showed that, after subcutaneous administration of α-methyldopa there was significant passage of the MA into brain tissue up to 2 weeks postpartum. α-Methylnorepinephrine penetrated the BBB in 2-week-old but not in 3-week-old rats. Both norepinephrine and α-methyldopamine were taken up by capillary cells and neurons in 1-week-old but not in 2-week-old animals. Similarly, an earlier study by Glowinski et al.[42] demonstrated closure of the rat BBB to tritiated norepinephrine by 2 weeks after birth.

7. DISCUSSION OF ONTOGENY OF BARRIER FUNCTION IN MAMMALS

7.1. Ontogeny of the BBB: Models and Questions

Is the BBB fully developed in newborns? This question has been vigorously debated for several decades. Sixty years ago, the caricature of the BBB in infants as immature and leaky was accepted on the basis of the model pertaining to the apparent pathogenesis of kernicterus in hyperbilirubinemic individuals. However, in 1941 Broman[15] challenged the prevailing thought by concluding from the extant literature that the BBB of neonates was fully developed. This alternative view was upheld recently in a review by Pardridge,[88] who emphasized that the current appreciation of nearly full development at birth has gradually replaced the common misconception of a leaky and incompletely developed barrier in perinatal organisms.

As with most controversies, evidence has been continuously forthcoming to support either side of the argument, gradually leading to a unitary theory. Thus, in the case of BBB ontogeny, the mammalian barrier in infants of some species is considerably more permeable to water-soluble nonelectrolytes (compared with the mature organism) but at the same time displays adult-like carrier transporters for nutrients (but with dissimilar kinetic properties). Perhaps, then, the more salient question is: What are the unique properties of the infant BBB? rather than is the BBB in infants fully mature? In other words, the BBB in infants is not necessarily immature, even though its permeability and transport properties differ markedly from adults. Selectivity and regulation of solute transfer in the infant BBB are not necessarily incompatible with incomplete development of the brain or organism per se.

To address appropriately the question about unique and specific properties, it is important to qualify differences among species in the developmental stage at birth and to provide quantitative descriptions of permeability/transport phenomena rather than anecdotal information. Since the rat has been the most popular model for investigating CNS barrier function in infants and adults, most of the following discussion is concerned with this species.

7.2. Permeability of the BBB in Infant Rats in in Vivo Experiments

It is difficult to distinguish between absolute and effective permeability (P) of the cerebral capillary wall. Interage comparison of the 1 hr V_d values for inulin and mannitol (Fig. 13), and for urea,[87] reveals greater permeation (by 50–100%) of these non-electrolytes into neonatal compared with adult brain. Since the augmented penetration of tracers in the youngest animals occurs in spite of lower values for both CBF (Fig. 4) and capillary density, i.e., surface area (Fig. 5), this suggests that the permeability of the cerebral capillary wall is greatest at the earliest stage of postnatal development. Because of the inertness of the tracers, it is possible to rule out carrier transport, metabolic sink factors, and macromolecular binding as complicating variables that affect distribution throughout development.

However, factors associated with fluid composition and turnover need consideration when comparing V_d data. The lesser density of parenchymal and interstitial elements (i.e., more water, less solids) in neonatal brain may slightly enhance net uptake of tracer from blood by minimizing tissue reflection (resistance to movement) of molecules and thus their backdiffusion into capillary lumina. Second, CSF secretion and thus sink action is less in the immature CNS,[55] and this factor would tend to promote the steady-state accumulation of tracers in CSF. However, in short-term experiments, say 1 hr or less, the CSF effect on nonelectrolyte distribution between plasma and brain in adults is negligible. (For example, acetazolamide in doses that inhibit CSF flow by 50% does not alter the 1 hr V_d of [^{14}C]urea in CSF and brain of adult rats.[53]) Overall, then, it is difficult to see how ontogenetic variation in either fluid turnover or composition could account quantitatively for substantial differences with advancing age in acute uptake by brain of tracer nonelectrolytes.

Ratio analysis of PS products also furnishes evidence for changing BBB permeability postpartum. On the assumption that, for a particular age, the capillary surface area, A, available for molecular diffusion is the same for inulin, mannitol, and urea, then the magnitude of the PS ratio of any two molecules should be identical to their P ratio. For example, $PS_{urea}/PS_{mannitol}$ would be equal to $P_{urea}/P_{mannitol}$. This facilitates permeability comparison among age groups, for which S is a developmental variable. As the BBB undergoes tightening, $P_{urea}/P_{mannitol}$ would be expected to increase. That there is a marked decrease in permeability after the second week can be demonstrated by the increasing $PS_{urea}/PS_{mannitol}$ values in Table VIII, and by the elevated mannitol-to-inulin ratios in Fig. 14 (see Section 4 about V_d ratios).

7.3. Regional Differences in Barrier Permeability in the Developing CNS

There is abundant evidence that adult CNS is characterized by regional differences in CBF, metabolism, and barrier permeability. On the other hand, regional differences in BBB and CP permeability in perinatal mammals has received less attention.[52,87,91,113] At all stages of development the effective permeability, i.e., PS product, for inulin, man-

Table VII. Summary of Permeability–Surface Area Products (PS) for Radioisotopic Ions and Urea in Various Regions of the Developing CNS[a]

	PS (ml/hr per g H_2O)[b,c]							
	Cerebral cortex		Cerebellum		CSF (fast)[d]		CSF (slow)[d]	
	1 wk	Adult	1 wk	Adult	1 wk	Adult	1 wk	Adult
^{22}Na	0.26	0.18	0.28	0.25	0.08	1.13	0.93	0.45
^{36}Cl	0.20[c]	0.14[c]	0.23[c]	0.18[c]	0.19	0.96	0.71[c]	0.58[c]
[^{14}C]Urea	0.48[c]	0.28[c]	0.48[c]	n.m.	0.56	0.30[c]	0.93	0.24[c]

[a]Mean values are compiled from refs. 56, 87, and 113. All data are for Sprague–Dawley rats. n.m., not measured.
[b]Standard errors are generally 5–8% of respective mean values, except for fast-component CSF ^{22}Na and ^{36}Cl, which have a SEM of 0.2–0.3 ml/hr per g.
[c]$p < 0.05$, ^{36}Cl or [^{14}C]urea versus ^{22}Na, by Student's t-test.
[d]CSF was sampled from cisterna magna. Fast and slow refer to components of tracer uptake, obtained by curve stripping procedures. The fast component is thought to represent tracer movement across choroid plexus.

nitol, Na$^+$, and Cl$^-$ (Tables VII and VIII; Figs. 12 and 13) is greater in CER than in CC. Similarly, in infant pigs, the *PS* for unbound bilirubin, and the *PS* for [^{125}I]albumin, is 2–3 times greater in CER than in four other regions tested.[66,67] The greater penetration of solutes into cerebellar tissue may be explained in part by higher density of capillaries (Fig. 6). However, regional differences in capillary wall structure in infants[118] cannot be ruled out because $PS_{urea}/PS_{mannitol}$ in 1-week-old rats is 5.9 in CER and 9.4 in CC (Table VIII).

PS products for ^{22}Na and ^{36}Cl in developing CER and CC can be dissected into their components of rate constant (k, hr^{-1}) and volume of distribution (%) (see Table II). For both ions and at different ages, the V_d is generally less in CER than in CC, whereas the penetration rate constants are 30–40% greater in CER than in CC.

Table VIII. Permeability–Surface Area Products (ml/hr per g H_2O) for Tracer Nonelectrolytes in the Developing Rat CNS, according to Region and Stage of Development[a]

Age	Cisternal CSF	Cerebral cortex	Cerebellum
1 week			
PS_{urea}	0.558[b,c]	0.480[c]	0.480
$PS_{mannitol}$	0.122[c]	0.051[c]	0.082[c]
$PS_{urea}/PS_{mannitol}$	4.6[c]	9.4[c]	5.9
2 weeks			
PS_{urea}	0.456[c]	0.402[c]	0.390
$PS_{mannitol}$	0.103[c]	0.039[c]	0.041[c]
$PS_{urea}/PS_{mannitol}$	4.4[c]	10.3[c]	9.5
Adults			
PS_{urea}	0.300	0.282	n.m.
$PS_{mannitol}$	0.021	0.012	0.015
$PS_{urea}/PS_{mannitol}$	14.3	23.5	—

[a]PS products were calculated from the slope over the initial hour of the respective uptake curves. PS_{urea} values for infant and adult rats were calculated from data in refs. 87 and 56. $PS_{mannitol}$ values were calculated from data in ref. 113. CSF uptake reflects penetration primarily across the blood–CSF barrier, i.e., the choroid plexuses.
[b]For a given region, slope values were compared by the method of Steel and Torrie.
[c]A significant difference ($p < 0.05$) between 1 and 2 weeks versus adults, by multiple range test. n.m., not measured.

The CP is the major transport interface by which many ions and molecules gain access to the CNS.[115] Following IP or intravenous IV injection, tracers rapidly penetrate the CSF; thus, the early uptake by CSF is thought to reflect solute movement across the plexuses (Table VIII). For samples at 1 hr from the cisterna magna, the $PS_{urea}/PS_{mannitol}$ is about 4.5 in 1-week-old and 2-week-old rats. Interestingly, this ratio is similar to that for the free diffusion coefficients, $D_{urea}/D_{mannitol}$, about 4. These observations indicate relatively unrestricted diffusion of small nonelectrolytes across the blood–CSF barrier in infant rats. (For discussion of unrestricted diffusion in mammalian CNS, see refs. 34, 57, and 84.) Between 2 weeks and adulthood, though, the blood–CSF barrier (like the BBB) tightens, as evidenced by the striking rise in $PS_{urea}/PS_{mannitol}$ to a value of 14 (Table VIII).

Two to 3 weeks after birth, the brisk increase in CSF secretion is signaled by the dramatic increase in the movement of Na^+ from plasma to CSF; thus, CSF PS ^{22}Na increases tenfold (Fig. 12). Concurrently, there is no sign of enhanced penetration of ^{22}Na across the vascular bed in either CC or CER (Fig. 12). This striking difference in Na-translocating ability, CP versus BBB, emphasizes that the fluid formation (secondary to Na^+ transport) is the function mainly of the plexus and not the cerebral capillary wall. Fluid elaboration by brain capillaries has been postulated by Bradbury[12]; if extant, it must be a small part of the total fluid production in the CNS.

7.4. Cation versus Anion and Nonelectrolyte Permeation

Table VII summarizes the accessibilities of Na^+, Cl^-, and urea to three CNS regions from plasma in infant and adult rats. ^{22}Na consistently penetrates CC and CER more rapidly than ^{36}Cl, and less fast than does [^{14}C]urea (60 M_r). Thus, the same relative permeativity ranking occurs for infants and adults.

CSF fast-component uptake (via CP) of both radioisotopic Na^+ and Cl^- is augmented by an order of magnitude after 1 week. Simultaneously, radio-urea permeation into CSF decreases by about one-half. Once again, this emphasizes the tightening of the blood–CSF barrier as the active ion transport associated with formation of CSF commences in the CP. With respect to the CSF slow-component of distribution, it is greater in infants than in adults for all three tracers. Less certain is the origin of the CSF slow component penetration (Table VII), one interpretation being that it represents solute movement across the BBB to ISF and eventually to CSF. The cerebral capillary origin of the CSF slow-component permeation is consistent with the fact that all three tracers have a greater PS value in BBB in infants than in adults. An alternative but not mutually exclusive interpretation is that the CSF slow component is associated with solute flux from the circumventricular organs into CSF.

7.5. Anatomic Substrates for More Permeable Interfaces

In the BBB of mature mammals, there is a lack of fenestrae, a paucity of vesicles, and the presence of continuous bands of tight junctions that seal the paracellular route. In the fetal rodent brain, fenestrae are also absent in the endothelia.[118] However, recent evidence has been found for developmental differences in both inter- and intraendothelial structure.[118] In highly permeable vessels outside the brain, blood-borne molecules traverse the capillary wall through leaky intercellular junctions and by tubulovesicular channels (or chains of vesicles in the endothelial cytoplasm). Prior to the postnatal period of 2–3 weeks, both the BBB and CP interfaces in Sprague–Dawley rats permit the ready access (compared with adults) of nonelectrolyte probes to neural tissue and CSF, respectively. Thus, the physical characteristics of each barrier in immature mammals may resemble those of extra-CNS capillaries. However, at the freeze-fracture microscopy level, the tight junction structure in cerebral capillaries, as well as choroidal epithelium, is similar in fetuses and adults.[79] This has led some investigators to conclude that the leakier transport interfaces cannot be attributed to greater patency of the zonulae occludentes; however, maturational differences in junctional configuration at the submicroscopic level are possible. Moreover, with routine electron microscopy and computer-assisted image analysis, Stewart and colleagues[118] recently observed in fetal mouse brain the existence of enlarged junctional clefts that decreased in number markedly by 2 weeks postpartum.

An additional explanation for leakiness is the presence of a specialized endoplasmic reticular (ER) network of tubules or canal-like structures in the perinatal barrier systems. Thus, evidence for transcellular vesicular transport of macromolecules has been presented by Lossinsky et al.[75] for cerebral endothelium in infant rats and by Møllgard for choroidal epithelium in fetal sheep.[80] The developmental disappearance of these vesicular transport mechanisms seems to coincide with the tightening of the barriers. Vesicular density is also higher in fetal versus adult BBB in mice.[118] It is of interest to determine whether such vesicles participate normally (i.e., nonpathophysiologic conditions) in the net transcellular transport of small water-soluble molecules as well as protein markers in the developing CNS.

Also promoting freer diffusion in perinatal animals are the incompletely developed elements (e.g., connective tissue, basement membrane, glial processes) in the interstitial space immediately surrounding the main cellular components of each barrier. In the case of the BBB, the less extensive covering of endothelium by glial foot processes in infants may permit less restricted exchange across the wall.[52,77] The progressive envelopment of capillaries by glial foot processes at 2–3 weeks in the rat coincides with the disappearance of vesicular transport;[75] this suggests an inductive effect by glia to reduce endothelial permeability as the animal matures. Sprouting is extensive between 1 and 3 weeks postpartum as the vessels elongate (Fig. 5); the tips of growing capillaries possibly have permeability properties different from the more proximal regions of the growing vessels. Overall, there is a plethora of anatomic evidence consistent with the physiologic information describing more permeable transport interfaces in perinatal rodent CNS.

7.6. Developmental Permeability: Neural versus Peripheral Tissues

Although the permeability of the BBB in infant rats is considerably greater than in adult counterparts, it is still much less than that in non-neural tissues. To place ontogenetic permeability in CNS in better perspective, it is helpful to compare tracer permeation into brain versus that in various peripheral tissues. In less than 1 hr, ^{36}Cl attains steady-state distribution in both submaxillary salivary gland and skeletal muscle (Fig. 11); this relatively rapid equilibrium distribution between plasma and tissue fluid is a reflection of the highly permeable capillary wall at all stages of development. On the other hand, ^{36}Cl steady-state distribution in CC requires 6–8 hr, even in 1-week-old animals; this suggests that the capillary wall is the rate-limiting step in Cl^- permeation, because at 1 week the glial cells have not yet proliferated (and should therefore not complicate the compartmental analysis of ^{36}Cl uptake).

Corroborating evidence for this postulate can be found from nonelectrolyte tracer permeation phenomena (Table IX). The 1-hr V_d of [^3H]mannitol (and inulin) measures

*Table IX. One-hr [³H]Mannitol Spaces (Volumes
of Distribution) in Neural versus Non-neural
Tissues in Postnatal Sprague–Dawley Rats[a]*

Age (weeks)	Brain	Heart	Liver	Lungs
0.5	6.4 (0.4)	24.6 (0.6)	66.3 (1.0)	36.0 (0.7)
1	5.5 (0.2)	24.6 (0.9)	67.8 (2.6)	32.7 (1.2)
2	5.0 (0.2)	25.4 (0.5)	68.2 (0.6)	35.2 (0.7)
3	4.6 (0.1)	22.8 (0.3)	66.6 (0.3)	36.0 (0.4)
>5	3.9 (0.2)	20.1 (0.4)	67.6 (0.7)	38.2 (0.9)

[a]Data were compiled from fig. 1 and table I in ref. 52. Brain data are for cerebral cortex. Means are for seven rats at each age. SEM are in parentheses. % Mannitol space = $100 \times$ dpm ³H/g wet tissue, divided by dpm ³H/g plasma H_2O.

ECFV not only in CP,[52,58] with its freely permeable capillaries, but also in peripheral tissues,[109] and at all stages of development.[51] However, even 3 days after birth (Table IX), the 1 hr V_d of mannitol in CC is one-fourth and one-sixth of that in heart and lungs, and only one-tenth of that in liver (where mannitol distributes in total tissue water). Since mannitol does not penetrate neurons, this comparative information is strong evidence that the cerebral capillary wall is the rate-limiting step in nonelectrolyte tracer uptake by the infant CNS.

Thus, even though the BBB in infant rats does not have adult level tightness, it apparently offers enough impediment to diffusion to permit relatively efficient operation of carrier transport systems for many organic solutes. Nevertheless, evidence is presented below that K_d (diffusion constant) for some organic nutrients is considerably greater in infants than adults.

7.7. Kinetic Constants for Carrier Transport in Immature Brain

Michaelis–Menten analysis of BUI data has furnished values for K_d, K_m, and V_{max} for several solutes in young rats. Although Daniel et al.[28] described a minor role for glucose diffusion into the CNS of suckling rats, Cremer and colleagues found a nonsaturable diffusion component (K_d) which they thought had physiologic significance, especially in hyperglycemia.[26] Sarna et al.[99] observed a 30% higher value of K_d for tryptophan in suckling versus mature rats. K_d for L-lactate is threefold greater in 2–3-week-old rats compared with adults.[26] Thus, for substrates transported by three different systems, there is considerable diffusional leakage of nutrient into the infant CNS.

Two laboratories have concluded that K_m for the BBB transport of glucose is age invariant.[26,28] For amino acids, K_m data indicate constancy as well as changes in postnatal development, depending upon species and type of transport system. The K_m for Trp (0.25 mM) is not different in suckling versus 6-month rats.[99] The K_m for Arg in neonatal rabbit BBB may correspond to adult affinity. On the other hand, K_m for the neutral amino acid system in newborn rabbits may be an age-related modulation.[14] Banos and colleagues[6] reported K_m values for Arg and Lys three- to fivefold less in 3-week than in adult rats. For the monocarboxylic acids, L-lactate and pyruvate, the K_m is 5 to 10 times greater in immature versus adult rats.[26] Overall, the limited developmental data indicate both lower and higher affinities of various nutrient carriers for their respective substrates in early development.[90]

Similarly, when comparison is made to adulthood, V_{max} is higher for some nutrients, lower for others, as the cerebral capillaries gradually undergo development. As a generalization for infants, V_{max} for glucose is less than the adult level; while the maximal transport capacity for many amino and monocarboxylic acids is greater. Particularly striking is the elevated V_{max} for β-hydroxybutyric acid in suckling rats[83]; this is apparently an inductive phenomenon,[40] since it is sensitive to plasma fluctuations in ketone bodies and acid–base balance secondary to dietary intake.

7.8. Interpretation of BUI Constants for Infants

Interage comparisons of K_m and V_{max} data (see Sections 7.7 and 7.11; see also Tables V, monocarboxylic acids,[98] and VI) must be done with considerable care. In ontogenetic studies, the BUIs obtained with the Oldendorf-type techniques may have variable degrees of error (yet undetermined), if there are differences with age in the extent of mixing of the injected bolus with arterial blood. The plasma composition of nutrients varies substantially with progression of age; thus, this factor could further complicate ontogenetic comparisons of transport constants, especially if there is greater percentage mixing of bolus with blood in infants. Since V_{max} is a function of the number of perfused capillaries as well as the extraction of the reference tracer, such factors have to be evaluated for each developmental stage before valid comparisons can be made with the adult transport capacities for various substrates.

The BUI approach in infant animals has furnished interesting information that needs corroboration or refinement by other methodologies. All ontogenetic techniques and models of the BBB face considerable challenge in sorting out and controlling the continual changes with age in the hemo- and hydrodynamic factors, and in the barrier transport mechanisms.

7.9. CNS Hemodynamic and Hydrodynamic Development

Between 1 and 3 weeks postpartum, the time course of the marked augmentation in cortical capillary surface area and length (per unit weight tissue) is matched by a corresponding rate of rise in CBF. These ontogenetically altered parameters can be depicted as S-shaped curves (Figs. 4 and 5). If cerebral metabolism is intimately coupled to CBF in infants, one would also expect a similar sigmoid curve for glucose transport into developing nervous tissue. Indeed, Moore et al. demonstrated such a relationship in rats between uptake of CH_3-glucose and advancing postnatal age (see Fig. 6 in ref. 82).

In the blood–CSF barrier over the same time period (i.e., second and third weeks after birth), there is a sharp increase in secretory activity by the CP.[54,55,113] Concurrently, there is a progressive increase with development in the ability of the plexus to respond to metabolic[91] and pharmacologic challenges. Also occurring during postnatal weeks 1–3 is the pronounced increase in CSF sink action on nonelectrolytes.[87] The more-or-less concurrent tightening of the blood–CSF barrier with that of the BBB results in the establishment of the CSF sink by 3 weeks postpartum. The similar time course of development of the two major barriers is likely a consequence of their common embryologic development.[10]

By 1 month after parturition, the greater CSF sink action, together with the smaller ECFV (Table I), have the combined effect to lower the V_d of water-soluble solutes penetrating from the blood. For certain charged substances such as iodide, their V_d is diminished even further throughout maturation as excretory systems (i.e., reabsorption oriented from brain or CSF to blood) attain full development in the cells of the barriers.[8,129] Bulk flow of CSF, across arachnoid villi into venous sinuses, increases concomitantly with CSF formation rate; this additional excretory factor facilitates nonselective removal of large water-soluble catabolites and proteins. The kinetics of distribution of proteins in developing CNS has been thoroughly reviewed by Saunders[100] and Saunders and Møllgard.[102]

7.10. Ontogeny of Acid–Base Modulation of Transport

The changes in blood pH that occur during postnatal development can alter BBB permeability to solutes, the activity of inorganic antiport systems, and the kinetic characteristics of organic nutrient transporters. In general, acidosis reduces cell membrane permeability. For example, NH_4Cl-induced metabolic acidosis decreases the rate of penetration of nonelectrolytes (e.g., raffinose) across the adult blood–CSF barrier (unpublished data). Greater acidity of blood can also affect exchange activity of the Na-H and Cl-HCO_3 exchangers in CP as well as the BBB[54,60]; this, in turn, has an effect on cell and CSF pH, fluid formation, and other factors. The chemical activity of H ion in arterial blood also determines the degree of protonation of numerous organic transport systems in the luminal membrane of cerebral capillaries; relationships between extracellular pH and the specific pK_a values of individual transport systems may change throughout development. Such problems deserve further quantification in infants, in whom systemic acid base homeostasis is vulnerable to disruption. When Oldendorf-type techniques are used in developmental studies, particular care should be exercised in matching bolus pH to that obtaining normally in vivo at a particular age.

Inside the brain there are developmental changes in intracellular pH (Fig. 10). Cell pH in neural as well as non-neural tissues is more alkaline than in adults, due mainly to elevated cell HCO_3^-.[61] The more alkaline pH in the cellular compartment is apparently conducive to growth and differentiation. Although CSF pH in infant rats is not well buffered in the face of systemic metabolic alkalosis[59,61] brain tissue pH is maintained during metabolic alkalosis or acidosis.[61] Barrier systems likely participate in brain pH homeostasis in the face of plasma perturbations, even early in development.

7.11. Pharmacologic and Toxicologic Considerations

The scope of this review on development deals primarily with pharmacokinetic (kinetics of distribution) rather than pharmacodynamic (mechanism of cellular function) phenomena in brain. For the fetus, the placental barrier serves as the first line of defense against perturbations in maternal plasma levels of water-soluble compounds, natural or foreign. For the neonate, the blood–milk barrier is the initial defense against hydrophilic drugs. However, once blood-borne in the perinatal vascular system, a pharmacologic agent or toxin will penetrate the BBB at a rate dependent on molecular size, charge, and lipid solubility[96] as well as the integrity (full development or not) of the BBB.

Although the data base for BBB mechanisms in humans is limited, the impression of some clinicians[104] is that human BBB permeability for various substances is significantly greater (compared with adults) in healthy term neonates, even though the growth spurt of CC occurs relatively early in fetal life (Fig. 1). In sick preterm infants (nonmeningitic), dopamine infused intravenously apparently penetrated the BBB readily to increase its concentration in CSF by almost twofold.[104] The substantial penetration was not related to patient illness type (oliguria, hypotension). Because the effective permeability of the BBB is a complex function of many systems, several interpretations are possible: e.g., there may have been leakage of dopamine through the CP; cerebral endothelial tight junctions in premature babies may lack complete sealing[118]; or the MAO degradative activity (i.e., the enzymic BBB) in endothelia may not exist at the adult level.[73] Because of limited comprehensive investigations for ethical reasons, it is difficult to draw sound conclusions about the integrity of the BBB in perinatal humans. Anecdotal information leads to clinical impressions, but more systematic studies will provide rational therapeutic strategies.

Valproic acid, an anticonvulsant, has a BUI of 47% and 17%, respectively, in newborn and adult rabbits.[24] This higher degree of extraction in neonates has been attributed to lower CBF at birth. With regard to plasma protein-binding effects, greater brain uptakes were observed when the unbound fraction of the drug was increased;[24] this is significant in the context that infant plasma protein levels are generally much lower than in adults.[3] Another interesting finding by Cornford et al. was that the BUI for [^{14}C]valproate (relative to [^{3}H]water) decreased with time after carotid injection during a 4-min washout period, suggesting that the antiluminal permeability (i.e., reabsorptive transport back to blood) was much greater than luminal membrane uptake. This result points to the need for careful pharmacologic interpretation of BUI data for drugs obtained in single-pass experiments where only unidirectional uptake is analyzed. Since therapeutic concentrations in neural tissue are also a function of excretory transport,[8] there is the need for reabsorptive and steady-state analyses in addition to uptake after a single-circulatory pass of bolus.

Cornford et al.[22] compared lipid-mediated BBB penetrability in neonates and adults. Uptake indices for 16 drugs were quantified in newborn rabbits, and a highly significant correlation was found between BUI and octanol–saline partition coefficients of the agents. Derived *PS* products revealed no difference in permeability between newborn and adult BBB. However, the lipoid properties of the neonatal BBB may differ from the adult, since the relationship between H-bond number and BUI determined for adult rat brain was not corroborated for the immature rabbit.[22]

The pharmacokinetics of lipid-soluble agents vary significantly with developmental stage. In adult rats, the peak concentration of apomorphine in brain is present at 5 min, and it declines with a half-life of 10 min.[121] In 1-week-old animals given the same dose (10 mg/kg), the peak concentration, attained at 10 min, decreases exponentially ($t_{1/2} = 28$ min). When immature rats are compared with mature ones, the delay in both onset and offset, respectively, of apomorphine effects on CNS function can likely be attributed to lower CBF and rate of hepatic enzyme degradation of drug.

Methadone, also lipid-soluble, penetrates the BBB readily in adults and infants.[92] However, even with stable and comparable levels in plasma, methadone attained a peak concentration in fetal and neonatal rats that was two- to fourfold greater than in adults.[92] The amount of methadone concentrating in brain from a single dose of 5 mg/kg increased with advancing postnatal age until day 15, and then decreased to mean adult levels by 1 month. The rapid deposition of cholesterol (greatest at 2 weeks) may promote the buildup of lipid-soluble drugs like methadone, with accompanying toxicity to infants. Chronic treatment of mothers during pregnancy resulted in offspring with delayed development of processes associated with the restrictive effects of the BBB.[92]

An increasingly important facet of neurotoxicology is the undesired accessibility of various pharmacologic agents and toxins to sensitive cellular elements in the developing brain. Although an extensive summary of neurotoxicology is beyond the scope of this review, examples can be cited for lead intoxication. During development, the CER is especially vulnerable to exposure to lead (Pb). Although pharmacodynamic differences among brain regions need to be considered, it is of pharmacokinetic interest that tightening of cerebellar capillaries lags that in CC.[52,118] Thus, more patent zonulae occludentes[118] in CER for a longer period may place this region at greater ontogenetic risk than other parts of the CNS. Exposure to lead may also alter the capillary wall per se. Postnatal administration of Pb^{2+} to young rats via maternal milk enhanced BBB permeability to Trypan blue.[126] When infant rats were exposed for 2 days to lead acetate (1 mg Pb/g wt per day), the endothelia of isolated brain capillaries showed grossly abnormal morphology; however, growing animals exposed for 3 weeks to high doses of Pb were able to recover from the initial capillary pathology.[125] The more recent study by Lefauconnier et al.[69] confirmed that relatively high doses of Pb^{2+} to infants caused hemorrhagic encephalopathy, especially in CER; the damage was reversible if Pb^{2+} administration was stopped. The toxicity at lower doses (10 μg) was attributed to a direct effect on brain cells, because they established that tissue $[Pb^{2+}]$ was elevated even when capillary transport of several metabolic substrates was unaffected.[69] Lead poisoning of the immature brain is a complex function of pharmacodynamic as well as pharmacokinetic factors.

7.12. Pathophysiology of the Developing Barriers

The effects of hypoxia, hypercapnia, or acidosis on cerebral hemodynamics, metabolism, and barrier permeability have been less extensively studied in infants than in adults. Recent investigations have established that the neonatal piglet is a convenient model for evaluating pathophysiologic phenomena in the unanesthetized state.[16] The stage of development of the porcine brain at birth is comparable to the human newborn at 36 weeks gestation. Preliminary studies indicate that in severe acute hypoxia, there is preferential protection of oxygenation of the CC at the expense of peripheral organs and that there is normal uptake of glucose by cortex.[43] This implies that the glucose transport capability of the neonatal BBB remains relatively intact when immature pigs are rendered severely hypoxic.

Does hypercapnia increase BBB permeability in infants? When arterial P_{CO_2} was elevated from 35 to 70 mm Hg for 1 hr, there was no change in the permeation of [125I]albumin in the cerebrum, CER, thalamus, midbrain, or brain stem of neonatal piglets.[16] However, in adult rats subjected to arterial P_{CO_2} of 100 mm Hg, there was a twofold enhancement of [125I]albumin concentration in the brain.[13] In explaining the different permeativities, one must consider species variation in the sensitivity of the vasculature to P_{CO_2} (threshold effect?) in addition to developmental differences.

Acidosis can affect several interacting variables (e.g., CBF, BBB permeability, plasma-protein binding, degree of ionization of drug), and so acid–base balance needs to be carefully assessed in studies of the developing BBB. This can be exemplified by bilirubin distribution. One risk factor thought to increase bilirubin toxicity is acidosis. In human neonates, greater binding of bilirubin to albumin is achieved upon correction of systemic acidosis. Induced respiratory and metabolic acidosis increased [14C]bilirubin uptake by brain in newborn guinea pigs.[31]

Respiratory acidosis has caused gross kernicterus in puppies.[71] The more recent systematic investigation by Bratlid et al.[13] in adult rats demonstrated that respiratory acidosis, but not metabolic acidosis, led to augmented permeation of bilirubin and albumin across the BBB; moreover, neither acidosis nor hypercapnia altered the level of apparent unbound bilirubin. Neonatal CBF is elevated in hypercapnia, and blood-flow limited distribution of bilirubin in several brain regions has been described by Burgess et al.[16] Kernicterus lesions are usually found in subcortical regions that receive the highest blood flow during asphyxia and hypercapnia. Overall, acidosis per se seems to have less effect on bilirubin permeation into CNS than does hypercapnia.

The longstanding controversy about the immature BBB and bilirubin penetration has generally not considered the functional implications of the incompletely developed choroid plexus–CSF system in neonates. Factors that reduce CSF formation (hypoxia) and increase blood–CSF barrier permeability (hypoxia and/or hypercapnia) also promote the uptake and retention of bilirubin in the CNS. The permeability of the blood–CSF barrier in dogs subjected to hypercapnia is much more markedly increased in infants than in adults.[71] Thus the immaturity of and alterations in CP function are significant factors that, together with hyperbilirubinemia, probably play a significant role in the production of bilirubin encephalopathy. Attenuated CSF sink action in infants could exacerbate retention of free and bound bilirubin in the brain. Again, this points to the necessity of considering blood–CSF barrier as well as BBB function when interpreting distributional phenomena in the developing CNS.

7.13. The Rat as an Ontogenetic BBB Model

A wide spectrum of developmental processes occurs in the rat between 1 and 3 weeks after birth.[129,130] During this interval, there is enhancement of blood supply to the CNS, rapid growth of the neuropil, proliferation of glia, reduction in ECFV, tightening of the BBB (Fig. 14) and blood–CSF barrier, and maturation of the CSF secretory process and sink action. Many of these events are functionally interdependent, so it is advantageous to study as many variables as possible in a single experiment.

Thus, the rat during infancy is ideal for analyzing endothelial–glial interactions in the immature BBB and for delineating unique and common transport systems in neurons, glia, and CP, because substantial changes take place (i.e., are condensed) in a relatively short period. By analyzing the series of events associated with the development of a particular physiologic parameter, a better understanding is gained about its operational significance in the mature system. The parallelism in the time course of development of the BBB and blood–CSF barrier strongly suggests an intimate functional relationship between them.

Sigmoid curves depict the rapidly building vascularity and perfusion of the CNS after the first week, undoubtedly to supply greater metabolic demands by the expanding cellular compartment. Cellular growth, by both number and volume, can be demonstrated microscopically as well as physiologically, by fluid and electrolyte concentration profiles (Fig. 15). Na–K–ATPase and carbonic anhydrase activities in various regions undergo a sharp rise between 1 and 3 weeks,[129,130] signaling the emergence of fluid and ion homeostatic transport mechanisms in association with the tightening of the barrier systems.

The developmental discontinuity in half-time values for the distribution of several solutes (Fig. 16) is a reflection of age-related dynamic changes in these physiologic factors. Thus, penetration kinetics is dependent upon the set of functional characteristics of the CNS at a particular stage of development.[87] The quantitative similarity between 1-week-old and adult half-time values for brain tissue and CSF (slow component), for various hydrophilic tracers (Fig. 16), probably relates to the condition that lesser vascularity, A, (and decreased CBF), together with greater permeability, P, at 1 week is

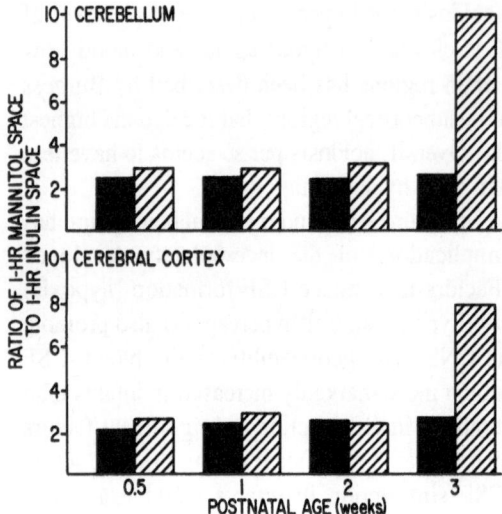

Figure 14. Ratio of 1-hr volumes of distribution, V_d, mannitol to inulin, in the cerebellum and cerebral cortex of developing Sprague–Dawley rats.[52] Each bar is the mean for five animals. Shaded bars represent V_d uncorrected, and diagonal-lined bars corrected, for residual blood[58] in sampled tissues. With advancing age, the progressively increasing blood content of tissues, together with steadily decreasing penetration of tracers across the developing BBB, make it critically important to correct for activity in residual blood of tissue samples. For example, note the importance of the correction, i.e., the resultant marked difference in ratio analysis, for the oldest animals.

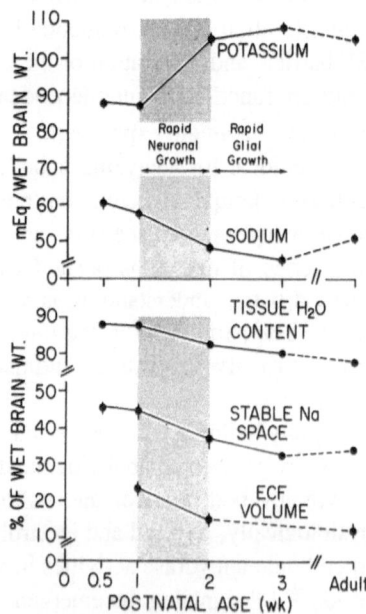

Figure 15. Dynamic phase of brain growth in postnatal Sprague–Dawley rats, as manifested by alterations of electrolyte and fluid profiles in brain tissue. Between 1 and 3 weeks postpartum, there is a substantial increase in [K], and a decrease in [Na], [H_2O], and extracellular fluid (ECF). Such changes reflect displacement of ECF by cellular fluid associated with growth of neuronal cell processes and glial proliferation/myelination. The curves are plots of original data.[58,113] Although there is obviously overlap between neuronal and glial growth spurts, the graph depicts demarcation between the two main growth periods in order to emphasize the predominant cellular activity at the respective stages of development. Carbonic anhydrase activity, a marker for glial growth, increases substantially between 12 and 24 days after birth.[129,130] Means ±SEM for five or more rats at each age.

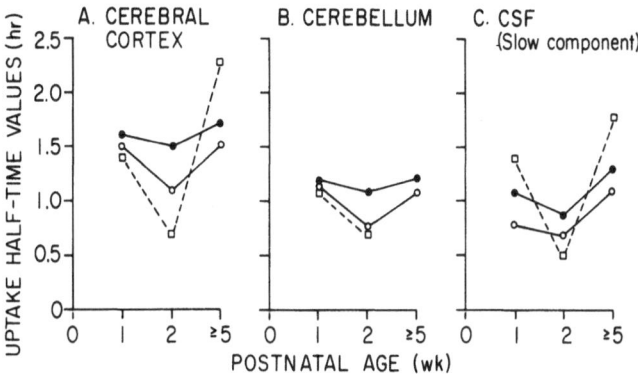

Figure 16. Half-time values for curves describing slow-component uptake by brain (and CSF) of tracers held at constant level in plasma of Sprague–Dawley rats. Infant animals (1 and 2 weeks) are compared with young adults (> 5 weeks). By curve-stripping procedures, the rate constant (k, hr^{-1}) was determined as the slope of the linear regression analysis of subtracted points (see Fig. 8 and ref. 108). Half-time ($t_{1/2}$,hr) = ln $2/k$. (□) [^{14}C]urea[60,87]; (●) ^{22}Na; and (○) ^{36}Cl.[113]

approximately countered (in terms of promoting permeation of tracer) by a larger A (and increased CBF) concurrent with the smaller P in adulthood. On the other hand, equilibration half-times are shortest at 2 weeks when A, CBF, and P collectively maximize uptake of water-soluble solutes by the brain.[87] Although the rapidly changing physiology in the postnatal rat can complicate kinetic analyses, it does represent an opportune model for delineating the respective factors affecting BBB permeation.

Another challenge to the kineticist is to evaluate geometric factors associated with distributional phenomena. The relative sizes and anatomic configurations of various CNS compartments undergo continual change in perinatal development. The size of the CP-CSF system, relative to that of brain, is considerably greater early in mammalian development than in adulthood. Theoretically, then, blood–CSF barrier function in fetuses should have a marked effect on distribution of solutes in brain. Factors such as the larger volume of ventricular CSF (and perhaps also CP secretion as well), relative to brain size, could effectively mask true BBB permeability.[101] Following the introduction of tracers to the vascular system, the rate of buildup of activity in CSF is determined by ventricular volume and CSF formation (itself related to blood flow).[87] The equilibrium distribution of tracer between CSF and brain is also a function of diffusion distances within the brain and the ependymal surface area and permeability, both of which are substantially altered throughout development.[36,103]

7.14. Future Research Directions

Although BBB and CP functions in infants continue to be the subject of considerable efforts, the physiology of the barrier systems in fetal mammals is still poorly understood. How does the relatively avascular brain in the fetus receive adequate nutrition? The effective permeability, PS, consists of components of capillary permeability and surface area. At early stages of development, the lower value of S (Fig. 5) may be compensated for by greater magnitude of P, e.g., more patent tight junctions,[118] or transendothelial tubulovesicular transport.[75] Indeed, postnatally (1–6 weeks), the cortical P_{Na} decreases in proportion to the increase in S; consequently, the PS is not significantly altered over time (Fig. 12). Natural adjustment of P and S in the BBB is one way to regulate flow of solutes into the developing CNS.

Moreover, since the turn of the century, when Goldman proposed that the flux of nutrients between blood and brain occurs at the CP, evidence has gradually accumulated in support of the hypothesis that transport at the blood–CSF barrier is vital to optimal function of the perinatal CNS. With respect to trophic function, the relative contribution of CP and cerebral capillaries to neuronal thriving is an important issue in neurobiology. Proteins secreted by the CP are thought to modulate brain cell growth.[103] Precursors for nucleic metabolism enter the CNS by way of CP.[115] Even in adults, the CP is the major transport interface for moving many hydrophilic solutes into the CNS (see Chapter 9). Yet, Michaelis–Menten analysis of transport systems in CP (and BBB) of perinatal mammals has rarely been done.[74] Overall, individual as well as correlative characteristics in the two major barrier systems need further analysis, especially for prenatal life.

The movement of hormones, polypeptides, and vitamins into the CNS of adult mammals has been investigated mainly by Pardridge and Mietus[89] and Spector.[114] Ontogenetic analysis in this field has received scant attention, but preliminary data indicate that neonatal rabbit CP can concentrate (and perhaps transport) thyroid hormone thyroxine (T4) and several water-soluble vitamins as effectively as the adult.[114,116] Proteins unique to immature animal plasma may modulate the movement of neurohumoral substances into the CNS. For example, neonatal rat serum contains a factor, presumably α-fetoprotein, that restricts estradiol influx into the developing brain.[78] The possibility of neurologic handicap has been linked to the penetration of maternal hormones (and antibodies) across a leaky blood–CSF barrier in human infants.[1] Such problems deserve pursuit because hormonal imbalances associated with pathophysiologic maternal states may be transmitted to fetal CNS at a time when cell differentiation is not yet complete, hence vulnerable. Additional insights need to be gained about the role of the BBB in mediating neuroendocrine and immunologic effects on developing neuronal networks. Clearly, there are many promising pharmacokinetic vistas in regard to future research on barrier properties in the immature CNS.

At the other end of the spectrum of life lie the physiologic phenomena associated with the aging BBB. In extremely advanced age, or in diseased geriatric patients, does the BBB break down and revert to the leakiness present at very early stages of development? For example, are the elevated levels of aluminum in the brains of Alzheimer patients the result of compromised BBB (or CP) function? These clinically relevant questions need to be addressed with animal models. To date, the experimental information obtained from old healthy Fischer-344 rats (24 months) indicates that the integrity of the BBB is maintained in respect to both permeability (PS Na)[111] and to carrier-mediated transport systems (K_m and V_{max} of large neutral amino acids).[122]

A particularly exciting prospect is the use of noninvasive techniques to evaluate solute movement into the CNS of perinatal and geriatric patients. The use of NMR and PET methodologies should help characterize the developing and aging BBB, while avoiding many of the ethical problems associated with invasive research on humans. Osmotic opening of the BBB may prove to be feasible in infants as well as adults, since acute urea-induced hyperosmolality has been shown to enhance albumin permeation into newborn piglet brain.[17]

8. RECAPITULATION AND PERSPECTIVE

This review has summarized evidence that the second and third weeks postpartum in the rat is the watershed period during which the BBB and blood–CSF barrier both undergo rapid and extensive evolution to adult-like characteristics. Especially intriguing are the apparent dynamic modulations of V_{max} and K_m for carrier transport systems in developing cerebral capillaries. Adaptation of the recently available in situ perfusion technique,[123] together with modified Oldendorf and arterial integral methodologies for application to neonatal rats,[70] should stimulate pursuit of developmental regional analyses of CBF and uptake of 2-DG and various metabolites.

The earliest studies of colloid and crystalloid substance penetration into perinatal CNS generated a wide range of interpretations about the nature of the BBB in the immature brain. When historical information about protein (or dye) permeation is evaluated, it is essential to consider probe dosage and binding, experimental duration, species, and specific region of brain analyzed. The lingering controversy of immature versus mature BBB in perinatal mammals is largely semantic in nature, and can be reconciled by more precise definition of terms (e.g., barrier) as well as to rigorous comparisons of experimental conditions. Cerebral capillaries in infant rodents do not allow ready access of proteins to brain substance. Even in the youngest fetuses, the tight junctions undoubtedly restrict penetration of proteins. Thus, the paracellular aspect of the barrier is probably established very early in development. The transcellular aspect of the barrier, however, is altered in ontogeny, probably by the appearance and then disappearance of complex endoplasmic-reticular systems. This transcellular route, particularly in the incompletely developed CP, seems to allow movement of plasma proteins into CSF, thereby rendering the plexus epithelial membrane permeable to macromolecules. The lack of development or maintenance of BBB (to protein) in transplanted fetal cortical tissue points to the need for assessing permeability characteristics of the vessels in the transplant as well as in the zone interfacing the tissues of the donor and the recipient.[98a] Although the BBB to lipid-insoluble molecules in infant rats is considerably more leaky than in adults, nevertheless, the cerebral capillary wall is rate limiting for the distribution of most ions, nonelectrolytes, and nutrients. To investigate highly permeable cerebral capillaries, it will be necessary to analyze fetal models in which, for example, inulin attains a V_d of 40% in brain.[36] Although the knowledge of capillary P has intrinsic value, the more practical goal is to understand the complex array of factors which govern the composition of brain extracellular fluid throughout development.

Raichle[95] hypothesized that three main cell groups (brain endothelium, astroglia, and choroidal epithelium) perform in a coordinated manner to regulate the internal environment of the brain. Ontogenetic analysis of such a central neuroendocrine system, for controlling brain volume, ion homeostasis, and metabolism, is one of the most fascinating and challenging problems in developmental neurobiology. Recently, tight junctions have been found between ependymal cells in the fetus,[103] whereas the more permeable gap junctions are found consistently in the ependymal wall of all adult mammals. Thus, the demonstration of a tight brain–CSF interface very early in prenatal development emphasizes the complexity and continually changing nature of another barrier system (in addition to the BBB and CP) for ensuring the requisite environment for growing neurons.

REFERENCES

1. Adinolfi M: Neurological handicap and permeability of the blood–cerebrospinal fluid barrier during fetal life to maternal antibodies and hormones. *Dev Med Child Neurol* 18:243–246, 1976.
2. Amtorp O: Transfer of ^{125}I-albumin from blood into brain and cerebrospinal fluid in newborn and juvenile rats. *Acta Physiol Scand* 96:399–406, 1976.

3. Amtorp O, Sorensen SC: The ontogenetic development of concentration differences for protein and ions between plasma and cerebrospinal fluid in rabbits and rats. *J Physiol (Lond)* 243:387–400, 1974.

4. Bakay L: Studies on blood–brain barrier with radioactive phosphorus III. Embryonic development of the barrier. *Arch Neurol Psychiatry* 70:30–39, 1953.

5. Bakay L: Studies in sodium exchange. *Neurology (NY)* 10:564–571, 1960.

6. Banos G, Daniel PM, Pratt OE: The effect of age upon the entry of some amino acids into the brain, and their incorporation into cerebral protein. *Dev Med Child Neurol* 20:335–346, 1978.

7. Bar T: *The Vascular System of the Cerebral Cortex.* Springer-Verlag, Berlin, 1980.

8. Bass NH, Lundborg P: Postnatal development of mechanisms for the elimination of organic acids from the brain and cerebrospinal fluid system of the rat: Rapid efflux of [^{3}H]Para-aminohippuric acid following intrathecal infusion. *Brain Res* 56:285–298, 1973.

9. Betz AL, Goldstein GW: Developmental changes in metabolism and transport properties of capillaries isolated from rat brain. *J Physiol (Lond)* 312:365–376, 1981.

10. Bradbury MWB: Ontogeny of mammalian brain barrier systems. In Cserr HF, Fenstermacher JD, Fencl V, (eds): *Fluid Environment of the Brain.* Academic, New York, 1975, pp. 81–103.

11. Bradbury M: The blood–brain barrier during the development of the individual and evolution of the phylum. In: *The Concept of a Blood–Brain Barrier.* Wiley, New York, 1979, pp. 289–322.

12. Bradbury MWB: The blood–brain barrier transport across the cerebral endothelium. *Circ Res* 57(2):213–222, 1985.

13. Bratlid D, Cashore WJ, Oh W: Effect of acidosis on bilirubin deposition in rat brain. *Pediatrics* 73:431–434, 1984.

14. Braun LD, Cornford EM, Oldendorf WH: Newborn rabbit blood–brain barrier is selectively permeable and differs substantially from the adult. *J Neurochem* 34:147–152, 1980.

15. Broman T: The possibilities of the passage of substances from the blood to the central nervous system. (Is there a blood–brain barrier and a blood–cerebrospinal fluid barrier?) *Acta Psychiatry Neurol* 16:1–25, 1941.

16. Burgess GH, Oh W, Bratlid D, et al: The effects of brain blood flow on brain bilirubin deposition in newborn piglets. *Pediatr Res* 19:691–696, 1985.

17. Burgess GH, Stonestreet BS, Cashore WJ, et al: Brain bilirubin deposition and brain blood flow during acute urea-induced hyperosmolality in newborn piglets. *Pediatr Res* 19:537–542, 1985.

18. Burton SA, Johanson CE: A new technique to evaluate regional blood–brain barrier permeability to zinc. *Soc Neurosci Abst* 6:830, 1980.

19. Burton SA, Johanson CE: Regional distribution of calcium in the CNS of infant and adult rats. *Soc Neurosci Abst* 7:88, 1981.

20. Caley DW, Maxwell DS: Development of the blood vessels and extracellular spaces during postnatal maturation of rat cerebral cortex. *J Comp Neurol* 138:31–48, 1970.

21. Cornford EM, Braun LD, Oldendorf WH: Developmental modulations of blood–brain barrier permeability as an indicator of changing nutritional requirements in the brain. *Pediatr Res* 16:324–328, 1982

22. Cornford EM, Braun LD, Oldendorf WH, et al: Comparison of lipid-mediated blood–brain-barrier penetrability in neonates and adults. *Am J Physiol* 243:C161–C168, 1982.

23. Cornford EM, Pardridge WM, Braun LD, et al: Increased blood–brain barrier transport of protein-bound anticonvulsant drugs in the newborn. *J Cereb Blood Flow Metab* 3:280–286, 1983.

24. Cornford EM, Diep CP, Pardridge, WM: Blood–brain barrier transport of valproic acid. *J Neurochem* 44:1541–1550, 1985.

25. Cremer JE, Braun LD, Oldendorf WH: Changes during development in transport processes of the blood–brain barrier. *Biochim Biophys Acta* 448:633–637, 1976.

26. Cremer JE, Cunningham VJ, Pardridge WM, et al: Kinetics of blood–brain barrier transport of pyruvate, lactate and glucose in suckling, weanling and adult rats. *J Neurochem* 33:439–445, 1979.

27. Cserr HF, Bundgaard M: Blood–brain interfaces in vertebrates: A comparative approach. *Am J Physiol* 246:R277–R288, 1984.

28. Daniel PM, Love ER, Pratt OE: The effect of age upon the influx of glucose into the brain. *J Physiol (Lond)* 274:141–148, 1978.

29. David S, Nathaniel EJH: Development of brain capillaries in euthyroid and hypothyroid rats. *Exp Neurol* 73:243–253, 1981.

30. Davson H: Ontogeny of the blood–brain barrier. In Boreus L (ed): *Fetal Pharmacology.* Raven, New York, 1973, pp. 75–88.

31. Diamond I, Schmid R: Experimental bilirubin encephalopathy: The mode of entry of bilirubin-^{14}C into the central nervous system. *J Clin Invest* 45:678–689, 1966.

32. Dobbing J: The blood–brain barrier. *Physiol Rev* 41:130–188, 1961.
33. Durbin GM, Dziegielewska KM, Evans CAN, et al: Penetration of labelled protein from blood into foetal brain and CSF. *Proc Physiol Soc* 25P–26P, 1975.
34. Dziegielewska KM, Evans CAN, Malinowaksa DH, et al: Studies of the development of brain barrier systems to lipid insoluble molecules in fetal sheep. *J Physiol (Lond)* 292:207–231, 1979.
35. Evans CAN, Reynolds JM, Reynolds ML, et al: The development of a blood–brain barrier mechanism in foetal sheep. *J Physiol (Lond)* 238:371–386, 1974.
36. Ferguson RK, Woodbury DM: Penetration of ^{14}C-inulin and ^{14}C-sucrose into brain, cerebrospinal fluid, and skeletal muscle of developing rats. *Exp Brain Res* 7:181–194, 1969.
37. Flexner LB: The development of the cerebral cortex: A cytological, functional, and biochemical approach. *Harvey Lect* 47:156–179, 1952.
38. Freedman LS, Samuels S, Fish I, et al: Sparing of the brain in neonatal undernutrition: Amino acid transport and incorporation into brain and muscle. *Science* 207:902–904, 1980.
39. Fries BA, Chaikoff IL: The phosphorus metabolism of the brain as measured with radioactive phosphorus. *J Biol Chem* 141:479–485, 1941.
40. Gjedde A, Crone C: Induction processes in blood–brain transfer of ketone bodies during starvation. *Am J Physiol* 229:1165–1169, 1975.
41. Gjedde A, Fuglsang A: Blood flow and hexose transfer to the newborn rat brain. *J Cereb Blood Flow Metab* 5 (suppl): (abstr), 1985.
42. Glowinski M, Axelrod J, Kopin IJ, et al: Physiological disposition of ^{3}H-norepinephrine in the developing rat. *J Pharm Exp Ther* 146:48–53, 1964.
43. Goldstein M, Oh W, Piva DL, et al: The effects of hypoxia (H) and hyperviscosity (HV) on cerebral cortex (CC) and gastrointestinal (GI) tract oxygenation and glucose metabolism. *Pediatr Res* 20:348(abstr), 1986.
44. Graziani LJ, Escriva A: Calcium exchange between brain and blood in cats and immature and adult rats. *Neurology (NY)* 19:314–315, 1969.
45. Gregoire NM, Gjedde A, Plum F, et al: Cerebral blood flow and cerebral metabolic rates for oxygen, glucose, and ketone bodies in newborn dogs. *J Neurochem* 30:63–69, 1978.
46. Hansen NB, Brubakk AM, Bratlid D, et al: The effects of variations in $PaCO_2$ on brain blood flow and cardiac output in the newborn piglet. *Pediatr Res* 18:1132–1136, 1984.
47. Hellmann J, Vannucci RC, Nardis EE: Blood–brain barrier permeability to lactic acid in the newborn dog: Lactate as a cerebral metabolic fuel. *Pediatr Res* 16:40–44, 1982.
48. Hernandez MJ, Brennan RW, Bowman GS: Autoregulation of cerebral blood flow in the newborn dog. *Brain Res* 184:199–202, 1980.
49. Himwich WA, Petersen JC, Allen ML: Hematoencephalic exchange as a function of age. *Neurology (NY)* 7:705–710, 1957.
50. Johanson CE: Choroid epithelial cell pH. *Life Sci* 23:861–868, 1978.
51. Johanson CE: Distribution of fluid between extracellular and intracellular compartments in the heart, lungs, liver and spleen of neonatal rats. *Biol Neonate* 36:282–289, 1979.
52. Johanson CE: Permeability and vascularity of the developing brain: Cerebellum vs. cerebral cortex. *Brain Res* 190:3–16, 1980.
53. Johanson CE: Differential effects of acetazolamide, benzolamide and systemic acidosis on hydrogen and bicarbonate gradients across the apical and basolateral membranes of the choroid plexus. *J Pharmacol Exp Ther* 231:502–511, 1984.
54. Johanson CE: Ontogenetic changes in the Na and H gradients across basolateral membrane of choroid plexus: A clue to the maturation of the CSF secretory process. *Soc Neurosci Abst* 11(2):839, 1985.
55. Johanson CE, Woodbury DM: Changes in CSF flow and extracellular space in the developing rat. In Vernadakis A, Weiner N (eds): *Drugs and the Developing Brain.* Plenum, New York, 1974, pp. 281–287.
56. Johanson CE, Woodbury DM: Uptake of ^{14}C-urea by the in vivo choroid plexus–cerebrospinal fluid–brain system: Identification of sites of molecular sieving. *J Physiol (Lond)* 275:167–176, 1978.
57. Johanson CE, Foltz FM, Thompson AM: The clearance of urea and sucrose from isotonic and hypertonic fluids perfused through the ventriculo–cisternal system. *Exp Brain Res* 20:18–31, 1974.
58. Johanson CE, Reed DJ, Woodbury DM: Developmental studies of the compartmentalization of water and electrolytes in the choroid plexus of neonatal rat brain. *Brain Res* 116:35–48, 1976.
59. Johanson CE, Woodbury DM, Withrow CD: Distribution of bicarbonate between blood and cerebrospinal fluid in the neonatal rat in metabolic acidosis and alkalosis. *Life Sci* 19:691–700, 1976.
60. Johanson CE, Parandoosh Z, Smith QR: Cl–HCO$_3$ exchange in choroid plexus: Analysis by the DMO method for cell pH. *Am J Physiol (Renal Fluid Electrolyte Physiol* 18) 249:F478–F484, 1985.

61. Johanson CE, Allen J, Withrow CD: Regulation of pH and HCO_3 in brain and CSF of the developing mammalian central nervous system. *Dev Brain Res* 466:255–264, 1988.
62. Katzman R, Leiderman PH: Brain potassium exchange in normal adult and immature rats. *Am J Physiol* 175:263–270, 1953.
63. Klosovskii BN: *The Development of the Brain and Its Disturbance by Harmful Factors.* Macmillan, New York, 1963, pp. 83–105.
64. Kuttner R, Sims JA, Gordon MW: The uptake of a metabolically inert amino acid by brain and other organs. *J Neurochem* 6:311–317, 1961.
65. Lajtha A: Amino acid and protein metabolism of the brain. II. The uptake of L-lysine by brain and other organs of the mouse at different ages. *J Neurochem* 2:209–215, 1958.
66. Lee C, Oh W, Stonestreet BS, et al: Maturation of blood–brain barrier (BBB) permeability for ^{125}I-albumin(ALB)-bound bilirubin (BR) in piglets. *Pediatr Res* 20:353(abstr), 1986.
67. Lee C, Stonestreet BS, Outerbridge E, et al: Postnatal maturation of the blood–brain barrier (BBB) for unbound bilirubin (BR) in piglets. *Pediatr Res* 20:353(abstr), 1986.
68. Lefauconnier JM, Bernard G: Evolution of cerebral blood flow and blood–brain clearance of phenylalanine and alanine during rat development. *J Cereb Blood Flow Metab.* 5 (suppl. 1), 1985.
69. Lefauconnier JM, Lavielle E, Terrien N, et al: Effect of various lead doses on some cerebral capillary functions in the suckling rat. *Toxicol Appl Pharmacol* 55:467–476, 1980.
70. Lefauconnier JM, Trouve R: Developmental changes in the pattern of amino acid transport at the blood–brain barrier in rats. *Dev Brain Res* 6:175–182, 1983.
71. Lending M, Slobody LB, Meston J: The relationship of hypercapnia to the production of kernicterus. *Dev Med Child Neurol* 9:145–151, 1966.
72. Levin E: Are the terms blood–brain barrier and brain capillary permeability synonymus? In Bito LZ, Davson H, Fenstermacher JD (eds): *The Ocular and Cerebrospinal Fluids.* Academic, London, 1977. (Supplement to *Exp Eye Res* 25:191–199, 1977.)
73. Loizou LA: Uptake of monoamines into central neurones and the blood–brain barrier in the infant rat. *Br J Pharmacol* 40:800–813, 1970.
74. Lorenzo AV, Smoly–Caruthers J, Greene E: Development of amino acid transport mechanisms in the choroid plexus. In Cserr H (ed): *Fluid Environment of the Brain,* Academic, London, 1975, pp. 167–180.
75. Lossinsky AS, Vorbrodt AV, Wisniewski HM: Ultracytochemical studies of transendothelial transport in cerebral micro-blood vessels during development. *J Cell Biol* 99:283(abstr), 1984.
76. Lou HC, Lassen NA, Friis–Hansen B: Impaired autoregulation of cerebral blood flow in the distressed newborn infant. *J Pediatr* 94:118–121, 1979.
77. Luciano DS: Sodium movement across the blood–brain barrier in newborn and adult rats and its autoradiographic localization. *Brain Res* 9:334–350, 1968.
78. McCall AL, Han SJ, Millington WR, et al: Nonsaturable transport of [³H]oestradiol across the blood–brain barrier in female rats is reduced by neonatal serum. *J Reprod Fertil* 61:103–108, 1981.
79. Møllgard K, Saunders NR: Complex tight junctions of epithelial and of endothelial cells in early foetal brain. *J Neurocytol* 4:453–468, 1975.
80. Møllgard K, Saunders NR: A possible transepithelial pathway via endoplasmic reticulum in foetal sheep choroid plexus. *Proc R Soc Lond [Biol]* 199:321–326, 1977.
81. Monin P, Daval JL, Vert P, et al: Study of blood–brain tissues [H^+] gradient during respiratory alkalosis and experimental seizures in newborn piglets: Effect on phenobarbital distribution. *Dev Pharmacol Ther* 7(suppl 1):185–189, 1984.
82. Moore TJ, Lione AP, Regen DM, et al: Brain glucose metabolism in the newborn rat. *Am J Physiol* 221:1746–1753, 1971.
83. Moore TJ, Lione AP, Sugden MC, et al: β-hydroxybutyrate transport in rat brain: Developmental and dietary modulations. *Am J Physiol* 230:619–630, 1976.
84. Murphy V, Johanson CE: Adrenergic-induced enhancement of brain barrier system permeability to small nonelectrolytes: Choroid plexus versus cerebral capillaries. *J Cereb Blood Flow Metab* 5:401–412, 1985.
85. Olsen OM, Sörensen SC: Stability of [Mg^{2+}] in cerebrospinal fluid during plasma changes and during hypercapnia in young and in adult rats. *Acta Physiol Scand* 82:466–469, 1971.
86. Olsson Y, Klatzo I, Sourander P, et al: Blood–brain barrier to albumin in embryonic new born and adult rats. *Acta Neuropathol (Berl)* 10:117–122, 1968.
87. Parandoosh Z, Johanson CE:. Ontogeny of the blood–brain barrier to, and cerebrospinal fluid sink action on, ¹⁴C-urea. *Am J Physiol* 243:R400–R407, 1982.
88. Pardridge WM: Brain metabolism: A perspective from the blood–brain barrier. *Physiol Rev* 63:1481–1535, 1983.

89. Pardridge WM, Mietus LJ: Transport of thyroid and steroid hormones through the blood–brain barrier of the newborn rabbit: Primary role of protein-bound hormone. *Endocrinology* 107:1705–1710, 1980.

90. Pardridge WM, Mietus LJ: Kinetics of neutral amino acid transport through the blood–brain barrier of the newborn rabbit. *J Neurochem* 38:955–962, 1982.

91. Pershing LK, Johanson CE: Acidosis-induced enhanced activity of the Na–K exchange pump in the in vivo choroid plexus: An ontogenetic analysis of possible role in cerebrospinal fluid pH homeostasis. *J Neurochem* 38:322–332, 1982.

92. Peters MA: Development of a blood–brain barrier to methadone in the newborn rat. *J Pharmacol Exp Ther* 192:513–520, 1975.

93. Purin VR, Syutkina EV: Local cerebral blood flow velocity in newborn rats in normocapnia and hypercapnia. *Bull Exp Biol Med* 84:139–141, 1977.

94. Purves MJ, James IM: Observations on the control of cerebral blood flow in the sheep fetus and newborn lamb. *Circ Res* 25:651–667, 1959.

95. Raichle ME: Hypothesis: A central neuroendocrine system regulates brain ion homeostasis and volume. In Martin JB, Reichlin S, Bick KL (eds): *Neurosecretion and Peptides*. Raven, New York, 1981, pp. 329–336.

96. Rapoport SI: *Blood–Brain Barrier in Physiology and Medicine*. Raven, New York, 1976.

97. Rapoport SI, Ohno K, Pettigrew KD: Blood–brain barrier permeability in senescent rats. *J Gerontol* 34:162–169, 1979.

98. Regen DM, Callis JT, Sugden MC: Studies of cerebral β-hydroxybutyrate transport by carotid injection; Effects of age, diet and injectant composition. *Brain Res* 271:289–299, 1983.

98a. Rosenstein JM: Neocortical transplants in the mammalian brain lack a blood–brain barrier to macromolecules. *Science* 235:771–774, 1987.

99. Sarna GS, Tricklebank MD, Kantamaneni BD, et al: Effect of age on variables influencing the supply of tryptophan to the brain. *J Neurochem* 39:1283–1290, 1982.

100. Saunders NR: Ontogeny of the blood–brain barrier. In Bito LZ, Davson H, Fenstermacher JD (eds): *The Ocular and Cerebrospinal Fluids*. Academic, London, 1977. (Supplement to *Exp Eye Res* 25:523–550, 1977).

101. Saunders NR, Bradbury MWB: The development of the internal environment of the brain. In Boreus L (ed): *Fetal Pharmacology*. Raven, New York, 1973, pp. 93–109.

102. Saunders NR, Møllgard K: The natural internal environment of the developing brain. *Trends Neurosci* 4:56–60, 1981.

103. Saunders NR, Møllgard K: Development of the blood–brain barrier. *J Dev Physiol* 6:45–57, 1984.

104. Seri I, Tulassay T, Kiszel J, et al: Effect of low-dose dopamine therapy on catecholamine values in cerebrospinal fluid in preterm neonates. *J Ped* 105:489–491, 1984.

105. Sershen H, Lajtha A: Capillary transport of amino acids in the developing brain. *Exp Neurol* 53:465–474, 1976.

106. Seta K, Sershen H, Lajtha A: Cerebral amino acid uptake in vivo in newborn mice. *Brain Res* 47:415–425, 1972.

107. Smith QR, Johanson CE: Active transport of chloride by lateral ventricle choroid plexus of the rat. *Am J Physiol (Renal Fluid Electrolyte Physiol. 18)* 249:F470–F477, 1985.

108. Smith QR, Johanson CE, Woodbury DM: Uptake of ^{36}Cl and ^{22}Na by the brain–cerebrospinal fluid system: Comparison of the permeability of the blood–brain and blood–cerebrospinal fluid barriers. *J Neurochem* 37:117–124, 1981.

109. Smith QR, Pershing LK, Johanson CE: A comparative analysis of extracellular fluid volume of several tissues as determined by six different markers. *Life Sci* 29:449–456, 1981.

110. Smith QR, Rapoport SI: Carrier-mediated transport of chloride across the blood–brain barrier. *J Neurochem* 42:754–763, 1984.

111. Smith QR, Takasato Y, Rapoport SI: Age-associated decrease in the rate of cerebrospinal fluid uptake of Na in the Fischer–344 rat. *Soc Neurosci Abs* 8:443, 1982.

112. Smith QR, Woodbury DM, Johanson CE: Uptake of ^{36}Cl and ^{22}Na by the choroid plexus–cerebrospinal fluid system: Evidence for active chloride transport by the choroidal epithelium. *J Neurochem* 37:107–116, 1981.

113. Smith QR, Woodbury DM, Johanson CE: Kinetic analysis of ^{36}Cl, ^{22}Na and ^3H-mannitol uptake into the in vivo choroid plexus–cerebrospinal fluid system: Ontogeny of the blood–brain and blood–CSF barriers. *Dev Brain Res* 3:181–198, 1982.

114. Spector R: Development of the vitamin transport systems in choroid plexus and brain. *J Neurochem* 33:1285–1287, 1979.

115. Spector R, Johanson CE: The mammalian choroid plexus: Structure, development and function. *Sci Amer*, accepted for publication.
116. Spector R, Levy P: Thyroxine transport by the choroid plexus in vitro. *Brain Res* 98:400–404, 1975.
117. Stern WE, Marshall C: Distribution of radioactive phosphorus in normal and diseased brain tissue: Experimental and clinical observations. *Proc Soc Exp Biol Med* 78:16–20, 1951.
118. Stewart PA, Hayakawa EM, Hayakawa K: Structural endothelial changes that underlie the maturation of the blood–brain barrier. *Soc Neurosci Abs*. 11(1):674, 1985.
119. Stonestreet BS, Nowicki PT, Hansen NW, et al: Effect of aminophylline on brain blood flow in the newborn piglet. *Dev Pharmacol Ther* 6:248–258, 1983.
120. Sturman JA: Taurine in developing rat brain: Changes in blood–brain barrier. *J Neurochem* 32:811–816, 1979.
121. Symes AL, Lal S, Sourkes TL: Time-course of apomorphine in the brain of the immature rat after apomorphine injection. *Arch Intern Pharmacodyn* 223:260–264, 1976.
122. Takasato Y, Momma S, Smith QR: Cerebrovascular transport of large neutral amino acids in mature and aging rats. *Soc Neurosci Abs* 10:791, 1984.
123. Takasato Y, Rapoport SI, Smith QR: An in situ brain perfusion technique to study cerebrovascular transport in the rat. *Am J Physiol* (Heart Circ Physiol 16) 247:H484–H493, 1984.
124. Tennyson VM, Pappas GD: Choroid plexus. In Minckler J (ed): *Pathology of Nervous System*. Vol. 1. McGraw-Hill, New York, 1968, pp. 498–518.
125. Toews AD, Kolber A, Hayward J, et al: Experimental lead encephalopathy in the suckling rat: Concentration of lead in cellular fractions enriched in brain capillaries. *Brain Res* 147:131–138, 1978.
126. Turnbull J, Brodeur J: Influence d'un retard de croissance sur la perméabilité de la barrière hémato-éncéphalique chez le rat nouveau-né exposé au plomb. *Can J Physiol Pharmacol* 62:142–145, 1984.
127. Vernadakis A, Woodbury DM: Cellular and extracellular spaces in developing rat brain. *Arch Neurol* 12·284–293, 1965.
128. Woodbury DM: Distribution of nonelectrolytes and electrolytes in the brain as affected by alterations in cerebrospinal fluid secretion. *Prog Brain Res* 29:297–313, 1967.
129. Woodbury DM: Maturation of the blood–brain and blood–CSF barriers. In Vernadakis A, Weiner N (eds): *Drugs and the Developing Brain*. Plenum, New York, 1974, pp. 259–280.
130. Woodbury DM, Johanson CE, Brondsted H: Maturation of the blood–brain and blood–CSF barriers and transport systems. In Zimmermann E, George R (eds): *Narcotics and the Hypothalamus*. Raven, New York, 1974, pp. 225–247.
131. Woodward DL, Reed DJ, Woodbury DM: Extracellular space of rat cerebral cortex. *Am J Physiol* 212:367–370, 1967.

8

The Blood–Nerve Barrier and the Pathologic Significance of Nerve Edema

Henry C. Powell and Robert R. Myers

1. INTRODUCTION

Edema occurs in peripheral nerves when excessive fluid accumulates either in the interstitium or within the myelin sheath, creating a disturbance in the endoneurial microenvironment. Nerve edema is concomitant with many peripheral neuropathies and can alter the endoneurial microenvironment by increasing pressure, reducing blood flow, or altering electrolyte concentrations in the endoneurial fluid (EF). Not only is it a complication of several diseases of peripheral nerve, but it can independently injure the nerve fiber. This is due to the vulnerability of the vasa nervorum to internal and external pressure. These vessels are characterized by numerous anastomoses that form the epineural and perineurial vascular plexuses (Figs. 1 and 2). Since they pierce the perineurium, pressure at this point causes ischemia, damaging nerve fibers. This chapter reviews new findings concerning the physiology of the endoneurial environment, mechanisms of increased endoneural fluid pressure (EFP), and their pathologic complications.

1.2. The Blood–Nerve Barrier and the Endoneurial Compartment

The endothelial lining of the vasa nervorum is formed by continuous, nonfenestrated endothelia in which individual cells are linked by tight junctions, rendering them impermeable to macromolecules carried in the circulation (Fig. 3). This blood–nerve barrier (BNB) and a similar mechanism in the innermost layers of the perineurial sheath isolate

Abbreviations: BNB, blood–nerve barrier; CNS, central nervous system; CSF, cerebrospinal fluid; EAN, experimental allergic neuritis; EF, endoneurial fluid; EFP, endoneurial fluid pressure; FITC, fluorescein isothiocyanate; HCP, hexachlorophene; HRP, horseradish peroxidase; PPD, purified protein derivative; TET, triethylin; TFA, transfascicular area.

Henry C. Powell • Department of Pathology, University of California, San Diego, La Jolla, California 92093.   *Robert R. Myers* • Departments of Anesthesiology and Neurosciences, Veterans Administration Medical Center, La Jolla, California 92093.

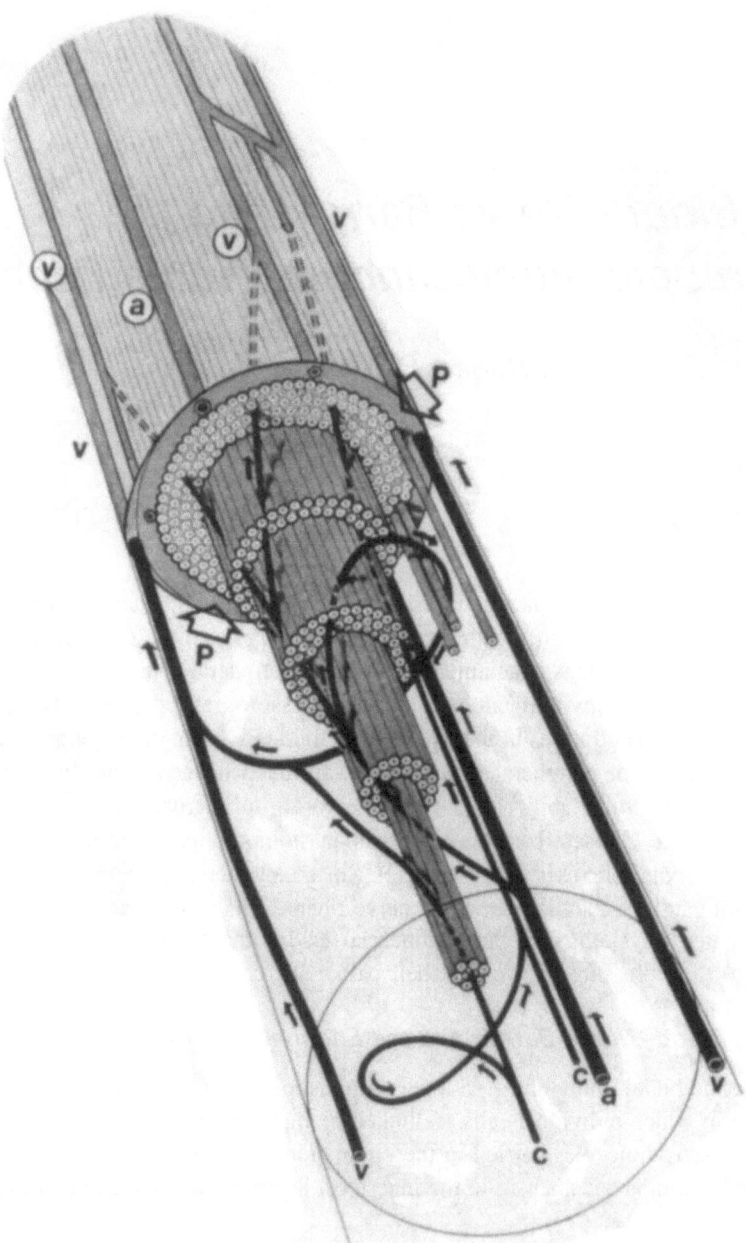

Figure 1. Schematic illustration of blood vessels in a normal peripheral nerve trunk. Veins (v) and arteries (a) traverse the perineurium (P) and form an endoneurial vascular plexus principally composed of capillaries (c). Note the capillary loops surrounding the nerve fibers. (From Lundborg.[23])

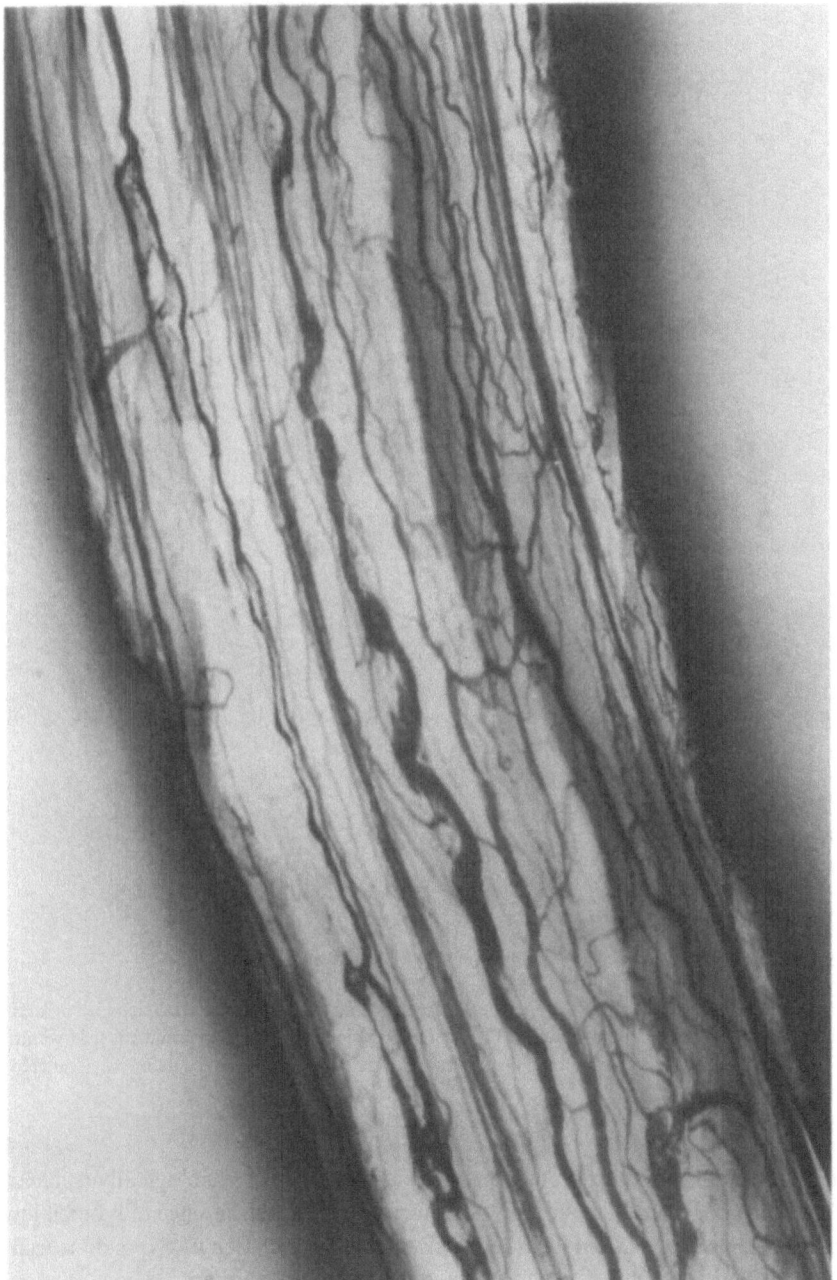

Figure 2. Gelatin cast of the sciatic nerve microcirculation in a rat. Liquid gelatin was infused into the distal aorta at systolic blood pressure. The nerve was subsequently excised and the tissue cleaned by immersion in methyl salicylate for 24 hr. The picture shows the richly anastomotic transaperineurial circulation which is vulnerable to pressure applied to the perineurial sheath. (From Powell and Myers.[43])

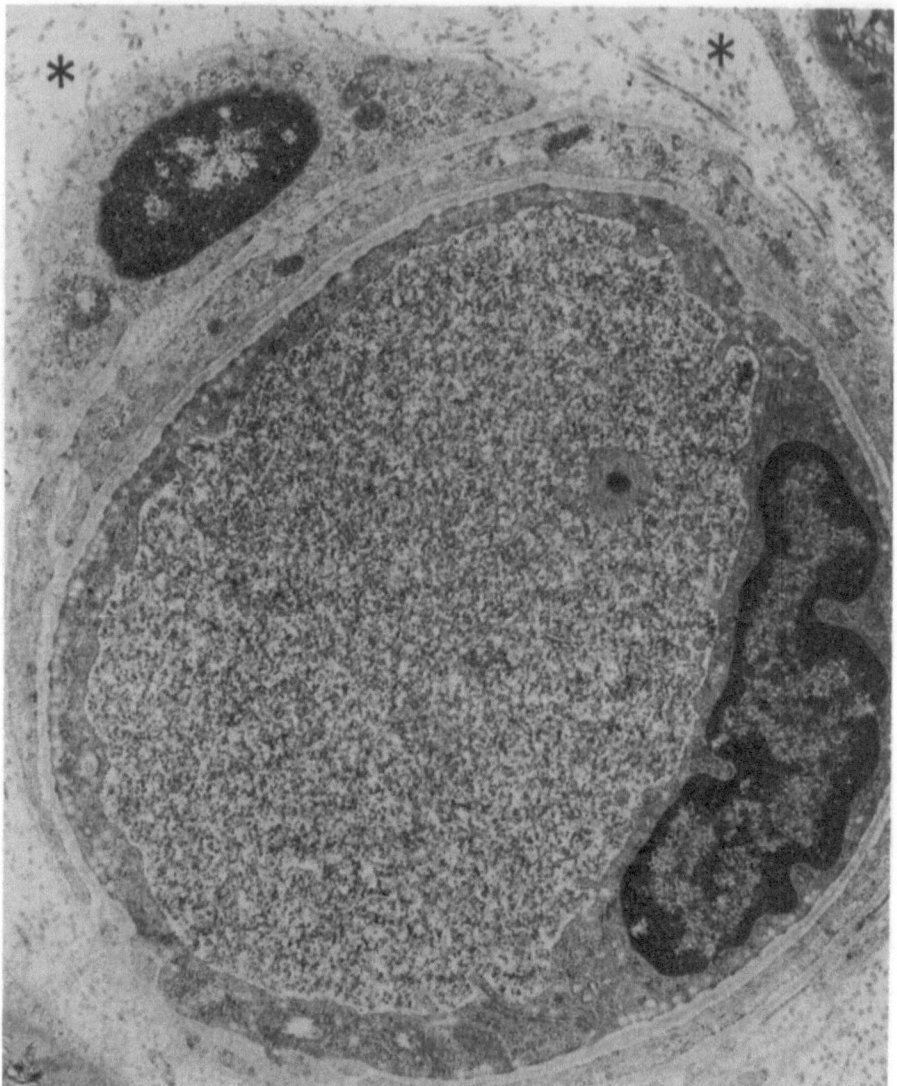

Figure 3. Blood–nerve barrier. Electron micrograph of an endoneurial capillary from a rat sciatic nerve after intravenous injection of horseradish peroxidase (HRP). The darkly stained HRP reaction product is held within the lumen of the vessel, while the surrounding interstitium (*) is free of any tracer and contains only scattered collagen fibrils.

the endoneurial interstitium, much in the same way as the brain is physiologically isolated by its endothelium and the choroid plexus. Other factors such as the absence of lymphatics and the distinctive electrolyte composition of EF reflect the isolation of the endoneurial compartment from the systemic circulation and extraneural interstitium.

2. ENDONEURIAL FLUID PRESSURE

Noting that endoneurial contents herniate through an incision in the perineurium, Sunderland[57] postulated that there is positive intrafascicular pressure in peripheral nerves.

This was confirmed by Low and Dyck[18,19] and by Myers et al.[31] using different techniques for measurement of endoneurial fluid pressure (EFP). Low and colleagues[17,19] implanted polyethylene matrix capsules in the peripheral nerves and allowed them to fill with EF. The capsule was connected to the surface by plastic tubing into which a micropipette could be introduced at a later date to determine EFP. Although this is an indirect method of EFP measurement, its results are in close agreement with the direct method[31] in which a glass micropipette, with a beveled, sharpened tip, 3 μm in diameter, is inserted through the perineurium. The direct method is less traumatic and allows for repeated measurements in the same fascicle. A third method by which EFP can be measured is by the wick catheter, a tube device with absorbent material packed in its orifice. After fluid equilibrates within the tube, the interstitial pressure can be determined. The wick catheters have been successfully employed in muscle tissue and in the carpal tunnel[9] but are less suited to small nerves in the experimental rat.

The mechanism by which normally positive fluid pressure is maintained is not completely understood. Both brain and peripheral nerve have constant positive interstitial fluid pressures, while the pressure in extraneural tissues is negative.[29] Donnan forces acting across selectively permeable barriers have been suggested.[14] These forces are generated by osmolytes sequestered in the endoneurial compartment by the BNB. The Donnan effect represents the net charges on protein molecules that cause unequal distribution of electrolytes across the semipermeable membranes of the BNB.[14] The absence of lymphatics from the endoneurial interstitium may also be contributory since lymphatic drainage is associated with the most negative pressures in the microcirculation. Other factors that might contribute to positive pressure include the sorbitol pathway, an obscure metabolic pathway that is active in nerve[28] and the contractile properties of the perineurial sheath.[52] The sorbitol pathway has recently been linked to sodium concentration in the endoneurium.[27] Endoneurial fluid is normally hypertonic[36] and the sorbitol pathway may play a role in the elevated sodium concentration seen in EF versus serum. The normal positive pressure in the EF is viewed as a mechanism of proximodistal drainage for the endoneurial interstitium, which lacks such mechanisms as lymphatic drainage or the cerebrospinal fluid (CSF) sink.

2.1. Mechanisms of Increased Endoneurial Fluid Pressure

To date, five mechanisms of endoneurial edema have been identified, each of them associated with increased EFP (Table I).

2.1.1. Neuropathies in Which There Is Altered Vascular Permeability

There are several experimental models of demyelinating and axonal disease in which altered vascular permeability precedes the development of edema and nerve fiber injury. Intraneural edema occurs when the internal homeostasis is disturbed, causing fluid to enter the endoneurial compartment. The first of the five mechanisms involved in this process is altered permeability of vascular endothelium, a change that has been documented in several neuropathies (Table I).

2.1.1a. Experimental Allergic Neuritis. The phenomenon of altered vascular permeability is readily appreciated in experimental allergic neuritis (EAN), in which sensitized cells,[15] as well as plasma, infiltrate the endoneurial space after passing between venular endothelial cells. The normally tight interendothelial cell junctions of venules

Table I. Mechanisms of Endoneurial Edema: Animal Models of Human Disease

Mechanism	Animal model	Human disease
Altered vascular permeability	Lead poisoning	Lead neuropathy
	Experimental allergic neuritis	Guillain-Barré syndrome
	Genetic, twitcher mouse	Globoid leukodystrophy
Intramyelinic edema	Hexachlorophene	Hexachlorophene neuropathy
	Triethyltin	Triethyltin neuropathy
Metabolic (hyperactive sorbitol pathway)	Galactose neuropathy	Diabetic neuropathy
	Streptozotocin	Diabetic neuropathy
Nerve fiber disintegration (Wallerian degeneration)	Crush injury (mechanical)	Carpal tunnel syndrome
	Laser injury (heat)	Burns, heat-induced neuropathy
	Compound 48/80 (chemical)	Chemical injury—inflammation
	Cryoprobe (cold, freeze)	Cold injury pain therapy
Altered perineurial permeability	Local anesthetics	Toxic neuropathies

loosen sufficiently to admit horseradish peroxidase (HRP) into the endoneurial interstitium.[49] Immunocytochemistry also reveals the edema fluid to be rich in immunoglobulin, consistent with its hematogenous origin. The significance of altered vascular permeability in EAN is emphasized by the fact that it is the earliest pathologic change in the extraganglionic nerves, presenting exactly 10 days after intravenous injection of sensitized lymphocytes.[11] Edema fluid is rich in inflammatory cells and may have a role in transporting them to their immunologic targets. Degranulation of mast cells is concomitant with edema formation,[11,49] and it is possible that these cells may assist in maintaining altered vascular permeability while the inflammatory process runs its course.[3] Mast cells possess receptors to certain classes of immunoglobulin and are also responsive to T cells during a delayed hypersensitivity reaction.[2] Their numbers increase during allergic neuritis.[3]

2.1.1b. Lead Neuropathy. A second example of altered vascular permeability occurs in experimental lead neuropathy, a condition in which there is primary evidence of endothelial cell injury. Electron-dense intranuclear inclusions, pathognomonic of lead intoxication, appear in endothelial cells (Fig. 4) within 1 week of lead intoxication[48] and there is a steady increase in endoneurial lead, maximal at day 30.[62] Paradoxically, no evidence of altered vascular permeability has been elicited until 7 weeks from the start of intoxication, when vessels become permeable to low-molecular-weight fluorescein isothiocyanate (FITC) dextran[33] and during subsequent weeks to high-molecular-weight dextrans and HRP (Figs. 5 and 6). It thus appears that lead gains access to the endoneurial compartment prior to the onset of endoneurial edema. Lead inclusions are visible inside Schwann cells (Fig. 7) within 2 weeks of commencing experimental intoxication.[48] While endoneurial edema (Fig. 8) does not appear to play a role in the pathogenesis of demyelination it has been linked to a disturbance in tonicity of EF.[36] Electron-probe microanalysis of EF has suggested it to be hypertonic.[36] During lead-induced endoneurial edema, it becomes isotonic with serum. While the significance of altered EF tonicity is hard to assess, it is a unique disturbance of the endoneurial environment and may accompany other forms of nerve edema.

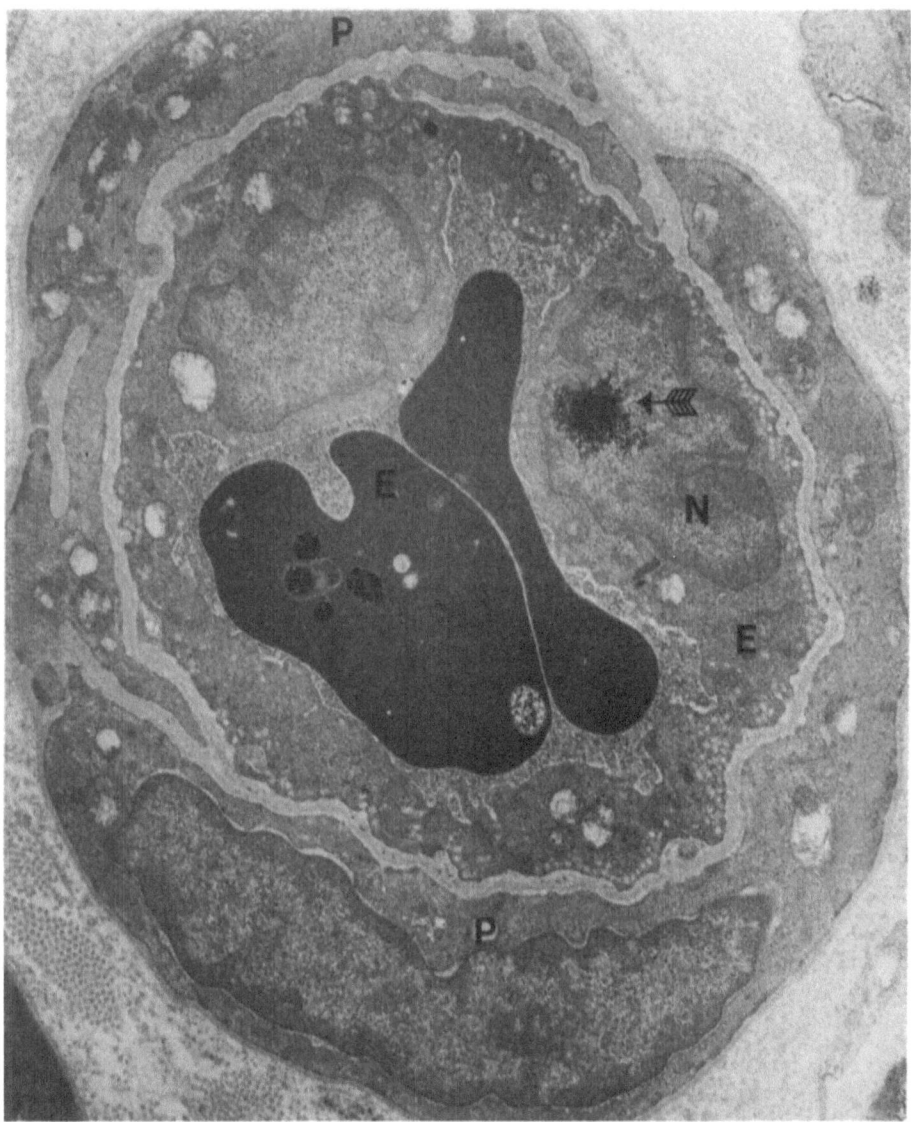

Figure 4. Microangiopathy in experimental lead neuropathy. In this capillary there is an electron-dense inclusion (arrow) in the nucleus (N) of an endothelial cell, a finding consistent with lead intoxication. Similar inclusions are found in liver, kidney, and astrocytic nuclei. Nuclei have the capacity to sequester lead, which combines with nucleoprotein and is thus prevented from injuring mitochondria of the cytosol. When this capacity is exhausted, the cell undergoes necrosis and a cycle of endothelial proliferation begins causing microangiopathy. E, endothelial cell; P, pericyte. (From Powell et al.[48])

2.1.1c. Murine Globoid Leukodystrophy. Altered vascular permeability has also been demonstrated in the twitcher mouse in which an inherited deficiency of galactosyl ceramidase causes globoid leukodystrophy resembling human Krabbe disease. Associated with a metabolic disturbance in Schwann cells, there is interendothelial cell leakage of HRP with endoneurial edema and increased EFP.[50] A rather unusual observation in twitcher nerves involves eosinophils, which are numerous throughout the endoneurial interstitium.[50] Although eosinophils may appear in local injuries such as foreign body

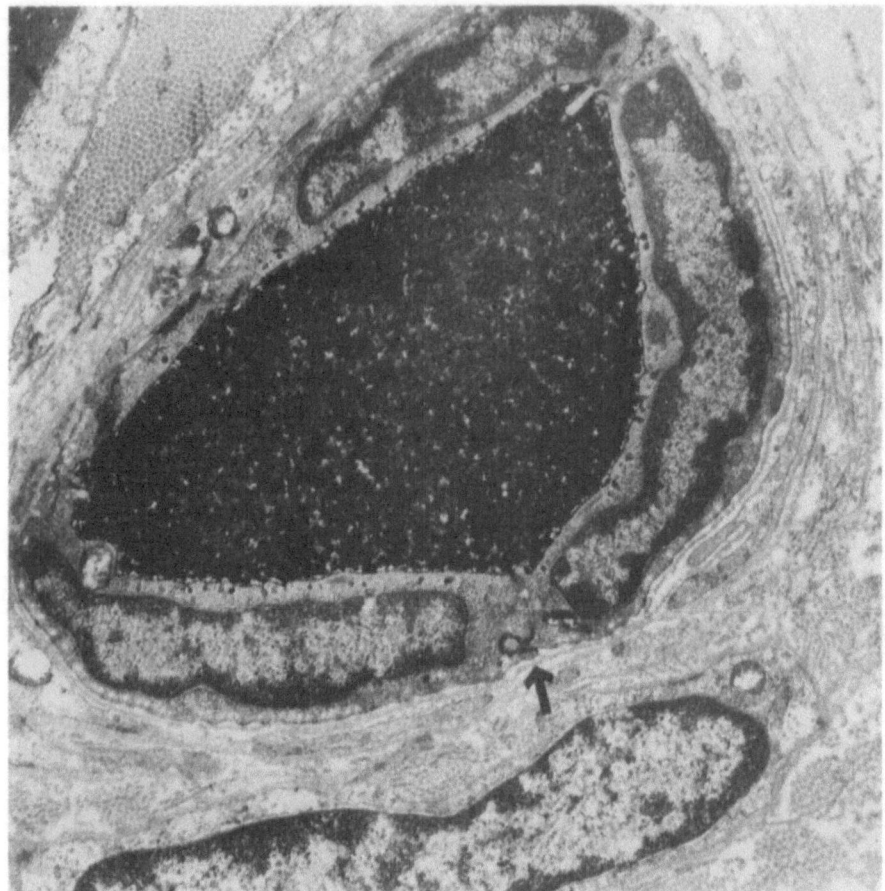

Figure 5. Endoneurial capillary from a rat sciatic nerve after intravenous injection of horseradish peroxidase (HRP) in a rat intoxicated with lead carbonate for 11 weeks. HRP reaction product can be seen leaking from the lumen through an endothelial cell junction (arrow).

reactions, in peripheral nerves 'their appearance in a symmetric polyneuropathy has not previously been described.

2.1.1d. Physical and Chemical Injury to Peripheral Nerves. Local disturbance of vascular permeability may occur in Wallerian degeneration following a crush injury or an analogous mechanical disturbance; it can be experimentally induced by interneural injection of mast-cell-degranulating agents such as histamine or Compound 48/80.[41] Intraneural injections of histamine caused local nerve swelling associated with mast-cell degranulation (Figs. 1 and 2), edema, and increased EFP.[46] On the other hand, heparin and serotonin failed to induce degranulation and had no significant effect on EFP. While histamine appears to act solely on the mast cell, Compound 48/80 affects mast cells, Schwann cells, and axons, causing Wallerian degeneration. Endoneurial mast cells are increased in many chronic neuropathies[41] and these cells may promote endoneurial drainage by modulating the amount of interstitial fluid in their immediate environment. Increased pressure promotes fluid flow along the nerve and thus plays an important role in interstitial drainage.

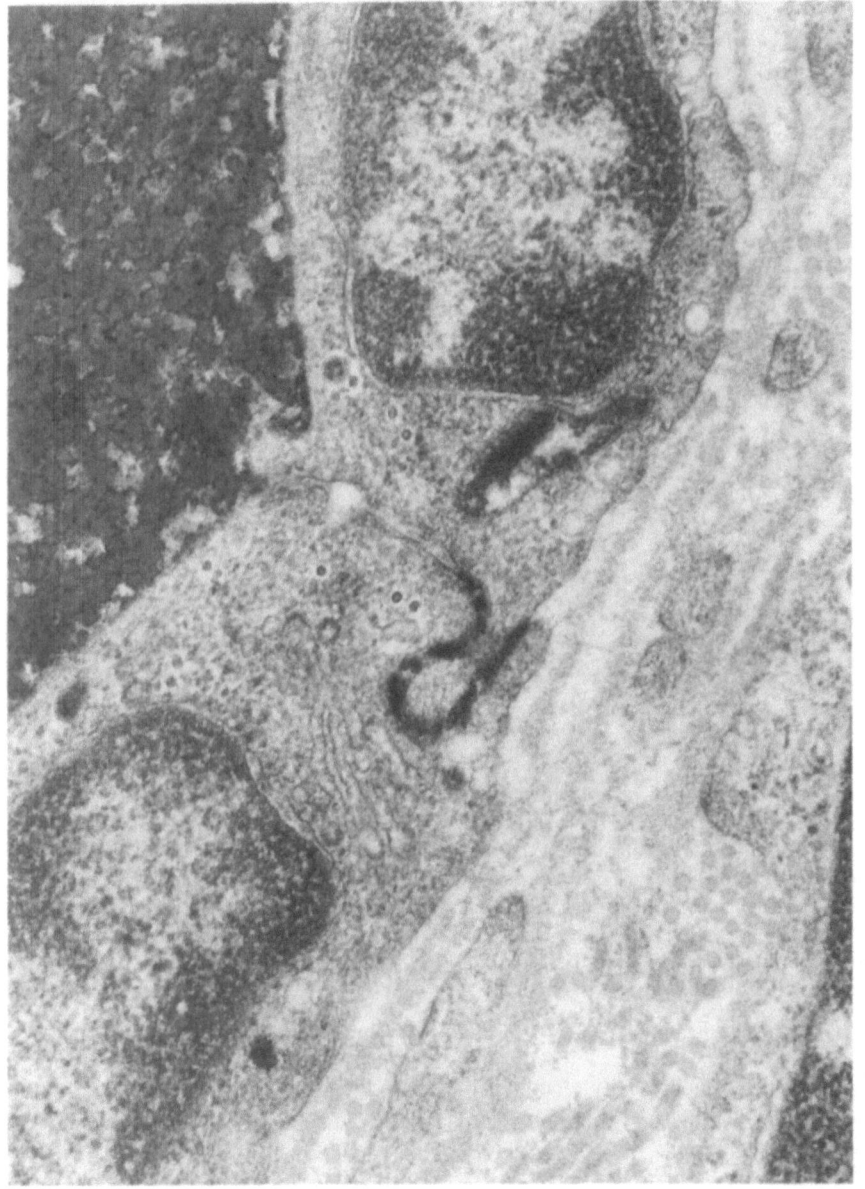

Figure 6. Higher-power electron micrograph demonstrating interendothelial cell leakage shown in Fig. 2. Leakage from lead-intoxicated capillaries is the mechanism of nerve edema in chronic lead neuropathy. (From Powell et al.[48])

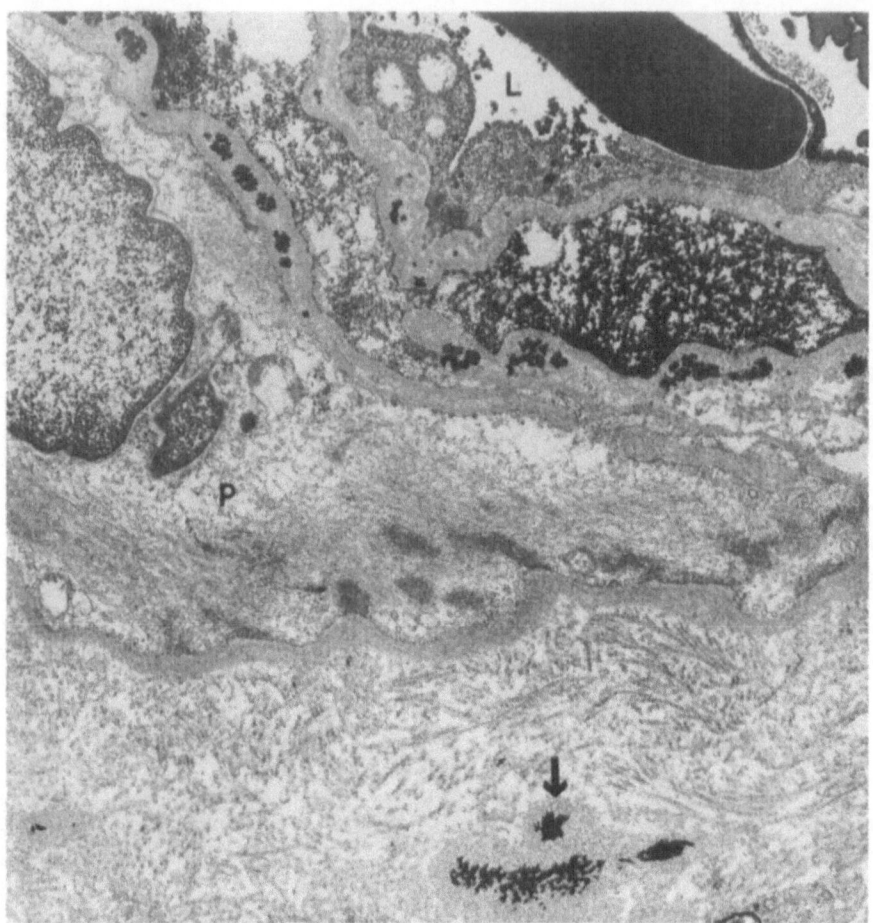

Figure 7. Leakage of horseradish peroxidase (HRP) into the subendothelial space of a rat fed 6% lead carbonate for 11 weeks. The vessel lumen (L) contains a red blood cell (RBC) as well as clumps of HRP reaction product, more of which is seen forming dense aggregates in the subendothelial space and in the interstitium subjacent to the vessel (arrow). Note the increased amount of cytoplasmic filaments in the pericyte (P). This is consistent with reaction to injury and contributes to the mural thickening characteristic of this microangiopathy. Leakage of plasma resulted in an increase in endoneurial fluid pressure in this rat to 11.7 cm H_2O, while the mean interstitial pressure in controls ($N = 6$) was 2.1 ± 1.0 cm H_2O.

2.1.2. Neuropathies in Which the Blood–Nerve Barrier Remains Intact

In certain types of neuropathy, edema is limited to the myelin sheath and is not found in the interstitium. However, enlargement of the myelinated fibers due to edema is sufficient to increase EFP.

2.1.2a. Intramyelinic Edema: Hexachlorophene and Triethyltin. Because of the lipid solubility of organic chemicals such as hexachlorophene (HCP) and triethyltin (TET), they can readily penetrate the blood–nerve barrier, causing a distinctive form of edema limited to the myelin sheath.[16,44] Within days of intoxication, there is conspicuous swelling of individual myelin sheaths while the surrounding interstitium remains unaffected and the BNB remains impermeable. Identical changes occur in central nervous system (CNS) myelin, which also remains impermeable to tracers. Thus, the toxin ap-

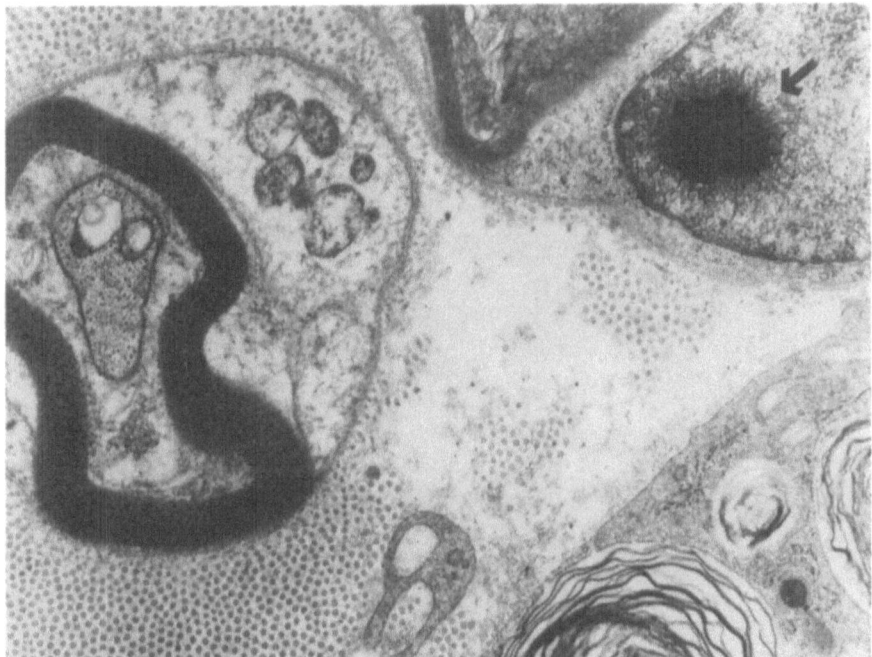

Figure 8. Electron micrograph of rat sciatic nerve after 10 weeks of lead intoxication. An electron-dense spicule, characteristic of lead intoxication, is present in a Schwann cell nucleus (arrow). Note the hydropic changes in the Schwann cell on the left and the large structureless space, representing edema at the center of the picture. (From Myers et al.[33])

pears to have a direct and highly specific effect on myelin, possibly through inhibition of enzyme systems in the myelin sheath.[4] The myelinopathy occurs throughout the nervous system, causing an increase in nerve water content both in brain[16] and nerve[5] and elevated EFP.[44] Because of the specificity of these morphologic lesions, HCP is a useful experimental model for complications of increased EFP and has been shown to be associated with reduced nerve blood flow and the pathogenesis of Wallerian degeneration.[26,35]

Intramyelinic edema can be caused by other neurotoxins, including organotin and acetylethyltetramethyltetralin.[56] Thus, HCP is the prototype for a class of neurotoxins having potent but restricted neuropathologic effects. It may also be worth noting that edema confined to the myelin sheath may be less easily treatable than interstitial edema associated with altered vascular permeability.

2.1.2b. The Sorbitol Pathway: Experimental Models of Diabetic Neuropathy. A third type of edema observed in peripheral neuropathy occurs when a hyperosmolar condition is induced by increased activity of the sorbitol pathway in experimental models of diabetes and galactose neuropathy. This type of edema is the result of both distinctive biochemical and permeability properties of the endoneurial compartment (Figs. 9 and 10). The enzyme aldose reductase is found in peripheral nerve and can convert sugar into sugar alcohols. Sorbitol is produced from glucose and accumulates in diabetic endoneurium, while galactitol is produced in correspondingly large quantities when galactose is fed to experimental animals.[7,54] The effects of increased activity in the sorbitol pathway (Fig.

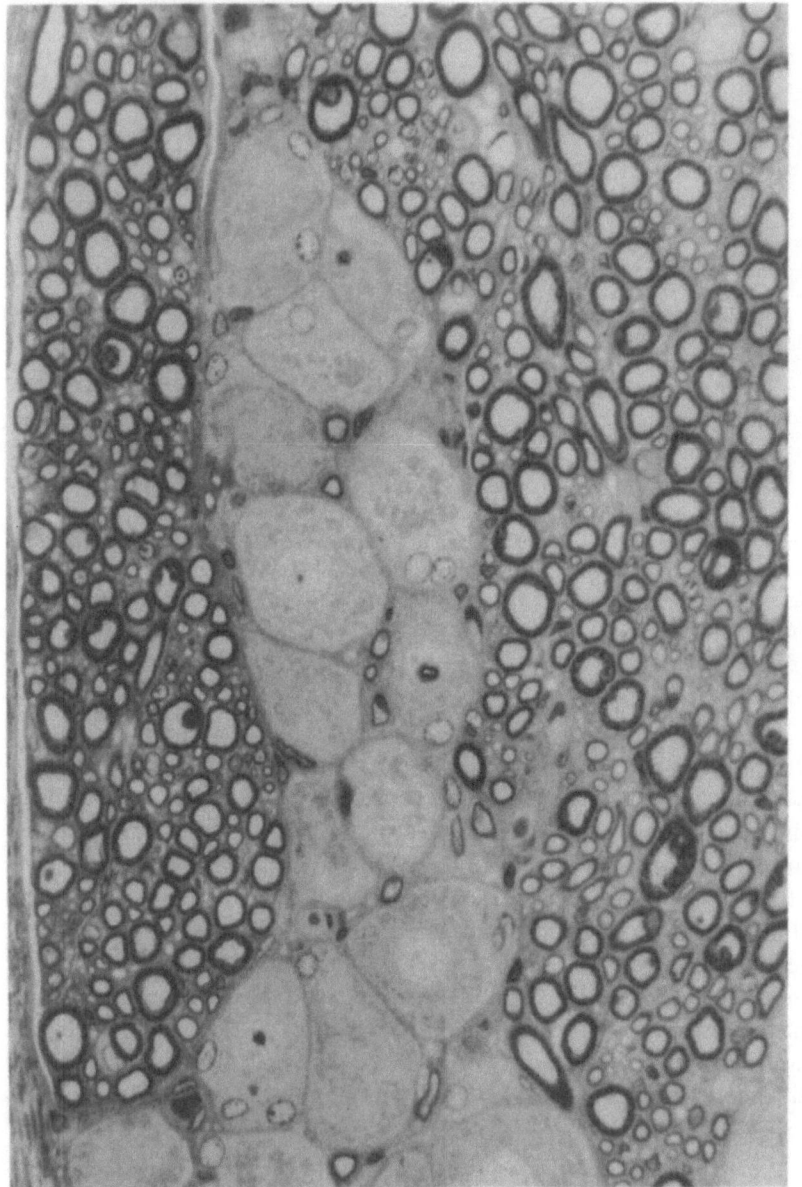

Figure 9. Galactose neuropathy. The pathogenesis of nerve edema in this metabolic neuropathy is unique. Galactitol and sodium accumulate within the endoneurial compartment because the blood–nerve barrier (BNB) remains intact. Thus, in the spinal ganglia and nerve roots in which the vessels are more permeable, no edema occurs (Fig. 9) while in the adjacent extraganglionic nerves (Fig. 10) massive edema fills the subperineurial space (E) and the endoneurial interstitium.

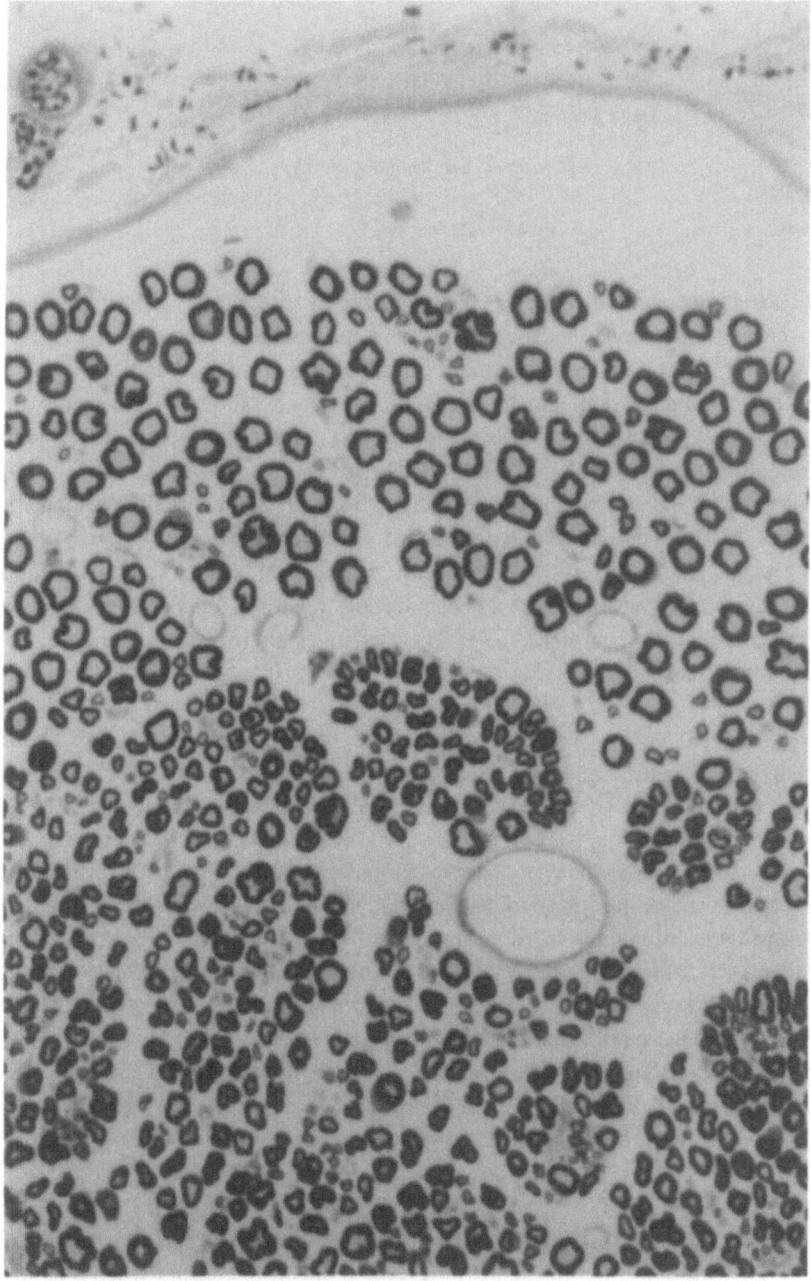

Figure 10. Galactose neuropathy. Rat sciatic nerve after administration of a diet containing 40% galactose by weight for several months. Massive edema is one of accumulation of both galactitol and sodium.[27,28] Edema is located in the extraganglionic nerves while the adjacent spinal ganglia (Fig. 9) are edema-free. This is one manifestation of the different properties of the blood–nerve barrier in the spinal ganglia and roots.

$$\begin{array}{lll}
\text{glucose} + \text{NADPH} & \rightarrow & \text{sorbitol} + \text{NADP}^+ \\
& \textit{aldose reductase} & \\
\text{galactose} + \text{NADPH} & \rightarrow & \text{dulcitol} + \text{NADP}^+
\end{array}$$

$$\begin{array}{lll}
\text{sorbitol} + \text{NAD}^+ & \rightarrow & \text{fructose} + \text{NADH} \\
& \textit{sorbitol dehydrogenase} & \\
\text{dulcitol} + \text{NAD}^+ & & \text{NR}
\end{array}$$

Figure 11. The sorbitol pathway.

11) are best appreciated in galactose neuropathy because of the inability of sorbitol dehydrogenase to metabolize galactitol further. Accumulation of osmotically active substances in the endoneurial compartment results in abnormal hydration,[54] increased EFP,[20,32] and ultimately demyelination and axonal degeneration.[17,42] As in the case of HCP neurotoxicity, increased EFP results in reduced nerve blood flow.[30] What is unique about the pathogenesis of this neuropathy is the paradoxical role of the BNB, which by restricting egress of osmotically active molecules from the endoneurium is responsible for the increase in EFP and ultimately for ischemic changes and nerve fiber injury. In addition, recent studies of EFP electrolytes reveal that abnormal sodium retention in galactose neuropathy is a major mechanism of disturbed endoneurial homeostasis in experimental diabetic neuropathy.[27,28] These changes are reversible by treatment with aldose reductase inhibitors.[28]

2.1.2c. Endoneurial Edema Associated with Wallerian Degeneration. The fourth mechanism of endoneurial edema is associated with myelinated nerve fiber degeneration following crush injury to nerve. Once again, edema is sufficient to increase EFP which rises immediately after injury and reaches a peak 6 to 7 days later.[45]

Various factors contribute to edema formation, including altered vascular permeability and breach of the perineurial sheath. However, these lesions occur at the time of the crush, while maximal pressure does not occur for several days. The probable explanation for this discrepancy is that the breakdown of myelin and of other tissues secondary to axonal degeneration is maximal after 4 to 6 days, and a hyperosmolar state exists during catabolism of the macromolecular constituents of the myelin sheath. This interpretation is supported by EFP studies in cryoprobe-induced lesions in which there is severe injury to the vasa nervorum at the site of freezing, resulting in a particularly steep rise in EFP within 24 hr.[34] However, EFP then declines, only to rise subsequently 6 to 7 days later (Fig. 12).

2.1.2d. Endoneurial Edema Associated with Extraneural Application of Local Anesthetics. The fifth and final mechanism to be described here occurs when a neurotoxin acts directly on the perineurium, causing a change in permeability with subperineurial edema. These abnormalities appear following extraneural injection of local anesthetics, such as chloroprocaine, tetracaine, and related agents. Although the perineurium is not breached mechanically, significant concentrations of local anesthetic in its vicinity can penetrate it and give rise to axonal degeneration and demyelination.[39] Evidence of the perineurial permeability change comes from tracer studies with HRP as well as ultrastructural findings of eosinophils in the endoneurium, dissociation of perineurial cells, and

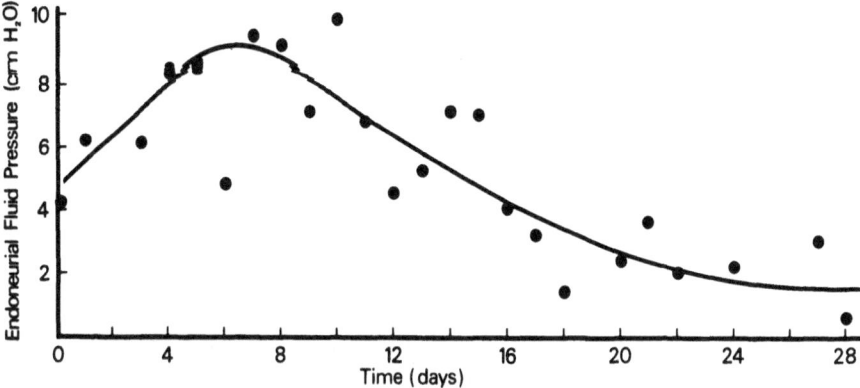

Figure 12. Endoneurial fluid pressure (EFP) in Wallerian degeneration. Analysis of EFP in 54 experimental rats following proximal crush injury to the sciatic nerve. FP increased within 90 min of crushing and reached a peak at 7 days, gradually returning to normal 3 to 4 weeks later. (From Powell et al.[45])

unusual morphologic changes in Schwann cells, whose cytoplasm becomes packed with lipid droplets.

3. SIGNIFICANCE OF ENDONEURIAL EDEMA

3.1. Topography of Endoneurial Edema

The distribution of EF follows consistent patterns in normal and diseased nerves. In most edematous neuropathies, excessive fluid accumulates in the interstitium and is most conspicuous in the subperineurial space and the perivascular spaces. Electron microscopic evaluation of streptozotocin diabetic nerve demonstrates that edema accumulates in structureless space[12] in the interstitium. Even in the most severe interstitial edema, the interfibrillar distance between collagen bundles appears undisturbed, the fluid instead accumulating outside the endoneurial collagen bed. In fact, edema accumulates preferentially in regions in which collagen is scarce: the subperineurial and perivascular areas. Compliance studies show that peripheral nerve tolerates a limited amount of swelling before EFP increases.[17,20,33] This would correspond with filling of available structureless space in these areas. Studies of the collagen bed in extraneural tissue[58] show the collagen bundles to be relatively resistant to interfibrillar fluid flow because of the organization of the collagen in which tightly coiled mucopolysaccharides between collagen fibrils reduce interstitial distance to 30Å between structures.

3.2. Impact of Endoneurial Edema

The effects of endoneurial edema include increased fluid pressure, reduced nerve blood flow, disturbances of EF electrolytes, endoneurial hypoxia, and altered nerve compliance in chronic edematous neuropathies. Compliance represents the inherent elasticity of peripheral nerve and is defined as the ratio of change in volume V of interstitial fluid to the simultaneous change in interstitial fluid pressure $P : C = dV/dP$. Nerve compliance can be determined indirectly by plotting changes in transfascicular area (TFA) due to endoneurial edema against EFP.[31] The normal positive EFP is due in part to the contractile properties of the perineurium, which resists expansion of the endoneurial contents. These physical properties were further discussed by Low and associates.[17,20]

3.2.1. Increased Endoneurial Fluid Pressure

Having discussed the mechanisms of increased EFP, it is now appropriate to examine its effects. The impact of increased pressure on the nerve fiber is perhaps best studied in neuropathies in which the BNB remains intact, minimizing homeostatic changes in the endoneurium. These two neuropathies are HCP myelinopathy, in which edema is intra-myelinic, and hyperosmolar neuropathy, in which endogenous fluid accumulates behind an intact barrier. In HCP neuropathy, increased EFP occurs within days as large vacuoles appear in the myelin sheaths. During the first 2 weeks, edema is exclusively intramyelinic and other structures remain intact. However, axonal degeneration appears in 4-week-old HCP-treated rats and was present in rats from whose diet HCP has been withdrawn after 2 weeks.[26,44] At that point, there was severe interstitial edema associated with disintegration of axons and their myelin sheaths. Since axonal degeneration in intramyelinic edema was also observed in TET neuropathy,[8] the putative mechanism was suggested to be ischemia; this hypothesis was tested by measuring nerve blood flow in HCP neuropathy. Nerve fiber injury was also observed as a late change in galactose neuropathy, in which the primary morphologic abnormality is edema. Microscopic evidence of injury to nerve fiber did not appear until edema had been present for several months[21] and was quite limited; however, animals that received galactose for 1 to 2 years had extensive nerve fiber damage, with both demyelination and axonal degeneration.[42] The predominant lesion was demyelination which occurred in association with swelling of Schwann cells, the cytoplasm of which was uniformly packed with glycogen granules. These changes occurred in the wake of a reduction in nerve blood flow[30] and appear to be the result of ischemia. The predominance of demyelinating over axonal injury in galactose may reflect the relatively slow onset of increased EFP in this neuropathy. Further evidence of the connection between edema and nerve fiber damage in galactose neuropathy is the absence of edema both from the spinal ganglia and from their roots.[47] The BNB is more permeable in the ganglia; thus, osmotically active material produced at these sites can escape from the endoneurial compartment.

3.2.2. Nerve Blood Flow

The importance of nerve blood flow in edematous neuropathy relates to peculiar aspects of the distribution of vasa nervorum, hence their vulnerability to ischemia. Peripheral nerves have a dual blood supply involving central axial vessels that perfuse the center of each fascicle,[10] while the circumference is served by an extensive anastomotic network of small blood vessels linked to the epineurial arteries by regional nutritive vessels.[24] These transperineurial vessels (Figs. 1 and 2) are affected by external compression[24,25,53] and internal pressure associated with endoneurial edema.[30,35] Nerve blood flow can be measured by a variety of techniques[35,55,59] both noninvasive and invasive.

3.2.3. Disturbances of Endoneurial Fluid Electrolytes

The specialized microenvironment within the endoneurial compartment is adapted to physiologic requirements of axons and their supporting cells. In 1955 Krnjevik reported that endoneurial electrolyte concentrations are greater than serum concentrations.[13] This observation was made in desheathed nerves and was subject to artifacts associated with rupture of blood vessels and damage to nerve fibers in the course of dissection. To confirm the hypertonicity of EF required pure EF uncontaminated by plasma and available in sufficient quantities for chemical analysis. Since direct aspiration of EF yields only microscopic droplets, we resorted to electron-probe microanalysis of samples aspirated in a picoliter pipette. Such methods confirm Krnjevik's observation that the sodium and chloride concentrations of EF are higher than plasma levels[27,28,36] and also demonstrate that when nerve edema is present, EF values can decline to plasma levels. This was demonstrated in lead neuropathy, in which endoneurial edema is secondary to increased permeability of the BNB. Endoneurial sodium concentrations can also increase, in galactose neuropathy, for example.[27,28] In this condition, endoneurial sodium concentrations almost double after several weeks of intoxication. Comigration of sodium with sugar into the endoneurial compartment appears to be the mechanism by which increased salt is retained within the endoneurial compartment; both the sorbitol pathway and passive glycosylation create a gradient mechanism by which greater than normal quantities of sugar can enter the endoneurial compartment. The process is reversible by administration of aldose reductase inhibitors.[28] Edema induced by this mechanism may be responsible for a number of pathophysiologic disturbances that constitute the earliest functional disturbance in this and other forms of experimental diabetic neuropathy. Future studies of endoneurial sodium concentration and its relationship to blood sugar concentration are necessary to elucidate the relationship between the sorbitol pathway and EFP since this enzyme system may contribute to maintenance of the normal positive pressure in peripheral nerve.

3.2.4. Endoneurial Hypoxia

Endoneurial hypoxia is a consequence of reduced nerve blood flow. Recently Tuck et al.[59] demonstrated that nerve blood flow and oxygen tension are reduced in streptozotocin diabetes and suggested that hypoxia in this condition may be responsible for nerve fiber injury. It has also been suggested that increased distance between capillaries in edematous neuropathies contributes to tissue hypoxia.[22] An alternative hypothesis views constriction of transperineurial vessels resulting from increased intrafascicular pressure in nerve edema as the causative mechanism. Support for the latter view comes from computer modeling of events in the perineurium and endoneurial interstitium in edematous neuropathy.[38]

4. NERVE FIBER LESIONS ASSOCIATED WITH INCREASED EFP

4.1. Wallerian Degeneration Secondary to Nerve Crush

When nerve fibers undergo Wallerian degeneration, endoneurial edema results (Table II). Measurement of EFP[45] and morphometric analysis[40] show that fluid accumulation reaches its peak 6–7 days after physical injury (Fig. 12). The pathogenesis of edema appears to involve the disintegration of myelin and other complex membranes as well as the activation of mast cells, whose granules release histamine and other agents that affect the BNB. When mast cells degranulate, histamine molecules exchange with sodium ions[60] and sodium bound to the granule matrix proteins may help maintain edema in the affected tissue.

*Table II. Endoneurial Edema
and Increased Endoneurial Fluid Pressure
Associated with Wallerian Degeneration
and Lesions of the Axon*

Traumatic injuries (crush)
Cold injury (cryoprobe)
Chemical injury (Compound 48/80)
Heat injury (laser)
Entrapment neuropathy (compression)

4.2. Cryoprobe Injury

Freezing a nerve, or part of a nerve, results in damage to myelinated axons[34] as well as necrosis of endoneurial blood vessels. Vascular injury results in a sharp increase in EFP, in contrast to the more gradual increase in EFP associated with Wallerian degeneration following crush injury.

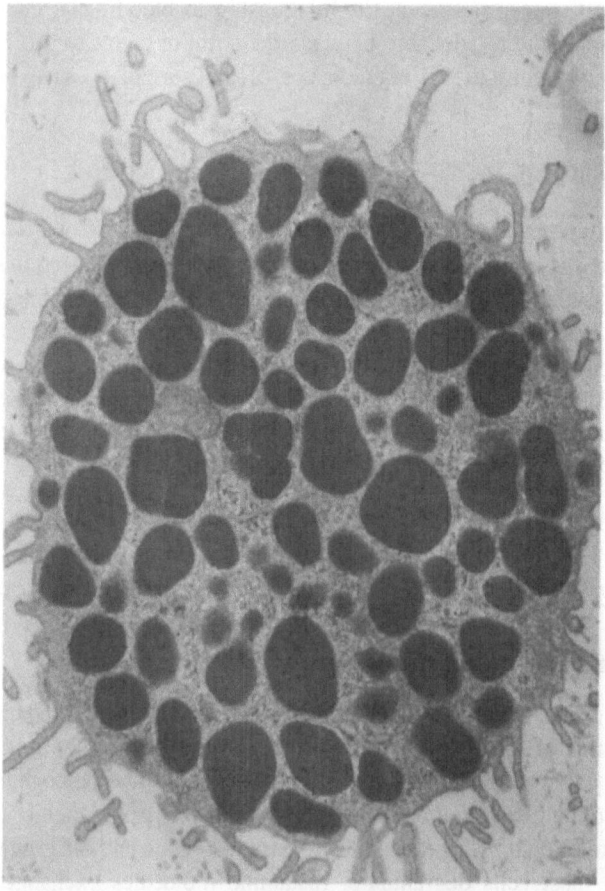

Figure 13. Mast cell degranulation, a mechanism of endoneurial edema. Normal-appearing rat mast cell in rat peripheral nerve after injection of phosphate buffered saline (control).

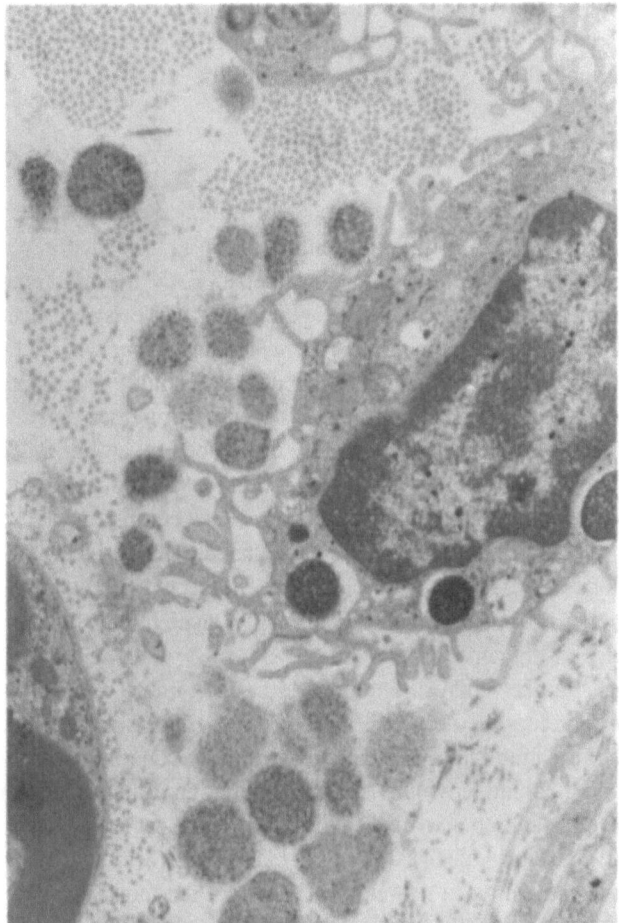

Figure 14. Mast cell degranulation, a mechanism of endoneurial edema, following endoneurial injection of histamine. Granules are being extruded into the adjacent interstitium. The structureless space represents edema.

4.3. Chemical Injury

Injection of the chemical agent Compound 48/80 is associated with extensive degranulation and lysis of mast cells as well as destruction of myelinated nerve fibers.[41,46] On the other hand, injection of histamine alone causes mast-cell degranulation (Figs. 13 and 14) leading to edema and increased EF; however, mast cells do not undergo lysis, and myelinated fibers remain undamaged.[46]

4.4. Laser Injury

Selective damage to the endoneurial parenchyma can result from laser injury.[37] Endoneurial edema and injury to myelinated fibers occurs in the damaged area.

4.5. Entrapment Neuropathy

Experimental nerve compression gives rise to endoneurial edema and nerve fiber injury. Compression of rat sciatic nerves for 2 hr with 80, 30, or 10 mm Hg causes edema and nerve fiber injury.[25,43] At high pressures, i.e., 80 mm Hg, axonal damage is predominant and edema is severe with a significant increase in EFP. At lesser pressures, i.e., 30 or 10 mm Hg, the principal injury is to the myelin sheath.

4.6. Secondary Axonal Injury

Axonal degeneration occurs in primary demyelinating neuropathies such as HCP myelinopathy and experimental allergic neuritis (EAN), in which severe edema and increased EFP occur. It is interesting to note that in EAN the most extensive axonal pathology occurs in guinea pigs.[1] Since the guinea pig nerve is rich in mast cells, the inflammatory process may be augmented by their degranulation, which appears to intensify the edema with increased EFP and consequent axonal injury. Intense axonal damage, presumably through the same mechanism, was observed in guinea pigs immunized with Freund's conjugated adjuvant and subsequently challenged by an intraneural injection of purified protein derivative (PPD). No demyelination was observed, but axonal damage was profound.[51] When the peripheral nerve is viewed as a compartment, the vulnerability of its contents to increased pressure can be appreciated. The perineurium is a semirigid sheath that envelops each fascicle as well as being the route of entry for the numerous transperineurial vessels. The vessels' vulnerability to tensile force in the perineurium is of primary importance to the nerve fiber.[38]

Last, external compression resulting in severe axonal injury may also occur when the muscle groups surrounding the nerve are affected by massive swelling in a muscle compartment. Infusion of saline into a muscle compartment is one mechanism of increasing pressure.[9] A second technique involves electric stimulation of a muscle group in anesthetized rabbits in which a limb has been encased in a plaster cast.[6] Repetitive electric stimulation of the muscle induces swelling, which when restricted by the plaster cast causes Wallerian degeneration in nerves located within the muscle.

In conclusion, edema produced through several mechanisms is a pathogenic factor in many types of neuropathy. Although the effects of edema in the brain are well known, nerve injury due to endoneurial edema is less clearly appreciated. Nerves passing through bony compartments are obviously vulnerable, but nerves in soft tissue are also subject to damage arising from edema, increased pressure, and ischemia. The perineurium, which plays a crucial role in isolating the nerve contents in a physiologic compartment, also plays a part in nerve injury through its distinctive pattern of vascularization. Edema may also be involved in functional disturbances of nerve through alterations of electrolyte concentrations that disturb the endoneurial microenvironment.

5. SUMMARY

Endoneurial edema occurs in many types of peripheral neuropathy (Fig. 15) and has independent pathologic effects on the nerve fiber. These are due to the compliance properties of peripheral nerve in which the perineurial sheath tends to resist swelling, hence the increase of EFP when edema develops. Experimental studies of nerve edema in rats and mice reveal five different pathogenic mechanisms of nerve edema (Table I). These include (1) altered vascular permeability, (2) intramyelinic edema due to lipid-soluble neurotoxins that pass across the BNB, (3) Wallerian degeneration, (4) edema due to accumulation of endogenously generated metabolites within the endoneurial compart-

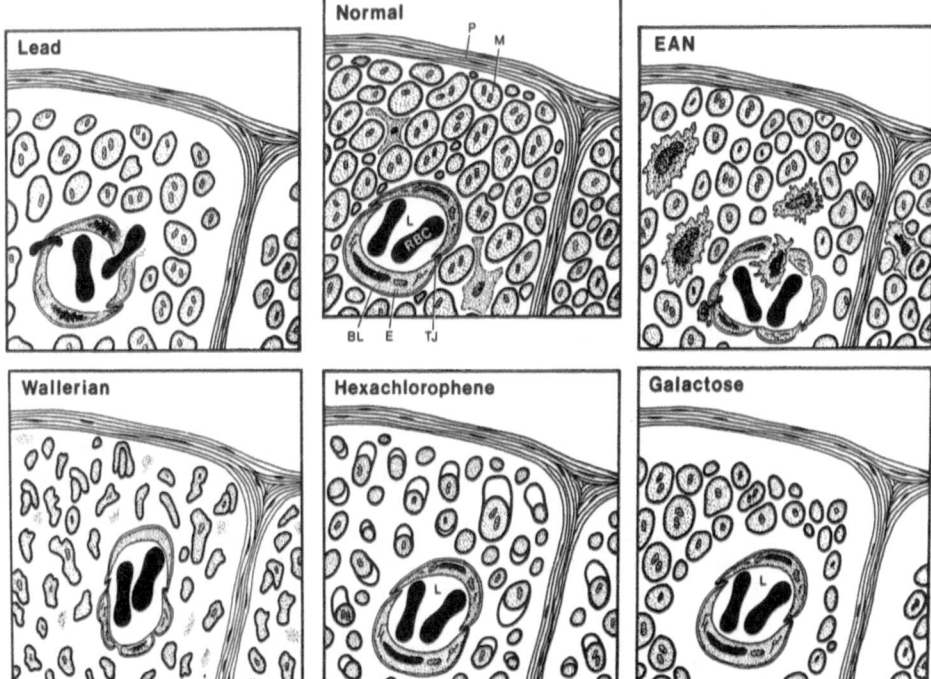

Figure 15. Endoneurial edema in different experimental models of peripheral neuropathy. Interendothelial cell leakage is the mechanism of altered permeability in lead neuropathy and experimental allergic neuritis (EAN), while tight junctions (TJ) remain sealed in hexachlorophene (HCP) and galactose neuropathy. In HCP neuropathy, edema is confined to the myelin sheaths, but in galactose neuropathy it is found in the interstitium, in which massive increase in sodium contents occurs due to comigration of electrolytes and sugar into the endoneurial compartment. P, perineum; E, endothelium; BL, basal lamina; L, lumen; RBC, red blood cell; M, myelinated nerve.

ment associated with increased activity of the sorbitol pathway, and (5) altered perineurial permeability associated with local anesthetics. These disturbances have different effects on the endoneurial environment: diminished nerve–blood flow, axonal degeneration and demyelination, disturbance of endoneurial electrolytes, and diminished endoneurial oxygen tension. Perhaps the single most important consequence of increased EFP is ischemia, which is responsible for degenerative changes of nerve fibers.

While the life-threatening effects of increased intracranial pressure are well known, damage to nerve fibers due to increased EFP has only recently been appreciated but has not been demonstrated in many types of neuropathy, including such clinically important conditions as diabetes mellitus, nerve entrapment, toxic neuropathies, and allergic neuritis.

ACKNOWLEDGMENTS

This work was supported in part by grants NS14162 and NS18715 from the National Institute of Neurological and Communication Disorders and Stroke and by the Veterans Administration.

REFERENCES

1. Allt G, Evans EM, Evans DHL: The vulnerability of immature rabbits to experimental allergic neuritis: A light and electron microscopic study. *Brain Res* 29:271–291, 1971.
2. Askenase PW, Bursztajn S, Gershon MD, et al: T-cell-dependent mast cell degranulation and release of serotonin in murine delayed-type hypersensitivity. *J Exp Med* 152:1358–1374, 1980.
3. Brosnan CF, Lyman WD, Tansey F, et al: Quantitation of mast cells in experimental allergic neuritis. *J Neuropathol Exp Neurol* 44:196–203, 1985.
4. Cammer W: Toxic demyelination: Biochemical studies and hypothetical mechanisms. In Spencer PS, Schaumburg HH (eds): *Experimental and Clinical Neurotoxicology.* Williams & Wilkins, Baltimore, 1980, pp. 206–219.
5. Costello ML, Powell HC, Myers RR: Microgravimetric analysis of nerve edema. *Muscle Nerve* 5:261–264, 1982.
6. Friden J, Lieber RL, Myers RR, et al: Myoneurial necrosis following high-frequency electrical stimulation of the cast-immobilized rabbit hindlimb. *Muscle Nerve,* (in press), 1988.
7. Gabbay KH, Snider JJ: Nerve conduction defect in galactose-fed rats. *Diabetes* 21:295–300, 1972.
8. Graham DI, Gonatas NK: Triethyl-tin sulfate-induced splitting of peripheral myelin in rats. *Lab Invest* 29:628–632, 1973.
9. Hargens AR: Introduction and historical perspectives. In Hargens AR (ed): *Interstitial Fluid Pressure and Composition.* Williams & Wilkins, Baltimore, 1981, pp. 1–11.
10. Hess K, Eames R, Darveniza P, et al: Acute ischemic neuropathy in the rabbit. *J Neurol Sci* 44:19–43, 1979.
11. Izumo S, Linington C, Wekerle H, et al: Morphologic study of experimental allergic neuritis mediated by T-cell lines specific for bovine P_2 protein in Lewis rats. *Lab Invest* 53:209–218, 1985.
12. Jacobsen J: Peripheral nerves in early diabetes: Expansion of the endoneurial space as a cause of increased water content. *Diabetologia* 14:113–119, 1978.
13. Krnjevik K: The distribution of Na and K in cat nerves. *J Physiol (Lond)* 128:473–488, 1955.
14. Landis EM, Pappenheimer JR: Exchange of substances through the capillary walls. In Landis EM and Pappenheimer JR (eds): *Handbook of Physiology.* Section 2: *Circulation.* American Physiology Society, Washington, DC, 1963, pp. 961–1034.
15. Lampert PW: Mechanism of demyelination in experimental allergic neuritis. *Lab Invest* 20:127–138, 1969.
16. Lampert PW, O'Brien JS, Garrett RS: Hexachlorophene encephalopathy. *Acta Neuropath (Berl)* 23:326–333, 1973.
17. Low PA: Endoneurial fluid pressure and microenvironment of nerve. In Dyck PJ, Thomas PK, Lambert EH, Bunge RP (eds): *Peripheral Neuropathy.* WB Saunders, Philadelphia, 1984, pp. 599–618.
18. Low PA, Dyck PJ: Increased endoneurial fluid pressure in experimental lead neuropathy. *Nature (Lond)* 269:427–428, 1977.
19. Low PA, Marchand G, Knox F, et al: Measurement of endoneurial fluid pressure with polyethylene matrix capsules. *Brain Res* 122:373–377, 1977.
20. Low PA, Dyck PJ, Schmelzer JD: Mammalian nerve sheath has unique responses to chronic elevations of endoneurial fluid pressure. *Exp Neurol* 70:300–306, 1980.
21. Low PA, Dyck PJ, Schmelzer JD: Chronic elevation of endoneurial fluid pressure is associated with low-grade fiber pathology. *Muscle Nerve* 5:162–165, 1982.
22. Low PA, Nukada H, Schmelzer JD, et al: Endoneurial oxygen tension and radial topography in nerve edema. *Brain Res* 341:147–154, 1985.
23. Lundborg G: Ischemic nerve injury. *Scand J Plast Reconstr Surg* (suppl. 6), 1970.
24. Lundborg G, Brancmark PI: Microvascular structure and function of peripheral nerves. Vital microscopic studies of the tibial nerve in the rabbit. *Adv Microcirc* 1:66–88, 1968.
25. Lundborg G, Myers RR, Powell HC: Nerve compression injury and increased endoneurial pressure: A "miniature compartment syndrome." *J Neurol Neurosurg Psychiatry* 46:1119–1124, 1983.
26. Maxwell IC, LeQuesne PM: Conduction velocity in hexachlorophene neuropathy. Correlation between electrophysiologic and morphologic findings. *J Neurol Sci* 43:95–110, 1979.
27. Mizisin AP, Myers RR, Powell HC: Endoneurial sodium accumulation in galactosemic rat nerves. *Muscle Nerve* 9:440–444, 1986.

28. Mizisin AP, Powell HC, Myers RR: Edema and increased endoneurial sodium in galactose neuropathy: Reversal with an aldose reductase inhibitor. *J Neurol Sci* 74:35–43, 1986.

29. Myers RR, Powell HC: Edema in peripheral neuropathy. In Hargens AR (ed): *Interstitial Fluid Pressure and Composition*. Williams & Wilkins, Baltimore, 1981, pp. 193–198.

30. Myers RR, Powell HC: Galactose neuropathy: Impact of chronic endoneurial edema on nerve blood flow. *Ann Neurol* 16:587–594, 1984.

31. Myers R, Powell HC, Costello ML, et al: Endoneurial fluid pressure: Direct measurement with micropipettes. *Brain Res* 148:510–515, 1978.

32. Myers RR, Costello ML, Powell HC: Increased endoneurial fluid pressure in galactose neuropathy. *Muscle and Nerve* 2:299–303, 1979.

33. Myers RR, Powell HC, Shapiro H, et al: Changes in endoneurial fluid pressure, permeability and peripheral nerve ultrastructure in experimental lead neuropathy. *Ann Neurol* 8:392–401, 1980.

34. Myers RR, Powell HC, Heckman HM, et al: Biophysical and pathological effects of cryogenic nerve lesion. *Ann Neurol* 10:478–485, 1981.

35. Myers RR, Shapiro HM, Mizisin AP, et al: Reduced nerve blood flow in hexachlorophene neuropathy. Relationship to elevated endoneurial fluid pressure. *J Neuropathol Exp Neurol* 41:391–399, 1982.

36. Myers RR, Heckman HM, Powell HC: Endoneurial fluid is hypertonic. Results of microanalysis and its significance in neuropathy. *J Neuropathol Exp Neurol* 42:217–224, 1983.

37. Myers RR, James HE, Powell HC: Laser injury of peripheral nerve—A model for focal endoneurial damage. *J Neurol Neurosurg Psychiatry* 48:1265–1268, 1985.

38. Myers RR, Murakami H, Powell HC: Reduced nerve blood flow in edematous neuropathies. *Microvascular Res* 32:145–151, 1986.

39. Myers RR, Kalichman MW, Reisner LS, et al: Neurotoxicity of local anesthetics: Altered perineurial permeability edema and nerve fiber injury. *Anesthesiology* 64:29–36, 1986.

40. Nichols PC, Dyck PJ, Miller DR: Experimental hypertrophic neuropathy: Change in fascicular area and fiber spectrum after acute crush injury. *Mayo Clin Proc* 43:297–305, 1968.

41. Olsson Y: Mast cells in the nervous system. *Int Rev Cytol* 24:27–70, 1968.

42. Powell HC, Myers RR: Schwann cell changes and demyelination in chronic galactose neuropathy. *Muscle Nerve* 6:118–127, 1983.

43. Powell HC, Myers RR: Pathology of experimental nerve compression. *Lab Invest* 55:91–100, 1986.

44. Powell HC, Myers R, Zweifach B, et al: Endoneurial pressure in hexachlorophene neuropathy. *Acta Neuropathol (Berl)* 41:139–144, 1978.

45. Powell HC, Myers RR, Costello ML et al: Endoneurial fluid pressure in Wallerian degeneration. *Ann Neurol* 5:550–557, 1979.

46. Powell HC, Myers RR, Costello ML: Increased endoneurial fluid pressure following injection of histamine and compound 48/80 into rat peripheral nerves. *Lab Invest* 43:564–571, 1980.

47. Powell HC, Costello ML, Myers RR: Galactose neuropathy. Permeability studies, mechanism of edema and mast cell abnormalities. *Acta Neuropathol (Berl)* 55:89–95, 1981.

48. Powell HC, Myers RR, Lampert PW: Changes in Schwann cells and vessels in lead neuropathy. *Am J Pathol* 109:193–205, 1982.

49. Powell HC, Braheny SL, Myers RR, et al: Early changes in experimental allergic neuritis. *Lab Invest* 48:332–338, 1983.

50. Powell HC, Knobler R, Myers RR: Peripheral neuropathy in the Twitcher mutant: A new experimental model of endoneurial edema. *Lab Invest* 49:19–25, 1983.

51. Powell HC, Braheny SL, Hughes RAC, et al: Antigen-specific demyelination and significance of the bystander effect in peripheral nerves. *Am J Path* 114:443–453, 1984.

52. Ross MH, Reith EJ: Perineurium: Evidence for contractile filaments. *Science* 165:604–606, 1969.

53. Rydevik B, Lundborg G: Permeability of intraneural vessels and perineurium following acute, graded experimental nerve compression. *Scand J Plastic Reconstr Surg* 11:179–187, 1977.

54. Sharma AK, Baker AN, Thomas PK: Peripheral nerve abnormalities related to galactose administration in rats. *J Neurol Neurosurg Psychiatry* 39:794–802, 1976.

55. Sladky JT, Greenberg JH, Brown MJ: Regional blood flow in normal and ischemic rat sciatic nerve. *Neurology (NY)* 38:101, 1983.

56. Spencer PS, Foster GN, Sterman AB, et al: Acetyl ethyl tetramethyl tetralin. In Spencer PS, Schaumburg HH (eds): *Experimental and Clinical Neurotoxicology*. Williams & Wilkins, Baltimore, 1980, pp. 296–308.

57. Sunderland S: The nerve lesion in the carpal tunnel syndrome. *J Neurol Neurosurg Psychiatry* 39:615–626, 1976.
58. Taylor AE, Gibson WH, Granger HJ, et al: The interaction between intracapillary and tissue forces in the overall regulation of interstitial fluid volume. *Lymphology* 6:192–208, 1973.
59. Tuck RR, Schmelzer JD, Low PA: Endoneurial blood flow and oxygen tension in the sciatic nerves of rats with experimental diabetic neuropathy. *Brain* 107:913–950, 1984.
60. Uvnas B: Histamine storage and release. *Fed Proc* 33:2172–2174, 1984.
61. Wiederheilm CA, Woodbury JW, Kirk S, et al: Pulsatile pressures in microcirculation of frog's mesentery. *Am J Physiol* 207:173–176, 1964.
62. Windebank AJ, McCall JT, Hunder HG, et al: The endoneurial content of lead related to the onset and severity of segmental demyelination. *J Neuropathol Exp Neurol* 39:692–699, 1980.

9

Potential for Pharmacologic Manipulation of the Blood–Cerebrospinal Fluid Barrier

Conrad E. Johanson

1. INTRODUCTION

Recent physiologic findings suggest the choroid plexus (CP) as a potentially important site for selective delivery of water-soluble agents and ions from blood into the adult mammalian central nervous system (CNS).[149] Because of structural and functional differences between the CP and the blood–brain barrier (BBB), it appears a priori that the unique properties of the blood–cerebrospinal fluid (CSF) barrier (i.e., mainly the CP) would afford pharmacotherapeutic opportunities not extant with cerebral capillaries.

Some previous reviews have relegated a relatively minor role, quantitatively, for CP transport, in comparison with the substantially more extensive BBB.[103–107] The main argument has been that the surface area of the CP is three to four orders of magnitude less than the total circumferential surface area of the brain capillaries; therefore, in the overall economy of CNS function, the choroidal epithelium has often been viewed as having a fine-tuning rather than a major function in brain extracellular fluid (ECF) homeostasis.

However, with respect to water-soluble solutes, evidence is accumulating that certain organic transport systems are present in CP, but apparently not in cerebral capillaries,[147] and that there are inorganic ion antiport and symport systems in CP with transport capacity far exceeding those in corresponding systems in the BBB.[135–138] Thus, there is need for reexamination of the question, Is it feasible therapeutically to exploit the unique properties of the CP tissues in the lateral, third, and fourth ventricles? This chapter reviews mainly hydrophilic solutes, because lipophilic agents generally do not penetrate the CP and BBB differentially. The severalfold purpose of this chapter is (1) to delineate characteristics of the rate-limiting membrane in CP; (2) to analyze sidedness of the choroidal epithelium; (3) to present a model for evaluating solute movement in vivo across the blood–CSF barrier; (4) to recapitulate recent basic transport studies on micronutrients, proteins, hormones, and ions; (5) to discuss how the choroidal membrane responds to osmotic and hydrostatic

Conrad E. Johanson • Department of Clinical Neurosciences, Brown University and Rhode Island Hospital, Providence, Rhode Island 02902

223

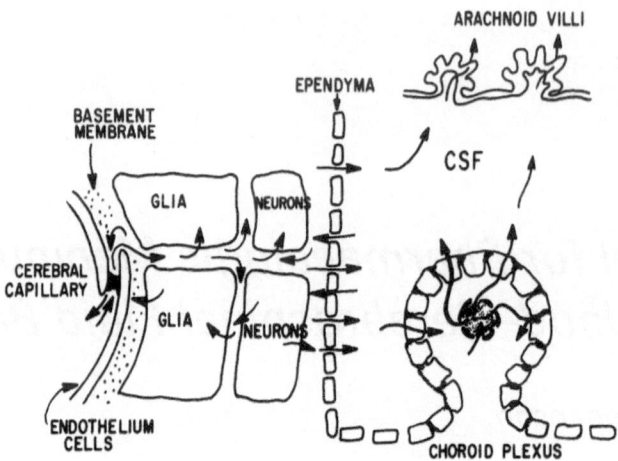

Figure 1. Schematic idealized representation of components of the blood–cerebrospinal fluid (CSF) barrier, i.e., the choroid plexus (CP), and its functional relationship to the other major transport interfaces in the CNS. The CP villus symbolizes collectively the tissues in the lateral, third and fourth ventricles. Solutes in choroidal plasma readily diffuse across capillaries in the core of each villus, and then into the interstitial fluid, which bathes the basolateral membrane. Thereafter, transmembrane transport occurs between and/or through the epithelial cells, depending on the size, charge, stereospecificity, or lipid solubility of the substance. Tight junctions exist at the apical pole of the cells, where the epithelia abut (darkened areas). The zonulae occludentes in the blood–CSF barrier are more penetrable than those in the BBB. Once into the ventricular fluid, a substance can then diffuse across the readily permeable ependymal lining into the brain substance, or can be carried out of the CNS, nonselectively, via bulk flow of fluid across the arachnoid villi into venous blood. (From Woodbury et al.[165])

pressure alterations; and (6) to project possible pharmacologic strategies for clinical problems.

The potential for pharmacologic manipulation of the CP is less well understood than that for the cerebral capillary wall. For therapeutic reasons, the ability to alter the blood–CSF barrier is predicated on a thorough knowledge of the physiology and pharmacology of the choroidal epithelium. Thus, this chapter describes current basic concepts within the context of possible alteration of certain transport and permeability characteristics of the plexus. Practical limitations are treated in Section 17 of this chapter.

2. CONSIDERATION OF GROSS STRUCTURE

The systemic relationship among the three major transport interfaces blood–CSF barrier, CSF–brain barrier, and blood–brain barrier (BBB) is depicted in Fig. 1. At the parenchymal level, functional comparison of the blood–CSF barrier with the BBB is facilitated by the fact that one cell type (choroidal epithelium or cerebral endothelium, respectively) comprises each barrier. Furthermore, comparison of transport/permeability phenomena in the two barriers is more straightforward, since each system consists of a single layer of circumferentially arranged parenchymal cells[70,94] (see Chapter 7, this volume). This is in contrast to multilayered systems such as skin, for which the pharmacokinetic phenomena are more complex and difficult to interpret.

3. BARRIER CELLS: PLEXUS EPITHELIUM VERSUS CEREBRAL ENDOTHELIUM

The parenchymal cells that make up the CP are of the typical epithelial genre. With a secretory as well as reabsorptive function, the choroidal epithelial cells thus resemble those found in the ciliary process, gallbladder, small intestine, proximal tubule, and other structures. On the other hand, the main cell type in the cerebral capillary is endothelial in nature. However, among capillaries in various organs, the endothelium in the brain is unique in that the cells have developed epithelial-like characteristics necessary for extensive net transcellular transport by active and facilitated mechanisms.[9] Thus, in many ways, the cerebral endothelium is functionally and structurally more akin to choroidal epithelium than to muscle (or other non-neural tissue) endothelium.

Nevertheless, there are important differences between the barrier cells of the blood–CSF and BBB interfaces. By taking advantage of dissimilarities in carrier specificity, transport capacity (V_{max}), tight junction permeability, vesicular transport capability, and fluid secretory ability, it should be possible to expedite the movement of certain solutes selectively across the CP or otherwise promote their retention in CSF. Once in the CSF, even macromolecules can migrate unhindered across the freely permeable ependymal wall surrounding the ventricles.[149]

4. POLARITY OF MEMBRANES IN CNS TRANSPORT INTERFACES

The epithelial membrane of the CP displays sidedness, i.e., the basolateral side of the choroid cell facing the plasma has structural and functional features strikingly different from those associated with the opposite, apical pole of the cell in contact with the CSF. Such polarization permits the net secretion of solutes and water from blood to CSF concurrently with net reabsorption of other compounds in the opposite direction.[63] By elucidating the physiologic nature of transepithelial fluxes of materials across the choroidal membrane, the pharmacologist and clinician should be better able to manipulate the movement of drugs, ions, and water between plasma and CSF. Many principles of renal pharmacology should apply to the CP, often referred to as the miniature kidney of the CNS. Indeed, there are many common denominators of ultrastructure when the epithelia of the CP and renal proximal tubule are compared.

The basolateral membrane of CP demarcates the base and two sides of the epithelium, which separates the outlying interstitial fluid from the cytoplasm within. The apical membrane interfaces the intracellular fluid (ICF) with ventricular CSF. There is similar sidedness in the BBB. Thus, the luminal and antiluminal (or abluminal) poles of the cerebral endothelium face the plasma and brain ECF, respectively.

The comparison of sidedness in the CP and BBB is important when analyzing analogies in barrier function. For example, there is preponderance of Na–K–ATPase in the CP apical membrane, which functions to keep CSF [K] relatively low.[133] Correspondingly, in the cerebral capillary wall, the polar distribution of Na–K–ATPase in the abluminal membrane likely serves to maintain brain ECF [K] lower than in plasma.[9] On the other hand, the differences in CP apical versus BBB abluminal membrane properties (or CP basolateral versus BBB luminal) may represent site-specific opportunities for regulating drug movements among CNS compartments.

5. EXPERIMENTAL PREPARATIONS OF CHOROID PLEXUS

Several different methods have been employed to characterize transport across the membranes of CP. Each has its own advantages and shortcomings. Collectively, the data procured from various in vitro, in situ and in vivo approaches have furnished an insightful overview of transporting capabilities.[64] The various preparations have been extensively critiqued in a recent review of CP methodologies.

5.1. In Vitro

Choroid plexuses incubated in synthetic CSF medium readily accumulate numerous substances.[134,148] Concentrations of drugs and inhibitors in the medium can be finely and widely altered.[70] Kinetic analyses of the rate and extent of uptake, expressed as the ratio of tissue to medium concentration of test substance, have furnished useful K_m and V_{max} parameters. However, the kinetic constants cannot be separately ascribed to basolateral and apical membranes.

The Ussing-type chamber preparation enables polarity of membrane transport to be maintained.[167] Thus, ionic and molecular fluxes of tracers can be determined advantageously for basolateral to apical movement, and in the reverse direction. As with any in vitro technique, the results obtained eventually have to be related to the natural environment of the CP in living organisms.

5.2. In Situ

Isolated plexus tissue secretes nascent CSF, which can be readily sampled by micropipette. Ames et al.[4] and, more recently, Miner and Reed[88] used their results to estimate fractional clearance of solute from blood per unit volume of CSF secreted. Extracorporeal perfusions of CP permit calculation of arteriovenous concentration differences.[111] All the in situ preparations are complex to set up and execute,[51] and they raise the question about the physiologic state of the animal after extensive surgery. However, because isolated CP preparations are potentially powerful techniques for sampling fresh CSF before it flows downstream, it is unfortunate that pharmacokinetic analyses of newly formed CSF have not been more commonly carried out in laboratory animals. Although in situ preparations of CP do not lend themselves to direct application in clinical investigations, their usage in basic experiments has provided valuable physiologic information.

6. IN VIVO ANALYSES OF CHOROID PLEXUS–CSF SYSTEMS

It would be ideal to analyze the function of the CP in its natural environment during noninvasive experimentation. The in vivo method comes close to achieving this aim.[67,68,71] Thus, the experiment is allowed to run its course in the intact animal; immediately after decapitation, the CPs are rapidly excised and analyzed for content of water, ions, tracers, and drugs by compartmentation analysis.[72] Choroid cell composition is then related to that in plasma and CSF at the time the animal is killed, and deductive conclusions can be made from concentration gradients, concerning polar transport across the external limiting membranes of the CP epithelium.[62,68,137] Although this approach lacks the precision found, for example, with direct microelectrode impalement of cells, it does furnish useful information about ion and drug distribution among the compartments in the plasma–CP–CSF system.

Scores of experiments have analyzed the time course of concentration of drugs in CSF following intravascular administration, but few have attempted to correlate the uptake of drugs by CP with that by CSF.[67] It is surprising that so few CSF investigators have analyzed CP tissue directly for content of drugs administered by in vivo protocols.

Can CSF composition be reliably related to transport phenomena in CP? First, it is important to sample CSF from a site near CP. This can be done by aspirating CSF from the cisterna magna and fourth ventricle region and concurrently removing CP from the fourth ventricle.[92] Second, the time of the sampling of CSF must be within 15–30 min after the plasma concentration of drug (or tracer) is at steady state. This ensures that the fast component of CSF uptake of solute is occurring.[92,135,137]

The literature is replete with information that CSF uptake curves resolve basically into the earlier fast component and the later slow component (Fig. 2). Three lines of correlative evidence strongly suggest that the fast component of solute uptake by CSF involves a transport pathway mainly including the CP rather than the brain:

1. Physically, the CP is suspended in CSF. The brisk blood flow to the CP along with the relatively short diffusion distance (in microns) between choroidal plasma and CSF, dictate that small (^3H_2O), lipid-soluble ([^{14}C]antipyrine), and carrier-transported (^{22}Na) agents rapidly (i.e., within a few minutes) build up in CSF.[67,68] By contrast, diffusion distances from blood to brain to CSF, on average, are in millimeters or centimeters.

2. Pharmacologically, enzyme inhibitors such as acetazolamide reduce CSF flow maximally by about 50%. ^{22}Na transport from plasma to CSF is directly proportional to CSF flow rate. Acetazolamide also inhibits ^{22}Na uptake by CP, and by CSF, by as much as 50%.[132] Acetazolamide does not significantly reduce ^{22}Na uptake by brain tissue.

3. Developmentally, as is the case of ^{22}Na, there is a fast component of uptake of ^{36}Cl by CSF for adult but not for immature animals (1-week-old Sprague–Dawley rats)[138] (see Fig. 2). Consistent with the above mentioned findings, the CSF formation rate by CP in immature rats is not nearly at adult-level capacity.[66,102]

7. RATE-LIMITING STEP IN TRANSCHOROIDAL TRANSPORT

The in vivo type of experiment just described has furnished information about the distribution of solutes from plasma in CP to fluid compartments outside the vessels. The capillaries in CP, unlike those in most regions of the brain, are readily permeable to water-soluble materials. The more-or-less equilibrium distribution of most compounds between plasma and interstitial fluid of CP is in contrast to the situation in brain. Thus, for permeation of most hydrophilic solutes from plasma into CSF, the rate-limiting step in distribution is not the CP endothelial–interstitial boundary.

Escaping readily from choroidal capillaries, most compounds diffuse rapidly through the interstitium before their movement toward CSF is thwarted by the basolateral membrane and tight junctions between epithelia. Evidence is presented in Section 8.2 that the basolateral membrane is the primary rate-limiting step for water-soluble agents. Although

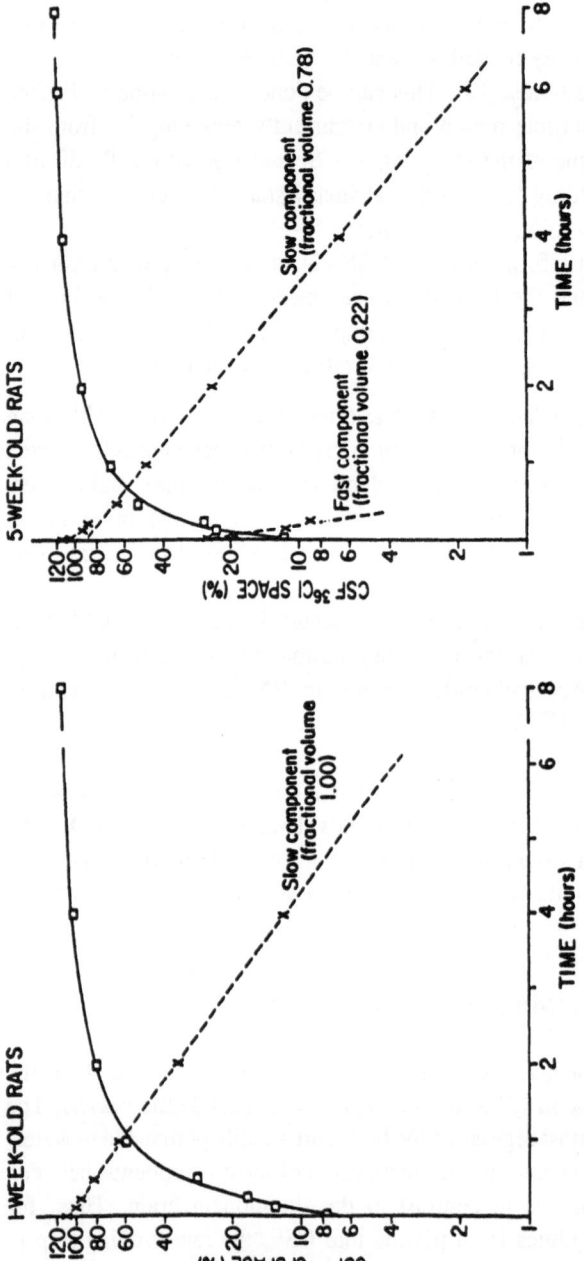

Figure 2. Resolution of cerebrospinal fluid (CSF) radioisotope curves into fast and slow components of uptake. ^{36}Cl penetration from plasma to CSF is depicted for infant and juvenile rats. ^{36}Cl was injected intraperitoneally, at various times prior to sacrifice, into 4-hr nephrectomized rats. Uptake curve: cisternal CSF radioactivity (unfilled squares) is expressed as a space (%) = $100 \times$ (dpm/g CSF)/(dpm/g plasma). CSF uptake of ^{36}Cl was resolved into a slow component by subtracting the observed spaces from the steady-state space, and plotting the difference (x····x) as a function of time after ^{36}Cl administration. The slope and y-intercept were obtained by fitting the linear portion of the graph to a straight line by the least-squares method. The fast component of curves for 5-week-old animals was resolved by subtracting the spaces predicted by the extrapolated slow component line from the spaces not fitted to the slow-component line (time = 0, 1/12, 1/6, and 1/4 hr), and plotting the difference (x····x) versus time. Again, a slope and y-intercept were obtained by linear regression. The fractional volume represents the ratio of the component volume to the steady-state space. Each point = three animals. Standard errors are generally < 5% of respective means. Resolution of the CSF uptake curve yielded a single statistically significant component (slow) for 1-week-old animals. However, the CSF uptake curve for the older rats resolved into two components: fast (constituting about one-fourth of the steady-state space) and slow (about three-fourths of the space). Similar kinetics were observed for CSF uptake of ^{22}Na (at corresponding ages). (From Smith et al.[138])

the tight junctions in CP are somewhat less restricting than those in cerebral endothelium, nevertheless they offer resistance to diffusing hydrophilic molecules above a certain critical molecular weight (see Section 8.1).

In respect to transport in the reverse direction, i.e., ventricular CSF to choroidal blood, the rate-limiting step is probably the apical membrane. Organic acids and bases are actively transported from CSF to CP epithelium, by systems with properties similar to those in the proximal tubule.[38,113,127] By increasing levels of stable organic cations and anions in ventricular fluid, it is possible to inhibit substantially the excretory clearance of the corresponding tracers from CSF to CP and blood. In the clinical evaluation of CNS levels of organic acid metabolites of certain neurotransmitters, the well-known CSF probenecid test takes advantage of competitive inhibition of anion transport in CP.

The concept of the rate-limiting step implies that once a solute is actively transported into the cell across the external limiting membrane, it can subsequently move readily out of the cell across the membrane at the other pole of the epithelium. The extrusion process involves movement down an electrochemical gradient by passive or facilitated diffusion.[149] The scope of the present review is more concerned with CP secretory transport (blood to CSF) than with reabsorptive transport (CSF to blood).

8. PERMEABILITY OF TIGHT JUNCTIONS AND MEMBRANE POLES

A thorough physiologic understanding of the CP is a prerequisite to effective pharmacologic manipulation of the blood–CSF barrier. Solutes can penetrate the choroid epithelial membrane either paracellularly, i.e., through tight junctions, and/or transcellularly, i.e., through the external limiting membranes (basolateral or apical) of the parenchymal cells.

8.1. Tight Junctions

The cells of transporting epithelia are joined to each other by a continuous band of tight junctions (zonulae occludentes) near the luminal (apical) surface. Tight junctions have been classified as either tight (stomach and urinary bladder) or leaky (intestine and gall-bladder). The zonulae occludentes in the CP are in the leaky category.[89] Discontinuities in the multiple strands of the tight junctions may well be the physical basis for permeability.

The close-to-isotonic nature of mammalian CSF is consistent with relatively permeable junctions in CP not permitting establishment of a large osmolality gradient between plasma and the elaborated fluid. Moreover, a substantial electrical gradient cannot be maintained. Wright and Zeuthen[169] summarized electrophysiologic evidence (i.e., transepithelial resistance, conductance, and current–voltage curves) for a significant paracellular pathway in CP.

Smaller solutes, like ions, readily penetrate the intercellular channels containing the tight junctions. There is a low degree of selectivity of passive ion permeation across frog CP epithelium compared with other physiologic systems at similar field strength.[169] Thus, the paracellular shunt accounts for at least 95% of transepithelial passive ion fluxes in the CP of amphibians, and presumably of mammalians as well. Electron microscopic analyses of trivalent lanthanum permeability have implicated the tight junction as the location for the shunt. Iodide, after being actively transported across the CP epithelium, diffuses back across the membrane by the paracellular route. Another piece of physiologic evidence for significant movement of ions, paracellularly, is the determination of a great passive unidirectional flux of sodium relative to the active transport of this cation. Overall, the freely penetrating ions studied have atomic weights in the range of 20–130.

The diffusion of much larger molecules, such as horseradish peroxidase (HRP: 43,000 M_r), is thwarted by tight junctions in CP, whether the proteinaceous probe is placed on the plasma or on the CSF side of the membrane.[15,16,156] Following intravascular administration, microperoxidase (14,000 M_r) and cytochrome c (13,000 M_r), as well as HRP, move readily through the basement membrane and intercellular clefts of the choroidal epithelium, from the connective tissue to the apical tight junctions where they are stopped.[87] Thus, the tight junctions, by effectively preventing further diffusion into the CSF, represent a significant part of the blood–CSF barrier to large lipid-insoluble molecules.

What, then, is the fate of intermediate-sized molecules, such as those with molecular weights approximating the size of pharmacologic agents? Many drugs are polar organic acids and bases having a molecular weight of roughly 150–600. Unfortunately, little attention has been paid to understanding the movement (or lack of movement) of water-soluble nonelectrolytes and organic electrolytes through tight junctions. In an experiment with intraventricularly perfused 5-hydroxydopamine (170 M_r), Richards[122] noted that the tight junctions in CP constitute a barrier to this amine. In general, catecholamine agents significantly penetrate neither the blood–CSF nor the BBB.

Transjunctional movement of nonelectrolytes, drugs, and even water is poorly understood, not only for CP but for epithelial systems in general. However, the basic studies conducted by Wright and Prather of nonelectrolyte permeation patterns in amphibian CP serve as a useful reference for future pharmacologic correlation of structure with penetrability.[168] These investigators used an in vitro preparation to quantify, by an osmotic method, the choroid epithelial permeability to 50 nonelectrolytes. Reflection coefficients, as an inverse measure of permeability, were determined by a rapid electrical procedure. Sucrose (342 M_r) and raffinose (504 M_r) were found to be essentially impermeant, i.e., with coefficients close to 1.0. For comparison, glycerol (92 M_r) and urea (60 M_r) had reflection coefficients of 0.81 and 0.56, respectively. The highly diffusible ethanol molecule had a coeffecient of zero. Although this early investigation was unable to distinguish between paracellular and transcellular permeation, it seems probable that the solutes with coefficients closest to one would encounter the most difficulty in penetrating the zonulae occludentes as well as the membrane surrounding the choroid cell.

8.2. Basolateral Membrane

This section describes in vivo distributional phenomena in experiments in which each tracer was administered either intraperitoneally (IP) or intravenously (IV) to attain a steady level in plasma. Polysaccharides and oligosaccharides have been used to estimate the ECF volume of CP, as these probes do not penetrate the basolateral membrane to gain access to epithelial cell water.[44,61,72] Inulin (5000–5500 M_r) either labeled with carbon-14 or tritiated, attains a steady-state volume of distribution of about 15% of wet tissue weight, in young adult rat CP in lateral and fourth ventricles (Fig. 3). Even though tritiated raffinose (504 M_r) has a molecular weight one-tenth that of inulin, this trisaccharide does not distribute into a larger volume, thereby, indicating confinement of marker to the extracellular compartment.[44,92]

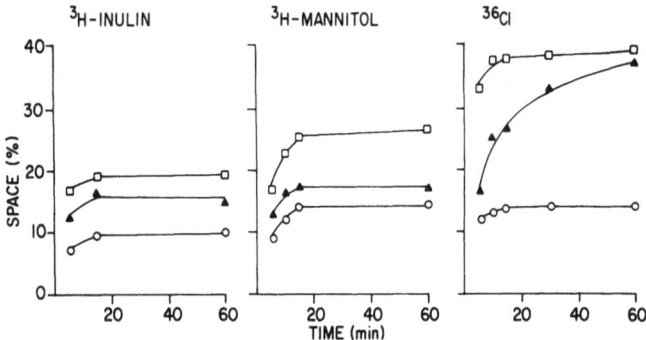

Figure 3. The rate and extent of uptake of tracers by fourth ventricle choroid plexus (triangles), compared with submaxillary salivary gland (squares) and skeletal muscle (circles), in adult Sprague–Dawley rats. Plasma level of isotopes was held steady by bilateral nephrectomy, following the intraperitoneal (IP) injection of tracers. Space is defined in Fig. 2. Each symbol is $n = 3$. Standard errors are generally <5% of respective means. Steady-state values for inulin and mannitol are estimates of extracellular fluid volume. ^{36}Cl gains access to the cellular water in the two secretory epithelia, but not significantly so in muscle fibers. (From Smith et al.[136])

Mannitol, a hexahydric alcohol of 182 M_r, does not cross the interstitial cell–membrane boundary in most tissues,[61] except in liver (see discussion of Table IX in Chapter 7). The steady-state volumes of distribution of [^{3}H]mannitol in CP, both in vitro and in vivo, are not appreciably greater than the corresponding values for radioinulin.[70] This finding strongly suggests that the basolateral membrane of CP is relatively impermeant to the six-carbon alcohol (Fig. 3).

Urea (60 M_r) penetrates the basolateral membrane, but not without impediment to diffusion.[68] Thus, in adult mammalian CP, urea does not attain equilibrium distribution between extracellular water and that inside the choroid cell.[102] Molecular sieving of urea in the basolateral membrane keeps the concentration of urea in intracellular water constantly lower than that in the interstitium (Fig. 4). At steady state, the level of urea in the choroid cell is about 70% that in plasma. Urea, with a molecular radius of 1.8 Å, can penetrate CP parenchymal cells, whereas mannitol (3.6 Å) evidently cannot (Fig. 3). This

	HOURS AFTER I.P. INJECTION OF ^{14}C-UREA						
ADULT RATS:	0.5	1	2	3	5	8	16
PLASMA H$_2$0	100	100	100	100	100	100	100
CHOROID CELL H$_2$0	2	14	38	52	65	72	71
CSF H$_2$0	17	30	43	52	62	71	72

Figure 4. Time-course analysis of the relative concentrations of [^{14}C]urea in cerebrospinal fluid (CSF) and choroid plexus epithelium, based on a plasma concentration of 100. For example, 2 hr after intraperitoneal (IP) injection of radiolabeled urea, the concentration of this tracer in the water of CSF, on average, was 43% that in plasma water. Each number is the mean concentration of [^{14}C]urea calculated from individual values for [^{14}C]urea spaces in choroidal tissues[68,102] and compartmentation data, i.e., tissue water content and residual erythrocyte volume[72] and extracellular space.[61] Standard errors for these derived values are generally in the range of 2–9% of the respective means; for steady-state values at 8 and 16 hr, errors are <5% of means. In the steady state, urea equilibrates between the choroid cellular and CSF compartments, i.e., across the apical membrane. By contrast, there is molecular sieving of urea across the plasma-facing membrane; as a result, the urea concentration in the epithelium is held below that in plasma. (From Parandoosh and Johanson.[102])

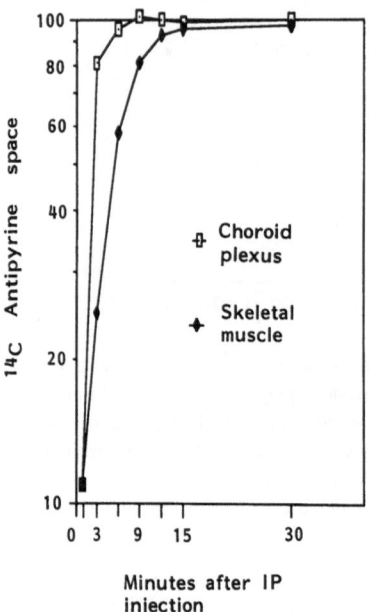

Figure 5. Lateral ventricle choroid plexus uptake from plasma of a lipid-soluble compound, the distribution of which is intimately related to blood flow. Skeletal muscle was from thigh. All Sprague–Dawley rats were anesthetized with ether. Tracer antipyrine was injected intraperitoneally, and its permeation quantified by relating radioactivity in tissue to that in plasma. Space is defined in Fig. 2. Each point is the mean for 3–5 animals. Standard errors are generally 2–8% of respective means. After allowing 3 min for distribution, the ^{14}C antipyrine space in muscle and CP was 25 and 82%, respectively, reflecting the much greater rate of perfusion of the plexus. The rapidly metabolizing and secreting CP has a blood flow 1–2 orders of magnitude greater than most regions of brain. (From Johanson and Woodbury.[67])

comparative observation gives a clue as to the size of the diffusion channels in the CP basolateral membrane.

The transport and distribution of two drugs, antipyrine and barbital, in the in vivo CP has been studied in adult rats.[67] [^{14}C]Barbital (184 M_r) penetrates the lateral ventricle CP at a rate about one-fourth that of tracer antipyrine, which has a comparable molecular weight of 188. Within 3–6 min, antipyrine distributes into the entire cell water of CP, due to its high degree of lipid solubility (Fig. 5). Barbital, which is considerably less lipophilic, undergoes molecular sieving in the basolateral membrane of CP; thus, the slow permeation of barbital into CSF correlates well with its sluggish rate of uptake by CP. The time course of uptake of barbital by CP is similar to that by cerebral cortex (see Table I in ref. 67). This is a manifestation of comparably slow uptakes of barbital across the blood–CSF barrier (basolateral membrane of CP) and the BBB (cerebral capillary wall in cortex).

8.3. Apical Membrane

There is considerable evidence that the CSF–facing membrane of CP has markedly different permeability properties compared with the basolateral counterpart. Both to ions and to nonelectrolytes, the apical pole of the cell is more permeable than the plasma-facing side.

Kinetic information obtained from in vivo mammalian studies reveals that during the approach to steady-state distribution, the tracer concentration in the CP cellular compartment closely matches that in the CSF. This is the case for both nonelectrolyte urea as well as for the ions, ^{22}Na and ^{36}Cl.[137] One interpretation is that the more permeable apical membrane promotes freer exchange of solutes than does the basolateral membrane. The assessment of absolute permeabilities of the two membranes is a difficult task, but it appears that the effective permeability of the apical membrane is greater. Since it is risky to use concentration gradient data to form conclusions deductively about relative membrane permeabilities, it is essential to pursue additional information by independent approaches.

In vitro experiments with amphibian CP have also furnished insight into the relative permeabilities of the epithelial poles. When plexuses are preloaded with radioactive amino acids, they subsequently release most of the isotope into the ventricular rather than the serosal solution.[166] This has been interpreted as evidence that the apical surface of bullfrog CP epithelium is more permeable to amino acids than the basolateral face. Electrophysiologic studies by Wright and Zeuthen[169] have provided evidence that the potassium conductivity of the apical membrane is greater than that of the basolateral membrane; this was determined in experiments in which intracellular [K] was altered upon passing an electrical current across the epithelium. Finally, the total conductance of the CP apical membrane is almost an order of magnitude greater than that of the basolateral membrane.[169]

Thus, from experiments with mammals and nonmammals, conducted in vivo and in vitro, using ionic, nonelectrolyte, and nutrient probes, the evidence collectively indicates that the apical membrane in contact with the CSF is more penetrable than the membrane facing the blood.

9. CARRIER TRANSPORT SYSTEMS IN CHOROID PLEXUS

Active transport systems in the membrane, in conjunction with its permselectivity, permit net translocation of solutes in both directions across the choroid cell. Thus, the asymmetry in permeability described above implies additional asymmetry (polarization) of the carrier transporters in the basolateral and apical membranes. Delineation of active and facilitated mechanisms in the CP basolateral membrane is an important initial step in regulating transchoroidal movement of water-soluble nutrients and drugs.

In general, research concerning the CP during the era of modern physiology has not been nearly as extensive as that of epithelial systems such as the kidney and gastrointestinal (GI) organs; therefore, the membrane mechanisms in mammalian CP are not as well understood. However, over the past decade there has been progressively increasing interest in the function of inorganic and organic transport in CP; these investigations have been carried out mainly with lateral ventricle tissues in rats, cats, and rabbits.

The most thoroughly analyzed primary active transport system in CP is the Na–K exchange pump, with its predominantly apical localization.[133] The Na–K exchanger is an energy-metabolizing system as it uses adenosine triphosphate (ATP) directly. By keeping choroid cell [K] high (150 mM) and [Na] low (30 mM), the Na–K pump has a cardinal role in helping to set up transmembrane gradients for electrical potential and for [Na].[71] Basolateral Na–H exchange (or antiport) is driven partly by the steep Na gradient directed inward, and by the cellular proton gradient oriented outward.[62,92,132] The secondary active transport systems, i.e., the Na–H and Cl–HCO$_3$ exchangers (Fig. 6), may be functionally coupled. Their operation is likely integral to fluid secretion as well as to choroid cell pH and volume regulation. Na–Ca antiport has been demonstrated in rat CP, but its polar distribution needs definition.[18] The physiologic coordination of the activities of the various inorganic antiport systems is poorly understood. Once gained, such knowledge will be invaluable in designing drugs to alter the flux of ions and water, hence CSF secretion, across the CP.

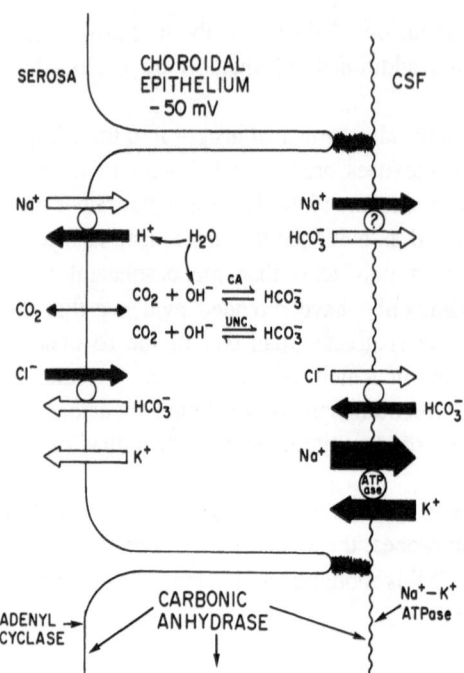

Figure 6. Model for antiport and symport ion translocation by basolateral and apical membranes of mammalian choroid plexus (CP) epithelium. Data for construction of transport scheme were obtained from adult rat lateral ventricle tissues; see refs. 18, 23, 60, 62, 70, 71, 92, 132–134, 137. Arrow widths are proportional to probable stoichiometric ion-transport exchange ratios. Filled and unfilled arrows, respectively, represent transport against and with electrochemical gradients. Because membrane density and turnover rate of the transport mechanisms await elucidation, the schema does not delineate the magnitude of net transepithelial ion fluxes. Normally, there is net secretion of Na, Cl, and HCO_3, and net reabsorption (cerebrospinal fluid [CSF] to blood) of K and H. CA, carbonic anhydrase; unc, uncatalyzed reaction. (From Johanson.[62])

Also present in CP are active systems for organic solutes that carry glucose, amino acids, anion metabolites, cationic bases, vitamins, and so forth.[6,149,167] Their function is complexly interdependent with that of the inorganic transporters that maintain cell [Na], pH, electrical potential, and so on. Vitamin and ribose-translocating systems move their substrates from plasma to CSF,[147] while the organic acid and base systems actively clear many metabolites from CSF to plasma.[113,127] Thus, micronutrient transfer is directed into CNS, while the movement of potentially deleterious catabolites is outward. The stereospecificity of the various carrier-protein molecules represents a convenient handle to exploit for therapeutically moving xenobiotic compounds into CSF.[149]

10. EXPLOITATION OF BARRIER TRANSPORT MECHANISMS IN THE BLOOD–CSF BARRIER

Several carrier-mediated transport systems present in CP appear to be absent or negligible in the BBB. Some of these micronutrients transferred across CP are ultimately actively taken up by brain cells. Thus, slight pharmaceutical modification of the structure of such endogenous compounds might retain capability both for transport into neurons/glia (to alter target cell metabolism) as well as for the initial rate-limiting transport across the blood–CSF barrier. Alternatively, knowledge of the kinetic properties (K_m, V_{max}, competitive inhibition) of these micronutrient transport systems could lead to more rational strategies for ameliorating states of deficiency (or excess) in the CNS; for example, hypovitaminoses or hypervitaminoses.

10.1. Ascorbic Acid

The criteria that establish carrier-mediated entry of a substance into brain, predominantly by way of the CP, were recently reviewed.[149] Regulation of ascorbic acid and isoascorbate in CSF, and thus effectively in brain parenchyma, depends primarily upon the rate-limiting transchoroidal transport of vitamins into ventricular fluid, followed by their unrestricted diffusion across the ependymal wall into the interstitial fluid, from which they are taken up by brain cells. Since ascorbic acid transport by the BBB is negligible, the nervous tissue is dependent on this specialized transport system in CP as its supply system.[151] The ascorbic acid carrier in CP is very efficacious, since it adequately furnishes this micronutrient for brain even in severe hypovitaminosis C.

10.2. Nucleosides and Vitamins

Deoxynucleosides, pyridoxine, and reduced folates are other examples of micronutrients that gain access to CNS predominantly by way of their respective stereospecific systems in the CP rather than by the BBB.[144,147,152] The active transport system in CP has a low affinity for endogenous nucleosides, and so it transfers them from blood to CSF as a function of their relative concentrations in plasma. In spite of relatively low affinity, however, there is some specificity in nucleoside transfer. It appears that the hydroxyl substituent at positions 2' (if present) and 3' of the ribose ring needs to have appropriate orientation for the active transport system in CP in order to have affinity for nucleosides. Since some chemotherapeutic compounds have arabinose substituted for the deoxyribose moiety, the therapeutic question presents itself whether or not the CP can actively transport cytosine arabinoside. Although the plexus does not happen to translocate this particular antineoplastic agent,[145] it is conceivable that the synthesis of comparable chemotherapeutic drugs could be successfully directed to achieving transport across CP as well as subsequently into target cells in the brain.

10.3. Purines

For several decades, many analogues of natural purine bases have been made and tested for treatment of malignancies in the CNS. In a basic study, the active accumulation of purines (and analogues) by in vitro rabbit CP, mediated by a stereospecific high-capacity system, was described by Berlin in 1969[8]; however, at that time the directionality of transport had not yet been determined. Covell et al.[21] recently presented an in vivo kinetic model for disposition of 6-mercaptopurine (6-MP) between plasma and CSF in monkeys. Their results demonstrated that the CP is the main port of entry of intravenously administered 6-MP into the CNS; thus, almost 90% of total penetration was through CP and about 10% across cerebral capillaries. Nascent CSF concentration of water-soluble 6-MP (152 M_r) was calculated as being equal to that in plasma, suggesting facilitated entrainment of drug in CP fluid. Although previous dosing strategies for the treatment of acute lymphocytic leukemia have focused on bypassing the BBB with such agents as 6-MP, this more recent multicompartmental kinetic analysis emphasizes the need to focus additional experimental attention on drug transport by choroidally secreted CSF.[21]

10.4. Amino Acids

Certain amino acids differentially permeate the CP and BBB. α-aminoisobutyric acid (AIB) slowly crosses the cerebral capillary wall, but it very rapidly penetrates into the CP–CSF system during in vivo experiments in rats,[93] presumably by carrier-mediated transport (Fig. 7). In vitro cat and rabbit CP concentrates AIB threefold (tissue-to-medium ratio) by a saturable mechanism (reviewed by Fenstermacher and Davson[33]). The chamber-mounted amphibian CP permits AIB permeation from the serosal to apical side, as well as in the reverse direction.[166] Thus, the relative perviousness and imperviousness of

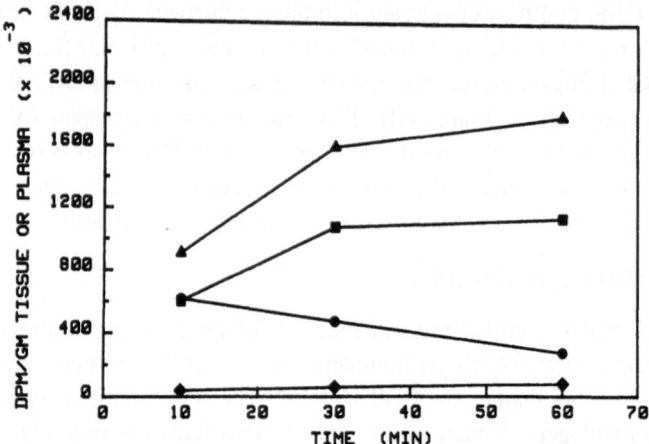

Figure 7. Blood–cerebrospinal fluid versus blood–brain barrier permeability to α-aminoisobutyric acid (AIB). Adult rats were anesthetized with ketamine, nephrectomized to maintain stable level of tracer in blood, and injected intraperitoneally (IP) with tritiated-AIB. Samples of lateral ventricle choroid plexus (CP) (squares), fourth ventricle CP (triangles), cerebral cortex (diamonds), and arterial plasma (circles) were taken at 10, 30, or 60 min after isotope injection. Standard errors for CP and cortical tissue are 5–10% of respective means. CP is able to accumulate AIB three- to fivefold above the plasma level. By contrast, AIB slowly penetrates the BBB. (From Murphy and Johanson.[93])

the CP and BBB, respectively, to water-soluble AIB (103 M_r), suggest that structurally altered amino acids could be used to advantage for selective penetration of the barrier systems. Indeed, anticancer agents resembling amino acids in structure have been used to ride the neutral amino acid system in the BBB. A similar strategy with the CP seems worth pursuing.

11. TRANSCELLULAR AND PARACELLULAR TRANSPORT OF PROTEINS

In spite of barrier systems, proteins are able to gain access slowly to the CSF.[49] Owing to permeability characteristics as well as transfer mechanisms, the movement of proteinaceous material appears to be more extensive in the blood–CSF barrier than in the BBB. Serum albumin crosses the cerebral capillaries so slowly that it has been used to quantify residual plasma volume in brain samples. However, in dogs, radioiodinated serum albumin enters CSF from plasma within 20 min, gradually attaining maximum concentration in 20 hr.[36] In humans, 3–4 days are necessary for radiolabeled proteins to achieve steady-state distribution between blood and CSF.[117] There is considerable micrographic evidence that the CP is the site for net transfer of many molecules, by way of either transcellular (vesicular transport) or paracellular (diffusion) routes, or both. The conventional markers (e.g., HRP) used to evaluate choroidal permeability are convenient for quantitative electron microscopic visualization, but they place constraints on interpreting the physiologic relevance of the distribution phenomena. Further studies are needed to evaluate the distribution of endogenous nonenzymic tracers of various sizes and electric charges.

11.1. Endothelial Vesicular Transport: Blood–Brain Barrier versus CP

It is likely that vesicular transport takes place in the cerebral endothelium (BBB) under certain circumstances, as it does under normal conditions in most noncerebral tissues. Some pathophysiologic stimuli induce the formation of vesicles in the BBB. Hypertension has been used as a model for studying vesicular blood-to-brain transport. Transport by way of vesicles is the best structural explanation for extravasation of protein tracer in acute hypertension experiments. Different mechanisms of extravasation may exist in long-term hypertension. With elevated hydrostatic pressure, vesicular transport likely occurs concurrently with paracellular opening of tight junctions in the cerebral capillaries.[94] However, in normal physiologic conditions, there are very low levels of pinocytotic activity in the BBB.

With respect to CP, there is evidence that vesicular transport occurs in both choroidal capillaries and in the epithelial cells. Transendothelial transport by means of vesicles is present in the thick parts of the endothelium.[158] Thus, the capillary endothelium in CP and cerebrum is dissimilar in that the former cells permit both vesicular transport and free diffusion of the smaller proteins. Transendothelial transport of macromolecules probably also occurs in the numerous vesicles in the choroidal arterioles.[158]

11.2. Transcellular Pathway (into or through Epithelium)

The extensive endocytotic activity of the choroidal epithelium has been widely analyzed in both secretory and reabsorptive studies. Starting with Becker and Almazon,[7] numerous investigators have observed that markers like microperoxidase (MRP), HRP, and cytochrome C are endocytosed from the interstitium by the basolateral surface of the epithelium.[87,120,157] The relatively large molecule of ferritin moves slowly across choroidal capillaries (through fenestrae), and it is also taken up by CP epithelium.[56] Subsequently, the lysosomal apparatus of the epithelia sequesters the markers. Thus, for some proteins, the lysosomes act as an enzymic barrier or intracellular sink, especially since vesicles with extravasated material usually do not discharge their content at the apical surface. Because the CP does not have a lymphatic drainage system, such a cellular digestive system helps prevent some macromolecules from gaining access to the interstitium from plasma.

What is the evidence for a truly transcellular pathway (entirely through the cell) in CP that could be therapeutically exploited to transport proteins into the adult CNS? Vesicular blood-to-CSF transport of macromolecules such as MRP and HRP does not seem to occur. Larger markers, such as Thorotrast, apparently become trapped in the basement membrane and therefore do not reach the cell (from blood) to be endocytosed. Yet some very large proteins, such as β-lipoprotein, gain access to the adult CSF system, presumably by choroidal and nonchoroidal systems. It is of interest that there is a transcellular pathway, which circumvents the tight junctions, in early fetal life. This is a tubular system of endoplasmic reticulum (ER) cisterns,[90] which may explain apparently unrestricted diffusion of molecules especially in early development. Albumin transport through cells was described by Jacobsen et al.[58] for third and fourth ventricle CP in fetuses. This ER system needs to be more extensively studied ontogenetically, in order to provide clues as to how the ER might be altered in the epithelium of adults to modify permeability of the blood–CSF barrier.

11.3. Paracellular Pathway (between Epithelia)

Alternatively, the tight junctions between the choroidal cells are a probable site of penetration of some substances because the zonulae occludentes are leakier than those between brain endothelia. Leakage of smaller proteins may occur via intercellular spaces between the CP epithelial cells, i.e., by the paracellular pathway. Rapoport and Pettigrew[116,117] quantitated protein movement through the blood–CSF barrier and calculated a mean size of 110–120 Å for pores in the CP. The inverse relationship between the CSF/blood concentration ratio and the hydrodynamic radii of proteins is evidence for diffusion, albeit restricted, across the blood–CSF barrier.

11.4. Factors Affecting CSF Protein Concentration

Therapeutically, it would be expedient to be able to modify the levels of certain enzymes or other proteins in the CSF. Fishman et al.[37] have discussed several parameters affecting the concentration gradient of protein in CSF.

11.4.1. Reabsorptive Sink of Choroid Plexus

Whatever the therapeutic strategy, it should be kept in mind that the CP has the capability of apically reabsorbing proteinaceous material from CSF into the choroidal cells.[99] Smith et al.[131] found that the CP can accumulate CSF proteins. Thus, the relatively high concentrations of β-trace proteins in CP as compared with other regions of brain have been attributed to a choroid cell reabsorptive sink for β-trace protein in the CSF, i.e., 3 mg/dl.[99] Ferritin injected into the subarachnoid space is incorporated retrogradely into CP epithelium for lysosomal digestion.[158] Thus, the CP apical transport mechanisms represent a reabsorptive sink for sequestering proteins from ventricular fluid.

11.4.2. Bulk Flow Sink of Cerebrospinal Fluid

On the other hand, the CSF flowing through the ventricles is hydrodynamically a generalized sink for receiving many proteins (and other macromolecules) diffusing from the brain interstitium across the ependymal wall. Inhibition of CSF formation by such agents as acetazolamide leads to retention of proteins in the CSF.[160] With reduction in the bulk flow of CSF across the arachnoid villi into venous blood, there is a subsequent nonselective buildup in the concentration of proteins in the ventricular fluid. There is also suggestive evidence that the influx of proteins, e.g., γ-globulin, secreted from blood to CSF (i.e., into ventriculocisternal perfusion fluid) is affected by the rate of CSF formation.[50] Clearly, the manipulation of CSF dynamics is one way in which to alter protein concentration, albeit nonselectively, in the ECF of the brain.

11.4.3. Protein Synthesis and Secretion of Choroid Plexus

Although some models assume that CSF proteins are derived from plasma by way of CP transport mechanisms,[91] more recent schemes have focused on the CP as a generator of certain proteins for CNS function.[29] Thus, another factor regulating the CSF concentration of protein is de novo synthesis in the CP epithelium.[29] It has been postulated that the CP synthesizes immunoglobulins.[129] Agnew et al.[1] provided direct evidence for protein synthesis by the choroidal epithelium. Saunders[128] used the ontogenetic argument to make the case that elevated levels of protein in fetal CSF arise from secretory phenomena rather than from leaks in the barriers.

Prealbumin is enriched in the plexus epithelial cells on the side proximate to CSF.[91] The concentration of prealbumin in CSF, and especially in the CP, greatly exceeds that in plasma.[162] The peroxidase–antiperoxidase method, as well as immunofluorescence, was used by Aleshire et al.[2] to localize prealbumin in human choroid epithelial cytoplasm, specifically on the ER. Prealbumin has a disproportionately high CSF concentration in view of its hydrodynamic radius and molecular weight. Although some prealbumin may derive from plasma (through transport of its monomers), it seems likely that the extensive Golgi staining for prealbumin in CP is indicative of choroidal synthesis.[2,25] Rat CP contains 100 times larger amounts of prealbumin RNA than does the rat liver; no pre-albumin messenger RNA (mRNA) is found in other regions of the brain.[25] Choroid plexus levels of transferrin mRNA are comparable to those in liver. Prealbumin is the main transporting protein in the bloodstream for thyroid hormone, and it possibly functions as the carrier for hormone transport across the blood–CSF barrier.[25] It appears that the synthesis of proteins in CP is a promising area of research for the control of macromolecu-lar flux from plasma to CSF.

11.5. Therapeutic Strategies

For pharmacotherapeutic efficacy, it is essential that the protein of choice reach the desired site of action at the concentration necessary to elicit the appropriate response. The problems can be exemplified by the distribution of the antiviral protein, α-interferon (IFN_α). Systemically administered recombinant IFN_α(rIFN_α) apparently crosses the BBB to a negligible extent, but it attains a concentration in primate CSF about 0.1% that in plasma.[121,140] The concentration of rIFN_α that is attainable in CSF is a sensitive function of the IV dose. Patients receiving 18×10^6 units displayed no measurable concentration in CSF, but 3 of 4 patients receiving 50×10^6 units of rIFN_α showed significant levels (17–70 pg/ml) in CSF sampled 1 hr after the start of IV infusion; in two cases, rIFN_α was measurable throughout the first 24-hr period.[140] The therapeutic index for IFN may be relatively low because in chemotherapeutic studies the dose of 100×10^6 units/day brought on disorientation, confusion, and hallucinations. Unfortunately, CSF levels of IFN do not permit estimation of concentrations attainable at specific populations of nerve or glial cells. In some cases, the relatively slow diffusion of this antiviral protein, from CSF into brain substance, over relatively long distances of several millimeters could be the limiting factor in its therapeutic effectiveness.

12. POLYPEPTIDE RECEPTORS AND TRANSPORT

The activity of many polypeptides markedly increases in the CP following either intravenous or intrathecal injection. Receptors and/or carrier transport of many hormones have been demonstrated in both in vivo and in vitro experiments. Theoretically, hormone receptors in the choroidal epithelium could be involved in either the regulation of CP function or transchoroidal transport between plasma and CSF. Some hormones that cross the CP have negligible permeability in the BBB. Thus, there is considerable interest in the ventricular route hypothesis that a hormone, once gaining access to the CSF (from either plasma or circumventricular organs), can be distributed by bulk flow of CSF to various regions of the brain.[75]

12.1. Oligopeptides

The tripeptide L-prolyl-L-leucyl-glycinamide, i.e., α-melanocyte-stimulating hor-mone (MSH)–release-inhibiting factor (MIF-I), potentiates the behavioral effects of L-dopa and reduces tremors. The suggestion of an antiparkinsonian effect has kindled interest in the kinetics of the distribution of MIF-I.[108] After intracarotid administration in rats, the MIF tripeptide tritiated radioactivity labels the cells of CP, ependyma, and circumventricular organs. The passage of radioactivity from blood to CSF is thought to occur at the level of the CP.[108] Diffusion of radioactivity over a short distance from the ventricles into underlying brain tissue is consistently observed.

The active accumulation of a peptide by the CP would be consistent with a regulatory role for the plexus in maintaining peptide concentrations and distribution in CSF. The tripeptide Tyr-D-Ala-Gly (TAG) is extensively concentrated by the in vitro rat CP. The tissue to medium ratio (T/M) for TAG is close to 5:1 after 1 hr of incubation.[52] TAG is useful as a test substance because it is minimally metabolized. The in vivo CP of the lateral and fourth ventricles actively accumulates TAG from ventriculocisternal perfusion fluid,[53] suggesting concentrative uptake across the apical membrane. The transport mechanism is sensitive to extracellular [K] (but not to [Na]), to amino acids, and to sympathectomy.

The dipeptide glycylglycine and the tripeptide TRH have been tested as competitive antagonists. Neither oligopeptide inhibits the concentrative uptake of methionine enkephalin.

12.2. Enkephalins

The methionine and leucine enkephalins are pentapeptides that exhibit morphine like activity. Pardridge and colleagues[107] discussed the restrictive transport of enkephalins at the BBB. Since the degradation of metenkephalin in tissue is almost complete within 7 min of incubation,[54] it has been expedient in CP studies to use the potent analogue enkephalinamide [D-Ala^2-Met-enkephalinamide (DAME) with tyrosyl ring 2–6], which displays similar action but resists tissue degradation.

DAME is accumulated against its concentration gradient in the in vitro CP of rats. T/M ratios of >15 : 1 are achieved by 15 min. Metabolic inhibitors such as N-ethylmaleimide reduce the enkephalin uptake by nearly 75% at 37°C.[54] The saturable accumulation process is stronger in CP than in slices of cerebral cortex. At 10^{-5} M, naloxone (but not morphine) reduces accumulation of enkephalinamide. Although various dipeptide residues of the enkephalinamide molecule do not affect accumulation, tyrosine does have an inhibitory effect.

Qualitatively similar findings have been reported for rabbit CP, but quantitative differences exist between species.[34,35] Ovine CP incubated in artificial CSF actively accumulates methionine–enkephalin (ME) to a T/M ratio of only 4 : 1 at 20 min. Similarly, the potent analogue DAME, not destroyed by tissue peptidases, attains a T/M distribution of 5 : 1 after 20 min of incubation. The uptake of endogenous opioid peptides is by a high-affinity system, competitively inhibitable by L-tyrosine. Saturation of DAME uptake occurs at concentrations above 10^{-5} M.[34]

Enkephalin as well as the larger β-endorphin have been implicated in various CNS pathophysiologies. The entry and exit of enkephalins into and out of the CNS are poorly understood; they may be affected by peptidase degradative activity in the cells of the barrier systems. Competitive studies with simple amino acids have revealed that DAME transport in the CP is probably not mediated by the A or L systems or by the acidic amino acid transporter. A reasonable working hypothesis is that enkephalins use the neutral amino acid system as the vehicle for transport in the CSF system. This is based on the observation from two laboratories that tyrosine competitively inhibits DAME uptake by CP.[34,35]

Bidirectional transport of enkephalins across the blood–CSF barrier is possible.

Excretory transport (i.e., CSF to choroidal blood) has been postulated, especially within the context of enkephalin as a neurotransmitter. By functional analogy, other amino acid neurotransmitters and their metabolites are actively removed by the intact CP (see Section 14). The instability of enkephalins in blood has hampered analyses of in vivo penetration from plasma to CSF.

12.3. Polypeptide Hormones

Central nervous system–endocrine interactions can be limited by the accessibility of hormones in plasma to the brain substance. As steroid hormones are lipid soluble, so they can readily penetrate the CNS barrier systems if they are not bound to globulin. However, polypeptide hormones are hydrophilic to varying degrees; thus many of them are able to permeate the CNS by way of carrier-mediated transport systems. Evidence is steadily accumulating for numerous polypeptide translocating systems in the CP epithelium of the blood–CSF barrier. Nearly 30 hormones have been found in the CSF.[164] However, this discussion is limited to several peptidergic substances that have receptors in CP or that have been isolated in the choroidal tissue or its neurosecretory innervation.

12.3.1. Insulin

Although there are insulin receptors in the cerebral capillary wall,[159] this hormone does not readily cross the BBB.[40] Yet, the CSF/plasma ratio of insulin is about 0.25. The polysaccharide inulin, which has a molecular weight of 5000 (comparable to insulin), has a CSF/plasma distribution tenfold less than 0.25. The relatively high concentration of insulin in CSF suggests that it penetrates the CNS capillaries in regions outside the BBB (e.g., CP, area postrema, subfornical organ) and is then selectively transported into CSF by ependymal-type systems.[103] Insulin-binding sites in rat brain have been auto-radiographically localized to CP as well as hypothalamic nuclei. In the developing rat brain, the insulin-binding sites are more heterogeneously distributed in the fetus than in adults, and they exhibit a waxing and waning, perinatally. The CP has somewhat elevated levels of insulin-binding sites.[171] Insulin may not only be translocated across the blood–CSF barrier, but it also modifies the function of the CP. Insulin enhances net secretion of Na into CSF,[126] an effect likely mediated by way of elevated levels of cyclic guanosine monophosphate (cGMP) and the Na–K pump activity in the apical membrane of the frog choroidal epithelium.[125] Stimulation of basolateral Na–H exchange in the rat CP was postulated by Murphy.[92]

12.3.2. Thyroid Hormones

Thyroid hormones are bound to plasma proteins.[106] The concentration of unbound thyroid hormones triiodothyronine (T3) and thyroxine (T4) in CSF is equal to or higher than in plasma.[43] In dogs, labeled T4 enters cerebral cortex slower than does T3, but it reaches steady-state distribution faster in CSF. Autoradiographs of adult rat brains demonstrate that nerve cells in cortical gray matter and in CP (but not in white matter of the corpus callosum) selectively concentrate and retain intravenously administered [^{125}I]-T3, by mechanisms susceptible to saturation with excess T3.[28] Using in vitro rabbit CP, Spector and Levy[150] found significant tissue accumulation of T4, but not T3, from CSF medium. The concentrative mechanism for T4 was not energy dependent and was one-half saturated at a concentration of unbound T4 of about 6 μg/liter, i.e., 600 times the normal level of unbound T4 in plasma. Thus, the facilitated transport of T4 by mammalian CP does not maintain T4 at a stable level in CSF; rather, the CSF levels reflect fluctuations of T4 in plasma. It appears that attempts to manipulate T4 concentration in CSF should be directed at titrations of the plasma level of the hormone.

12.3.3. Melatonin

Melatonin, a hormonal product of the pineal gland, can attain concentrations in CSF up to five times greater than in blood.[45] However, cerebral extraction of melatonin from blood is only about 12–14%, far less than the 92–97% unidirectional clearance from a single pass through the liver.[105] On the other hand, the rapid uptake of [³H]melatonin by the adult rat CP and CSF points to carrier transport in the blood–CSF barrier.[85] The uptake by CP (corrected for residual blood) of intravenously administered melatonin is 3–12 times higher than by hypothalamus and mid-brain, and CSF peak uptake occurs by 5 min. Pineal-synthesized melatonin may reach the choroidal cells directly by way of venous blood flow from the pineal gland to the plexus.[114] With species differences in mind, it should be mentioned that frog CP can actively reabsorb melatonin from the apical side of the epithelium, and thus CSF clearance into choroidal blood may contribute to the circadian rhythm of melatonin concentration in CSF.[141]

12.3.4. Vasopressin and Oxytocin

Can neurohypophysial hormones cross the blood–CSF barrier? The source of antidiuretic activity in the CSF, which is elevated following hemorrhage or vagal stimulation, has been long debated. Earlier, it was not clear whether arginine vasopressin (AVP) reaches the CSF from blood and/or is directly secreted into the CSF by juxtaependymal processes of neurosecretory neurons. It is now appreciated that one cell population secretes AVP into CSF and another releases the hormone into peripheral blood.

Within 2 min of IV injection of vasopressin into rabbits, this peptide appeared to permeate the blood–CSF barrier.[46] However, the substantial rise in arterial pressure probably disrupted the integrity of the blood–CSF barrier; in this regard, Parandoosh and Johanson[101] observed substantially augmented penetration of urea into lateral ventricle CP and CSF of adult rats made hypertensive with AVP. Another study by Zaidi and Heller[172] in 1974 used tracer amounts of [³H]oxytocin to avoid concurrent vascular effects, and found negligible penetration of the nonapeptide into rabbit CSF. More recent investigations have confirmed that the AVP in CSF is functionally isolated from blood by a bidirectionally impervious blood–CSF barrier.[59,81] Even though there is not free exchange across the CP, the concentration of AVP in CSF is similar to that in plasma.

More than three decades ago, oxytocic activity was isolated from extracts of CP.[154] Subsequently, Rodriguez and Heller[123] demonstrated antidiuretic principle in toad CP, the activity in third ventricle CP being four times greater than in fourth ventricle tissue. More recently, Rudman and Chawla[124] isolated an antidiuretic peptide in bovine, porcine, and human CP, with a molecular weight of 750–3500. The choroidal epithelium has the same embryologic origin as the neurohypophysis, so possible cellular origin can be postulated for CP synthesis of peptides related to neurohypophyseal hormones. An alternative explanation for the CP content of AVP or arginine vasotocin (AVT) is that the cells accumulate hormone being transported in one direction or another across the epithelial lining. Still another possibility is that the neurohypophyseal hormones are transported from hypothalamic nuclei directly to CP by neurosecretory fibers as traced by the immunoperoxidase method.[17] Whatever the origin of the ADH, this hormone can modulate the fluid–transport capabilities of the blood–CSF barrier. Antidiuretic hormone (ADH) involvement in brain interstitial–ventricular dynamics has been discussed by Schultz et al.,[130] and by Rap[115] within the context of cerebral edema.

12.3.5. Prolactin and Vasoactive Intestinal Polypeptide

Unlike most peptide hormones, prolactin readily penetrates the blood–CSF barrier. The CSF levels of prolactin run parallel to those in plasma.[79] It has been considered that prolactin gains access to CSF by way of the CP.[5] Posner et al.[112] characterized lactogen binding sites in CP. Prolactin receptors in CP have also been described by Lorenzo et al.,[80] who postulated a role for these receptors in the water balance of the perinatal brain. CSF prolactin concentration can be altered by pregnancy, drugs, and stressful states. The use of CSF/plasma ratios of prolactin has been expedient in clinical evaluations. CSF prolactin concentrations may modulate the secretion of prolactin.[97] Electrolyte functions of prolactin in the CP–CSF system need to be elucidated.

Vasoactive intestinal polypeptide (VIP), a highly basic 28 amino acid peptide, was initially isolated from pig intestine. Since its discovery in 1970, VIP has subsequently been identified in many regions of the CNS,[32,74] including the CP.[77] VIP distribution in rat brain is similar to that in the human.[74] The ability of VIP to induce vasodilation and to stimulate exocrine secretion suggest that this hormone is a modulator of ion and fluid movement across the epithelium of the blood–CSF barrier.[155] However, the kinetics of distribution of VIP across the CP is unknown. The CSF level of VIP exceeds that in blood by several fold, and there is no correlation between the concentrations of VIP in the two compartments.[32] Altered CSF levels of VIP have been linked to hydrocephalic disorders and dementias, on the basic assumption that the lowered concentrations of VIP in these disease states primarily reflect the brain levels. In this regard, however, it will also be necessary to evaluate whether CP transport of this polypeptide has an important influence on the CSF level.

13. PROSTAGLANDINS AND LEUKOTRIENES

Many detrimental effects have been ascribed to the buildup of·prostaglandins (PGs) and leukotrienes in the CSF and interstitial fluid of the brain. The movement of these arachidonic acid metabolites across the barrier systems is by carrier-mediated transport. The existing data suggest greater permeation of the blood–CSF barrier in the direction of excretion (CSF to blood) than in the reverse route, i.e., inward to the ventricular fluid.

$[^3H]$-PGF$_{2\alpha}$ can attain a tenfold concentrative accumulation in the isolated rabbit CP.[11] At 0.1 mM concentration, both PGE$_1$ and PGA$_1$, as well as the precursor arachidonic acid, caused substantial inhibition of the uptake of $[^3H]$-PGF$_{2\alpha}$ by choroidal tissue. Probenecid competed for PG transport, but neither perchlorate nor iodide inhibited it. In vivo analysis of PG movement has also been done by Bito and colleagues.[10] $[^3H]$-PGF$_{2\alpha}$ clearance from synthetic CSF perfused through the ventricular system in rabbits was reduced by stable PGF$_{2\alpha}$, PGF$_{2\beta}$, probenecid, and iodipamide, but not by perchlorate.

Prostaglandins not only are translocated by CP,[24] but they also are synthesized in this tissue and can modulate transport functions. Cultured bovine CP responds to PGE$_1$ by elevating its content of cAMP.[22] Prostacyclin (PGI$_2$) is extensively formed in rat CP[39];

PGE_2 and $PGF_{2\alpha}$ are also made, but to a lesser extent. PGE_2 substantially accelerates chloride transport from blood to CSF.[23] In in vivo experiments with rats, Deng[23] quantified the dose–response relationship between ^{36}Cl uptake by CSF and the concentration of PGE_2 instilled intraventricularly.

Intravenously injected leukotrienes achieve very low levels in brain. Leukotriene transport and effects at the blood–CSF barrier were recently analyzed in vitro by Spector and Goetzl.[148] After ruling out binding and metabolic factors, they postulated that leukotriene C4 is actively transported from CSF to blood. The system that accumulates LTC4 is probenecid and temperature sensitive, specific, and energy dependent.

14. NEUROTRANSMITTERS

Many neurotransmitters and their metabolites are transported at the blood–CSF barrier.[153] Separate transport systems probably exist for amines and acids.[83] Because these same biogenic amines and their anion metabolites can modulate functional processes in the CP, there must be compartmentalization of these substances in and around the choroidal epithelium as they are being translocated. Otherwise, the transported amines could disrupt the normal functioning of the barrier cells.

In general, the CSF and brain are protected against fluctuations in the concentration of neurotransmitters in plasma. The limited lipid solubility of most monoamine neurotransmitters, and especially of their hydrophilic metabolites, thwarts diffusive penetration from blood to CSF. Degradative enzymes in the CP form a second line of defense against neurotransmitter permeation; this chemical barrier is represented enzymically as hydrolytic activities of MAO and catechol-*O*-methyltransferase in the epithelium.[78] Still other factors regulating CSF concentration of neurotransmitters and metabolites are the active transport systems for reabsorbing acid anions and basic cations from CSF into the choroidal blood.[113,127] The well-known clinical test involving probenecid usage takes advantage of the ability of this drug competitively to inhibit the active clearance of anion metabolites such as 5-HIAA from ventricular fluid.[38]

Because most drugs are either organic acid or base electrolytes, there are theoretically many possible kinetic effects of pharmacologic agents on CSF levels of neurotransmitters. High concentrations of naloxone inhibit 5-HIAA and HVA transport by rat CP[55]; however, morphine potentiates 5-HIAA accumulation. When analyzing mechanisms of drug action, pharmacodynamic versus pharmacokinetic explanations, it is essential to consider transport interactions in the blood–CSF barrier between endogenous substances (e.g., neurotransmitters) and xenobiotic agents (e.g., narcotics). In one such investigation, Huang and Wajda[55] deduced that the increase of HVA and 5-HIAA in brain after morphine can not be attributed to the inhibition of the excretion of these neurotransmitter metabolites by the CP.

The involvement of neurogenic mechanisms in CSF formation and transport has been widely investigated by Edvinsson and colleagues.[30] The adrenergic innervation to the mammalian CP regulates not only blood flow but cellular enzymic and ion-transport mechanisms as well. CP blood vessels are covered with both alpha and beta adrenergic receptors, or adrenoceptors. The tiny endings of adrenergic fibers, originating in the superior cervical ganglion, abut against the choroidal epithelium. Nathanson's studies of broken cell preparations of dissociated CP have characterized β-adrenergic receptors intimately associated with adenylate cyclase.[95]

Perfusion of the cerebral ventricles with norepinephrine in incremental concentrations (10^{-11}–10^{-3} M) resulted in a dose-related curtailment of CSF production. Lower and higher doses, respectively, inhibited CSF secretion, likely by beta effects on the choroidal epithelia and alpha effects (vasoconstrictive) on tissue blood flow.[30] Sympathectomy can increase CSF flow by about 30%, whereas stimulation of sympathetic nerves leads to a 30% reduction in fluid production. Resection of sympathetic nerve fibers to the CP brings about alterations in the activities of Na, K–ATPase, and carbonic anhydrase, both of which are key transport enzymes. One week after unilateral sympathectomy, the CP on the operated side has a threefold greater ability to accumulate choline.[163] This interesting finding suggests that organic solute movement across the blood–CSF barrier may be modulated through a functional influence of neural tone on the choroidal epithelium.

15. ALTERED FLUXES OF INORGANIC IONS ACROSS CHOROID PLEXUS

Many pharmacotherapeutic implications are associated with drug-induced modification of ion transport across CP; among these are considerations of volume of brain ECF, cell and interstitial pH regulation, CSF secretion, and ion homeostasis. For many ions, their main route of entry into the CNS is by way of the blood–CSF barrier, not extensively via the BBB. Thus, ion-transport mechanisms in the CP represent a quantitatively important site for drug action.

15.1. Sodium and Chloride

Na and Cl are the predominant cation and anion species, respectively, that are transported from plasma to CSF. With data for permeability surface–area products and plasma ion concentrations, it can be calculated that two-thirds to three-fourths of the Na and Cl gaining access to the rat CNS do so by transport across the CP.[138] Since Na and Cl transport are integral to the CSF secretory process, it is obvious that attempts to modify substantially the fluid dynamics in the CNS have to focus on Na and Cl transport mechanisms in the basolateral membrane of CP. The CP is the workhorse for fluid production in the CNS; by comparison, the cerebral capillary wall has low fluid-secreting capacity.[14]

Na and Cl distribution into tissues and transcellular fluids like the CSF are closely linked (Fig. 8). Thus, the radioisotopic uptake curves for ^{22}Na and ^{36}Cl are similar.[135,137,138] It appears that the transport of Na is obligatorily coupled to that of Cl, in order to maintain charge and osmotic balance across membranes. This means that CSF production can be slowed down by directly inhibiting the transport of either Na *or* Cl. Such a scheme has been presented by this author, i.e., Na–H and Cl–HCO$_3$ antiport systems operating in parallel on the plasma-facing membrane of CP (Fig. 6). Because the antiporters are more-or-less loosely coupled functionally, CSF secretion can be reduced by amiloride inhibition of Na–H exchange,[92] or by disulfonic stilbene inhibition of Cl–HCO$_3$ exchange.[23] Alternatively, a tightly coupled NaCl cotransport (symport) would be inhibitable by loop diuretics such as furosemide or bumetanide. However, these high-ceiling diuretics do not significantly alter ^{36}Cl transport in CP, or the flow of CSF, in nephrectomized rats with stable plasma ionic composition. Moreover, Vogh and Langham[161] found that bumetanide did not alter CP secretion as estimated by ^{22}Na turnover from plasma to CSF in cats. Still, there are reports that the potent loop diuretic agents alter

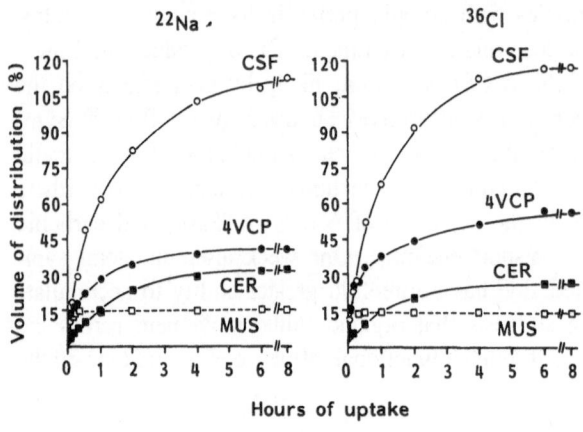

Figure 8. A comparison of the penetrability of Na and Cl in various transport interfaces. Tissue radioactivity is expressed as a space (%) = 100 × (dpm/g tissue or CSF)/(dpm/g interstitial fluid or plasma). Either Na^{22} or Cl^{36} (0.1 μC/g) was injected at various times prior to sacrifice. Each point is the mean uptake for 3 adult rats anesthetized with ether. Central nervous system (CNS) data are replotted from Smith et al.[135] (CSF, cisternal cerebrospinal fluid; 4VCP, fourth ventricle choroid plexus; CER, cerebellum). Data for skeletal muscle (MUS) from C. Johanson and Q. Smith, unpublished data. Na penetration kinetics are similar to that of Cl, not only in the blood–CSF barrier, but also in the blood–brain barrier (BBB) and in a non-neural tissue, i.e., skeletal muscle.

mammalian CSF production;[84] thus, species and methodologic differences have to be carefully evaluated before it becomes clear which agents are the most promising for arresting CSF elaboration in humans.

Most pharmacologic attempts to alter CSF secretion have sought to inhibit flow. Such experimental focus on reducing fluid output by CP relates to the clinical need for diminishing fluid retention in the CNS, e.g., in pathophysiologic states of hydrocephalus and cerebral edema. On the other hand, far less concentration has been directed to enhancing CSF production. The demonstration by Epstein et al.[31] that cholera toxin boosts CSF flow in cats served to fix attention on possible facilitation of secretory mechanisms in CP mediated by elevated cAMP in the epithelium. Recent findings indicate that agents that promote cAMP accumulation (adenosine, PGE_2, dibutyryl cAMP plus theophylline) increase CP transport of ^{36}Cl from plasma to CSF; moreover, PGE_2 enhances CSF formation.[23] However, the role of cAMP in CSF secretion is still not well understood. Acceleration of fluid and ion transport by CP has practical and theoretical implications. Augmented CSF sink action associated with increased CSF flow is expected to promote clearance of water-soluble drugs in the brain, say, in the case of drug overdosing. From the theoretical point of view, the elucidation of mechanisms underlying enhanced fluid turnover will probably provide insight not previously gained by the widely used inhibitory analyses.

15.2. Bicarbonate

Anionic bicarbonate, generated intracellularly in CP, plays a fundamental role in formation of CSF and in the regulation of pH in the CP–CSF system.[63] The abundant supplies of H_2O and CO_2 in choroidal epithelium are indispensable underlying substrates in CSF secretion. Interference by carbonic anhydrase inhibitors with the cellular hydration of CO_2 can abruptly halt fluid movement across the choroidal membrane. Protolysis of H_2O furnishes not only OH for converting CO_2 to HCO_3, but also H for exchange with extracellular Na.[62] Thus, basolateral Na–H exchange activity furnishes Na directly (and HCO_3 indirectly) to apical extrusion mechanisms.[92]

Traditionally, the carbonic anhydrase inhibitor acetazolamide has had immense popularity as the agent of choice by basic and clinical researchers trying to curtail CSF production. Therapeutically, acetazolamide has limited efficacy. The pharmacologically related agents benzolamide and methazolamide have not yielded the desired selectivity in CP that has been documented for other secretory systems. However, the prolific work by Maren and colleagues has furnished numerous biochemical and physiologic insights about the kinetic properties of carbonic anydrase in the mammalian CP–CSF system.[82,161]

The nature of HCO_3 and H transport by CP is undergoing intensive investigation.[60,62,70,73,82] Over the past three decades the ventriculocisternal perfusion preparation has been extensively employed in several mammalian species to analyze HCO_3 movement between blood and CSF and between CSF and brain. CSF $[HCO_3]$ and plasma $[HCO_3]$ are about equal.[73] More recent endeavors, now at the cellular level, have tried to delineate the distribution of HCO_3 across the basolateral and apical membranes of CP,[62] as well as the choroid cell content of HCO_3.[70,92] As a result, proposals have been made for $Cl–HCO_3$ antiport,[23,70] $NaHCO_3$ symport,[62] and HCO_3 conductance pathways. Since the disulfonic stilbene agents (DIDS and SITS) alter HCO_3 transport and distribution in CP, by inhibiting $Cl–HCO_3$ exchange and possibly $Na–HCO_3$ symport as well, this class of agents may prove a significant new addition to the armamentarium of drugs for controlling fluid production in the CNS. An appreciation of the functional relationship between the pH–PCO_2–HCO_3 system and transepithelial Na transport is a fundamental physiologic problem that is gradually becoming better understood.[62,92,132]

15.3. Potassium

Normally there is net transport of K from CSF to blood, in both the CP and BBB. Thus, one of the distinguishing characteristics of mammals is that their CSF [K] is usually kept at a concentration of 0.5–0.75 less than that in plasma.[72] CSF [K] is maintained well, if not perfectly, in the face of severe hypo- or hyperkalemia.[71] The CP has a primary role in CSF [K] homeostasis.[57,133] It has been postulated that carrier inward transport of K in the basolateral membrane works in concert with Na–K–ATPase apically to regulate CSF [K]. For example, in sustained hypokalemia induced by diet or desoxycorticosterone, the reduction in plasma [K] is reflected in a sizeable decrease in CP cell [K] but not in CSF [K].[71] Hyperkalemia stimulates accumulation of CP [K] proportionally more than CSF [K].[71]

Some pharmacologic agents alter CSF [K] through a modulating effect on the CP Na–K pump. Fluctuations in CSF [K] can lead to changes in neuronal excitability. Cardiac glycosides like ouabain are not lipophilic enough to penetrate from plasma through the choroidal membrane to affect Na–K–ATPase activity on the CSF face of the epithelium.[133] Distribution studies have revealed negligible permeation of [³H]ouabain in CSF 1 hr after IP administration to adult rats.[165] Several investigators noted that hydrophilic digitalis glycosides do not alter liquor production unless the drugs are applied on the ventricular side of the CP.

More lipid-soluble glycosides like digoxin are gradually taken up by CP.[76] Initially, the CP concentration may be low, thereby stimulating CSF production; later, the greater tissue accumulation of digoxin likely reduces CSF formation. Attenuated CSF production may explain some of the digoxin-induced side effects associated with the CNS of geriatric patients.[76] Recently, the cardiac glycosides have been the subject of renewed interest for arresting CSF production.[110] For therapeutic usefulness, it must be convincingly demonstrated that digitalis negligibly increases CSF [K] while reducing fluid output by CP. Fortunately, there is considerable evidence that alterations in bulk formation of CSF are independent of K exchange between plasma and CSF. Histamine does not alter CSF flow, but it does accelerate the transport of tracer ^{42}K from blood to CSF.[27] Moreover, Domer was able substantially to decrease CSF formation in cats by use of triamterene or chlorothiazide, without affecting K exchange at the blood–CSF barrier.

Chemical composition studies have shown that acute treatment (1 hr) with the anticonvulsant drug phenytoin (40 mg/kg) markedly elevates CP cell [K]/[Na] while decreasing CSF [K].[65] Thus, Na–K pump activity in CP, like that in salivary gland epithelium, may be facilitated by phenytoin treatment. A more recent investigation, chronically with rats, has confirmed that phenytoin treatment leads to a reduction in CSF [K]. With regard to antiepileptic mechanisms of drug action, phenytoin undoubtedly has multiple effects upon ion distribution across membranes. One working hypothesis deserving further attention is the involvement of the CP Na–K pump to lower CSF [K], thereby indirectly reducing CNS excitability through alterations of the transmembrane gradient for K between brain ISF and neurons.

15.4. Lithium

The pharmacology of lithium has been a topic of considerable interest ever since its effectiveness was demonstrated in affective disorders. The delay in the onset of antimanic action was initially attributed anecdotally to slow permeation of Li across the BBB. It is now known that Li readily penetrates the CP and distributes into CSF with a steady–state concentration of 0.6 that in plasma; this distribution ratio is maintained over a wide range of plasma [Li].[170] Li movement from blood into the CP–CSF system mimics that of urea,[68] i.e., is consistent with molecular sieving of the diffusing solute. The rapid penetration of Li into the ventricles is not counteracted by active extrusion of this monovalent cation back into CP and its blood supply. Thus, there is no evidence for active transport of Li either out of CSF in vivo or concentrative uptake by CP in vitro.

Attention has been directed to the interaction between Li and ^{22}Na transport in CP, due to implications for CSF secretion in patients undergoing lithium salt therapy. Reed and Yen[119] demonstrated that Li administered intravenously did not alter ^{22}Na transport actively into in situ CSF of their chamber preparation of CP. Saito and Wright[125] also noted lack of effect of Li on ^{22}Na efflux from in vitro frog CP. Yet the evidence for lack of interactive transport of Li and Na in CP is equivocal. Hesketh[47] reported that elevated Li in blood caused a marked decrease in the fast component of Na entry to CSF. And Yen and Reed[170] suggested that Li is translocated by the Na system responsible for CSF secretion in cats. Yet Li does not alter CSF secretion rate, at least by the in situ CP. In addition to transcellular movement of Li, there is probably paracellular diffusion as well. Li has the small atomic weight of 7 and an ionic radius similar to divalent Mg; thus this monovalent cation undoubtedly penetrates tight junctions.

15.5. Calcium and Magnesium

Alterations in CSF [Ca] have been closely linked to changes in CNS excitability in animals and to mental state in humans. Normally CSF [Ca] is regulated within the narrow range of 1.2–1.5 mmoles/liter, presumably by carrier-mediated transport primarily in the CP.[18] Ca entry into CNS is substantially greater across the CP than the BBB.[42,139] Therefore, since transport at the blood–CSF interface is the dominant factor in pharmacokinetic modeling of Ca distribution among CNS compartments,[26] it is important to understand the physiology of Ca transport systems in CP.[20] It would also be expedient to know the effects of Ca-blocking agents on CSF Ca homeostasis for therapeutic as well as toxicologic reasons.[18]

Ca moves from CP ISF, i.e., ultrafiltrate of plasma, into ventricular fluid by a process consistent with passive diffusion. By use of electron microscopy and a precipitation technique, it has been demonstrated that Ca moves between the choroid epithelial cells in transit to CSF.[86] CSF [Ca] is likely regulated by a Ca carrier mechanism in CP[18] that extrudes Ca to blood at a rate proportional to CSF Ca concentration. Burton characterized the Michaelis–Menten kinetics of Na–Ca exchange in the in vitro CP, and has putatively ascribed the apical membrane as the site for the Ca antiport system. Na–Ca exchange in rat CP is inhibitable by amiloride and ouabain (competitively) but not by verapamil (a Ca-channel blocker). Plasma lead interferes with ^{45}Ca penetration across the barrier systems in vivo.[19]

Several investigators have observed that the major route of entry of tracer ^{45}Ca to some brain regions is by way of the CP to the ventricular system, and then to paraventricular brain tissue.[19,26,139] The transchoroidal flux of ^{45}Ca (and many other ions including Mg) into CSF is one to two orders of magnitude greater than that through cerebral capillaries into cortical tissue; thus, a concentration gradient from CSF to brain is rapidly established for radioisotopic Ca (and Mg).[139] Especially since the CP is a quantitatively and homeostatically significant interface for Ca translocation, it is pertinent to consider that a compromise in its Ca-transporting ability could be detrimental to Ca metabolism in brain. Postischemic elevation in total [Ca] is rapid in CP and of considerable magnitude, i.e., fourfold increase 5 hr after the start of ischemia. Ca accumulation by cells is often a reflection of morbidity, and so the blood–CSF barrier may be particularly vulnerable to ischemia.

Magnesium is the other major divalent cation of interest here. For a long time there has been consensus among CSF physiologists that Mg is actively transported from plasma to CSF. In a chemical composition study, Reed and Yen[118] have used the in situ CP to characterize CSF [Mg] regulation. In vivo kinetic analyses have shown that tracer ^{28}Mg in plasma rapidly gains access to CSF.[139] The blood-to-CSF pathway for Mg appears to involve a transcellular carrier, i.e., active pump, more than paracellular diffusion, because the choroidal tissue has a real but limited permeability to Mg. Pharmacologic analyses of CSF [Mg] have been rare. Mg deficiency in rats decreases CP [Mg] by 40% with attendant changes in weight (loss), color (milky white to brick red), and gross morphology (epithelial cell separation).[142] Such pronounced changes must alter the homeostatic capability of CP to maintain CSF ion concentrations. Stroke patients exhibit deficits in CSF as well as serum Mg.[3] The Alturas have discussed the prophylactic effects of Mg salts for brain ischemia and cerebrovasospasm. The lowering of extracellular Mg in situ and in vitro leads to enhanced uptake of Ca ions in the brain and cerebral arteries. Transient ischemic attacks may be related to a true deficiency of Mg in CSF and brain.[3]

16. HYPERTENSIVE AND OSMOTIC OPENING OF THE CHOROID PLEXUS

Many laboratories have systematically analyzed the opening of the BBB by acutely elevated arterial pressure. Until recently scant attention has been focussed on possible hypertensive effects on the permeability of the blood–CSF barrier.

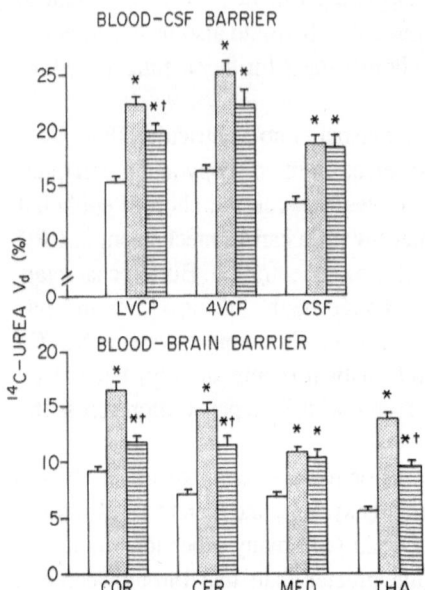

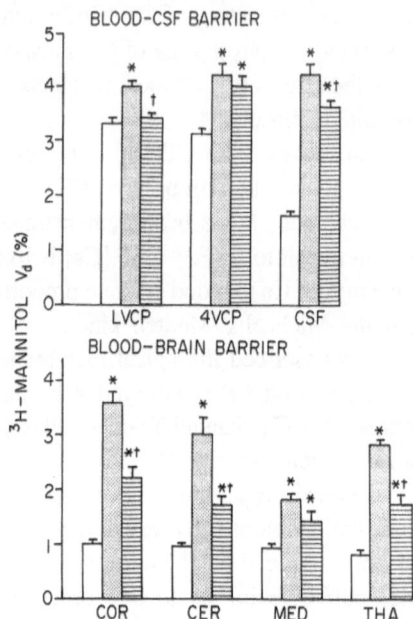

Figure 9. Acute hypertension opens up the blood–cerebrospinal fluid (CSF) barrier as well as the blood–brain barrier (BBB) to nonelectrolyte tracer molecules urea and mannitol, with molecular weights of 60 and 182, respectively. Arterial blood pressure in adult rats was elevated by either intravenous (IV) epinephrine (stippled bars) or phenylephrine (striped bars) to a peak of 160 mm Hg from a baseline of 95–100. Unfilled bars are controls receiving saline vehicle. The tracer was injected simultaneously with the sympathomimetic agent, and 10 min was allowed for the urea or mannitol to distribute from blood into various regions of CNS. Vd, volume of distribution of isotope. Means ± SEM for 5–6 animals. LVCP, 4VCP, lateral or fourth ventricle choroid plexus; CSF, cerebrospinal fluid from cisterna magna; COR, cortex from parietal lobe; CER, cerebellum; MED, medulla; THA, thalamus. *, $p < 0.05$, control versus epinephrine. †, $p < 0.05$, phenylephrine versus epinephrine, by one-way analysis of variance and Bonferroni t statistic. (From Murphy and Johanson.[94])

Sudden elevation of MABP was found by Murphy and Johanson[94] to disrupt the permeability of the rat CP to tracer nonelectrolytes and protein. We used epinephrine or phenylephrine to induce a MABP rise to 160 mmHg. Epinephrine enhanced the permeation of tritiated mannitol and [14C]urea into both the lateral and fourth ventricle CP (Fig. 9); and this sympathomimetic agent augmented the 10-min volume of distribution (from plasma) of urea, mannitol and [125I]bovine serum albumin in the CSF by 1.5-, 2.7-, and 30-fold, respectively. It was postulated that the elevated arterial pressure distended the cores of CP villi, with resultant stretching and opening of the tight junctions between the epithelia; thus, with apparently less hindrance to diffusion, the small nonelectrolytes were cleared across the blood–CSF barrier at rates closer to free diffusion.[94]

Osmotic modification of cerebral capillary permeability has been a useful approach in facilitating drug entry into the brain. Few efforts have been directed at hyperosmotically disrupting the relatively impervious blood–CSF barrier. However, the involvement of the CP in the osmotically induced passage of antibody (to measles) from plasma to CSF (but not brain) of immunized monkeys was initially suggested by Hicks et al.[48] a decade ago. Ventricular perfusion experiments have demonstrated that the choroidal and ependymal cells act as osmometers when the osmolality of artificial CSF is raised above 300 mOsm.[69] Bouchaud and Bouvier[13] described the fine structure of altered tight junctions between rat choroidal cells, after barrier opening induced by hyperosmotic urea or sucrose. More recently, Neuwelt et al.[96] reported that with the higher-molecular-weight compounds infused with osmotic bolus, there is a differential permeability effect on the blood–CSF versus BBB. CSF concentration of methotrexate, monoclonal antibodies, and Evans blue–albumin all increased markedly after CNS barrier opening.[96] There was a sixfold greater concentration of IgM monoclonal antibody in CSF when the barrier was modified via the vertebral compared with the carotid artery.[96] It was postulated that the brisk vascular supply of the fourth ventricle CP is responsible for the improved delivery of high-molecular-weight substances into the CSF.[96] Moreover, the mechanism of osmotic alteration of zonulae occludentes between choroidal epithelia may be dissimilar to the disruption of the tight junctions between cerebral endothelial cells.

17. FUTURE OUTLOOK AND PRACTICAL CONSIDERATIONS

The transport data obtained for the blood–CSF barrier over the last decade have provided substantial evidence along many lines that the functional properties of the CP transport systems have numerous features that distinguish this interface from the cerebral capillary wall. Some previously narrow views of the CP have implied a relatively minor role for this secretory tissue in the overall economy of the CNS. It is now clear that the CP is more than a fine tuner of CSF composition. Rather, it is an epithelium capable of translocating proteins, immune complexes, hormones, water-soluble drugs, and inorganic ions into the CNS; for many solutes, their main, if not exclusive, route of entry into the brain is by way of the blood–CSF barrier. Indeed, some substances that apparently are not carried across the cerebral endothelium can be rapidly deployed from blood to CSF by stereospecific saturable carriers. Thus, the functional uniqueness of the CP presents opportunities for pharmacologically manipulating the transfer of micronutrient and therapeutic agents into the highly protected cerebral environment.

Compared with transport proteins in other epithelial systems, the carriers in CP have been much less extensively analyzed at the molecular level. However, the prolific work of Spector and colleagues has furnished considerable insight into the stereospecific and saturable nature of water-soluble vitamin transport from blood to CSF. The molecular structure requirements for vitamin penetration of the plexus have been thoroughly delineated in in vitro and in vivo experiments. Significantly, a folate-binding transport protein in CP has been isolated and partially characterized. Pronase, a proteolytic enzyme, has been used to furnish evidence that transport mechanisms (carriers) for folate, iodide, leucine, and glucose consist of proteins existing on the surfaces of the choroidal epithelium.[143]

The issue of bidirectional movement of protein across the CP is complex, but there is weighty evidence that the movement of proteinaceous material is more extensive in the blood–CSF barrier than in the BBB. Clinical as well as basic studies indicate that normally about two-thirds of CSF protein has entered ventricular fluid at the CP[116]; even β-lipoprotein, with a molecular weight of 2.2×10^6, enters CSF from blood. Peress and Tompkins[109] provided an extensive description of immune-complex penetration into CP; and in other promising studies, they used cationized ferritins to evaluate the effects of molecular charge on protein permeativity in the blood–CSF barrier. The significant permeability to macromolecules not only affords therapeutic opportunities with the CP for the distribution of blood-borne enzymes and monoclonal antibodies into CSF, but also presents pathophysiologic complications because some intravascular viruses, protein toxins and detrimental antibodies are able to leak into the CNS. Potentially CNS-reactive antibodies present in blood may penetrate brain via the CSF route and subsequently destroy neurons, especially in senescent individuals. Choroidal aposomes merit further consideration as a mechanism of secreting protein into CSF.[41] The liposomal approach to enhance delivery of water-soluble agents has not been fruitful in BBB analyses,[104] but these lipid spheres deserve scrutiny in the CP.

An exciting vista in CSF physiology is the role of the CP in regulating hormone and eicosanoid traffic into and out of the CNS. Many polypeptide hormones penetrate the cerebral capillaries very slowly, if it all. The circumventricular organs, particularly the CP, probably have a significant role in the transport of enkephalins, insulin, prolactin, and melatonin. However, more information is needed to clarify the relative roles of the blood–CSF barrier and BBB in regulating extracellular concentrations of these polypeptides in various regions of the CNS. The existence of polypeptide transport systems, possibly uniquely located in the CP, enhances the prospect of covalently linking poorly penetrable polypeptides to those endogenous peptides with carriers in the plexus; thus, the piggy back approach would appear to be more natural than various latentiation procedures previously directed at cerebral capillaries.[104] Prostaglandin and leukotriene transport by the CP may be important components of homeostatic systems to maintain fluid balance between intracranial compartments during ischemia and hemorrhagic states. Inflammatory processes and certain organic acid drugs affect the choroidal removal of arachidonic acid metabolites from the ventricular system.[24]

The CP is the "workhorse" for fluid production in the CNS. Therefore, pharmacologic regulation of CSF secretion will come under finer control as agents are identified for modulating the activity of ion-transport systems in the choroidal epithelium (Fig. 8). Na–H antiport appears to be the main system for moving Na from plasma into the choroid cells, seemingly the initiating step in driving the CSF secretory process. It would be desirable to find amiloride-like agents for which the CP has a high and selective affinity, in order to be able to reduce the elaboration of CSF effectively. Another Na-transport problem begging solution is the nature of the apical Na-extrusion step, which is integral to fluid movement. The Na–K pump certainly has at least a permissive role in setting up the steep transmembrane gradient for Na, but is there another Na-extruding system (e.g., Na–HCO_3) that under girds the process of CSF secretion? To test for Na–HCO_3 cotransport, such disulfonic stilbene agents like 4,4'-diisothiocyano-2,2'-disulfonic acid stilbene (DIDS) and 4-acetamide-4'-isothiocyano-stilbene-2,2'-disulfuric acid (SITS) will likely be useful. However, other drugs are needed as tools because the stilbene agents also interfere with Cl–HCO_3 exchange. A formidable challenge will be to engage in a systematic biochemical analysis of the effects of cytosolic modulators (e.g., pH, Ca, cAMP) on the activity of the ion transport systems integral to CSF secretion. With respect to receptor coupling to the transport processes, it will also be important to identify more precisely those endogenous neurohumoral substances having the most prominent effects on the turnover of CSF.

The drug-metabolizing abilities of the blood–CSF barrier have not been frequently studied. This problem needs further attention, especially in the context of failures experienced with the use of intrathecal antibiotics. Cephalothin is often unsatisfactory for the therapy of bacterial meningitis, probably due to the inactivation of the parent compound when it is metabolized by the CP to the less potent desacetylcephalothin.[98] Thus, the barrier cells of the blood–CSF interface may exert a pharmacokinetic influence on antibiotic availability to periventricular tissue. The weak acid excretory transport system in

CP has minimal affinity for the third-generation cephalosporin, ceftriaxone; thus, bacteriocidal concentrations are achieved in CSF because the clearance of ceftriaxone, CSF to blood, is less than that for penicillin.[146] In general, more pharmacologic knowledge is needed to evaluate the drug-inactivating capabilities and transport affinities of the choroidal epithelium. Toxicologic analyses of CP function have rarely been carried out in vivo. The initial evidence indicates that the plexus acts as a permeability barrier and a transport sink, both functions keeping heavy metals at a low concentration in CSF.[100]

To recapitulate, the epithelial cells of the CP have an important role to play in determining the concentration of many natural and foreign substances in CSF. The key practical question awaiting resolution is what substances can have their transport directed across the blood–CSF barrier, ultimately to enhance concentration in what regions of the brain? The findings with ascorbate suggest that this vitamin does not penetrate cerebral capillaries, yet via the CP–CSF system it can widely and sufficiently reach target cells presumably deep in the parenchyma. The experimental scientist as well as the clinician is interested in finely regulating the flow of water-soluble materials to specific regions of the the CNS. At the very least, once having permeated the CP, a substance has ready access to the brain tissue lining the ventricular and subarachnoid spaces.

Such distribution can be typified by the ability of Purkinje neurons to extract selectively both small and large molecules from CSF flowing by the cerebellum. Propidium iodide (PI), although apparently unable to cross the BBB, reaches the CSF and imparts an orange color to it; 1 hr after IV administration, PI is observed in the dendrites and somata of Purkinje neurons.[12] PI binds selectively to nucleic acids, which gain access to the CNS by transport through the CP.[144,147,149] It has been postulated that the Purkinje neurons extract PI from CSF by a physiologic mechanism designed to supply neurons with nucleic acids. A general model for this 3-step transport sequence—i.e., CP blood to CSF, CSF to brain interstitial fluid (ISF), and ISF into brain cells—was set forth by Spector and Johanson.[149] Such findings hold promise that there must be a variety of ways for basic and clinical scientists noninvasively to introduce an array of water-soluble agents to brain.

Practical issues deserve comment. The relatively small size and the interior ventricular location of the plexuses are factors that discourage previous clinical efforts to devise transport strategies with the CP. However, there are compelling reasons for intensifying efforts to delineate barrier function. It is becoming widely appreciated that the plexuses are substantially more involved than the BBB in directing the flow of ions, water, polypeptides, vitamins, proteins, and immunologic substances from plasma into CNS. For such water-soluble materials, the pharmacokinetic models of distribution in the brain need to encompass the marked influence of the CP–CSF system. Improved resolution of the circumventricular organs in brain scans should expedite the interpretation of distributional phenomena. Quantitative autoradiographic techniques need to be applied more extensively to the plexuses. Basic transport studies of CP will continue to lay the groundwork for the therapeutic exploitation of the unique properties of the blood–CSF barrier. The emerging concepts of the CP as an immune organ, and as having neuroendocrine integrative functions, should increasingly attract the attention of neuropathologists.

REFERENCES

1. Agnew WF, Alvarez RB, Yuen TGH, et al: Protein synthesis and transport by the rat choroid plexus and ependyma. *Cell Tissue Res* 208:261–281, 1980.

2. Aleshire SL, Bradley CA, Richardson LD, et al: Localization of human prealbumin in choroid plexus epithelium. *J Histochem Cytochem* 31:608–612, 1983.
3. Altura BT, Altura BM: The role of magnesium in etiology of strokes and cerebrovasospasm. *Magnesium* 1:277–291, 1982.
4. Ames A III, Sakanoue M, Endo S: Na, K, Ca, Mg and Cl concentrations in choroid plexus fluid and cisternal fluid compared with plasma ultrafiltrate. *J Neurophysiol* 27:672–681, 1964.
5. Assies J, Schellekens APM, Touber JL: Protein hormones in cerebrospinal fluid: Evidence for retrograde transport of prolactin from the pituitary to the brain in man. *Clin Endocrinol (Oxf)* 8:487–491, 1978.
6. Barany EH: Organic anion and cation transport in vitro by dog choroid plexus: Effects of neuroleptics and tricyclic antidepressants. *Acta Pharmacol Et Toxicol* 44:146–155, 1979.
7. Becker NH, Almazon R: Evidence for the functional polarization of micropinocytotic vesicles in the rat choroid plexus. *J Hist Cytochem* 16:278–280, 1968.
8. Berlin RD: Purines: Active transport by isolated choroid plexus. *Science* 163:1194–1195, 1969.
9. Betz AL: Epithelial properties of brain capillary endothelium. *Fed Proc* 44:2614–2615, 1985.
10. Bito LZ, Davson H, Hollingsworth JR: Facilitated transport of prostaglandins across the blood–cerebrospinal fluid and blood–brain barriers. *J Physiol (Lond)* 256:273–285, 1976.
11. Bito LZ, Davson H, Salvador EV: Inhibition of in vitro concentrative prostaglandin accumulation by prostaglandins, prostaglandin analogues and by some inhibitors of organic anion transport. *J Physiol (Lond)* 256:257–271, 1976.
12. Borges LF, Elliott PJ, Gill R, et al: Selective extraction of small and large molecules from the cerebrospinal fluid by Purkinje neurons. *Science* 228:346–348, 1985.
13. Bouchaud C, Bouvier D: Fine structure of tight junctions between rat choroidal cells after osmotic opening induced by urea and sucrose. *Tissue Cell* 10:331–342, 1978.
14. Bradbury MWB: The blood–brain barrier-transport across the cerebral endothelium. *Circ Res* 57(2):213–222, 1985.
15. Brightman MW: The intracerebral movement of proteins injected into cerebrospinal fluid of mice. *Prog Brain Res* 29:19–40, 1968.
16. Brightman MW, Reese TS: Junctions between intimately apposed cell membranes in the vertebrate brain. *J Cell Biol* 40:648–677, 1969.
17. Brownfield MS, Kozlowski GP: The hypothalamo-choroidal tract. I. Immunohistochemical demonstration of neurophysin pathways to telencephalic choroid plexuses and cerebrospinal fluid. *Cell Tissue Res* 178:111–127, 1977.
18. Burton SA: Sodium–Calcium Exchange in the Rat Choroid Plexus. Doctoral thesis, University of Utah, 1982.
19. Burton SA, Johanson CE: Pentobarbital, morphine and lead decrease Ca-45 uptake in rat central nervous system. *Fed Proc* 41:1734(abstr), 1982.
20. Burton SA, Johanson CE: Calcium influx into the in vitro rat CP is inhibited by sodium, potassium, amiloride and ouabain. *Fed Proc* 42:989(abstr), 1983.
21. Covell DG, Narang PK, Poplack DG: Kinetic model for disposition of 6–mercaptopurine in monkey plasma and cerebrospinal fluid. *Am J Physiol* 248:R147–156, 1985.
22. Crook RB, Farber MB, Prusiner SB: Hormones and neurotransmitters control cylic AMP metabolism in CP epithelial cells. *J Neurochem* 42:340–350, 1984.
23. Deng QS: *Drug Modification of Chloride Transport in the Choroid Plexus–Cerebrospinal Fluid System of the Rat.* Doctoral thesis, University of Utah, 1986.
24. DiBenedetto FE, Bito LZ: Transport of prostaglandins and other eicosanoids by the choroid plexus: Its characterization and physiological significance. *J Neurochem* 46:1725–1731, 1986.
25. Dickson PW, Aldred AR, Marley PD, et al: High prealbumin and transferrin mRNA levels in the choroid plexus of rat brain. *Biochem Biophys Res Com* 127:890–895, 1985.
26. Dienel GA: Regional accumulation of calcium in postischemic rat brain. *J Neurochem* 43:913–925, 1984.
27. Domer F: Effects of diuretics on cerebrospinal fluid formation and potassium movement. *Exp Neurol* 24:54–64, 1969.
28. Dratman MB, Futaesaku Y, Crutchfield FL, et al: Iodine-[125]labeled triiodothyronine in rat brain: Evidence for localization in discrete neural systems. *Science* 215:309–312, 1982.
29. Dziegielewska KM, Evans CAN, New H, et al: Synthesis of plasma proteins by rat fetal brain and choroid plexus. *Int J Dev Neurosci* 2:215–222, 1984.
30. Edvinsson L, Lindvall M, Owman C, et al: Autonomic nervous control of cerebrospinal fluid production and intracranial pressure. In Wood JH (ed): *Neurobiology of Cerebrospinal Fluid.* Vol. 2. Plenum, New York, 1983, pp. 661–676.

31. Epstein MH, Feldman AM, Brusilow SW: Cerebrospinal fluid production: Stimulation by cholera toxin. *Science* 196:1012–1013, 1977.

32. Fahrenkrug J, Schaffalitzky de Muckadello O, Fahrenkrug A: Vasoactive intestinal polypeptide (VIP) in human cerebrospinal fluid. *Brain Res* 124:581–584, 1977.

33. Fenstermacher JD, Davson H: Distribution of two model amino acids from cerebrospinal fluid to brain and blood. *Am J Physiol* 242:F171–F180, 1982.

34. Firemark HM: Choroid-plexus transport of enkephalins and other neuropeptides. In Wood JH (ed): *Neurobiology of Cerebrospinal Fluid.* Vol. 2. Plenum, New York, 1983, pp. 77–81.

35. Firemark H, Brand N: Enkephalin transport by choroid plexus. *Clin Res* 27:52A, 1979.

36. Fishman RA: Exchange of albumin between plasma and cerebrospinal fluid. *Am J Physiol* 175:96–98, 1953.

37. Fishman RA, Ransohoff J, Osserman EF: Factors influencing the concentration gradient of protein in cerebrospinal fluid. *J Clin Invest* 37:1419–1424, 1958.

38. Forn J: Active transport of 5-hydroxyindoleacetic acid by the rabbit choroid plexus in vitro. *Biochem Pharmacol* 21:619–624, 1972.

39. Goehlert UG, Ng Ying Kin NMK, Wolfe LS: Biosynthesis of prostacyclin in rat cerebral microvessels and the choroid plexus. *J Neurochem* 36:1192–1201, 1981.

40. Goodner JC, Berrie MA: The failure of rat hypothalamic tissues to take up labeled insulin in vivo or to respond to insulin in vitro. *Endocrinology* 101:605–612, 1977.

41. Gudeman D, Nelson SR, Merisko EM: Apocrine secretion by choroid plexus: Isolated apical fragments synthesize protein in vitro. *Fed. Proc* 45:892(abstr), 1986.

42. Graziani LJ, Escriva A: Calcium exchange between brain and blood in cats and immature and adult rats. *Neurology (NY)* 19:314–315, 1969.

43. Hagen GA, Solberg LA: Brain and cerebrospinal fluid permeability to intravenous thyroid hormones. *Endocrinology* 95:1398–1410, 1974.

44. Harbut RH, Johanson CE: Third ventricle choroid plexus function and its response to acute perturbations in plasma chemistry. *Brain Res* 374:137–146, 1986.

45. Hedlund L, Lischko MM, Rollag MD, et al: Melatonin: Daily cycle in plasma and cerebrospinal fluid of calves. *Science* 195:686–687, 1977.

46. Heller H, Hasan SH, Saifi AQ: Antidiuretic activity in the cerebrospinal fluid. *J Endocrinology* 41:273–280, 1968.

47. Hesketh JE: Effects of potassium and lithium on sodium transport from blood to cerebrospinal fluid. *J Neurochem* 28:597–603, 1977.

48. Hicks JT, Albrecht P, Rapoport SI: Entry of neutralizing antibody to measles into brain and cerebrospinal fluid of immunized monkeys after osmotic opening of the blood–brain barrier. *Exp Neurol* 53:768–779, 1976.

49. Hochwald GM, Wallenstein M: Exchange of albumin between blood, cerebrospinal fluid, and brain in the cat. *Am J Physiol* 212:1199–1204, 1967.

50. Hochwald GM, Wallenstein MC: Exchange of γ-globulin between blood, cerebrospinal fluid and brain in the cat. *Exper Neurol* 19:115–126, 1967.

51. Howarth F, Jowett A: A technique for surgical encapsulation of a canine choroid plexus. *J Physiol (Lond)* 162:20P(abstr), 1962.

52. Huang JT: Accumulation of peptides by choroid plexus in vitro: Tyr-D-Ala-Gly as a model. *Neurochem Res* 6:681–689, 1981.

53. Huang JT: Accumulation of peptide Tyr-D-Ala-Gly by choroid plexus during ventriculocisternal perfusion of rat brain. *Neurochem Res* 7:1541–1548, 1982.

54. Huang JT, Lajtha A: The accumulation of [³H] enkephalinamide (2-D-alanine-5-methioninamide) in rat brain tissues. *Neuropharmacology* 17:1075–1079, 1978.

55. Huang JT, Wajda IJ: The effects of morphine on the accumulation of homovanillic and 5-hydroxyindoleacetic acids in the choroid plexus of rats. *Br J Pharmacol* 60:363–367, 1977.

56. Hurley JV, Anderson RMcD, Sexton PT: The fate of plasma protein which escapes from blood vessels of the choroid plexus of the rat—An electron microscopic study. *J Pathol* 134:57–70, 1981.

57. Husted RF, Reed DJ: Regulation of cerebrospinal fluid potassium by the cat choroid plexus. *J Physiol (Lond)* 259:213–221, 1976.

58. Jacobsen M, Clausen PP, Jacobsen GK, et al: Intracellular plasma proteins in human fetal choroid plexus during development I. Developmental stages in relation to the number of epithelial cells which contain albumin in telencephalic, diencephalic and myelencephalic choroid plexus. *Dev Brain Res* 3:239–250, 1982.

59. Jenkins JS, Mather HM, Ang V: Vasopressin in human cerebrospinal fluid. *J Clin Endocrinol Metab* 50:364–367, 1980.
60. Johanson CE: Choroid epithelial cell pH. *Life Sci* 23:861–868, 1978.
61. Johanson CE: Permeability and vascularity of the developing brain: Cerebellum vs. cerebral cortex. *Brain Res* 190:3–16, 1980.
62. Johanson CE: Differential effects of acetazolamide, benzolamide and systemic acidosis on hydrogen and bicarbonate gradients across the apical and basolateral membranes of the choroid plexus. *J Pharmacol Exp Ther* 231:502–511, 1984.
63. Johanson CE: Choroid plexus. In Adelman G (ed): *Encyclopedia of Neuroscience.* Vol. 1. Birkhauser, Cambridge, Massachusetts, 1987, pp. 236–239.
64. Johanson CE: The choroid plexus-arachnoid membrane: Cerebrospinal fluid system. In Boulton AA, Baker GB (eds): *Neuromethods,* Vol. VII, Humana Press. Clifton, New Jersey, 1987, pp. 33–104.
65. Johanson CE, Smith QR: Phenytoin-induced stimulation of the Na–K pump in the choroid plexus–cerebrospinal fluid system. *Soc Neurosci Abs* 3:316, 1977.
66. Johanson CE, Woodbury DM: Changes in CSF flow and extracellular space in the developing rat. In Vernadakis A, Weiner N (eds): *Drugs and the Developing Brain.* Plenum, New York, 1974, pp. 281–287.
67. Johanson CE, Woodbury DM: Penetration of ^{14}C-barbital and ^{14}C-antipyrine into the choroid plexus and cerebrospinal fluid of the rat. *Exp Brain Res* 30:65–74, 1977.
68. Johanson CE, Woodbury DM: Uptake of C-^{14} urea by the in vivo choroid plexus–cerebrospinal fluid–brain system: Identification of sites of molecular sieving. *J Physiol (Lond)* 275:167–176, 1978.
69. Johanson CE, Foltz FM, Thompson AM: The clearance of urea and sucrose from isotonic and hypertonic fluids perfused through the ventriculo-cisternal system. *Exp Brain Res* 20:18–31, 1974.
70. Johanson CE, Parandoosh Z, Smith QR: Cl–HCO$_3$ exchange in choroid plexus: Analysis by the DMO method for cell pH. *Am J Physiol* (Renal Fluid Electrolyte Physiol 18) 249:F478–F484, 1985.
71. Johanson CE, Reed DJ, Woodbury DM: Active transport of sodium and potassium by the choroid plexus of the rat. *J Physiol (Lond)* 241:359–372, 1974.
72. Johanson CE, Reed DJ, Woodbury DM: Developmental studies of the compartmentalization of water and electrolytes in the choroid plexus of neonatal rat brain. *Brain Res* 116:35–48, 1976.
73. Johanson CE, Woodbury DM, Withrow CD: Distribution of bicarbonate between blood and cerebrospinal fluid in the neonatal rat in metabolic acidosis and alkalosis. *Life Sci* 19:691–700, 1976.
74. Johansson BB, Fahrenkrug J, Wikkelso C, et al: Vasoactive intestinal polypeptide in human cerebrospinal fluid. *Front Hormone Res* 9:189–197, 1982.
75. Kozlowski GP: Ventricular route hypothesis and peptide-containing structures of the cerebroventricular system. *Front Hormone Res* 9:105–118, 1982.
76. Krakauer R, Steiness E: Digoxin concentration in choroid plexus, brain, and myocardium in old age. *Clin Pharmacol Ther* 24:454–458, 1978.
77. Lindvall M, Alumets J, Edvinsson L, et al: Peptidergic (VIP) nerves in the mammalian choroid plexus. *Neurosci Lett* 9:77–82, 1978.
78. Lindvall M, Hardebo JE, Owman CH: Barrier mechanisms for neurotransmitter monoamines in the choroid plexus. *Acta Physiol Scand* 108:215–221, 1980.
79. Logins IS, Macleod RM: Prolactin in human and rat brain, serum and CSF. *Brain Res* 132:477–483, 1977.
80. Lorenzo AV, Winston KR, Welch K, et al: Evidence for prolactin receptors in the CP and a possible role in water balance in neonatal brain. *Z Kinderchir* 38(suppl II):68–70, 1983.
81. Luerssen TG, Robertson GL: Cerebrospinal fluid vasopressin and vasotocin in health and disease. In Wood JH (ed): *Neurobiology of Cerebrospinal Fluid.* Vol. 1. Plenum, New York, 1980, pp. 613–623.
82. Maren TH, Broder LE: The role of carbonic anhydrase in anion secretion into cerebrospinal fluid. *J Pharmacol Exp Ther* 172: 197–202, 1970.
83. Mattaliano VJ, O'Brien RA: The uptake of dopamine by the isolated choroid plexus. *Fed Proc* 39:529(abstr), 1980.
84. Melby JM, Miner LC, Reed DJ: Effect of acetazolamide and furosemide on the production and composition of cerebrospinal fluid from the cat choroid plexus. *Can J Physiol Pharmacol* 60:405–409, 1982.
85. Mess B, Trentini GP: ^{3}H–melatonin level in cerebrospinal fluid and choroid plexus following intravenous administration of the labeled compound. *Acta Physiol Acad Sci Hung* 45:225–231, 1974.
86. Milhorat TH, Davis DA, Hammock MK: Use of calcium ion as an electron microscopic tracer for injection into blood and cerebrospinal fluid. In Arceneaux C, Bailey G (eds): *Proceedings of the Thirty-second Annual Meeting of the Electron Microscopy Society of America.* Claitor's Publishing, Baton Rouge, 1974, pp. 82–83.

87. Milhorat TH, Davis DA, Lloyd BJ Jr: Two morphologically distinct blood–brain barriers preventing entry of cytochrome c into cerebrospinal fluid. *Science* 180:76–78, 1973.

88. Miner LC, Reed DJ: Composition of fluid obtained from choroid plexus tissue isolated in a chamber in situ. *J Physiol (Lond)* 227: 127–139, 1972.

89. Møllgard K, Saunders NR: Complex tight junctions of epithelial and of endothelial cells in early foetal brain. *J Neurocytol* 4: 453–468, 1975.

90. Møllgard K, Saunders NR: A possible transepithelial pathway via endoplasmic reticulum in foetal sheep choroid plexus. *Proc R Soc Lond B* 199:321–326, 1977.

91. Møllgard K, Jacobsen M, Jacobsen GK, et al: Immunohistochemical evidence for an intracellular localization of plasma proteins in human foetal choroid plexus and brain. *Neurosci Lett* 14:85–90, 1979.

92. Murphy VA: Sodium–Hydrogen Exchange in the Rat Choroid Plexus. Doctoral thesis, University of Utah, 1984.

93. Murphy VA, Johanson CE: Uptake of $H-^3$ α-aminoisobutyric acid (AIB) into central and peripheral tissues of the rat. *Fed Proc* 43:673(abstr), 1984.

94. Murphy V, Johanson CE: Adrenergic-induced enhancement of brain barrier system permeability to small nonelectrolytes: Choroid plexus versus cerebral capillaries. *J Cereb Blood Flow Metab* 5:401–412, 1985.

95. Nathanson JA: Adrenergic-receptor mechanisms in mammalian choroid plexus. In Wood JH (ed): *Neurobiology of Cerebrospinal Fluid.* Vol. 2. Plenum, New York, 1983, pp. 677–686.

96. Neuwelt EA, Barnett PA, McCormick CI, et al: Osmotic blood–brain barrier modification: Monoclonal antibody, albumin and methotrexate delivery to cerebrospinal fluid and brain. *Neurosurg* 17:419–423, 1985.

97. Nicholson G, Greely GH, Humm J, et al: Prolactin in cerebrospinal fluid: A probable site of prolactin autoregulation. *Brain Res* 190:447–457, 1980.

98. Nolan CM, Ulmer WC: A study of cephalothin and desacetylcephalothin in cerebrospinal fluid in therapy for experimental pneumococcal meningitis. *J Inf Dis* 141:326-330, 1980.

99. Olsson J-E, Link H: Distribution of serum proteins and beta-trace protein within the nervous system. *J Neurochem* 20:837–846, 1973.

100. O'Tuama LA, Kim CS, Gatzy JT, et al: The distribution of inorganic lead in guinea pig brain and neural barrier tissues in control and lead-poisoned animals. *Toxicol Appl Pharmacol* 36:1–9, 1976.

101. Parandoosh Z, Johanson CE: Effect of vasopressin on the penetration of C-14 urea into brain compartments protected by barrier systems. *Soc Neurosci Abs* 5:308, 1979.

102. Parandoosh Z, Johanson CE: Ontogeny of the blood–brain barrier to, and cerebrospinal fluid sink action on, C^{14}-urea. *Am J Physiol* 243:R400–R407, 1982.

103. Pardridge WM: Transport of nutrients and hormones through the blood–brain barrier. *Diabetologia* 20:246–254, 1981.

104. Pardridge WM: Strategies for drug delivery through the blood–brain barrier. In Borchardt RT, Repta AJ, Stella VJ (eds): *Directed Drug Delivery.* Humana Press, Clifton, New Jersey, 1985, pp. 83–96.

105. Pardridge WM, Mietus LJ: Transport of albumin-bound melatonin through the blood–brain barrier. *J Neurochem* 34:1761–1763, 1980.

106. Pardridge WM, Mietus LJ: Transport of thyroid and steroid hormones through the blood–brain barrier of the newborn rabbit: Primary role of protein-bound hormone. *Endocrinology* 107:1705–1710, 1980.

107. Pardridge WM, Frank HJL, Cornford EM, et al: Neuropeptides and the blood–brain barrier. In Martin JB, Reichlin S, Bick KL (eds): *Neurosecretion and Brain Peptides.* Raven, New York, 1981, pp. 321–328.

108. Pelletier G, Labrie F, Kastin AJ, et al: Radioautographic localization of radioactivity in rat brain after intraventricular or intracarotid injection of ^{3}H-L-prolyl-L-leucyl glycinamide. *Pharmacol Biochem Behav* 3:675–679, 1975.

109. Peress NS, Tompkins D: Effect of molecular charge on choroid-plexus permeability: Tracer studies with cationized ferritins. *Cell Tissue Res* 219:425–431, 1981.

110. Pollay M, Reynolds E, Tompkins P, et al: Inhibition of choroid plexus $Na^+–K^+$ ATPase and cerebrospinal fluid formation. *Soc Neurosci Abs* 10:791, 1984.

111. Pollay M, Stevens A, Estrada E, et al: Extracorporeal perfusion of choroid plexus. *J Appl Physiol* 32:612–617, 1972.

112. Posner BI, van Houten M, Patel B, et al: Characterization of lactogen binding sites in choroid plexus. *Exp Brain Res* 49:300–306, 1983.

113. Pullar IA: The accumulation of [^{14}C]5-hydroxyindol-3-yl-acetic acid by the rabbit choroid plexus in vitro. *J Physiol (Lond)* 216: 201–211, 1971.

114. Quay WB: Retrograde perfusions of the pineal region and the question of pineal vascular routes to brain and choroid plexuses. *Am J Anat* 137:387–402, 1973.
115. Rap ZM: Inhibitory effect of antidiuretic hormone on outflow of the cerebrospinal fluid in vasogenic brain edema induced by cold lesion. In Cervos-Navarro J, Fritschka ER (eds): *Cerebral Microcirculation and Metabolism.* Raven, New York, 1981, pp. 171–175.
116. Rapoport SI: Passage of proteins from blood to cerebrospinal fluid. Model for transfer by pores and vesicles. In Wood JH (ed): *Neurobiology of Cerebrospinal Fluid.* Vol. 2. Plenum, New York, 1983, pp. 233–245.
117. Rapoport SI, Pettigrew KD: A heterogenous, pore-vesicle membrane model for protein transfer from blood to cerebrospinal fluid at the choroid plexus. *Microvasc Res* 18:105–119, 1979.
118. Reed DJ, Yen M-H: The role of the cat choroid plexus in regulating cerebrospinal fluid magnesium. *J Physiol (Lond)* 281:477–485, 1978.
119. Reed DJ, Yen M-H: The effect of lithium on electrolyte transport by the in situ choroid plexus of the cat. *J Physiol (Lond)* 309:329–339, 1980.
120. Reese TS, Feder N, Brightman MW: Electron microscopic study of the blood–brain and blood–cerebrospinal fluid barriers with microperoxidase. *J Neuropathol Exp Neurol* 32:137–138, 1971.
121. Riccardi R, Kramer RJ, Trown PW, et al: Serum and cerebrospinal fluid (CSF) pharmacokinetics of recombinant leucocyte A interferon (IFLrA) in monkeys. *Proc Am Assoc Cancer Res* 23:203(abstr), 1982.
122. Richards JG: Permeability of intercellular junctions in brain epithelia and endothelia to exogenous amine: Cytochemical localization of extracellular 5-hydroxydopamine. *J Neurocytol* 7:61–70, 1978.
123. Rodriguez EM, Heller H: Antidiuretic activity and ultrastructure of the toad choroid plexus. *J Endocrinol* 46:83–91, 1970.
124. Rudman D, Chawla RK: Antidiuretic peptide in mammalian choroid plexus. *Am J Physiol* 230:50–55, 1976.
125. Saito Y, Wright EM: Kinetics of the sodium pump in the frog choroid plexus. *J Physiol (Lond)* 328:229–243, 1982.
126. Saito Y, Wong H, Wright EM: Insulin effects on sodium transport by choroid plexus. *Soc Neurosci Abs* 7:86, 1981.
127. Sampath SS, Neff NH: The elimination of 5-hydroxyindoleacetic acid from cerebrospinal fluid: Characteristics of the acid transport system of the choroid plexus. *J Pharmacol Exp Ther* 188:410–414, 1974.
128. Saunders NR: Ontogeny of the blood–brain barrier. *Exp Eye Res* 25(suppl):523–550, 1977.
129. Schliep G, Felgenhauer K: Serum-CSF protein gradients, the blood–CSF barrier and the local immune response. *J Neurol* 218:77, 1978.
130. Schultz WJ, Brownfield MS, Kozlowski GP: The hypothalamochoroidal tract. II. Ultrastructural response of the choroid plexus to vasopressin. *Cell Tissue Res* 178:129–141, 1977.
131. Smith DE, Streicher E, Milkovic K, et al: Observations on the transport of proteins by the isolated choroid plexus. *Acta Neuropathol (Berl)* 3:372–386, 1964.
132. Smith QR, Johanson CE: Effect of carbonic anhydrase inhibitors and acidosis on choroid plexus epithelial cell sodium and potassium. *J Pharmacol Exp Ther* 215:673–680, 1980.
133. Smith QR, Johanson CE: Effect of ouabain and potassium on ion concentrations in the choroidal epithelium. *Am J Physiol* 238:F399–F406, 1980.
134. Smith QR, Johanson CE: Active transport of chloride by lateral ventricle choroid plexus of the rat. *Am J Physiol (Renal Fluid Electrolyte Physiol* 18) 249:F470–F477, 1985.
135. Smith QR, Johanson CE, Woodbury DM: Uptake of Cl and Na by the brain–cerebrospinal fluid system: Comparison of the permeability of the blood–brain and blood–cerebrospinal fluid barriers. *J Neurochem* 37:117–124, 1981.
136. Smith QR, Pershing LK, Johanson CE: A comparative analysis of extracellular fluid volume of several tissues as determined by six different markers. *Life Sci* 29:449–456, 1981.
137. Smith QR, Woodbury DM, Johanson CE: Uptake of Cl[36] and Na[22] by the choroid plexus–cerebrospinal fluid system: Evidence for active chloride transport by the choroidal epithelium. *J Neurochem* 37:107–116, 1981.
138. Smith QR, Woodbury DM, Johanson CE: Kinetic analysis of Cl[36], Na[22] and H[3]-mannitol uptake into the in vivo choroid plexus–cerebrospinal fluid system: Ontogeny of the blood–brain and blood–CSF barriers. *Dev Brain Res* 3:181–198, 1982.
139. Smith QR, Tai C-Y, Rapoport SI: Brain capillary permeability to inorganic ions. *Soc Neurosci Abs* 9:161, 1983.

140. Smith RA, Norris F, Palmer D, et al: Distribution of alpha interferon in serum and cerebrospinal fluid after systemic administration. *Clin Pharmacol Ther* 37:85–88, 1985.

141. Smulders AP, Wright EM: Role of choroid plexus in transport of melatonin between blood and brain. *Brain Res* 191:555–558, 1980.

142. Sparks DL, Greene WB, Powers JM, et al: Magnesium deficiency: Its effect on choroid plexus. *Fed Proc* 40:933(abstr), 1981.

143. Spector R: The effect of Pronase on choroid plexus transport. *Brain Res* 134:573–576, 1977.

144. Spector R: Nucleoside transport in choroid plexus: Mechanism and specificity. *Arch Biochem Biophys* 216:693–703, 1982.

145. Spector R: Pharmacokinetics and metabolism of cytosine arabinoside in the central nervous system. *J Pharmacol Exp Ther* 222:1–6, 1982.

146. Spector R: Ceftriaxone pharmacokinetics in the central nervous system. *J Pharmacol Exp Ther* 236:380–383, 1986.

147. Spector R, Eells J: Deoxynucleoside and vitamin transport into the central nervous system. *Fed Proc* 43:196–200, 1984.

148. Spector R, Goetzl EJ: Leukotriene C4 transport by the choroid plexus in vitro. *Science* 228:325–327, 1985.

149. Spector R, Johanson CE: The mammalian choroid plexus: Structure, development and function. *Sci Amer*, accepted for publication.

150. Spector R, Levy P: Thyroxine transport by the choroid plexus in vitro. *Brain Res* 98:400–404, 1975.

151. Spector R, Lorenzo AV: Specificity of ascorbic acid transport system of the central nervous system. *Am J Physiol* 226:1468–1473, 1974.

152. Spector R, Lorenzo AV: Folate transport by the choroid plexus in vitro. *Science* 187:540–542, 1975.

153. Tochino Y, Schanker LS: Transport of serotonin and norepinephrine by the rabbit choroid plexus in vitro. *Biochem Pharmacol* 14:1557–1566, 1965.

154. Tramezzani JH, Negreiros de Paiva CE, et al: Oxytocic activity of the toad's brain. *Acta Endocrinol (Stockh)* 23:175–184, 1956.

155. Uddman R, Fahrenkrug J, Malm L, et al: Neuronal VIP in salivary glands: distribution and release. *Acta Physiol Scand* 110:31–38, 1980.

156. Van Deurs B: Choroid plexus absorption of horseradish peroxidase from the cerebral ventricles. *J Ultrastruct Res* 55:400–415, 1976.

157. Van Deurs B: Microperoxidase uptake into the rat choroid plexus epithelium. *J Ultrastruct Res* 62:168–180, 1978.

158. Van Deurs B: Structural aspects of brain barriers, with special reference to the permeability of the cerebral endothelium and choroidal epithelium. *Int Rev Cytol* 65:117–191, 1980.

159. Van Houten M, Posner BI, Kopriwa BM, et al: Insulin-binding sites in the rat brain: In vivo localization to the circumventricular organs by quantitative autoradiography. *Endocrinology* 105:666–673, 1979.

160. Van Wart CA, Dupont JR, Kraintz L: Effect of acetazolamide on passage of protein from cerebrospinal fluid to plasma. *Proc Soc Exp Biol Med* 106:113–114, 1961.

161. Vogh BP, Langham MR: The effect of furosemide and bumetanide on cerebrospinal fluid formation. *Brain Res* 221:171–183, 1981.

162. Weisner B, Kauerz U: The influence of the choroid plexus on the concentration of prealbumin in CSF. *J Neurol Sci* 61:27–35, 1983.

163. Winbladh B, Edvinsson L, Lindvall M: Effect of sympathectomy on active transport mechanisms in choroid plexus in vitro. *Acta Physiol Scand* 102:85A(abstr), 1978.

164. Wood JH: Physiology and pharmacology of peptide, steroid, and other hormones in cerebrospinal fluid. In Wood JH (ed): *Neurobiology of Cerebrospinal Fluid*. Vol. 2. Plenum, New York, 1983, pp. 43–65.

165. Woodbury DM, Johanson CE, Brondsted H: Maturation of the blood–brain and blood–CSF barriers and transport systems. In Zimmermann E, George R (eds): *Narcotics and the Hypothalamus*. Raven, New York, 1974, pp. 225–247.

166. Wright EM: Accumulation and transport of amino acids by the frog choroid plexus. *Brain Res* 44:207–219, 1972.

167. Wright EM: Transport processes in the formation of the cerebrospinal fluid. *Rev Physiol Biochem Pharmacol* 83:1–34, 1978.

168. Wright EM, Prather JW: The permeability of the frog choroid plexus to nonelectrolytes. *J Membrane Biol* 2:127–149, 1970.

169. Wright EM, Zeuthen T: The paracellular pathway across the frog choroid plexus. In Bradley SE, Purcell EF (eds): *The Paracellular Pathway*. Josiah Macy Foundation, New York, 1982, pp. 323–332.
170. Yen M-H, Reed DJ: Regulation of lithium in cerebrospinal fluid of the cat by the choroid plexus isolated in situ. *Arch Int Pharmacodyn et de Therap* 251:217–227, 1981.
171. Young WS III, Kuhar MJ, Roth J, et al: Radiohistochemical localization of insulin receptors in the adult and developing rat brain. *Neuropeptides* 1:15–22, 1980.
172. Zaidi SMA, Heller H: Can neurohypophysial hormones cross the blood–cerebrospinal fluid barrier? *J Endocrinol* 60:195–196, 1974.

10

The Blood–Brain Barrier and the Immune System

Marianne Juhler and Edward A. Neuwelt

1. BASIC CONCEPTS OF IMMUNOLOGY

The immune system is a broad term covering elements involved in antimicrobial defense, autoimmune reactions, hypersensitivity/graft rejection reactions, and antineoplastic reactions.[63,123] Conventionally, immunity is divided into cellular and humoral components.

Cellular immunity is principally a function of T lymphocytes, which are activated upon contact with the appropriate antigen. T-cell activation requires antigen to be presented in conjunction with the major histocompatibility complex (MHC). T-helper lymphocytes recognize foreign antigens presented by macrophages or other cells in the presence of class II MHC antigens. These antigens in the murine systems are designated Ia and in the human systems as HLA-DR and are present on the cell surface. By contrast, the MHC antigens for cytotoxic T-lymphocyte presentation probably consist of class I MHC gene products. The restriction pattern for suppressor cell lineages has been less clearly elucidated but may—at least in part—be class II restricted.[61,167] As illustrated in Fig. 1, the requirement of antigen presentation that must occur in conjunction with the appropriate MHC antigen occurs on the surface of antigen-presenting cells (APC).[175,179]

Monocytes/macrophages are thought to be the main source of APC. Under pathologic conditions brain endothelial cells may function as APC.[139,143,179,188] In addition, recent evidence suggests that astrocytes within the central nervous system (CNS) may function as accessory cells for T-cell activation and thereby be an integral part of the afferent arm of the cellular immune system. This hypothesis is supported by the fact that class II MHC (HLA-DR) antigens have been observed on large numbers of astrocytes and endothelial cells in areas of gliosis surrounding metastases and abscesses. Indeed, using single- and double-labeled avidin, biotin (ABC) immunoperoxidase techniques employing monoclonal antibodies for HLA-DR, macrophages, T lymphocytes, and glial fibrillary acidic

Marianne Juhler • Department of Neurology, Rigshospitalet, Copenhagen, Denmark. *Edward A. Neuwelt* • Divisions of Neurosurgery and Biochemistry, School of Medicine, Oregon Health Sciences University, and Neurosurgery Section, Veterans Administration Medical Center, Portland, Oregon 97201.

protein (GFAP), Frank et al.[61] were able to demonstrate the presence of HLA-DR on a large number of GFAP-positive astrocytes. In addition, cultured astrocytes can be induced to produce cell surface HLA-DR by exposure to γ-interferon (IFN$_\gamma$). The fact that astrocytes cannot only function as APC but also release interleukin-1 (IL-1), a factor that induces interleukin-2 (IL-2) receptors on lymphocytes as demonstrated by Fontana et al.[56–58] gives strong evidence that astrocytes as well as endothelial cells may be very important in immune processes involving the CNS. Indeed, circulating intravascular T lymphocytes may first encounter foreign antigen in the presence of HLA-DR antigen on the surface of brain endothelial cells. Later, after these cells have been activated and have penetrated the endothelial barrier, they may be further stimulated by antigen-presenting astrocytes.

When activated, the T cells proliferate, undergo blastogenesis, and secrete substances referred to as lymphokines.[114,140] The lymphocytes via lymphokines promote an inflammatory response by their effects on macrophage behavior and by increasing microvasculature permeability.[15,16,36,91,114,140] Tissue damage secondary to these events may cause release of tissue kinins and a secondary cascade of events resulting in increased local capillary permeability.[16,97,114,138,140]

Humoral immunity is a result of systemically circulating antibodies secreted by B lymphocytes in response to antigenic challenge.[85,96,123] The secreted antibodies are specifically directed against a given antigen, but cross reactions with other substances bearing antigenic resemblance to the challenging antigen can occur.[21,96,110] Antibodies may directly combine with and neutralize the antigen or may require complement to deal with the antigen.[96,123] Complement may be defined as an enzymatic cascade causing (1) lysis of foreign cells, (2) attraction of lymphocytes and macrophages, and (3) increase in microvasculature permeability.[119]

Although cellular and humoral immunity are classically described as entirely different entities, they share common features and often interact (Fig. 1). This makes the actual distinction less sharp. In many clinical and experimental settings, it may therefore be less obvious to attribute a situation mainly or entirely to one of the two. It may be appreciated from this brief description of both systems that the final event, which may be triggered by either limb of the immune system, can be very similar, i.e., increased vascular permeability, cellular infiltration, and tissue destruction.

The division of the immune system into a strict B-cell-dependent compartment and a strict T-cell-dependent compartment becomes even less rigid when considering immune-regulatory mechanisms.[179,180] Not infrequently, antigenic stimulation is not enough to cause B-cell production of antibodies, and T-cell assistance is required.[63,96,123,175] Helper and suppressor T cells are specialized T-cell subsets that interact with MHC and immune-regulatory genes to modulate and regulate the immune response.[63,175,180] Helper T cells proliferate in response to antigens and by secretion of lymphokines stimulate B cells, other T cells, and macrophages (Fig. 1). Suppressor T cells have been shown to decrease or abolish graft rejection reactions.[96] The change in circulating suppressor/helper cell ratio in diseases such as acquired immunodeficiency syndrome (AIDS),[92,98] and in multiple sclerosis (MS) exacerbations[4,7,75,153] seem to suggest a central regulation and coordination of immune response by helper–suppressor interaction.

It has also been proposed that the CNS itself may modulate immune reactivity. It has been shown that electrolytic lesions in the anterior hypothalamus, amygdala, hippocampus, and mammillary bodies alter number and function of spleen- and thymus-derived cells.[31,161] The mechanisms involved in this neuroimmunomodulation are unresolved.

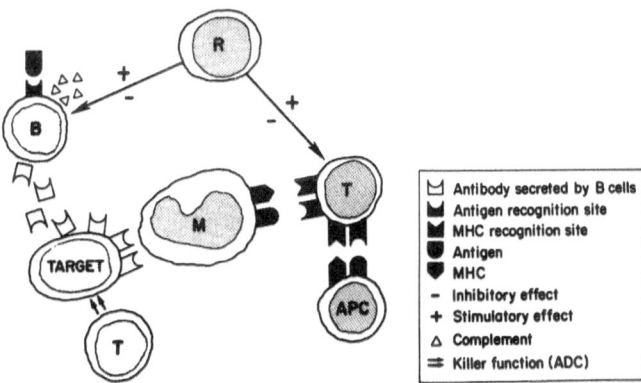

Figure 1. The activity of the immune system is centrally regulated by immunoregulatory cells (R) that may subserve a suppressing (−) or a helping (+) function by the secretion of mediators known as lymphokines. T cells require that antigen is presented on the surface of antigen-presenting cells (APC) together with the major histocompatibility complex (MHC). B cells often require T-cell help to produce antibodies and may require complement to be able to lyse cellular antigens. B cells and T cells may also cooperate in lysis of foreign cells in antibody-dependent cytotoxicity (ADC). Macrophages (M) may act both as effector cells by phagocytosis and as APC.

The lesions have been thought to exert their effect by interruption of central catecholamin-ergic pathways that may be important for maintaining immunologic homeostasis and readiness.[39,162]

2. IS THE CENTRAL NERVOUS SYSTEM AN IMMUNOLOGICALLY PRIVILEGED SITE?

From the introduction to some of the basic concepts of immunology, it can be appreciated that an immune reaction elicited by the presence of an antigen requires the arrival of either immunocompetent cells or specific antibodies at the site. In most organs, the capillary endothelium allows passage of macromolecules through fenestrae or inter-cellular junctions. The vasculature of the CNS, however, is characterized by the existence of the blood–brain barrier (BBB). The BBB can be regarded as both an anatomic and a physiologic phenomenon. The anatomic substrate for the BBB is the interendothelial tight junction[23,26,51,121] that forms a continuous sealing of the interendothelial spaces. Thus, the intravascular compartment, which includes the circulating immune system, is anatom-ically entirely separated from the brain extracellular fluid (ECF).

The circulating immune system is separated from the cerebrospinal fluid (CSF) by the blood–CSF barrier. This barrier is anatomically situated in the epithelial lining of the choroid plexus. Cells and antibodies may easily escape through the fenestrated choroidal capillaries, but the blood–CSF barrier between the epithelial cells of the choroid plexus (as opposed to choroidal capillary endothelial cells) greatly limits permeation of these elements into the CSF.[52]

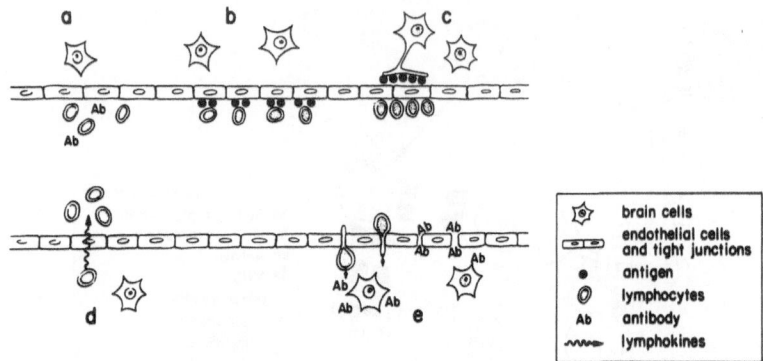

Figure 2. Potential ports of entry for the immune system into the CNS. (a) The BBB, which consists of endothelial cells (EC) with tight junctions between them, greatly impedes the entry of lymphocytes (Ly) and antibodies (Ab) into the CNS. (b) EC may act as antigen-presenting cells and thus attract Ly. (c) Astrocytes may act as antigen-presenting cells and through interaction with EC attract Ly. (d) Ly involved in immune surveillance may encounter CNS antigen and secrete lymphokines and thereby attract other Ly. (e) Ab may enter, e.g., through some of the leaky areas in the BBB; on interaction with their antigen counterpart in the CNS, they may cause damage to EC, making the BBB permeable. Ab and Ly can enter through the leaks.

Teleologically, the junction of these barriers may be explained as keeping potentially noxious substances away from the brain. Highly polar molecules and macromolecules have very poor access to the brain.[23,27,52,128,131,143,151] In an immunologic context, this means that both the cellular and humoral immune systems are exposed to brain antigens to a very limited degree (Fig. 2). Keeping this in mind, it may be readily understood why the brain may be regarded as a so-called immunologically privileged site.[41,123,150] In fact, transplants of noncompatible tissues in brain may survive and grow with little or no graft rejection,[8,120,178] viruses may linger for long periods, causing slow tissue destruction and CNS dysfunction, the so-called slow-virus diseases, with almost no inflammatory response.[89,149,174]

Much of the experimental evidence for clarifying the CNS as an immunologically privileged site dates back to the 1920s when Murphy and Sturm[120] showed that mouse sarcoma could survive in rat brain without graft rejection. If an autograft of splenic tissue was implanted in the rat brain together with the mouse sarcoma, the tumor failed to grow. On the other hand, systemic immunization against the graft failed to inhibit growth of an intracerebral implant of mouse sarcoma in rat brain.[120] The conclusion of this series of experiments is that a barrier separates the peripherally circulating lymphocytes from the brain, thereby preventing a cellular immune reaction, such as graft rejection, in the CNS.

The excellent survival of graft and the ability to establish anatomically normal neuronal circuitry between graft and appropriate host regions after implantation of hippocampal, nigrostriatal, or pituitary/hypothalamic donor tissue,[172,190] is additional evidence in favor of immunologic privilege of the CNS.

This conclusion is supported by another series of experiments in which human lymphocytes were introduced into the subarachnoid space of experimental animals.[124] If the animal had been sensitized by a similar injection previously, it developed pulmonary edema and cellular infiltration of the choroid plexus without any evidence of parenchymal CNS lesions (Figs. 3–5).

However, immunologic reactions do occur in the CNS,[182] so the seclusion of the brain from the circulating immune system may be less than perfect. It has been observed that graft survival is quite short after intracerebral implantation of highly antigenic tumors or tissues.[155,186] However, the graft rejection can be prevented by pretreating the recipient with antilymphocyte serum.[154] Raju and Grogan[150] reported on a series of experiments demonstrating that recipients of intracerebral skin grafts exhibit significant hemagglutinin titers and that rejection of a second implant occurs once the first implant was established. They also observed that the ability of recipient lymphocytes to undergo blastogenesis when cocultured with donor lymphocytes was altered (increased in ≈75% and decreased in ≈25%). They concluded that the afferent arc, i.e., input, of immune response in the CNS seems intact and that the immunologic privilege may be due to a weak efferent arc, i.e., output.

Acceptance of the quoted experiments as evidence of the ability of circulating immunocompetent cells to cross the BBB and initiate immune reactions meets one main objection: the mechanical disruption of the BBB by the surgical trauma obligatory for the experimental protocol.

Another piece of evidence is the fact that it is known that systemic immunization can

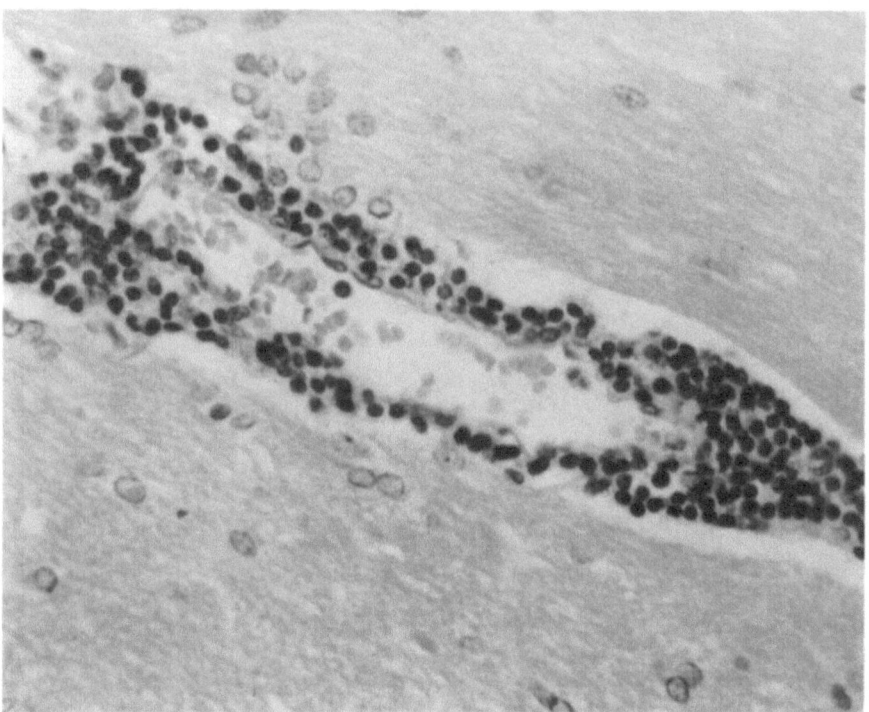

Figure 3. Photomicrograph showing lymphocytic infiltration of the subarachnoid space of a New Zealand white rabbit 36 hr after intrathecal infusion of 2×10^9 viable human lymphocytes. The CSF cell count just before autopsy was 2000 cells/mm³, all of which were viable lymphocytes. (H & E.) ($\times$ 350.) (From Neuwelt and Doherty.[124])

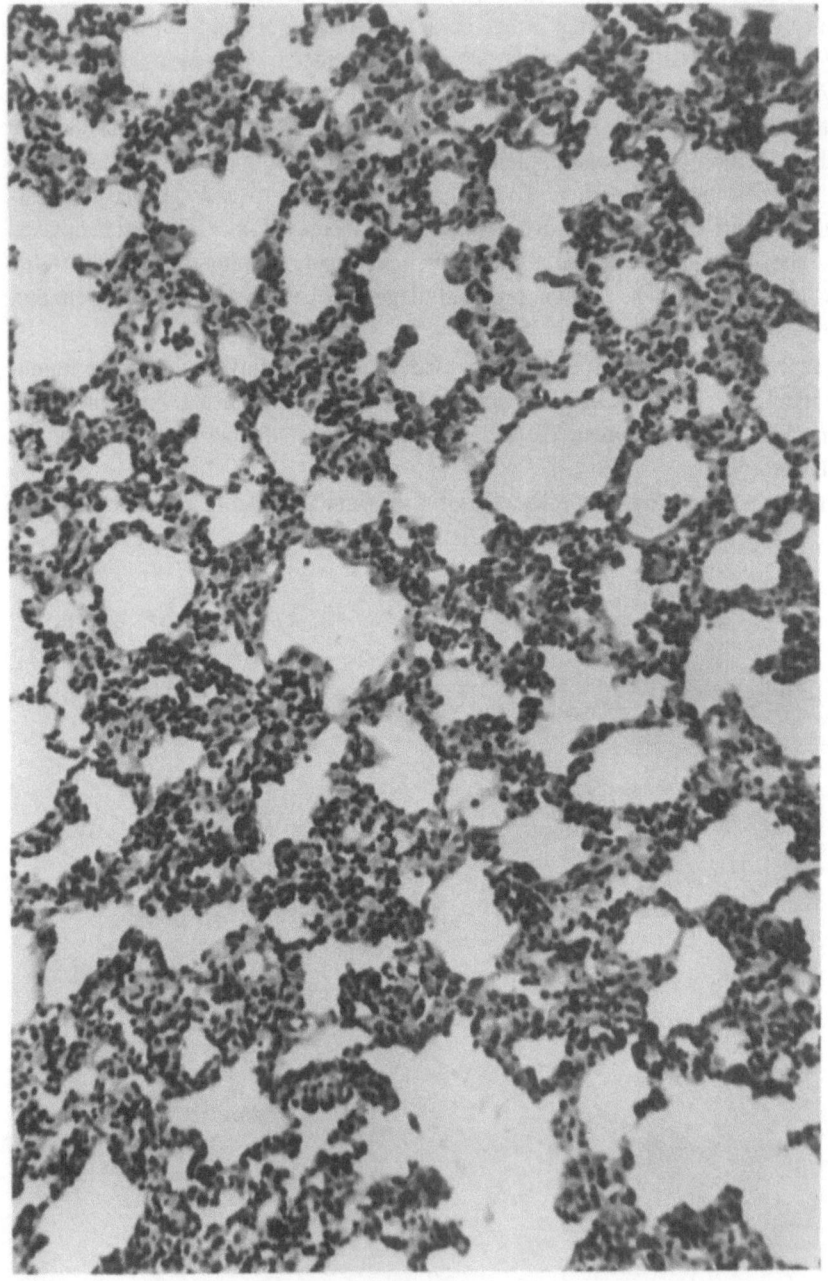

Figure 4. (a) Absence of pulmonary edema 48 hr following an initial intrathecal (IT) infusion of xenogeneic lymphocytes into a New Zealand white rabbit. The above histologic section was taken from the same rabbit whose normal choroid plexus is illustrated in Fig. 5a. (H & E.) (× 134)

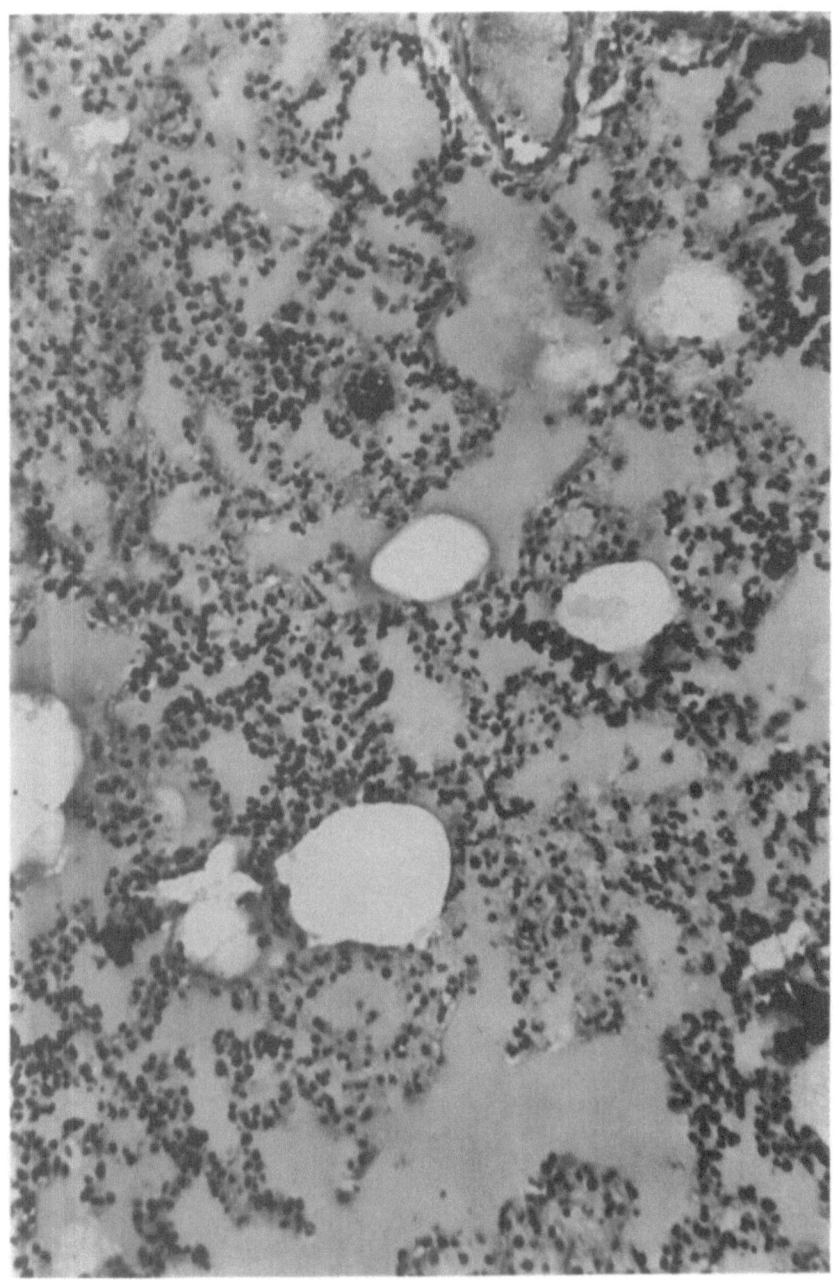

(b) Pulmonary edema 48 hr following a second IT infusion of xenogeneic lymphocytes into a New Zealand white rabbit. The above histologic section was taken from the same rabbit that developed choroid plexitis, as illustrated in Fig. 5b, and received the same lymphocyte preparation as the rabbit in Fig. 5a. (From Neuwelt and Doherty.[124])

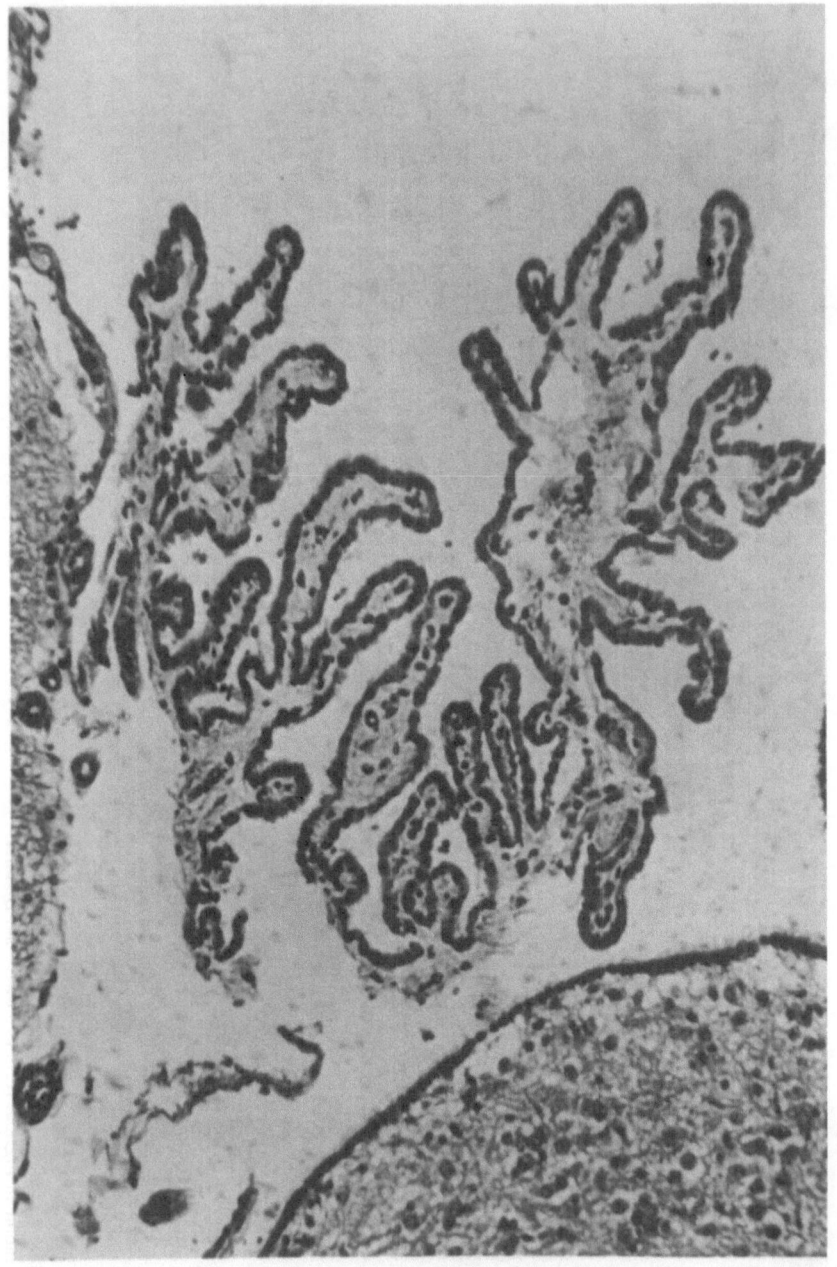

Figure 5. Choroid plexus following IT infusion of normal human lymphocytes into New Zealand white rabbits. (a) Normal choroid plexus 48 hr post-IT infusion of 1×10^9 viable human lymphocytes from donor H. C. Host rabbit had no prior exposure to human cells. CSF cell count at time of necropsy was 2900 lymphocytes/mm^3. (H & E.) ($\times$ 34)

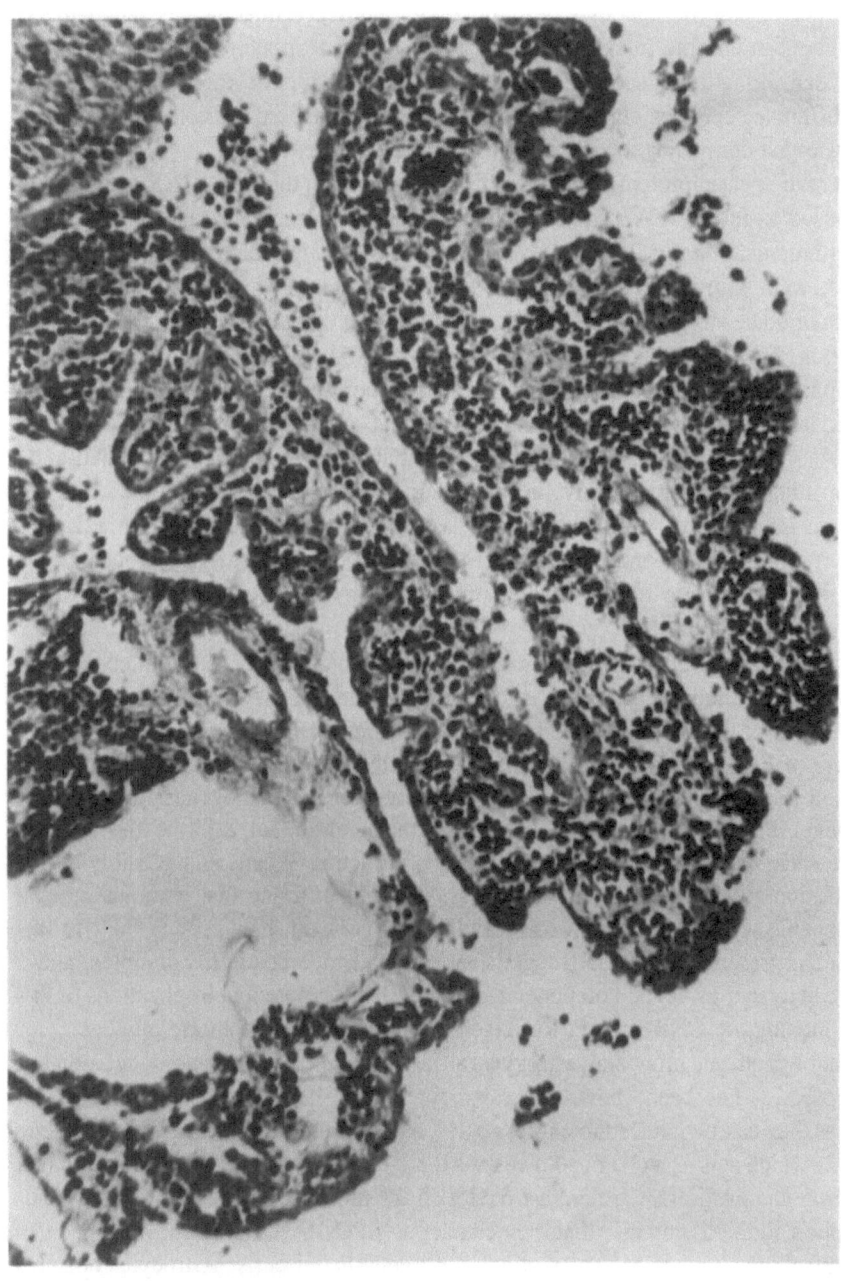

(b) Marked lymphocytic infiltration of choroid plexus 48 hr post-IT infusion of 1×10^9 viable human lymphocytes from donor H. C. Same donor and cell preparation as in A. The host rabbit had a prior human lymphocyte infusion 6 weeks previously from donor L. B. CSF cell count at the time of necropsy was 900 lymphocytes/mm³. Brain parenchyma, due to an intact BBB, was uninvolved by this immune response. (H & E.) ($\times$ 34) (From Neuwelt and Doherty.[124])

cause immune reactions in the CNS. Experimental allergic encephalomyelitis (EAE) is a classic example of systemic immunization resulting in an immune reaction in the CNS.[3,145,156] EAE also creates an excellent setting for studying the individual antigenic substances of the CNS.[32,87,189] The neuropathology of EAE is remarkable for BBB damage,[122] perivascular lymphocytic inflammation, and demyelination.[97,99] The latter is more prominent in chronic relapsing EAE,[99,189,192] whereas in acute monophasic EAE, only slight demyelination is seen in the areas covered by lymphocytic infiltrates.[188] The observed difference between chronic and acute EAE suggests that the cellular inflammatory response and demyelination may be two separate entities.

When myelin basic protein (MBP) was first isolated from the crude CNS homogenate originally used to induce EAE, it was found to possess encephalitogenic properties.[87] However, immunization to MBP alone results in only very slight demyelination, so other antigens had to be implicated to explain the full-blown picture of EAE.[32,189] It has been shown that antibodies to galactocerebroside—a glycolipid component of myelin—caused demyelination, but only in the presence of a cellular exudate consisting of lymphocytes and macrophages.[32] The explanation for these experimental data seems straightforward. Galactocerebroside is a surface antigen of myelin that is readily accessible to attack by antibodies.[32,105] Inflammatory cells, most notably macrophages, cause demyelination by lysis or unrolling of the lamellar myelin sheath and thus expose the deeply buried MBP, which, as an integral protein of myelin,[53,105] is necessary for holding the lamellae together. The joint action of antigalacrocerebroside antibodies, anti-MBP antibodies, and cells thus seems to bring about total disruption of myelin, both biochemically and structurally.

The more puzzling question to answer is how systemic induction of immunocompetent cells and/or antibodies can induce an allergic reaction against antigenic determinants on the other side of the BBB. It is well known that increased BBB permeability is a very early feature of EAE preceding the occurrence of cellular infiltrates by several days.[83,84] The question may therefore be rephrased to searching for the pathogenesis of the early BBB damage. Part of the explanation may be that endothelial cells of the cerebral vasculature seem capable of acting as APCs, thus attracting T cells and possibly aiding their penetration across the BBB as well. Thus, it has been found that brain endothelial cells are capable of expressing histocompatibility antigens on their surface.[61,177] In this way, antigen is presented at the BBB together with the histocompatibility complex, and T cells can be activated by actual components of the BBB. The potential implications of this theory for making the CNS a target for attack by the immune system are obvious.

Another hypothesis proposing astrocytes as key cells in CNS immunomodulation has been proposed. It has been shown that astrocytes may meet T-cell requirements for activation as they can be induced to express MHC antigens on their surface and are able to secrete IL-1, which acts as a chemical mediator for T-cell activation. The observation that T cells clump and proliferate around astrocytes in the presence of specific antigen lends further support to the hypothesis that they can act as APCs.[55,61] The astrocytes may also serve a controlling function via interactions with endothelial cells. Astrocyte foot processes are closely opposed to the endothelium being separated only by the matrix of a basement membrane. It is possible that astrocytes are involved in bidirectional transfer of antigens between endothelium and brain tissue thereby promoting antigen presentation by endothelial cells. Recognition of antigen on endothelial cells by circulating immunocompetent cells could then create the initial attraction of immunocompetent cells into the CNS. Furthermore, T cells sensitized against MBP secrete endoglycosidase, an enzyme capable of breaking down the subendothelial basement membrane,[122] which could conceivably facilitate their penetration into the CNS.

Another—not necessarily exclusive—explanation might be found in the ability of the damaged brain to generate vasoactive substances.[15,16,36,118] Brain lipids are rich in arachidonic acid (AA)[15,16] which may be released by trauma to the brain tissue, by neoplastic invasion or by ischemia.[15,16,118] It has been shown experimentally that AA and leukotrienes (the latter are formed from AA) can increase BBB permeability when injected directly into the rat brain.[15] The leukotriene content of brain tissue correlates significantly with the amount of edema surrounding various CNS neoplasms,[16] and it is conceivable that leukotrienes released from the damaged brain contribute to BBB disruption and vasogenic edema in CNS neoplasia. Similarly, inflammation of brain tissue in immune-mediated CNS disease could possibly cause release of AA and leukotrienes which could open the BBB and might propagate any initial tissue damage.

However, it is not necessary to postulate a physical disruption of the BBB to explain immunologic phenomena in the setting of systemic immunization. There are some potential flaws in the perfect seclusion of the brain from the circulating immune system, all of which could represent a possible port of entry for immunocompetent cells and/or antibodies into the CNS.

One such possible flaw may be immune surveillance. Immune surveillance is thought to be a major antineoplastic defense of the body to ensure early destruction of malignant cell transformation.[35,63,123] It is a function of lymphocytes and/or macrophages, which leave the vascular space at random to enter various organs (possibly including the brain), where they may elicit an immune response if foreign antigen is encountered.

A few leaky areas in the CNS—hypothalamus, area postrema, and the spinal and cranial nerve roots[23,26,83,84,139,151]—provide another potential port of entry for the immune system into the CNS. Exchange between circulating plasma solutes and brain ECF occurs more readily in these areas.[23,83,84]

The brain itself lacks lymphatics. However, subarachnoid and ventricular CSF, including macromolecular solutes from the extracellular fluid, may drain via subarachnoid space surrounding the olfactory nerve bundles into the submucous spaces of the nose and from there to the retropharyngeal lymph nodes.[24] Thus, fluorescein-labeled dextrans, Evans blue or ^{131}I-labeled albumin, horseradish peroxidase (HRP), and corpuscular elements (erythrocytes, lymphocytes, macrophages) can be retrieved from the retropharyngeal lymph nodes after intraventricular or intracerebral injection.[24,104,135] It is obvious that direct movement of brain antigens via this route into the lymphatic system may provide contact between the CNS and the circulating immune system.

3. CEREBROSPINAL FLUID COMPOSITION

The total protein concentration of the cerebrospinal fluid (CSF) is approximately 1% of the protein concentration of plasma.[73,176] The protein fractions of the CSF are mainly the same as those found in plasma[27,40,73] and their entry into the CSF is limited by their diffusion across the BBB. Thus, the plasma/CSF ratio for a low-molecular-weight protein is ≈ 200, whereas the ratio for immunoglobulin G (IgG) ($\approx 150,000 \ M_r$) is 300–600[134] (Fig. 6).

Analysis of the protein composition of the CSF can give useful clinical information. Increased BBB permeability is seen in many diseases of the CNS (e.g., infections,

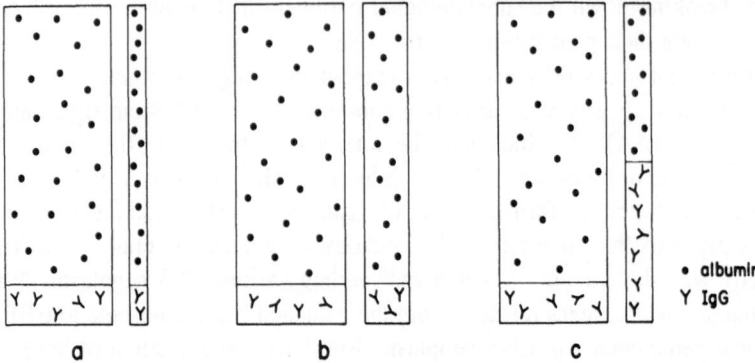

Figure 6. (a) Ratio between albumin and IgG in the cerebrospinal fluid (CSF) and in plasma under normal circumstances. The concentration of each molecule is much less in the CSF than in plasma. (b) If the BBB is damaged, the concentration of both albumin and IgG increases in the CSF. (c) If there is local production of IgG in the CSF, IgG accounts for a much larger fraction of the total CSF protein.

inflammatory disorders, neoplasms, ischemia)[54,176] and can be demonstrated by a decrease in the plasma/CSF albumin ratio[176] (Fig. 6).

Because the relative plasma protein concentration of IgG (about 20%) is so much higher than the CSF concentration, modest changes in BBB permeability can give significantly increased relative CSF IgG concentrations. Increased CSF IgG may reflect (1) diffusion from plasma to CNS, (2) local production of IgG in the CNS, or (3) a combination of both. The contribution to the total CSF IgG from plasma can be corrected by comparing the CSF IgG concentration to the CSF albumin concentration. The CSF IgG/albumin ratio is normally approximately 0.30[72,176] and a higher ratio indicates that there is relatively more IgG in the CSF than permeation from the plasma can account for, i.e., IgG is probably also being synthesized locally in the CNS[72,134,176] (Fig. 3). The so-called IgG index[72,176] is a more sensitive and accurate correction:

$$\text{IgG index} = \frac{\text{CSF IgG} \times \text{plasma albumin}}{\text{plasma IgG} \times \text{CSF albumin}}$$

The upper limit for this index in normal CSF is 0.77.[72] The index is elevated mainly in inflammatory, demyelinating, and infectious diseases.[54,72,109,176] The correction is more accurate when the BBB damage is modest, hence the contribution from plasma to CSF of IgG, is moderate. When the BBB permeability is high, i.e., total protein is more than 100–200 mg %, the plasma-derived IgG in the CSF is abundant compared with any possible contribution from local CNS production, and the index becomes quite difficult to interpret.

When CSF is subjected to electrophoresis on agarose gel, the different IgG fractions become separated into a bandlike pattern.[72,103,133,147,176] Discrete bands that emigrate as γ-globulins in the CSF, which are not detected on simultaneous electrophoresis of the patient's serum, are called oligoclonal bands.[103] The presence of oligoclonal bands indicates intrathecal synthesis of immunoglobulins and may be found in demyelinating/inflammatory disorders such as multiple sclerosis and in chronic infections such as neurosyphilis.[176]

Preferential elevation of certain IgG fractions may provide a useful diagnostic tool, e.g., in determining a specific infectious etiology. Thus, in a patient with acute measles encephalitis, the ratio of antibody titers for measles in the CSF to serum change from 1 : 160 to 1 : 2 during the illness, whereas the titers for reference viruses (adeno-virus and poliomyelitis virus) did not change.[133] Similar results can be obtained for cases of sub-acute sclerosing panencephalitis (SSPE),[133,176] strongly suggesting that measles virus is the etiologic agent of this disease.[133,169] The serologic diagnosis of neurosyphilis can likewise be performed on CSF by demonstration of a positive VDRL, *Treponema pallidum* immobilization (TPI), or fluorescent treponemal antibody absorption (FTA-ABS), the latter two being the more specific.[54,141] A positive VDRL, reaction in the CSF is sometimes obtained in the absence of a positive serum reaction. TPI and FTA-ABS generally remain positive in the serum and cannot reliably be used as an indicator of disease activity.[139]

4. CIRCULATING ANTIBODIES IN NEUROLOGIC DISEASE

Several clinical entities with CNS pathology and pathophysiology have been found to coexist with circulating antibodies directed against various CNS components, e.g., Huntington chorea,[76] paraneoplastic syndromes,[9] ataxia telangiectasia,[86] kuru[168] and Creutzfeld–Jakob disease.[168] When attempting to explain how such antibodies might cause damage to the CNS, one is first forced to explain how they reach their antigenic counterparts across the BBB. It may be appreciated from the previous discussion that theories on the initial pathogenesis of EAE may be relevant for interpreting such clinical situations. Systemic lupus erythematosus (SLE) is often complicated by CNS involvement.[46,47,67] The heterogeneous neurologic manifestations of SLE include encephalopathy, seizures, different focal signs of cerebral dysfunction, myelitis, and peripheral neuropathy. CNS involvement in SLE will be used as an example of considerations on the pathogenic significance of peripherally circulating antibodies in neurologic disease (see also Chapter 25, Vol. 2).

Both circulating immune complexes and antineuronal antibodies are demonstrable in SLE,[21,194] and theoretically both may be implicated in causing damage to the CNS. The pathogenesis of lupus nephritis is generally thought to be trapping or precipitating circulating immune complexes in the glomeruli, with a subsequent activation of complement, leading to local vasculitic damage.[82] The arrangement of fenestrated vascular loops in the choroid plexus is somewhat similar to the vascular structure of glomeruli; the finding of immunoglobulin deposits in the choroid plexus of SLE patients similar to the deposits found in the kidneys[6,46,66] suggests that a similar pathogenesis might account for the CNS involvement in SLE. Furthermore, immunologic cross-reactivity between the choroidal and glomerular basement membranes has been found.[110]

Experimental immune-complex disease in rabbits seems to be associated with a selective increase in the permeability of the blood–CSF barrier leaving the BBB intact,[74] thus lending support to the theory that immune-complex disease can create a break in the immunologic seclusion of the CNS. However, the existence of choroidal deposits of immune complexes does not seem to be sufficient to explain the CNS manifestations of SLE: (1) the choroidal deposits are often found in SLE patients without CNS disease,[22] and (2) CNS lupus may occur in patients who are serologically and otherwise in systemic clinical remission.[187]

Antineuronal antibodies and lymphocytotoxic antibodies cross-reacting with neural antigens have been found both in the serum and CSF of SLE patients.[21] IgG antineuronal activity in the CSF is reported to be higher in patients with active CNS disease than in patients without CNS manifestation of lupus.[21] The presence of antibodies in the CSF of SLE patients is frequently not associated with a similar increase in CSF albumin.[187,194] This provides strong evidence of intrathecal IgG synthesis, i.e., the antibodies result from local synthesis within the CNS rather than from permeation across the BBB.[21,187,194] The fact that IgG found in the CSF in these cases is often oligoclonal, i.e., directed against a few antigens only, lends further support for a local CNS synthesis of antibodies directed against a few CNS antigens.

The various manifestations of CNS lupus have been found to differ in their association with IgG antineuronal antibody in the CSF. More diffuse CNS disease, such as psychosis or generalized seizures, is more often associated with antibodies in the CSF than are cases with focal neurologic deficits.[21] By contrast, patients with focal neurologic symptoms or encephalopathy often have increased BBB permeability as evidenced by their CSF/serum albumin ratio.[187] The latter observation may reflect vasculitic foci in the brain or spinal cord parenchyma which at the same time damage the BBB locally with a resulting increase in its permeability to macromolecules and gives rise to focal tissue damage with corresponding symptoms of focal nature.[46,47,82,194]

Whether the BBB damage and the antineuronal antibodies in the CSF are the cause of, or merely a reflection or the result of, tissue damage in the CNS is an unresolved question. Since minute amounts of serum proteins, including antibodies, can cross the normal BBB and blood–CSF barrier, it is tempting to suggest that the presence of such small amounts of immune-reactive substances could be sufficient to cause at least the initial damage to both neurons and CNS microvessels.

5. DEMYELINATING DISEASES OF THE CENTRAL NERVOUS SYSTEM

Demyelination is a characteristic feature of several immunopathologic disorders of the CNS.[3,94,97,112,138,158] It frequently occurs in the setting of ongoing or antecedent contact with viral antigen.[144]

Progressive multifocal leukoencephalopathy (PML) is caused by infection with the human JC papovavirus.[142] It occurs almost exclusively in persons with impaired cellular immunity. PML has been reported in patients with various malignancies, AIDS, and severe congenital immune deficits.[95] The diagnosis can be established by brain biopsy and subsequent determination of virus in brain tissue by immunofluorescence, immune agglutination, or culture of the virus.[95,142] Most likely, demyelination is the direct result of destruction of oligodendroglia infected with the virus.

Postvaccinal or postinfectious demyelination constitutes another group of demyelinating diseases. This entity was first described in 1888, when it was found that vaccination with rabies virus attenuated by several serial passages in rabbit spinal cord could result in paralytic accidents due to acute disseminated encephalomyelitis (ADE) in humans. That this was due to an immunologic reaction against the foreign CNS tissue used for the culture, rather than a result of the rabies virus itself, was demonstrated in 1933 by Rivers et al.[156] These workers observed that monkeys receiving multiple injections of rabbit brain extracts developed encephalomyelitis with myelin destruction closely resembling the pathology seen in ADE. This experimental analogue is known as experimental allergic encephalomyelitis (EAE). ADE has also been reported following a number of spontaneous viral infections, including smallpox, rubella, rubeola, and varicella.[152] The pathogenesis of these postinfectious encephalomyelitides has not been proved. One possible explanation is that either the virus or parts of it inserts into the host CNS cell membranes, creating a non-self surface antigen amenable to attack by the immune system due to a compromised BBB. An alternative explanation is that the viral genome incorporates itself into the host genome, thereby coding for a new antigen, which, if expressed as a surface antigen on the infected cells, could be recognized as non-self.[144]

The third group of demyelinating diseases consists of a group of spontaneously occurring neurologic disorders in which no definite relationship has been established to ongoing or previous viral infections. Multiple sclerosis (MS) and variants of the MS disease process (e.g., monosymptomatic optic neuritis or transverse myelitis) are by far the most common representatives of this group.

Efforts to detect active or latent viral infections as the cause of MS have not been conclusive. Parainfluenza, mumps, rubella, canine distemper, and measles virus have all been reported as possible etiologic agents of MS.[2,133,134] The evidence for a possible association of these viral agents with MS is all indirect, i.e., serologic.[2,133,134] Immunologic reactivity against measles virus has been associated with MS far more consistently than any other virus.[133] Adams and Imagawa[1] in 1962 were the first to report that measles antibody titers are increased in the serum of many MS patients. Later it was shown that antibodies against measles virus are also increased in the CSF of MS patients.[133,134] Evidence that the antibody in the CSF is due to local production in the CNS, rather than to leakage across the damaged BBB, can be obtained by comparing the CSF concentration of IgG and albumin with the serum concentration of these two components.[72,176] Preferential elevation of IgG can be demonstrated in 60–80% of patients with a clinical diagnosis of MS.[54,72,176] By comparing the serum/CSF ratio of antibodies with that of several viral antigens, Norrby et al.[134] demonstrated that about 60% of MS patients have evidence of local production of measles antibodies in the CNS compared with rubella in 20%, mumps in 15%, and herpes in 10%, for example. He also reported that in 7% of the patients, local production of antibodies against three or more viral antigens could be found.

The presence of oligoclonal bands in the CSF of MS patients, i.e., IgG produced by only a few clones of antibody-producing cells, provides further support to the concept of local CNS IgG production in the CNS in this disease. Oligoclonal bands are reported to occur in 80–95% of MS patients,[72,103] but only a fraction of the oligoclonal IgG seems to be specifically directed against measles virus antigen.[133] Thus, the significance for the MS disease process of antibodies produced in the CNS—whether directed against measles or other known or unknown antigens—is unclear. They could be directly associated with the etiology of the disease; they might be important for the pathogenesis once the disease process has been initiated by some other etiologic factor, or they might be mere epiphenomena.

Blood–brain barrier damage in MS has been demonstrated by increased amounts of albumin in the CSF[176] and by enhancement of plaques on CT scan by radiocontrast media in acute exacerbations[101] (Fig. 7). The possible significance of BBB damage for the evolution of disease has been discussed in the section of this chapter dealing with EAE, often used as an experimental model for MS.[3,180] Whether the increased BBB permeability is a phenomenon of true etiologic/pathogenetic importance remains to be seen.

a

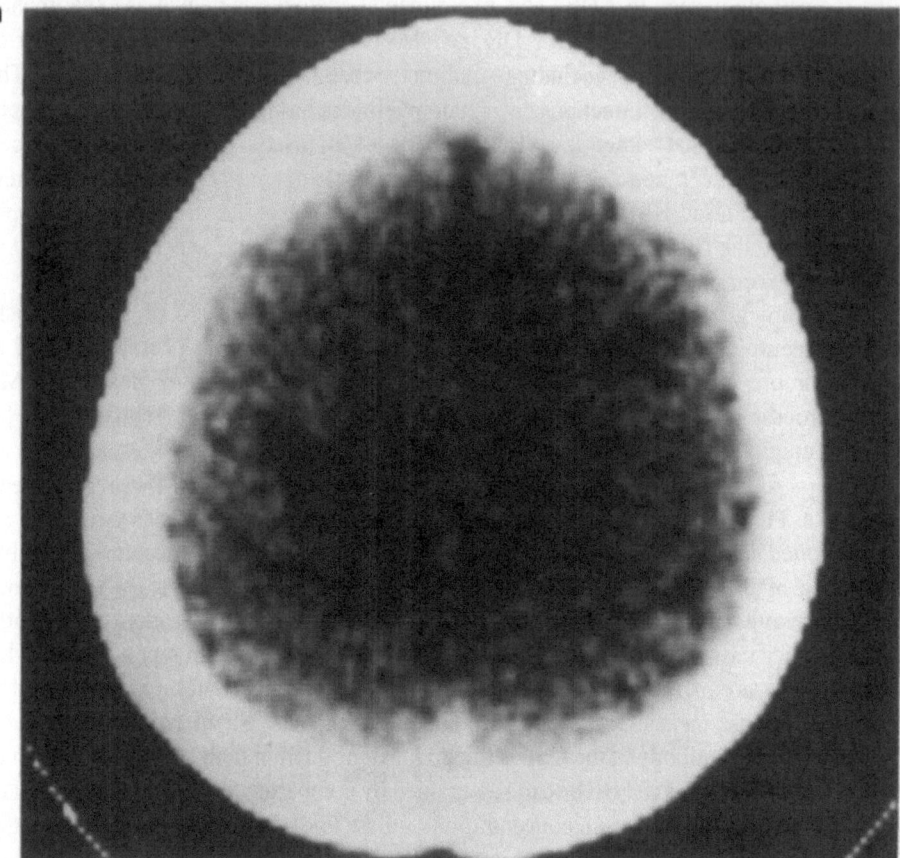

Figure 7. (a) The hypodense area on the noncontrast computed tomography (CT) image is a multiple sclerosis (MS) plaque. (b) Blood–brain barrier (BBB) damage in the MS plaque allows the intravenously injected radiocontrast medium to escape into the brain parenchyma, which produces the faint ring enhancement.

6. NEOPLASTIC DISEASE OF THE CENTRAL NERVOUS SYSTEM

Most studies of immunity and CNS neoplastic disease pertain to gliomas. It has been known for several years that interaction between the immune system and gliomas takes place. It is generally accepted that cellular immunity is far more important than humoral immunity in this context.[5,157] Thus, there is lymphocytic infiltration of the tumor in 30% of cases.[34,71,185] Conversely, the demonstration of impaired cellular immunity in glioma patients[34,107,148,185] could indicate that the presence of the tumor exerts a modulating influence on the immune system (Figs. 8–10). Both observations suggest a potential benefit of manipulating an immune response in glioma patients by enhancing desired immune functions and leaving others suppressed.

A prerequisite for creating/enhancing an immune response against a neoplasm is the existence of specific neoplastic antigens that may be recognized and interacted with by the immune system. The existence of glioma-specific antigens could ensure a very specific immune reaction, resulting in tumor rejection without any damage to, or interference

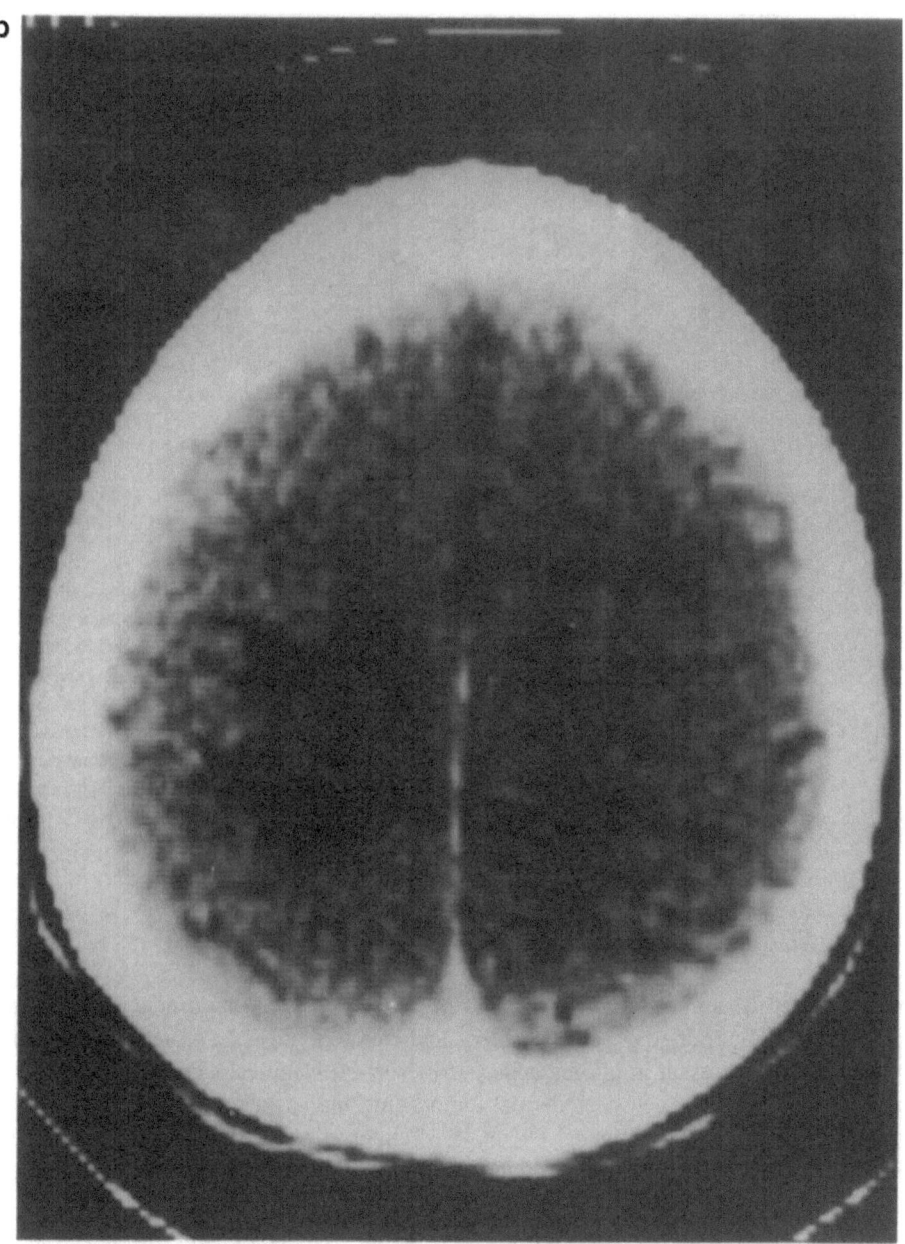

Figure 7. (Continued)

with, the rest of the brain and/or other organs. One of the problems in this approach has been the demonstration of such specific antigens. The search is complicated by (1) genotypic and phenotypic variability between tumors of the same histopathologic appearance and even between cells of the same tumor,[11,12,14] and (2) growth phase-dependent expression of surface antigens.[5]

The discovery of monoclonal antibodies has been an invaluable instrument in defining glioma-associated antigens.[33,34,148,183,185] Gliomas share antigenicity with a

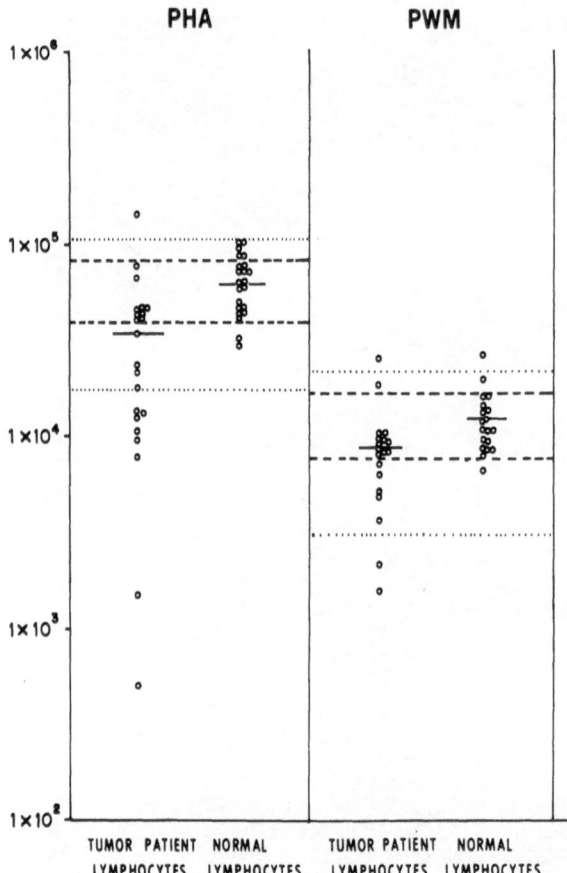

Figure 8. [³H]Thymidine incorporation of normal and tumor patient lymphocytes in response to phytohemagglutinin (PHA) and pokeweed mitogen (PWM). Horizontal bars indicate mean [³H]thymidine incorporation for each group. Dashed line indicates 1 SD from the mean level of incorporation of normal lymphocytes. Dotted line indicates 2 SD from that mean. (From Neuwelt et al.[127])

number of other tissues and tumors (see also Chapter 17, Vol. 2). Thus, antigen present in normal human adult brain, e.g., glial fibrillary acidic protein (GFA), a protein specific for astrocytes, may be present in glioma cells.[5,33,34,60,148,185] Neuroectodermal antigens are shared by gliomas, other neuroectodermal tumors (melanoma, neuroblastoma), and human fetal brain (oncofetal antigens).[33,34,60,148,185] Interestingly, a number of lymphoid differentiation antigens found on T cells and leukemic cells are also found in normal brain and neuroectodermal tumors.[33,34,60,148,185] The ontogenetic significance of this observation and its implications for regarding the CNS as an immunologically privileged site secluded from the circulating immune system are unclear.

Recently, two different anti-glioma-specific monoclonal antibodies (MAb) have been reported (BF7 and GE2), which react almost exclusively with glioma cells when tested against a panel of glioma cells, other neuroectodermal tumor cells, and metastatic tumor cells from systemic cancer in tissue culture.[148] Coakham and et al.[38] reported a panel of six MAb that bind to the vast majority of human gliomas.

Another necessity for cell-mediated immune response against a tumor is the presence of histocompatibility antigen (HLA) on the cell surface.[175] HLA antigens have been demonstrated on the surface of glioma cells,[148,185] and on astrocytes around brain metastases.[61] These antigens are known to be important for antigen presentation to T cells.

If specific tumor antigens and the HLA complex are present on glioma cells, then

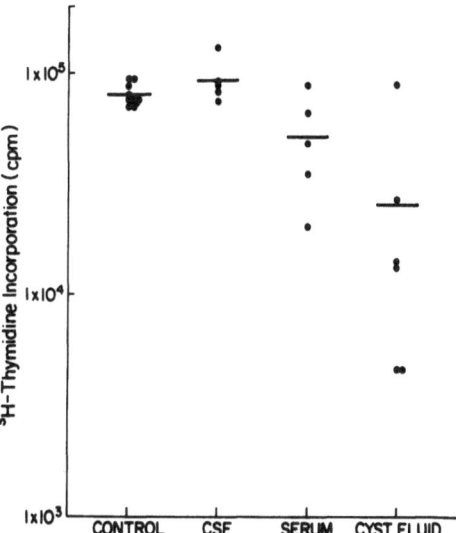

Figure 9. Phytohemagglutinin (PHA)-stimulated [³H]thymidine incorporation of normal lymphocytes in the presence of tumor-cyst fluid, cerebrospinal fluid (CSF), and serum from brain-tumor patients. Horizontal bars indicate mean counts per min (cpm) for each group. In control cultures, 40 μl normal pooled human serum was added in each well. Each experimental culture received either tumor-cyst fluid, CSF, or serum (40 μl). Statistical analysis of data using the unpaired Student's *t*-test shows that both serum and tumor-cyst fluid, but not CSF, are statistically different from control (*p* > 0.01). (From Kikuchi and Neuwelt.[88])

why doesn't the patient's immune system recognize them and reject the tumor? One possibility is that the tumor is hidden behind a protective BBB, preventing components of the immune system from reaching the tumor. However, it has been demonstrated that the BBB of malignant brain neoplasms can be leaky[18,69] and can permit passage of substances across the blood–tumor barrier (BTB), which are excluded from normal brain by the intact BBB. Altered immune function of the glioma patient and escape mechanisms whereby the tumor avoids contact with the immune system may be other factors that

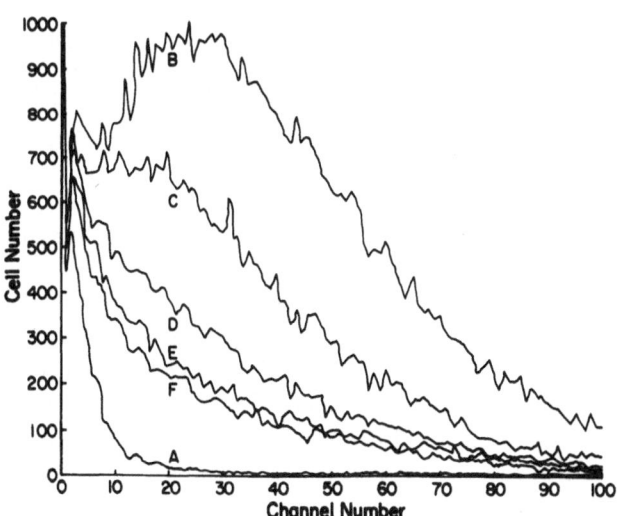

Figure 10. Effect of tumor cyst fluid obtained from a glioblastoma patient on mitogen-induced increase in normal lymphocyte cell size. Channel number is a measure of cell size. (Curve A) Control but no phytohemagglutinin (PHA) and no cyst fluid. (Curve B) Control + PHA but no cyst fluid. (Curve C) PHA + cyst fluid (0.05 ml/tube). (Curve D) PHA + cyst fluid (0.10 ml/tube). (Curve E) PHA + cyst fluid (0.15 ml/tube). (Curve F) PHA + cyst fluid (0.20 ml/tube). (From Kikuchi and Neuwelt.[88])

prevent rejection of brain tumors. It has already been mentioned briefly that many glioma patients exhibit impaired cellular immunity (Fig. 8). This may be seen as impaired or absent skin hypersensitivity reactions after intradermal injections of such common antigens as *Candida*, PPD, mumps, streptokinase, and *Trichophyton*.[28,29,107] There is also a relative lymphopenia, with a decreased number of T cells and a relatively increased number of suppressor T cells in glioma patients.[25,30,34,60,107]

Glioma cells seem to be able to escape the immune system by a number of different mechanisms. It has been demonstrated that glioma cells in vitro can synthesize a coat of glycosaminoglycans by interaction with lymphocytes in mixed culture.[64] It is possible that tumor surface antigens may be hidden behind such a coat. Furthermore, glioma cells are able to secrete factors that directly interfere with T-cell function. IL-2, also known as T-cell growth factor (TCGF), is a mediator produced by activated lymphocytes, which can induce proliferation and secretion of interferon by previously activated T cells.[59] Human glioblastoma cells in tissue culture can release a blocking factor that prevents IL-2-induced T-cell activation.[59] This effect is not reversed by the addition of IL-2 to the medium. The presence of T-cell inhibitory factors in tumor cyst fluid and serum from glioma patients[59] suggests that blocking factors may be of importance in the immune-deficient status of these patients and may effectively prevent an immune attack on, and rejection of, the tumor (Figs. 9 and 10).

The lymphocyte population found to infiltrate 30% of glial tumors has been classified as belonging mainly to the suppressor subset.[71] It is not known how the tumor specifically attracts suppressor T cells, but it is easy to imagine that this could provide yet another barrier against cellular attack on the tumor. Indeed, tumor infiltrating lymphocytes in glioma, when stimulated in vitro with potent mitogens such as PHA, have been shown to have a markedly diminished proliferative response (N. deTribolet, preliminary results).

Immunotherapy against brain tumors can be classified in four main categories: (1) passive, (2) adoptive, (3) active nonspecific, and (4) active specific.[5,33,34,106]

Passive immune therapy is provided by administration of monoclonal antibodies (MAbs) against tumor antigens. The impact of humoral immune mechanisms against tumor is probably quite ineffective, but potentially, monoclonal antibodies might stimulate antibody-dependent cellular cytotoxicity (ADCC).[5] Another possible use of MAb is a carrier of radioactive particles for local tumor irradiation, chemotherapeutic agents, radio enhancers, and radioactive molecules for imaging purposes by, e.g., positron emission tomography (PET)[33,34,48,106] (see Chapter 17, Vol. 2).

Adoptive immunotherapy uses the transfer of specifically activated immunocompetent cells to the immunodeficient brain tumor patient. Such cells have been administered systemically[157,191] as well as inoculated directly into the tumor bed[80,125,173,191,193] in order to circumvent the BBB and obtain maximal local concentration of cytotoxic cells in the tumor (see Chapter 1). Administration of IL-2 together with autologous T cells specifically activated by coculture with tumor cells has been attempted in various types of systemic cancer with promising results[159,160] but might not be expected to work in glioma cases since gliomas may secrete blocking factors preventing IL-2-dependent T-cell activation. However, it has been demonstrated in vitro that IL-2-activated lymphocytes (LAK) can lyse glioma cells without damaging normal cells.[79] Jacobs et al.[80] reported on a series of nine glioma patients who received LAK cells in the tumor bed following surgical resection. None of the patients exhibited systemic or brain toxic symptoms. Their estimated life expectancy was 6 weeks, but the combination of surgery, radiation, and administration of LAK cells into the tumor bed made a follow-up period of 3–11 months possible, although there is no mention of actual survival. No controlled studies of LAK cell administration versus conventional therapy are yet in existence. After LAK cell infusion IL-2 must be administered to maintain LAK cell cytotoxicity. Since IL-2 does not cross the BBB and diffusion in the extracellular space of the CNS is poor, LAK cell therapy in the CNS will be difficult. Intracerebral injection of IL-2 may also incite cerebral edema (R. Merchant, personal communication).

Studies in vitro have demonstrated that the addition of allogenic lymphocytes to lymphocyte–glioma coculture elicits a cytolytic response against the tumor cells.[65] It has been speculated that the allogenic lymphocytes may somehow enhance immunogenicity of the tumor cells. It is also possible that the tumor-induced block of IL-2-activated T cells could be circumvented in this way. Clonal expansion of tumor infiltrating lymphocytes may be very useful in such studies.

Active nonspecific immunotherapy is nonspecific stimulation of the immune system by repeated bacille Calmette-Guérin (BCG) or *corynebacterium* injections.[42,166] This has been demonstrated to alter neither general immune competence nor clinical outcome.

Active specific immune therapy represents an attempt at specific immunization against the tumor. Subcutaneous implants of tumor tissue obtained by resection or biopsy,[20,68] as well as repeated injections of lethally irradiated tumor cells with or without adjuvant (i.e., nonspecific immune stimulation), have been tried.[19,34,108] Repeated injections with adjuvant seemed to induce specific antibodies and to prolong survival in one study,[34] but so far patient groups have been small. Some of the studies are without adequate control groups.[19,20,34,68] A potential danger of this type of therapy is the induction of EAE, which has been demonstrated experimentally[13] as well as in some patients in clinical trials.[28] If a specific tumor (glioma) antigen were present, an anti-idiotype vaccine would avoid this problem.

7. IS THERE A SEPARATE IMMUNE SYSTEM IN THE CENTRAL NERVOUS SYSTEM?

The possible existence of an immune system inherent to the CNS separated from the systemic immune system obviously offers a ready explanation for immune reactions in the CNS. The problem in accepting this explanation has been defining cells which could constitute such a CNS-specific immune system. Microglia seem to be the most likely candidates.[123]

As immune reactions in the CNS are very often accompanied by increased permeability of the BBB at some time point, it may be very difficult in the actual clinical or experimental situation to determine, whether a response is primarily due to the systemic immune system leaking into the CNS or to the actions of an inherent immune system.

The microglial cell was described in 1919 by del Rio Hortega[43] who discovered a cell that differed from the remainder of the glial population in its affinity for silver carbonate stain. In comparative light microscopic studies, he described two types of microglia: a resting microglia in the normal undamaged CNS and an activated or reactive microglia that could be observed in the CNS after trauma or inflammation and that was phagocytic and migratory.[44,45] As both types showed affinity for silver carbonate, they were assumed to be identical. del Rio Hortega described a number of transition forms between resting and reactive microglia. He also described argyrophilic cells in the pia and concluded that these were precursors of microglia.

The origin and function of microglia and the question of identity between resting and reactive cells have since been intensely studied and debated. Recent reviews give a detailed account of the controversies of this debate in a historic setting as different methods for study successively became available.[102,136,137] The light microscopic studies conducted by del Rio Hortega have already been mentioned. A methodologically similar, contemporary study by Rydberg differed only as to the origin of microglia: a neuroectodermal origin from the subependymal origins of the lateral ventricles was proposed contrary to del Rio Hortega's concept of a pial (mesodermal) origin of the cells.[163] The pericyte, which is found adjacent to CNS capillaries surrounded by capillary basement membrane, has also been suggested as a precursor cell for microglia.[102,115]

With the advent of electron microscopy, the ultrastructure of microglia has been described. The resting cells have only a few organelles and dense chromatin in the nucleus and thus appear to be functionally relatively inert.[102,137] The reactive microglia, however, have larger nuclei with fine chromatin and more abundant organelles, including ribosomes, rough endoplasmic reticulum, a prominent Golgi apparatus, and lysosomes,[102,137] and thus appear to be actively synthesizing cells with a potential for digesting engulfed material.

Modern techniques have enabled studies of functional capacity, surface markers/receptors, and cytokinetics in addition to the above-mentioned purely structural studies. In summary, reactive microglia have enzymes and surface markers that are identical to monocytes, whereas resting microglia and pericytes have none of these characteristics.[102,136,137] Cytokinetic studies initially used radiolabeled thymidine (TdR) as a DNA precursor (i.e., marker for mitosis), whereby successive labeling of immature glial cells, microglia, and astrocytes was observed.[137,164] Recently Schelper and Adrian[164] pointed out that TdR may be redistributed and reused by other cells after the death of the originally labeled cells. Instead, they used radiolabeled deoxyuridine (UdR), which cannot be reused, and found that only reactive microglia and no other cells were labeled by this procedure.

The evidence of modern dynamic studies thus strongly indicate that resting microglia and reactive microglia are entirely different cell populations and that one does not develop into the other or into astrocytes or any other kind of cell.[136,137,164] Reactive microglia are most likely derived from monocytes, which extravasate by diapedesis in response to CNS injury to become brain macrophages.[164] This interpretation is supported by the occurrence of transfused labeled bone marrow cells, lymphocytes, and peritoneal macrophages in the CNS following neural injury.[135,137] When their task is completed, these macrophages migrate back to the bloodstream or lyse locally[111] but do not seem to involute to become resting microglia.[137,164] True microglial proliferation, however, can be seen in response to indirect neuronal damage following, e.g., peripheral axonotomy and Wallerian degeneration and even ischemia[126,164,170] (see Fig. 1, Chapter 1), but must not be confused with the macrophage response to direct CNS injury.[164]

AIDS is a viral infection resulting in severe impairment of cellular immunity.[50] Apart from a multitude of systemic opportunistic infections, similar CNS infections are often encountered in AIDS victims. Toxoplasmosis, fungal infections, TB, progressive multifocal leukoencephalopathy, and cytomegalovirus (CMV) are examples of CNS infections that occur far more frequently in AIDS than in immunologically intact persons.[49,50,116,117,184] Primary CNS lymphoma—a rare disease—also occurs with increased frequency in AIDS.[49,50,146] Other immunocompromised individuals, e.g., transplant patients on immunosuppressant therapy and patients with congenital immunodeficiency syndromes, are also at risk of acquiring similar CNS infections and CNS lymphoma.[37,95,113,165] These observations seem to suggest that an intact systemic immune apparatus is important for anti-infectious and antineoplastic surveillance on both sides of the BBB. Conversely, simultaneous impairment of both the systemic and CNS immune system is also a possibility.

Primary CNS lymphoma is histologically identical to systemic lymphoma and shares the affinity for silver stains with microglia[123]; it has also been referred to as microglioma. However, this staining affinity is not consistent, and recent research has established that these CNS tumors share surface antigens with systemic lymphomas,[100,130] suggesting that they are no different immunologically from other primary lymphomas. It is therefore remarkable that they only rarely metastasize outside the CNS.[123,130] It is possible that these intracerebral tumors may be shielded from the systemic immune apparatus by the BBB and that, although deficient, immune surveillance on the other side of the BBB might be adequate to prevent proliferation, if not being able to kill, the neoplastic cells. On the other hand, the frequent occurrence of Kaposi sarcoma in AIDS patients[62] seems to contradict adequate systemic antineoplastic surveillance in these cases. Thus, it remains a possibility that CNS lymphomas are true primary brain tumors and not simply lymphomas that happen to arise in the CNS; thus as with other primary brain tumors they do not metastasize outside the CNS.

8. THERAPEUTIC IMPLICATIONS OF THE BLOOD–BRAIN BARRIER

The CNS may be considered an immunologically privileged site. Thus, the interaction between the systemic immune system, cellular or humoral, is restricted by the existence of the BBB and blood–CSF barrier. The kinetics across these barriers must be considered for any substance administered for the treatment of CNS diseases. Immune therapy for CNS diseases is no exception. The significance of a number of variables (CBF, barrier kinetics, water–lipid solubility of the permeating compound, systemic distribution, and clearance of the compound) for the delivery of pharmacotherapy to the CNS is reviewed in Chapter 6 and was recently reviewed specifically for the delivery of pharmacotherapy to brain tumors.[70]

Interferon therapy of multiple sclerosis is one example of immunotherapy against a CNS disease.[10,78,81,93] Interferons (IFNs) are immunologically potent substances with potent effects on the immune response. IFNs are subdivided into three major subtypes: IFN-α, which is produced by leukocytes; IFN-β, which is produced by fibroblasts; and IFN-γ, which is produced by activated mononuclear cells. Both IFN-α and IFN-γ may influence immune regulation by modifying B cell function through enhancement of B-cell interaction with helper cells and by changing the density of histocompatibility antigens on cell membranes, respectively. IFNs have antiviral properties and may also be involved in the pathogenesis of autoimmune diseases.[77,171] The BBB permeability is very low to these proteins.[78] Systemic administration (intramuscular injection) of interferon has been reported to decrease the number of exacerbations in MS patients,[93] but the mechanism of action seems difficult to explain in view of the poor BBB permeability, unless increased BBB permeability in the plaque early during an exacerbation[101] can be viewed as a port of entry. In order to circumvent the BBB, intrathecal introduction of the compound via a lumbar subarachnoid puncture has been proposed.[78] IFN has been reported to penetrate the blood–CSF barrier well in monkeys,[78,171] but no similar human data are available. It seems unlikely that therapeutic parenchymal concentrations in the CNS can be obtained at a distance of more than a few millimeters away from the ependymal surface because the extracellular space is so small that even small molecules do not penetrate far by diffusion.[90]

Another example of immunotherapy for CNS diseases is the administration of monoclonal antiglioma antibodies to patients with gliomas.[5,33,34,48,106,128,131,132] The evaluation of the efficacy of this treatment is complicated by complex tumor–host interactions, the ability of the tumor to evade immune surveillance, and the immunosuppressed state in glioma patients.[5,34,185] Permeability of antibodies across the normal BBB is limited. Permeability of the blood–tumor barrier (BTB) is often heterogeneous and unpredictable within the same tumor.[18,69] Generally speaking, the BTB permeability to small water-soluble molecules is approximately 10 times the undisturbed BBB permeability, e.g., to α-aminoisobutyric acid (AIB), a small nonmetabolizable neutral amino acid with a molecular weight of 103.[17] This BTB permeability is still probably much too low to permit entry of significant amounts of antibodies, which are much larger molecules, i.e., IgG $\simeq 150,000$ M_r, and procedures to increase BBB permeability or to circumvent the BBB to increase access of the antibodies to the tumor will be necessary. Delivery of higher molecular-weight compounds such as immunoglobulins seems to depend mostly on molecular size rather than on lipid/water solubility, a primary determining factor for the permeability of small molecules, such as many chemotherapeutic agents.[52,128,129] It has been shown that reversible osmotic disruption of the BBB both in experimental animals and in patients with brain tumors can increase delivery of monoclonal antitumor antibodies to the brain parenchyma as much as 25–100 times the delivery to undisturbed brain or to the hemisphere contralateral to the manipulated one[128,131,132] (see Chapter 17, Vol. 2).

9. SUMMARY AND CONCLUSIONS

This chapter reviews the relationship between the immune system and the BBB and discusses the significance of this relationship. Probably the most important and still controversial aspect of this association is the question: Is the CNS an immunologically privileged site? The answer to this question is that under normal circumstances with an intact BBB, the CNS is, to a large extent, immunologically privileged. However, under many circumstances of pathologic significance, the BBB is at least partially disturbed; thus, the immune privilege of the BBB is partially disturbed as well.

Clearly, the BBB is responsible for the dramatic difference in the composition of CSF versus serum. Under normal circumstances, this fluid is essentially acellular; the only immunoglobulins present are those that are passively derived by diffusion from the systemic circulation. However, in pathologic situations (i.e., demyelinating diseases, infections, and possibly even tumors) a local immune response occurs within the CNS and can result in the local production of immunoglobulins. Detection of such local immune responses in CSF is predominantly of diagnostic value only. However, these local CNS immune responses almost certainly play a role in the pathogenesis of the diseases in which they occur.

In some disease processes, antibodies, such as antineuronal antibodies in systemic lupus erythematosus or sensitized lymphocytes in the case of experimental allergic encephalomyelitis, appear to gain access to the CNS across the BBB. Exactly how this process takes place is an area of an intense study, and the answers have not been clearly delineated. However, substances that can increase vascular permeability, such as leukotrienes, may play a role. Primary brain tumors, which appear to originate in the brain, may damage the barrier sufficiently to permit some access of the systemic immune system to the CNS. Indeed, at least 30% of gliomas are infiltrated by tumor-infiltrating lymphocytes (TIL). These sensitized cells may be particularly important with regard to adoptive immunotherapy of brain tumors.

Since there are no true lymphatics in the brain and the microglial cell is the only endogenous brain cell that under normal circumstances has characteristics of immune function, the next question is whether there is a separate immune system in the CNS. There probably is not a separate immune system, but under pathologic circumstances astrocytes can be stimulated to become an APC and can be activated to release IL-1. Thus, under pathologic circumstances, certain aspects of the immune system can function within the CNS.

Finally, the BBB poses a major obstacle to immunotherapy. For instance, the delivery of a macromolecular protein such as interferon, which may have efficacy in multiple sclerosis, is greatly impeded by the BBB, as are monoclonal antibodies directed against antigens on the surface of gliomas. Similarly, if activated tumor-infiltrating lymphocytes or LAK cells are to have any role in the treatment of CNS tumor as a form of adoptive immunotherapy, the problem of circumventing the BBB needs to be addressed.

REFERENCES

1. Adams JM, Imagawa DT: Measles antibodies in multiple sclerosis. *Proc Soc Exp Biol Med* 111:562–566, 1962.
2. Altes M: Is multiple sclerosis caused by a virus? In Boese A (ed): *Search on the Cause of Multiple Sclerosis and Other Chronic Diseases of the Central Nervous System. First International Symposium of the Heitie Foundation.* Verlag Chemie, Frankfurt, 1980, pp. 374–391.
3. Alvord EC, Kies MW, Suckling AG (eds): Experimental allergic encephalomyelitis, a useful model for multiple sclerosis. A satellite conference of the International Society of Neurochemists. Seattle, Washington, July 16-19, 1983. *Prog Clin Biol Res* 146:1–554, 1984.
4. Antel JP, Arnason BGW, Medof ME: Suppressor cell functions in multiple sclerosis: Correlation with clinical disease activity. *Ann Neurol* 5:338–342, 1979.
5. Apuzzo MLJ, Mitchell MS: Immunological aspects of intrinsic glial tumors. *J Neurosurg* 55:1–18, 1981.
6. Atkins GJ, Kondon JJ, Quismorio FP, et al: The choroid plexus in systemic lupus erythematous. *Ann Intern Med* 76:65–72, 1972.
7. Baach MA, Phan-Din-Toy F, Tournie E, et al: Deposit of suppressor T-cells in active multiple sclerosis. *Lancet* 2:1221–1223, 1980.
8. Barker CF, Billingham RE: Immunologically privileged sites. *Adv Immunol* 25:11–15, 1977.
9. Bell CE, Seetharam S: Identification of the Schwann cell as a peripheral nervous system cell possessing a differentiation antigen expressed by a human lung tumor. *J Immunol* 118:826–831, 1977.
10. Berry CC: Intrathecal interferon for multiple sclerosis. (Lett.) *Science* 217:269–270, 1982.
11. Bigner DD: Biology of gliomas: Potential clinical implications of glioma cellular heterogeneity. *Neurosurgery* 9:320–326, 1981.
12. Bigner DD, Bigner SH, Pontén J: Heterogeneity of genotypic and phenotypic characteristics of fifteen permanent cell lines derived from human gliomas. *J Neuropathol Exp Neurol* 40:201–229, 1981.
13. Bigner DD, Pitis OM, Wikstrand CJ: Induction of lethal experimental allergic encephalomyelitis in nonhuman primates and guinea pigs with human glioblastoma multiforme tissue. *J Neurosurg* 55:32–42, 1981.
14. Bigner SH, Bjerkvig R, Laerum OD: DNA content chromosomal composition of malignant human gliomas. In Vick NA, Bigner DD (eds): *Neurologic Clinics.* Vol. 3/4: *Neurooncology.* WB Saunders, Philadelphia, 1985, pp. 769–784.

15. Black KL, Hoff JT: Leukotrienes increase blood–brain barrier permeability following intraparenchymal injections in rats. *Ann Neurol* 18:349–351, 1985.

16. Black KL, Hoff JT, McGillicuddy JE, et al: Increased leucotriene C4 and vasogenic edema surrounding brain tumors in humans. *Ann Neurol* 19:592–595, 1986.

17. Blasberg RG, Fenstermacher JD, Patlak CS: The transport of alpha-amino isobutyric acid across brain capillaries and cellular membranes. *J Cereb Blood Flow Metab* 3:8–32, 1983.

18. Blasberg RG, Molnar P, Groothuis DR: Concurrent measurements of blood flow and transcapillary transport in ASV-induced experimental brain tumors: Implications for brain tumor chemotherapy. *J Pharmacol Exp Ther* 231:724–735, 1984.

19. Bloom HJ, Peckham MJ, Richardson AE: Glioblastoma multiforme: A controlled trial to assess the value of specific active immune therapy in patients treated by radical surgery and radiotherapy. *Br J Cancer* 27:253–267, 1973.

20. Bloom WH, Carstarb KC, Crompton MR: Autologous glioma transplantation. *Lancet* 2:77–78, 1960.

21. Bluestein HG, Woods VL: Antineuronal antibodies in systemic lupus erythematosus. *Arthritis Rheu* 24:773–778, 1982.

22. Boyer RS, Sun NC, Verity MA, et al: Immunoperoxidase staining of the choroid plexus in systemic lupus erythematosus. *J Rheumatol* 7:645–650, 1980.

23. Bradbury M: *The Concept of a Blood–Brain Barrier*. Wiley, New York, 1979.

24. Bradbury M: Lymphatics and the central nervous system. *Trends Neurosci* 4:100–101, 1981.

25. Braun DP, Penn RD, Harris JE: Regulation of natural killer cell function by glass-adherent cells in patients with primary intracranial malignancies. *Neurosurgery* 15:29–33, 1984.

26. Brightman MW, Reese TS: Junctions between intimately opposed cell membranes in the vertebrate brain. *J Cell Biol* 40:648–677, 1969.

27. Brightman MW, Luatzo I, Olsson Y, et al: The blood–brain barrier to proteins under normal and pathological conditions. *J Neurol Sci* 10:215–239, 1970.

28. Brooks WH, Roszman TL: Immunobiology of intracranial tumors. In Waters H (ed): *Immunotherapy. The Handbook of Cancer Immunology*. Garland, New York, 1978, pp. 345–355.

29. Brooks WH, Caldwell HD, Mortara RH: Immune responses in patients with gliomas. *Surg Neurol* 2:419–423, 1974.

30. Brooks WH, Roszman TL, Rogers AS: Impairment of rosette-forming T-lymphocytes in patients with primary intracranial tumors. *Cancer* 37:1869–1873, 1976.

31. Brooks WH, Cross RJ, Roszman TL, et al: Neuroimmuno-modulation: Neural anatomical basis for impairment and facilitation. *Ann Neurol* 12:56–61, 1982.

32. Brosnar CF, Traugott U, Raine CS: Analysis of humoral and cellular events and the role of lipid haptens during CNS demyelination. *Acta Neuropathol (Berl)* 9:(suppl)59–70, 1983.

33. Bullard DE, Bigner DD: Applications of monoclonal antibodies in the diagnosis and treatment of primary brain tumors. *J Neurosurg* 63:2–16, 1955.

34. Bullard DE, Gillespie GY, Mahaley MS, et al: Immunobiology of human gliomas. *Semin Oncol* 13:94–109, 1986.

35. Burnet FM: Immunological aspects of malignant disease. *Lancet* 1:1171–1174, 1967.

36. Chen ST, Hsu CY, Hogan EL, et al: Thromboxane, prostacyclin and leucotrienes in cerebral ischemia. *Neurology (NY)* 36:466–470, 1986.

37. Cleary ML, Sklar J: Lympho proliferative disorders in cardiac transplant recipients are multifocal lymphomas. *Lancet* 2:489–493, 1984.

38. Coakham HB, Garson JA, Allan PM, et al: Immunohistological diagnosis of CNS tumors using monoclonal antibody panel. *J Clin Pathol* 38:165–173, 1985.

39. Cross RJ, Jackson JC, Brooks WH, et al: Neuroimmunomodulation: Impairment of humoral immune responsiveness by 6-hydroxydopamine treatment. *Immunology* 57:145–152, 1986.

40. Cutler RWP, Watteis GV, Hammerstad JP. The origin and turn over rates of cerebrospinal fluid, albumin and gamma globulin in man. *J Neurol Sci* 10:259–268, 1970.

41. Darling JL, Hoyle NR, Thomas DGT: Self and non-self in the brain. *Immunol Today* 2:176–181, 1981.

42. de Carvalho S, Kaufman A, Pineda A: Adjuvant chemo-immunotherapy in central nervous system tumors. In Salmon SE, Josen SE (eds): *Adjuvant Therapy in Cancer*. North-Holland, Elsevier, 1977, pp. 495–502.

43. del Rio Hortega P: El tercei elemento de los antros resviosos. I: La microglia en estado normal. II. Intevencion de la microglia enlos procesos patologicos. III. Naturalenza probable de la microglia. *Bol Soc Esp Biol* 9:20–68, 1919.

44. del Rio Hortega P: Microglia. In Penfield W (ed): *Cytology and Cellular Pathology of the Nervous System.* Vol. 2. Hafner, New York, 1965, pp. 483–534.

45. del Rio Hortega P, Penfield W: Cerebral cicatrix. The reaction of neuroglia and microglia to brain wounds. *Bull Johns Hopkins Hosp* 41:41–68, 1927.

46. Denburg JA, Temesvari P: The pathogenesis of neuropsychiatric lupus. *Can Med Assoc J* 128:257–260, 1983.

47. Ellis SG, Verity MA: Central nervous system involvement in systemic lupus erythematosus: A review and neuropathologic findings in 57 cases. *Semin Arthritis Rheum* 8:212–221, 1979.

48. Epenetos AA, Courtnay-Luck N, Pickering D, et al: Antibody-guided irradiation of brain glioma by arterial infusion of radioactive monoclonal antibody against epidermal growth factor receptor and blood group A antigen. *Br Med J Clin Res* 290:1463–1466, 1985.

49. Fauci AS: Immunologic abnormalities in the acquired immunodeficiency syndrome (AIDS). *Clin Res* 32:491–499, 1984.

50. Fauci AS, Henry M, Gelmann EP, et al: The acquired immunodeficiency syndrome: An update. *Ann Intern Med* 102:800–813, 1985.

51. Fenstermacher JD, Rapoport SI: The blood–brain barrier. In Renkin EM, Michel CC (eds): *Handbook of Physiology: The Cardiovascular System.* Vol IV. American Physiology Society, Bethesda, 1984, pp. 969–1000.

52. Fenstermacher JD, Blasberg RG, Patlak CS: Methods for quantifying the transport of drugs across brain barrier systems. *Pharmacol Ther* 14:217–248, 1981.

53. Fidelio GD, Maggio B, Cumai FA: Interaction of soluble and membrane proteins with monolayers of glycosphingolipids. *Biochem J* 203:717–725, 1982.

54. Fishman RA. *Cerebrospinal Fluid in Diseases of the Nervous System.* WB Saunders, Philadelphia, 1980.

55. Fontana A, Fierz W: The endothelium–astrocyte immune control system of the brain. *Springer Semin Immunopathol* 8:57–70, 1985.

56. Fontana A, Kristensen R, Dubs D, et al: Production of prostaglandin E and interleukin-1 like factors by cultured astrocytes and C-6 glioma cells. *J Immunol* 129:2413–2419, 1982.

57. Fontana A, McAdam KPWJ, Kristensen F, et al: Biological and biochemical characterization of an interleukin-1 like factor from rat C-6 glioma cells. *Eur J Immunol* 13:685–689, 1983.

58. Fontana A, Fierz W, Wekerle H: Astrocytes present myelin basic protein to encephalitogenic T-cell lines. *Nature (Lond)* 307:273–276, 1984.

59. Fontana A, Hengastner H, de Tribolet N, et al: Glioblastoma cells release interleukin-1 and factors inhibiting interleukin-2 mediated effects. *J Immunol* 132:1837–1844, 1984.

60. Frank E, de Tribolet N: Immunobiology of brain tumors. *Neurosurg Rev* 9:31–37, 1986.

61. Frank E, Pulve M, de Tribolet N: Expression of class II major histocompatibility antigens on reactive astrocytes and endothelial cells within the gliosis surrounding metastases and abscesses. *J Neuroimmunol* 12:29–36, 1986.

62. Friedman-Kien AE (ed): *AIDS, the Epidemic of Kaposi's Sarcoma and Opportunistic Infections.* Masson, New York, 1984.

63. Gallin JI, Fauci A (eds): *Advances in Host Defense Mechanisms.* Vol 2: *Lymphoid Cells.* Raven, New York, 1983, pp. 1–43, 101–143, 241–275.

64. Gately CL, Muul LM, Greenwood MA, et al: In vitro studies on the cell-mediated immune response to human brain tumors. II. Leucocyte induced coats of glycosaminoglycan increase the resistance of glioma cells to cellular immune attack. *J Immunol* 133:3395–3397, 1984.

65. Gately MK, Glaser M, Dick SJ, et al: In vitro studies on the cell mediated immune response to human brain tumors. I. Requirement for third-party stimulatory lymphocytes in the induction of cell mediated cytotoxic responses to allogenic cultured gliomas. *J Natl Cancer Inst* 69:1245–1254, 1982.

66. Gerschwin ME, Hyman LR, Steinberg AD. The choroid plexus in CNS involvement of systemic lupus erythematosus. *J Pediatr* 87:588–590, 1975.

67. Gibson T, Myers AR: Nervous system involvement in systemic lupus erythematosus. *Am Rheum Dis* 35:398–406, 1976.

68. Grace JT, Perese DM, Metzgar RS: Tumor autograft responses in patients with glioblastoma multiforme. *J Neurosurg* 18:159–167, 1961.

69. Groothuis DR, Molnar P, Blasberg RG: Regional blood flow and blood-to-tissue transport in five brain tumor models. In Rosenblum M, Wilson C (eds): *Prog Exp Tumor Res.* 27:132–153, 1984.

70. Groothuis DR, Blasberg RG: Rational brain tumor chemotherapy: The interaction of drug and tumor. In

Vick NA, Bigner DD (eds): *Neurologic Clinics.* Vol. 3/4. *Neurooncology.* WB Saunders, Philadelphia, 1985, pp. 801–816.

71. Hanwehr RI von, Hofman FM, Taylor CR, et al: Mononuclear lymphoid populations infiltrating the microenvironment of primary CNS tumors. *J Neurosurg* 60:1139–1147, 1984.

72. Hershey LA, Trotter JL: The use and abuse of the cerebrospinal fluid IgG profile in the adult: A practical evaluation. *Ann Neurol* 8:426–434, 1980.

73. Hochwald GM: Influence of serum proteins and their concentration in spinal fluid along the neuraxis. *J Neurol Sci* 10:269–278, 1970.

74. Hoffman SA, Arbogast DN, Day TT, et al: Permeability of the blood–cerebrospinal fluid barrier during acute immune complex disease. *J Immunol* 130:1695–1698, 1983.

75. Huddlestone JR, Oldstone MBA: T-suppressor (T_g) lymphocytes fluctuate in parallel with changes in the clinical course of patients with multiple sclerosis. *J Immunol* 123:1615–1618, 1979.

76. Husby G, Williams RC, Besin RM, et al: Antineuronal antibodies in diseases affecting the basal ganglia particularly Sydenham's and Huntington's chorea. In Rose FC (ed): *Clinical Neuroimmunology.* Blackwell, Oxford, 1979, pp. 90–105.

77. Husby G, Williams RC, Ramirez F, et al: Absence of interferon-α and -γ in renal lesions of systemic lupus erythematosus and membraneous glomeruls nephritis. *Clin Immunol Immunopathol* 39:68–80, 1986.

78. Jacobs L, O'Malley J, Freeman A, et al: Intrathecal interferon reduces exacerbations of multiple sclerosis. *Science* 214:1026–1028, 1981.

79. Jacobs SK, Wilson DJ, Kornblith PL: In vitro killing of human glioblastoma by interleukin-2-activated autologous lymphocytes. *J Neurosurg* 64:114–117, 1986.

80. Jacobs SK, Wilson DJ, Kornblith PL, et al: Interleukin-2 and autologous lymphokine-activated killer cells in the treatment of malignant glioma. *J Neurosurg* 64:743–749, 1986.

81. Johnson KP: Systemic interferon therapy for multiple sclerosis. Design of a trial. *Arch Neurol* 40:681–682, 1983.

82. Johnson RT, Richardson EP: The neurologic manifestations of systemic lupus erythematosus. A clinical pathological study of 24 cases and review of the literature. *Medicine (Baltimore)* 47:337–369, 1968.

83. Juhler M, Blasberg RG, Fenstermacher JD, et al: A spatial analysis of the blood–brain barrier damage in experimental allergic encephalomyelitis. *J Cereb Blood Flow Metab* 5:534–553, 1985.

84. Juhler M, Laursen H, Barry DI: The distribution of immunoglobulins and albumin in the central nervous system in acute experimental allergic encephalomyelitis. *Acta Neurol Scand* 73:119–124, 1986.

85. June NK: The natural selection theory of antibody formation. *Proc Natl Acad Sci (USA)* 41:840–851, 1955.

86. Kaufman DB, Miller HC: Ataxia telangiectasia: An autoimmune disease associated with a cytotoxic antibody to brain and thymus. *Clin Immunol Immunopathol* 7:288–299, 1977.

87. Kies MW: Chemical studies on an encephalitogenic protein from guinea pig brain. *Ann NY Acad Sci* 122:161–169, 1965.

88. Kikuchi K, Neuwelt EA: Presence of immunosuppressive factors in brain-tumor cyst fluid. *J Neurosurg* 59:790–799, 1983.

89. Kimberlin RH (ed): *Slow Virus Diseases of Animal and Man.* North-Holland/American Elsevier, New York, 1976.

90. Kimmelberg HK, Kung D, Watson RE, et al: Direct administration of methotrexate into the central nervous system of primates. Part 1: Distribution and degradations of methotrexate in nervous systemic tissue after intraventricular injection. *J Neurosurg* 48:883–894, 1978.

91. Kirby JA, Suckling AJ, Rumsby MG:. Chronic relapsing experimental allergic encephalomyelitis. The presence of the cerebrospinal fluid of factors chemotactic for monocytes. *J Neuroimmunol* 5:271–281, 1983.

92. Klatzman D, Barre-Sinoussi F, Nugeyre MT: Selective tropism of lymphadenopathy associated virus for helper/inducer T-lymphocytes. *Science* 224:59–63, 1984.

93. Knobler RL, Panitch HS, Braheny SC, et al: Systemic alpha-inteferon therapy of multiple sclerosis. *Neurology (NY)* 34:1273–1279, 1984.

94. Kono R, Kuroiwa Y: Subacute myelo-optic neuropathy is not a special form of multiple sclerosis. *Lancet* 2:267, 1982.

95. Krupp LB, Lipton RB, Swedlow ML, et al: Progressive multifocal leukoencephalopathy: Clinical and radiographic features. *Ann Neurol* 17:344–349, 1985.

96. Lachman PJ, Peters DK (eds): *Clinical Aspects of Immunology.* Blackwell Scientific, Oxford, 1982.

97. Lampert PW: Pathological implications of immunological disease in the central nervous system. In Waksman BH (ed): *Immunoneuropathology.* WB Saunders, Philadelphia, 1982, pp. 347–369.

98. Lane HC, Deppes JM, Greene WC: Qualitative analysis of immune function in patients with the acquired immune deficiency syndrome. *N Engl J Med* 313:79–84, 1985.

99. Lassmann H: Comparative neuropathology of chronic experimental allergic encephalomyelitis and multiple sclerosis. *Schriftenr Neurol* 25:1–135, 1983.

100. Li C-Y, Witzig TE, Phyliky RL, et al: Diagnosis of B cell non-Hodgkins lymphoma of the central nervous system by immunocytochemical analysis of cerebrospinal fluid lymphocytes. *Cancer* 57:737–744, 1986.

101. Lidegard O, Gyldensted C, Juhler M, et al: CT findings in acute MS. *Acta Neurol Scand* 68:77–86, 1983.

102. Ling E-A: The origin and nature of microglia. *Adv Cell Biol* 2:33–82, 1979.

103. Link H: Comparison of electrophoresis on agar gel and agarose in the evaluation of gamma globulin abnormalities in cerebrospinal fluid and serum in multiple sclerosis. *Clin Chim Acta* 46:393–399, 1973.

104. Love JA, Leslie RA: The effects of raised ICP on lymph flow in the cervical lymphatic trunks in cats. *J Neurosurg* 60:577–581, 1984.

105. Maggio B, Cumar FA, Roth GA, et al: Neurochemical and model membrane studies in demyelinating diseases. *Acta Neuropathol (Berl)* 9(suppl):71–85, 1984.

106. Mahaley MS, Gillespie GY: Immunotherapy of patients with glioma: Fact, fancy, future. *Prog Exp Tumor Res* 28:118–135, 1984.

107. Mahaley MS, Brooks WH, Roszman TL: Immunobiology of primary intracranial tumors. I. Studies of the cellular and humoral general immune competence of brain tumor patients. *J Neurosurg* 46:467–476, 1977.

108. Mahaley MS, Bigner DD, Dudka LF: Immunobiology of primary intracranial tumors. Part 7. Active immunization of patients with anaplastic human glioma cells. A pilot study. *J Neurosurg* 59:201–207, 1983.

109. Murkowitz H, Kokmen E: Neurologic diseases and the cerebrospinal fluid immunoglobulin profile. *Mayo Clin Proc* 58:273–274, 1983.

110. McIntosh RM, Groswold WR, Chenack WB, et al: The choroid plexus; a possible role in autoimmune nephritis. *Clin Res* 21:324–325, 1973.

111. McKeever PE, Balentine JD, Paris DV: Macrophage migration through the brain parenchyma to the perivascular space following particle ingestion. *Am J Pathol* 93:153–165, 1978.

112. Miller RG: The Guillian-Barré syndrome. *Postgrad Med* 77:57–64, 1985.

113. Mirra SS, Check IJ, Porter JD, et al: Rapid evolution of central nervous system lymphoma in renal transplant recipient. *Lancet* 1:868–869, 1981.

114. Mizuno D, Cohn ZA, Takaya K, et al. (eds): *Self-Defense Mechanisms. Monokines and Lymphokines.* University of Tokyo Press, Tokyo, 1983.

115. Mori S, Leblond CP: Identification of microglia in light and electron microscopy. *J Comp Neurol* 135:57–80, 1969.

116. Moskowitz LB, Hensley GT, Chan JC, et al: The neuropathology of acquired immune deficiency syndrome. *Arch Pathol Lab Med* 108:867–872, 1984.

117. Moskowitz LB, Gregorius JB, Hensley GT, et al: Cytomegalovirus. Induced demyelination associated with acquired immune deficiency syndrome. *Arch Pathol Lab Med* 108:873–877, 1984.

118. Moskowitz MA, Kiwak KJ, Hekimium K, et al: Synthesis of compounds with properties of leukotrienes C4 and D4 in gerbil brains after ischemia and reperfusion. *Science* 224:886–888, 1984.

119. Müller-Eberhard HJ, Schreiber RD: Molecular biology and chemistry of the alternative pathway of complement. *Adv Immunol* 29:1–39, 1980.

120. Murphy JB, Sturm E: Conditions determining the transplantation of tissues in the brain. *J Exp Med* 39:183–197, 1923.

121. Nagy Z, Peters H, Huttner I: Fracture facets of cell junctions in cerebral endothelium during normal and hyperosmotic conditions. *Lab Invest* 50:313–322, 1984.

122. Naparstek Y, Cohen IR, Fuks Z, et al: Activated T-lymphocytes produce a matrix-degrading heparan sulphate endoglycosidase. *Nature (Lond)* 310:241–244, 1984.

123. Neuwelt EA, Clark WK: *Clinical Aspects of Neuroimmunology.* Williams & Wilkins, Baltimore, 1978.

124. Neuwelt E, Doherty D: Toxicity kinetics and clinical potential of subarachnoid lymphocyte infusions. *J Neurosurg* 47:205–217, 1977.

125. Neuwelt EA, Clark WK, Kirkpatrick JB, et al: Clinical studies of intrathecal autologous lymphocyte infusions in patients with malignant glioma: A toxicity study. *Ann Neurol* 4:307–312, 1978.

126. Neuwelt EA, Garcia JH, Mena H: Diffuse microglial proliferation after global ischemia in a patient with aplastic bone marrow. *Acta Neuropathol (Berl)* 43:259–262, 1978.

127. Neuwelt EA, Kikuchi K, Hill S, et al: Immune responses in patients with brain tumors. Factors such as anti-convulsants that may contribute to impaired cell-mediated immunity. *Cancer* 51:248–255, 1983.

128. Neuwelt EA, Barnett P, McCormick CI, et al: Osmotic blood–brain barrier modification: Monoclonal antibody, albumin and methotrexate delivery to CSF and brain. *Neurosurgery* 17:419–423, 1985.

129. Neuwelt EA, Specht HD, Hill SA: Permeability of human brain tumor to ^{99m}Tc–glucoheptonate and ^{99m}Tc-albumin: Implications for monoclonal antibody therapy. *J Neurosurg* 65:194–198, 1985.

130. Neuwelt EA, Frenkel EP, Gumerlock MK, et al: Developments in the diagnosis and treatment of primary CNS lymphoma: A prospective series. *Cancer* 58:1609–1620, 1986.

131. Neuwelt EA, Minna J, Frenkel E, et al: Osmotic blood–brain barrier opening to IgM monoclonal antibody in the rat. *Am J Physiol* 250:R875–R883, 1986.

132. Neuwelt EA, Specht HD, Barnett PA, et al: Increased delivery of tumor-specific monoclonal antibodies to brain after osmotic blood–brain barrier modification in patients with melanoma metastatic to the CNS. In preparation, *Neurosurgery* 20:885–895, 1987.

133. Norrby E: Characterization of the virus antibody activity of oligoclonal IgG produced in the central nervous system of patients with multiple sclerosis. In ter Meulen V, Katz M (eds): *Slow Virus Infections of the Central Nervous System*. Springer-Verlag, New York, 1977, pp. 159–164.

134. Norrby E, Salmi AA, Link H, et al: The measles virus antibody response in subacute sclerosing panencephalitis and multiple sclerosis. In Zwman W, Hemmette EH, Brunson JG (eds): *Slow Virus Diseases*. Williams & Wilkins, Baltimore, 1974, pp. 72–85.

135. Oehmichen M, Grüninger H, Weithölte H, et al: Lymphatic efflux of intracerebrally injected cells. *Acta Neuropathol (Berl)* 45:61–65, 1979.

136. Oehmichen M: Functional properties of microglia. In Smith WT, Cavanaugh JB (eds): *Recent Advances Neuropathol*. Vol 2. Churchill Livingstone, London, 1982, pp. 83–107.

137. Oehmichen M: Are resting and/or reactive microglia macrophages? *Immunobiology* 161:246–254, 1982.

138. Oldstone MBA: Immunopathologic disease of the central nervous system. In Brockes J (ed): *Neuroimmunology*. Plenum, New York, 1982, pp. 125–139.

139. Olsson Y: Topographical differences in the vascular permeability of the peripheral nervous system. *Arch Neuropathol (Berl)* 10:26–33, 1968.

140. Oppenheim JJ, Cohen S (eds): *Interleukins, Lymphokines and Cytokines*. Proceedings of the Third International Lymphokine Workshop. Academic, New York, 1983.

141. Pacner AR: Spirochetal diseases of the CNS. In Booss J, Thornton GF (eds): *Neurologic Clinics*. Vol. 4/1: *Infectious Diseases of the Central Nervous System*. WB Saunders, Philadelphia, 1986, pp. 207–222.

142. Padgett BH: Progressive multifocal leucoencephalopathy—A viral disease. In Boese A (ed): *Search on the Cause of Multiple Sclerosis and Other Chronic Diseases of the Central Nervous System. First International Symposium of the Heitie Foundation*. Verlag Chemie, Frankfurt, 1980, pp. 280–283.

143. Pardridge WM: Blood–brain barrier: Interface between internal medicine and the brain. *Ann Intern Med* 105:82–85, 1986.

144. Paterson PY: Immune responses implicated in immunopathological disorders of the central nervous system. In Boese A (ed): *Search on the Cause of Multiple Sclerosis and Other Chronic Diseases of the Central Nervous System. First International Symposium of the Heitie Foundation*. Verlag Chemie, Frankfurt, 1980, pp. 173–183.

145. Paterson PY, Drobish DG, Hansen MA, et al: Induction of experimental allergic encephalomyelitis in Lewis rats. *Int Arch Allergy* 27:26–40, 1970.

146. Payan MJ, Gambarelli D, Routy JP: Primary lymphoma of the brain associated with AIDS. *Acta Neuropathol (Berl)* 64:78–80, 1984.

147. Pearl GS, Check IJ, Hunter RL: Agarose gel electrophoresis and immunonephelometric quantitation of cerebrospinal fluid immunoglobulins: Criteria for application of the diagnosis of neurologic disease. *Am J Clin Pathol* 81:575–580, 1984.

148. Piguet V, Diserens A-C, Card S, et al: The immunobiology of human gliomas. *Springer Semin Immunopathol* 8:111–127, 1985.

149. Prusiner SB, Hadlow WY (eds): *Slow Transmissible Diseases of the Nervous System*. Academic, New York, 1978.

150. Raju S, Grogan JB: Immunologic study of the brain as a privileged site. *Transpl Proc* 9:1187–1191, 1977.

151. Rapoport SI: *Blood–Brain Barrier in Physiology and Medicine*. Raven, New York, 1976.

152. Reik L: Disorders that mimic CNS infections. In Booss J, Thornton GF (eds): *Neurologic Clinics*. Vol. 4/1: *Infectious Diseases of the Central Nervous System*. WB Saunders, 1976, pp. 223–248.

153. Reinhesz EL, Weiner HL, Hansen SL, et al: Loss of suppressor T-cells in active multiple sclerosis: Analysis with monoclonal antibodies. *N Engl J Med* 303:125–129, 1980.

154. Ridley A: Survival of guinea pig skin grafts in the brains of rats under treatment with antilymphocytic serum. *Transplantation* 10:86–91, 1970.

155. Ridley A, Cavanaugh JB: The cellular reactions to heterologous, homologous and autologous skin implanted into brain. *J Pathol* 99:193–203, 1969.

156. Rivers TM, Sprunt DH, Berry GP: Observations on attempts to produce acute disseminated encephalomyelitis in monkeys. *J Exp Med* 58:39–53, 1933.

157. Romani L, Nardelli B, Bianchi R, et al: Adoptive immune therapy of intracerebral murine lymphomas. Role of different lymphoid populations. *Int J Cancer* 35:659–665,1985.

158. Ropper AH, Miett T, Chiappa KH: Absence of evoked potential abnormalities in acute transverse myelopathy. *Neurology (NY)* 32:80–92, 1980.

159. Rosenberg SA: Immunotherapy of cancer by systemic administration of lymphoid cells plus interleukin-2. *J Biol Res Mod* 3:501–511, 1984.

160. Rosenberg SA, Lotz M, Muhl L, et al: Observations on the systemic administration of autologous lymphokine-activated killer cells and recombinant interleukin-2 to patients with metastatic cancer. *N Engl J Med* 313:1485–1492, 1985.

161. Roszman TL, Brooks WH: Neural modulation of immune function. *J Neuroimmunol* 10:59–69, 1985.

162. Roszman TL, Jackson JC, Cross RJ, et al: Neuroanatomic and neurotransmitter influences on immune function. *J Immunol* 135:769–772, 1985.

163. Rydberg E: Cerebral injury in newborn children consequent on birth trauma, with an inquiry into the normal and pathological anatomy of the neuroglia. *Acta Pathol Microbiol Scand* 10(suppl):1–247, 1932.

164. Schelper RL, Adrian EK: Monocytes become macrophages; they do not become microglia: A light and electron microscopic autoradiographic study using ^{125}I-deoxyuridine. *J Neuropathol Exp Neurol* 45:1–19, 1986.

165. Schenck SA, Penn I: De-novo brain tumors in renal transplant recipients. *Lancet* 1:983–986, 1971.

166. Selker RG, Wolmark N, Fisher B: Preliminary observations on the use of corynebacterium parvum in patients with primary intracranial tumors. *J Surg Oncol* 10:299–303, 1978.

167. Simpson E, Lieberman R, Ando I, et al: How many class II immune response genes. A reappraisal of the evidence. *Immunogenetics* 23:302–308, 1986.

168. Sotelo J, Gibbs CJ, Guydusek DC: Autoantibodies against axonal neurofilaments in patients with Kuru and Creutzfeld–Jakob disease. *Science* 210:190–193, 1980.

169. Sotrel A, Rosen S, Ronthal M, et al: Subacute sclerosing panencephalitis: An immune complex disease? *Neurology (NY)* 33:885–890, 1983.

170. Steinweg AE: The origin of brain macrophages in traumatic lesions, Wallerian degeneration, and retrograde degeneration. *J Neuropathol Exp Neurol* 31:696–704, 1972.

171. Stiehm ER, Kronenberg LH, Rosenblatt HM, et al: UCLA conference on interferon: Immunobiology and clinical significance. *Ann Intern Med* 96:80–93, 1982.

172. Sunde N, Hamberg S, Zimmer J: Brain grafts can restore irradiation damaged neuronal corrections in newborn rats. *Nature (Lond)* 310:51–53, 1984.

173. Takakura K, Miki Y, Kubo O, et al: Adjuvant immunotherapy for malignant brain tumors. *Jpn J Clin Oncol* 12:109–120, 1972.

174. ter Meulen V, Katz M (eds): *Workshop on Slow Virus Infections. Slow Virus Infections of the Central System: Investigational Approaches and Pathogens of These Diseases.* Springer-Verlag, Berlin, 1977.

175. Tilney NL, Strom TB, Milford EL: The role of suppression of the immune responses. In Stuart FP, Fitch FW (eds): *Immunologic Tolerance and Enhancement.* University Park Press, Baltimore, 1979, pp. 125–149.

176. Tourtelotte W: On cerebrospinal IgG quotients in multiple sclerosis and other diseases. A review and a new formula to estimate the amount of IgG synthesized per day by the central nervous system. *J Neurol Sci* 10:279–304, 1970.

177. Traugott U, Reinheiz EL, Raine CS: Multiple sclerosis: Distribution of T-cell subsets and Ia-positive macrophages in lesions of different ages. *J Neuroimmunol* 4:201–221, 1983.

178. Tse WJ, Tai J: Intracerebral allotransplantation of purified pancreatic endocrine cells and pancreatic islets in diabetic rats. *Transplantation* 39:107–111, 1984.

179. Unanue ER, Beller DI, Lu CY, et al: Antigen presentation: Comments on its regulation and mechanism. *J Immunol* 132:1–14, 1984.

180. Vandenbark AA, Raus JCMY (eds): *Immunoregulatory Process in Experimental Allergic Encephalomyelitis and Multiple Sclerosis.* Elsevier, New York, 1984.

181. Vass K, Lassmann H, Wisniewski H, et al: Ultracytochemical distribution of myelin basic protein after injection into the cerebrospinal fluid. Evidence for transport through the blood–brain barrier and binding to the luminal surface of cerebral veins. *J Neurol Sci* 63:423–433, 1984.

182. Weiner HL, Hauser SL: Neuroimmunology. I. Immunoregulation in neurological disease. *Ann Neurol* 11:437–449, 1982.
183. Weiner HL, Hauser SL: Neuroimmunology. II. Antigenic specificity of the nervous system. *Ann Neurol* 12:499–509, 1982.
184. Welch K, Finkbeine W, Alpus CE, et al: Autopsy findings in the acquired immune deficiency syndrome. *JAMA* 252:1152–1159, 1984.
185. Wikstrand CJ, Bigner DD: Immunobiologic aspects of the brain and human gliomas. *Am J Pathol* 98:517–567, 1980.
186. Wilson CB, Barke M: Intrathecal transplantation of human neural tumors to the guinea pig. *J Natl Cancer Inst* 41:1229–1240, 1968.
187. Winfield JB, Shaw M, Silverman LM, et al: Intrathecal IgG synthesis and blood–brain barrier impairment in patients with systemic lupus erythematosus and central nervous system dysfunction. *Am J Med* 74:837–844, 1983.
188. Wisniewski HM, Lassmann H: Etiology and pathogenesis of monophasic and relapsing inflammatory demyelination—Human and experimental. *Acta Neuropathol (Berl)* 9(suppl):21–31, 1983.
189. Wisniewski HM, Lassmann H, Brosman CF, et al: Multiple sclerosis: Immunological and experimental aspects. In Matthews WB, Glase CH (eds): *Recent Advances in Neurology*. Vol. 3. Churchill Livingstone, New York, 1982, pp. 95–124.
190. Wolberger JE: Transplantation of central nervous tissue. *Neurosurgery* 13:90–94, 1983.
191. Yamasaki T, Handa H, Yamashita J, et al: Specific adoptive immunotherapy with tumor specific cytotoxic T-lymphocyte clone for murine malignant gliomas. *Cancer Res* 44:1776–1803, 1984.
192. Yonezawa T: Circulating myelinotoxic factors in human and experimental demyelinating tissue. *Acta Neuropathol (Berl)* 9(suppl):47–58, 1983.
193. Young HF, Kaplan AM, Regelson W: Immunotherapy with autologous white cell infusions (lymphocytes) in the treatment of recurrent glioblastoma multiforme. *Cancer* 40:1037–1044, 1977.
194. Zwaifler NY, Bluestein HG: The pathogenesis of central nervous system manifestations of systemic lupus erythematosus. *Arthritis Rheum* 25:862–866, 1982.

11

Cerebral Edema and the Blood–Brain Barrier

Hanna M. Pappius

1. INTRODUCTION

Brain edema accompanies a wide variety of pathologic processes and contributes to the morbidity and mortality of many neurologic diseases.[42] The early literature on this subject, dating back more than 50 years, was both confused and confusing for two main reasons: increased intracranial pressure (ICP) was often equated with the presence of edema, and there was no appreciation that different types of edema exist with different mechanisms of formation and resolution and thus presumably open to different therapeutic approaches.

1.1. Definition of Edema

By definition, cerebral edema is a condition in which the volume of brain tissue is increased due to an increase in its water content.[42] Increased intracranial content, hence ICP, may be due not only to increased water content, or edema, but also to increased cerebral blood volume (CBV), increased cerebrospinal fluid (CSF) volume, or a combination of the three. When the precise nature of the processes involved is not clear, brain swelling is a more appropriate term to use, whereas edema should be restricted to describe conditions in which actual increase in tissue water is known to have occurred.

This is not a question of pure semantics. If one wants to prevent the catastrophic consequences of increased ICP, the chances of success are much greater if the underlying mechanisms are delineated as clearly as possible.

1.2. Types of Cerebral Edema

Much of the progress in our understanding of the whole subject of cerebral edema has derived from experimental studies in which various contributory factors can be controlled

Hanna M. Pappius • The Goad Unit of The Donner Laboratory of Experimental Neurochemistry, Montreal Neurological Institute, McGill University, Montreal, Quebec H3A 2B4, Canada.

and water content of tissue measured, something that is rarely possible in the clinical setting. The following discussion is based exclusively on results obtained with various animal models. Although there is every reason to suppose that conclusions derived from such investigations apply to comparable human material, it is fully appreciated that the clinical situations are much more complicated than the average experimental setup.

The modern era in the field of brain edema started with the classic experimental studies of Klatzo, which led in 1967 to his definition of vasogenic edema as associated with gross damage to vascular elements and consequent breakdown of the blood–brain barrier (BBB).[45] Klatzo, at that time, also coined the term cytotoxic edema to describe all conditions not covered by the term vasogenic. In the intervening years, our understanding of the processes involved has permitted a delineation of three additional types of cerebral edema, i.e., ischemic, osmotic, and interstitial.

1.2.1. Vasogenic Edema

Vasogenic edema is induced by damage to vascular elements in the brain regardless of the type of injury sustained.[42,45] Clinically, it is the edema seen as a result of head injury, in association with tumors, with even successful surgery. It is now also quite clear that it also accompanies infarction. Experimentally, a variety of insults have been used to induce vasogenic edema. The freezing lesion is the most common model.[42]

Peritumoral edema must be considered as a special case of vasogenic edema. Recent evidence shows that the major site of extravasation of this edema fluid is the hyperpermeable vasculature that proliferates in the tumor, rather than the damaged vessels in the surrounding brain tissue.[21,22,86]

The salient characteristics of vasogenic edema are a breakdown of the BBB or its absence in the case of tumors, extravasation of a plasma-like fluid that penetrates the tissue under the force of systemic pressure and spreads because of pressure gradients that develop within the tissue. The extravasated fluid accumulates mainly in white matter. This is the type of cerebral edema that is primarily associated with events at the BBB.

1.2.2. Ischemic Edema

A completely different type of cerebral edema occurs as a consequence of ischemia.[44] It affects cerebral cortex, not white matter, and consists, at least initially, of intracellular accumulation of water and sodium, with loss of potassium occurring at later stages. The BBB remains intact at first. However, continuing flow reduction leads to cellular damage, extravasation of serum proteins, and the development of vasogenic edema. Edema associated with ischemia, although not a consequence of changes at the BBB, is considered briefly below to underline the differences between the two edematous processes of greatest clinical interest. It should be pointed out that hypoxia per se does not appear to induce cerebral edema.[42,56]

1.2.3. Osmotic Edema

Osmotic edema occurs when an osmotic gradient exists between plasma and cerebral tissue, as in water intoxication or rapid hemodialysis.[42,73] Acutely, the edema fluid consists of water unaccompanied by electrolytes; it affects all tissue elements. This is a transient phenomenon, however; with time, brain tissue loses electrolytes instead of gaining water, and as a result, osmotically induced changes in brain tissue volume are not sustained. Thus, it was not possible to demonstrate significant edema in an experimental model of inappropriate secretion of ADH[23,24] and increased ICP is not a consistent finding in patients with this syndrome.[24,42] Neurologic disturbances seen under such conditions may in fact reflect loss of potassium from the tissue, apparently in response to the accompanying hyponatremia.[42,60]

1.2.4. Interstitial Edema

Interstitial edema is the term used to describe increased brain fluid content associated with hydrocephalus.[29] It affects periventricular white matter but differs from true vasogenic edema inasmuch as the fluid involved is similar in composition to CSF (not plasma) and that the BBB remains intact.

1.2.5. Cytotoxic Edema

Cytotoxic edema is often used to denote intracellular swelling, e.g., following ischemia. It was a useful term when mechanisms underlying most cellular swelling were not understood. This is no longer the case, however, and its use should be restricted to conditions under which noxious factors are involved, such as triethyltin and hexachlorophene poisoning, or when mechanisms underlying the cellular swelling are not clear.

2. DEVELOPMENT AND RESOLUTION OF VASOGENIC EDEMA

Three phases can be distinguished in vasogenic edema, each having somewhat different characteristics.[11] These phases are formation and spread from the area of injury, followed by a period of equilibrium when the exudation of fluid into the tissue is balanced by its removal from the affected areas and, finally, resolution when no more fluid is extravasated while the process of clearance of the edema fluid continues. The information available on this subject has come primarily from experimental studies in which the freezing lesion was used to induce edema, but essentially similar results have also been obtained with tumor models.[33,39]

2.1. Formation and Spread of Vasogenic Edema

2.1.1. Breakdown of BBB

Focal disruption of the normal permeability pattern in the area of injury, commonly referred to as a breakdown of the BBB, is the starting point in the chain of events that lead to development of vasogenic edema.[44,45,75] In the case of tumors, the vessels within the tumor are the main site of focal leakage. It must be emphasized that the breakdown of BBB is restricted to the area of injury, while permeability characteristics of the edematous tissue at a distance from the lesion remain unchanged. This was originally shown in experiments in which the spread with time from the area of injury of a fluorescent marker, given at the time a freezing lesion was made, occurred together with the edema fluid.[47] However, when the marker was given not at the time of the lesion but at various times after the lesion, always exactly 1 hr before the animal was killed, its spread was limited and constant, irrespective of the degree of edema that had previously developed (Fig. 1).

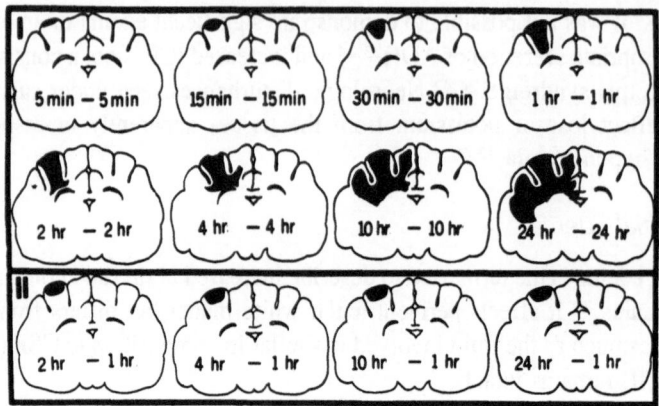

Figure 1. Diagrammatic representation of the movement of fluorescein-labeled serum albumin in brain of two series of cats in which cerebral edema was induced by a freezing lesion. In each case, time on the left refers to time since the lesion, on the right to time during which the marker was present in the circulation at the time the animals were killed. Series I: protein injected just before lesion was made. Series II: protein injected always 1 hr before end of experiment. (Modified from Klatzo et al.[47] Original plates made available by courtesy of I. Klatzo.)

This was a clear demonstration that the increased permeability was limited to the area of the lesion and that the permeability of blood vessels in the edematous tissue remained unaltered at a distance from the lesion. These findings have been confirmed more recently with quantitative autoradiographic techniques.[8]

The exact pathogenesis of the focal breakdown of the BBB resulting from injury is still under discussion.[49,81] The most obvious mechanism, namely extravasation of fluid through ruptured endothelial cell membranes in the damaged tissue, has been demonstrated[5,13,49] but appears to be only a part of the overall process. In the periphery of the lesion area, passage of BBB markers has been observed through apparently intact, but obviously damaged, endothelial cells[49] as well as through leaky interendothelial clefts[5,13,35]; involvement of pinocytosis has also been implicated.[5,13,35]

A pivotal role of damaged vascular bed in formation of vasogenic brain edema was confirmed by experiments in which microexcision of the cold lesion resulted in complete prevention or arrest of the edematous process, depending on the time of the surgical intervention.[1] Furthermore, it was shown that injured brain does not contain other factors important in the genesis of edema, since lesioned area isolated from its capillary bed did not give rise to edema.[1] A role in induction of cerebral vasogenic edema for fatty acid cyclo-oxygenase products of arachidonic acid metabolism and for biogenic amines, both known to be increased in injured brain, had been previously ruled out. Thus pretreatment with indomethacin, a known inhibitor of prostaglandin synthetase in brain,[87,88] was shown to have no effect on the amount of fluid accumulating in response to a standardized freezing lesion.[68] Similarly, manipulations of serotonin levels in brain with reserpine and tryptophan loading were equally without effect.[26] Contradictory results have been reported regarding the role of lipoxygenase products of arachidonic acid metabolism in BBB opening and edema formation.[7,57,83] Edema-inducing properties of various other agents have also been either postulated or demonstrated.[32,40] To date, none of these has been unequivocally shown to be involved in mechanisms of breakdown of BBB and formation of edema resulting from trauma. Nevertheless, evidence is accumulating for a possible mediator role of the kallikrein–kinin system in vasogenic edema.[3,84] In contrast, while oxygen-derived free radicals generated from exogenous xanthine oxidase/hypoxanthine/ADP–Fe³⁺ system have been shown to induce brain injury and edema,[14,15] involvement of endogenously formed free radicals has not been demonstrated.

2.1.2. Extravasation and Bulk Flow of Edema Fluid

Increased capillary hydraulic conductance resulting from disruption of BBB leads to extravasation of plasma-derived fluid, while tissue hydrostatic and oncotic pressure is the force that drives the extravasated fluid through the tissue.[27,28,75] Substances of diverse molecular weights, ranging from sucrose to protein molecules, were shown to spread from the site of injury at the same speed, indicating bulk movement of edema fluid.[8,11,48,76] Increased brain tissue pressure was demonstrated in the vicinity of the lesion and a decrease in this pressure with distance from the lesion (Fig. 2).[52,76] This provided evidence that tissue pressure gradients existed that could account for the bulk movement of fluid within the edematous brain from the area of the lesions to the ventricles. That this was an important route of removal of the extravasated fluid was shown by recovery of labeled plasma proteins and other markers in the CSF when the edema front reached the ventricles.[11,54,76,77]

2.1.3. Factors Determining Spread of Edema Fluid

Several factors determine the extent to which vasogenic edema spreads from the area of the lesion. Obviously, the actual surface area of barrier damage is the first of these. The larger the area affected, the greater will be the volume of extravasated fluid.[66] Arterial pressure is the hydrostatic pressure that drives the fluid into the tissue; the higher it is, the greater the speed and extent of edema formation. This was shown clearly when in animals made hypotensive by use of appropriate drugs, the rate of spread and the extent of edema were considerably diminished, while drug-induced hypertension caused both to increase.[18,48] Furthermore, delivery of edema fluid into CSF was shown to be dependent on pressure gradients between the edematous tissue and the CSF. Increasing the gradient enhanced the clearance of edema fluid, while decreasing it had the opposite effect.[77]

2.1.4. Distribution of Edema Fluid

The bulk of the edema associated with experimental brain injury is confined to the white matter extracellular space of the lesioned hemisphere. This has been demonstrated

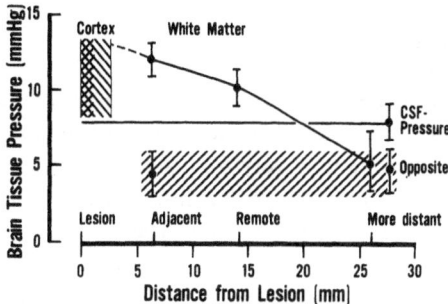

Figure 2. Distribution of interstitial fluid pressure and cerebrospinal fluid pressure in white matter of the injured and opposite hemisphere 6 hr after cold injury. (From Reulen et al.[76])

Table I. Edematous Changes Induced in Cat Brain by Cortical Freezing Lesion[a,b]

	Difference in weight[c] (right vs. left) (g)	Dry weight[d]	Na[e] per fw	Na[e] per dw	K[e] per fw	K[e] per dw
		White matter				
Normal (9, 14)	−0.01 ± 0.16	32.4 ± 1.5	57 ± 7	173 ± 20	83 ± 6	256 ± 11
After freezing lesion						
3 hr (8)	0.20 ± 0.17	—	—	—	—	—
7 hr (10)	0.28 ± 0.08	—	—	—	—	—
24 hr						
Left (11)		31.9 ± 1.7	60 ± 6	189 ± 26	80 ± 3	251 ± 27
Right[c] (29, 11)	0.65 ± 0.15[f]	21.4 ± 2.4[f]	90 ± 8[f]	421 ± 80[f]	57 ± 7[f]	266 ± 31
48 hr right (40, 11)	0.72 ± 0.20[f]	24.2 ± 2.1[f]	77 ± 7[f]	318 ± 33[f]	66 ± 10[f]	273 ± 48
72 hr right (7, 6)	0.54 ± 0.20[f]	25.0 ± 2.2[f]	75 ± 12[f]	304 ± 71[f]	69 ± 8[f]	274 ± 16
		Cerebral cortex				
Normal (9, 14)		18.8 ± 0.6	64 ± 6	340 ± 20	96 ± 7	511 ± 31
After freezing lesion						
24 hr						
Left (11)		19.3 ± 1.3	66 ± 7	342 ± 52	98 ± 7	508 ± 31
Right[c] (29, 11)		18.3 ± 1.0	74 ± 11[g]	404 ± 54[f]	89 ± 9[g]	486 ± 26[g]
48 hr right (40, 11)		19.7 ± 0.9	65 ± 4	329 ± 44	94 ± 7	477 ± 29[f]
72 hr right (7, 6)		19.1 ± 0.5	67 ± 8	350 ± 35	94 ± 7	492 ± 50

[a]Data from Pappius and Gulati,[66] Pappius and McCann,[67] and unpublished observations.
[b]Averages ±SD. Numbers in parentheses indicate, respectively, the number of animals in the total edema group and, where appropriate, in the water and electrolyte content group.
[c]The lesion was always made on right hemisphere. Average weight of normal hemisphere: 10 g.
[d]Dry weight in milligrams per 100 mg fresh weight.
[e]Na and K in milliequivalents per kilogram fresh weight (fw) or dry weight (dw).
[f]Significantly different from normal $p < 0.01$.
[g]$p < 0.05$.

in both morphologic and chemical studies (Table I).[4,9,34,39,43,51,66,67,74] The difference between white matter and cortex has generally been ascribed to differences in tissue architecture. The parallel arrangement of fibers in white matter presumably allows opening up of potential extracellular spaces, while interwoven cell processes in the cortex present considerable mechanical resistance to expansion of intercellular volume and accumulation of the extravasated fluid.

2.2. Equilibrium Phase

The next stage of vasogenic edema is reached when the amount of fluid extravasated in the area of lesion is balanced by the clearance of water from the area of edema.[11,75] The precise time at which this occurs may vary from study to study[11,18,45,67] and depends on several factors, among them the size of the lesion and the level of systemic blood pressure. At that point the BBB remains open. The extracellular spaces have become enlarged and, as a result, pressure gradients within the tissue are minimized.[52,54] There is still flow of fluid into the edematous areas, but this is matched by clearance of edema fluid, so there is no further spread of edema. The clearance into the CSF is diminished as tissue pressure gradients dissipate,[53] and this stage occurs by diffusion rather than by bulk flow of the edema fluid.[11] Evidence is now accumulating that there is transfer of ions and molecules across intact capillaries from the edematous tissue into the blood[11] although total clearance by brain vasculature appears to be quantitatively relatively small.[52] A strong case has been made for protein clearance from the extracellular space by the astrocytes as a basis for redistribution of water according to the Starling hypothesis.[46,75] Removal of large molecules across the capillary wall by pinocytosis has also been proposed as one mechanism involved in removal of edema fluid[11,85] but the significance of this process remains a subject of controversy.[12]

2.3. Resolution of Vasogenic Edema

Finally, if BBB is repaired, whether by regeneration of normally impermeable microvessels in the area of injury[13] or excision of abnormally permeable capillaries in the tumor, there is no further extravasation of fluid. Clearance of edema fluid continues until all the edema is resolved, presumably by the various mechanisms operating during the equilibrium phase. These may include transport processes for ions and small molecules, protein clearance by astrocytes, redistribution of water in accordance with the Starling hypothesis, and possible pinocytosis of large molecules.

2.4. Chemical Characteristics

The determination of chemical composition of edematous brain has been instrumental in furthering our understanding of the different types of edema. In the first place, measurement of water content can definitively demonstrate the presence of edema. Chemical analysis of the swollen tissue has provided information about the nature of the edema fluid, such as its electrolyte and protein content. This is vital to establishing the source of fluid taken up by the tissue, e.g., whether it is derived from plasma, is a plasma ultrafiltrate, or is not related to plasma at all. Attempts at isolating and analyzing edema fluid itself have also been made.[19,31]

2.4.1. Methods to Measure Edema

Measurement of percentage dry weight, which is the reciprocal of percentage water content, or determination of specific gravity are the simplest and most generally accepted methods for establishing whether brain-tissue sample is edematous and to what degree.[10,55,66] Sample dry weight data can be converted to sample volume changes (percentage swelling) by the use of an appropriate formula[25] but for any degree of accuracy this requires knowledge of the dry weight of the edema fluid.[16,59] Such measurements do not give any information, however, as to the total amount (volume) of edema formed in response to a lesion, which can only be determined from increases in total fresh weight or total water content of the whole brain or of a hemisphere.[66,67] Unfortunately, this most direct measure of total edema is not very sensitive, especially when the lesion and the resulting edematous zone are relatively small and localized.

2.4.2. Total Edema and Changes in Water and Electrolyte Content

The degree and time course of edematous changes resulting from injury depend on a number of factors and will vary with experimental conditions. The results summarized in Table I (see also Fig. 3) represent the changes characteristic of vasogenic edema in general, but their extent and evolution are specifically the consequence of the standardized freezing lesion used.

Total edema, estimated as the difference in weight between the intact and the lesioned hemisphere, was measurable within 3 hr after the lesion and continued to increase for 48 hr, with a decrease noted at 72 hr. At its peak the edema was equivalent to a 7–8% increase in the volume of the traumatized hemisphere, while at that point the increase in volume of the edematous white matter, calculated from the dry weight data using the Elliott–Jasper formula amounted to between 30–40%.[59]

The edematous changes in water and electrolyte content were essentially restricted to the white matter of the lesioned hemisphere, where throughout the experimental period of 3 days a sharp decrease in percentage dry weight, or increase in water content, was accompanied by a rise in sodium and a fall in potassium per unit fresh weight. The increase in white matter sodium content represented a net gain, while the decrease in potassium reflected a dilution of normal tissue content, as could be seen when the results were calculated in terms of dry weight. The observed changes were compatible with the uptake of high-sodium, low-potassium fluid, similar in electrolyte content to plasma.

The marginal decrease in the dry weight of cerebral cortex at 24 hr and the accompanying small but statistically significant changes in electrolytes, which are not always demonstrable,[58,67] may reflect sampling of necrotic tissue in the lesion area.

2.4.3. Protein Content of Edema Fluid

Presence of serum proteins in the extravasated fluid was demonstrated in earliest studies of vasogenic edema.[45] Initial efforts at quantitating the amount of protein in the edematous brain tissue involved systemic application of exogenous tracers, such as radio-iodinated [^{131}I]human serum albumin.[16,67] More recently, an immunochemical method was developed that permits detection of endogenous serum proteins in the affected tissue.[9]

Regardless of the method used, the serum protein extravasation into brain tissue was partly dissociated from changes in water content.[9,16,67] At the height of edema it was less than would be expected on the basis of increase in water (see Fig. 3). This could be due to

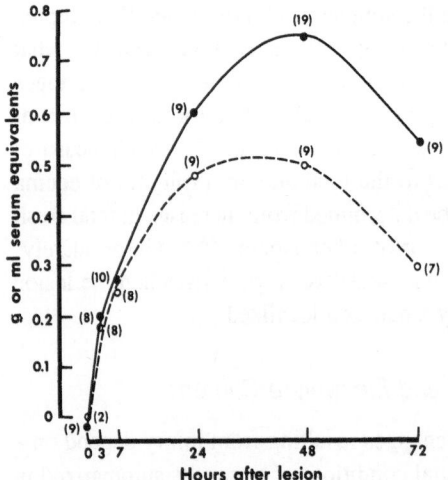

Figure 3. Difference in weight (closed circles), as a measure of total edema and in radiolabeled iodinated serum albumin (RISA) content in serum equivalents (open circles) between right and left hemisphere of cat brain after a freezing lesion on the right side. Number of animals is shown in brackets. The lines would be superimposed if protein content of edema fluid was the same as in serum. (Modified from Pappius and McCann.[67])

metabolic removal of the protein[67] or some intracellular swelling involving water only[9,16] or a combination of the two processes. Whatever the underlying cause, the discrepancy invalidates the use of protein extravasation as a quantitative method of estimation of accumulated edema fluid.

3. CHARACTERISTICS AND REVERSIBILITY OF ISCHEMIC EDEMA

The pathophysiology of ischemic brain swelling is complex, involving intercompartmental fluid shifts, net changes in brain water and electrolytes, and delayed changes in BBB permeability. Different processes are operative during the ischemic and the postischemic periods. Experimental work on the subject has been recently reviewed in depth.[38]

3.1. Ischemic Changes

When considering the consequences of ischemia, it is important to remember that both flow and time thresholds exist for disturbances associated with diminished brain circulation. Cerebral blood flow must be considerably decreased before any effects are seen.[2,6,78] Movement of water and sodium into the cellular compartments and changes in electrical function occur first, followed by loss of potassium from the cells and uptake of calcium, and eventually cell death.

Once the threshold for edema has been reached, further decrease in blood flow correlates quite well with decreases in the specific gravity of the tissue, another way of determining water content.[20] However, when ischemia is complete, and there is no blood flow, edema does not develop, as under such conditions no extra fluid can enter the ischemic tissue.[20,36,38] At the same time, it has been clearly demonstrated that shifts of ECF into the cellular compartments occur very rapidly under these conditions, indicating that severe disturbances in ion and water homeostasis have occurred.[36,37,78]

3.2. Postischemic Edema

Devastating edema can develop on recirculation of previously completely ischemic areas; this is referred to as postischemic brain edema.[36,38] Actually, similar changes are associated with prolonged, incomplete ischemia, i.e., when below threshold blood flows persist in affected areas.[37]

During complete circulatory arrest, there is a severe disturbance of water and ion homeostasis and a shift of fluid into intracellular spaces, but no edema develops. Immediately upon resumption of circulation, although the BBB remains intact, the disturbances in ionic homeostasis persist, and now edema develops. There is net influx of fluid into the tissue from the vascular compartment.[36,38]

Subsequent events depend on the duration of ischemia and the quality of reoxygenation achieved during the postischemic period. The longer the period of ischemia, the less likely the reversal of changes that accompanied it, although we now know that the critical period is not as short as previously assumed. The other key factor is the extent to which blood flow is resumed. In the absence of postischemic flow disturbances, when recirculation occurs within a time limit compatible with recovery of metabolic activity, ion homeostasis is normalized and edema is resolved in a matter of hours.[37]

3.3. Delayed Breakdown of Blood–Brain Barrier

When either the ischemic period is prolonged or postischemic recirculation remains deficient, there is progression of edema even during the postischemic period. The changes that occur are similar to those seen with below-threshold, but incomplete, ischemia. ICP increases and can be fatal. With time, BBB breakdown occurs and irreversible injury develops. These delayed changes have been called the maturation phenomenon by Klatzo and his collaborators,[30,41] according to which the time course of the development of various parameters of ischemic injury, including BBB damage, is inversely related to the intensity and/or duration of the ischemic insult. The mechanisms involved have not been elucidated.

Thus, the edematous changes associated with ischemia are quite complex, consisting initially of shifts of water and sodium into the intracellular compartment in the cortex, while BBB remains intact. These changes are reversible as long as ischemia is not prolonged and recirculation complete. Prolonged ischemia and poor recirculation lead to permanent injury, breakdown of BBB, and uncontrolled development of vasogenic edema (see Chapter 26, Vol. 2).

4. FUNCTIONAL DISTURBANCES IN INJURED BRAIN MAY NOT BE RELATED TO CEREBRAL EDEMA

It is well established that brain injury causing gross damage to vascular elements results in opening of the BBB and an extravasation of fluid, giving rise to vasogenic edema. Nevertheless, the general assumption that this in turn leads to functional disturbances has not been validated, although it has been questioned.[66,72,82] Brain injury is associated with other events, such as release of arachidonic acid from membrane phospholipids, release of neurotransmitters, and formation of prostaglandins, thromboxanes, and other eicosanoids, processes that can be envisaged as leading to disturbances of neuronal function, independent of development of cerebral edema.[72] These relationships are shown schematically in Fig. 4. The following discussion of the mechanisms by which injury to the brain may cause functional neurologic disturbances is pertinent to the consideration of cerebral edema. In most instances, involvement of edema as the major cause of functional abnormalities is tacitly assumed, and other possible causes of neurologic malfunction are rarely taken into account.

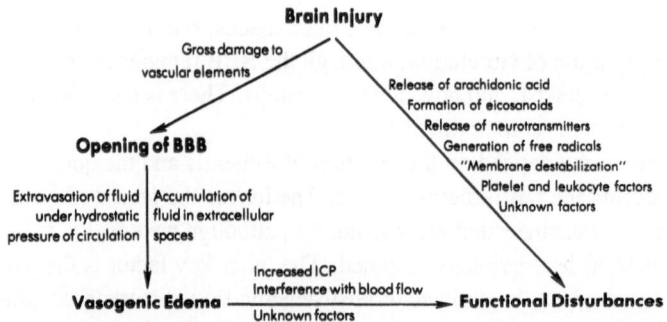

Figure 4. Schematic representation of the processes that may be involved in development of functional disturbances resulting from brain injury.

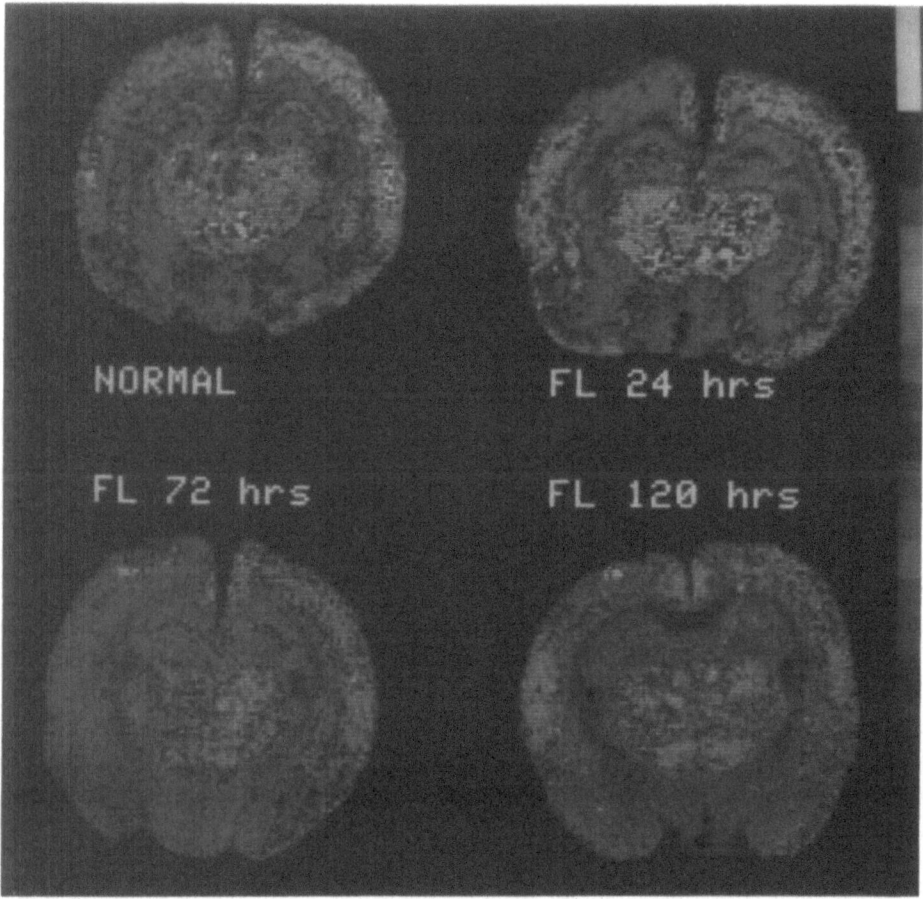

Figure 5. Cerebral glucograms prepared from selected [¹⁴C]deoxyglucose autoradiographs. LCGU in micro-moles/100 g per min. Sections through the lesion area are shown from a normal animal and representative animals at 24, 72, and 120 hr after a freezing lesion. A color reproduction of this figure appears following p. xxviii. (Data from Pappius.[62])

Until the late 1970s, lack of a good method for assessing cerebral function in animals was a major obstacle to the study of mechanisms underlying functional disturbances in traumatized brain.[61] This difficulty was overcome by the development of the deoxy-glucose technique for measurement of local cerebral glucose utilization[80] and by valida-tion of the use of this method for mapping of cerebral functional activity in awake animals.[79]

Using this approach, it was shown that with time after a focal freezing lesion, a widespread depression of local cerebral glucose utilization (LCGU) developed in cortical areas of the lesioned hemisphere.[62] This is clearly seen in Fig. 5, in which glucograms from corresponding sections of normal and traumatized brain can be directly compared. In the normal animal, heterogeneous rates of glucose utilization were seen throughout, but no side-to-side differences were present. At 24 hr after a superficial freezing lesion in the parietal cortex (upper left of each section), LCGU in the cortical areas of the traumatized hemisphere was decreased and became obviously depressed 72 hr after the lesion. At 120 hr there was a return toward normal. The depression of LCGU was not restricted to regions surrounding the lesion but involved the cortical areas of the whole hemisphere where at its peak at 3 days postlesion, the glucose metabolism was only 50% of normal.

The changes in glucose utilization did not appear to be mediated by cerebral edema, as their spatial distribution and time course were different from those demonstrated previously for edema. Since blood flow was not affected,[62] there was no evidence of a direct effect on cerebral metabolism. In keeping with the hypothesis that the functional state of cerebral tissues is closely coupled to their metabolism, the results were interpreted as representing a manifestation of cerebral dysfunction independent of the edema process.[62] This interpretation implies that the energy needs of cerebral cortex in injured brain are diminished as a result of a functional depression. Elucidation of the mechanisms involved in the postulated functional depression has been the goal of subsequent investigations.

Further studies showed that the post-traumatic depression of cortical glucose utilization could be modified by dexamethasone,[63] nonsteroidal anti-inflammatory drugs (NSAIDs),[69] and drugs inhibiting biogenic amine synthesis.[71] NSAIDs were previously shown to be without effect on vasogenic edema[68]; manipulation of the monoamines was similarly ineffective.[26] The discrepancy between the effects of the various drugs on the decrease of cortical metabolism and on edema is further evidence that the two processes are unrelated consequences of brain injury.

The effects of indomethacin and dexamethasone on arachidonic acid and prostaglandin (PG) content in injured brain were also determined.[70] The arachidonic acid content of the cortical lesion area increased sharply within 60 sec of injury, but this release was not affected by either dexamethasone or indomethacin treatment. The accumulation of $PGF_{2\alpha}$, the only PG measured in the two treated groups, was also not affected by dexamethasone treatment but, as might be expected, was more than 90% inhibited in the indomethacin-treated rats. Thus, dexamethasone in the dose that modified cortical glucose utilization in injured rat brain did not affect the immediate arachidonic acid release from phospholipids in the lesion area. Its effects in traumatized brain must be mediated independently of the PG cascade. On the other hand, the effects of indomethacin on LCGU and on the PG content in injured brain implicate these eicosanoids in the chain of events leading from injury to functional disturbances.

Amelioration of the depression of cortical glucose utilization in injured brain by inhibition of serotonin and catecholamine synthesis[71] indicated that biogenic amine neu-

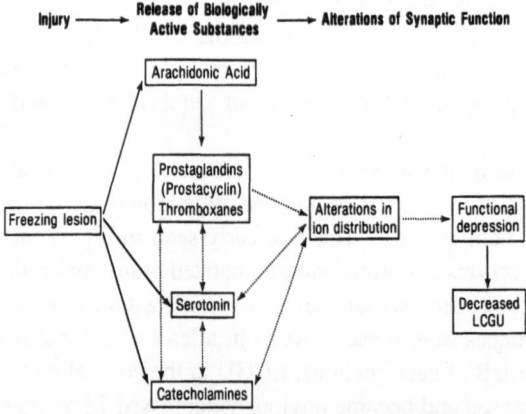

Figure 6. Schematic representation of the working hypothesis linking brain injury to alteration in function.

rotransmitter systems were involved in the underlying mechanisms. Accordingly, serotonin, dopamine, and their metabolites, as well as norepinephrine, were determined in cerebral cortex between 4 hr and 10 days after lesion.[64] The results suggested that in focally injured rat brain, the serotonergic system was activated unilaterally (ipsilaterally to the lesion), and norepinephrine levels decreased bilaterally, while dopamine and its metabolites were unaffected. Thus, the changes in these transmitter systems were complex; further work will be required for the elucidation of their contribution to the functional disturbances resulting from injury.

On the basis of these and other results, a working hypothesis was developed according to which the widespread functional changes, as reflected by the observed metabolic depression in focally traumatized brain, are mediated through neurotransmitter system or systems and alterations in ion distribution (Fig. 6).[72] Prostaglandins (and dexamethasone) could then act by modifying the response of these systems to injury or directly on the membrane and its ionic channels. Furthermore, changes in the neurotransmitters could be interrelated and in turn influence the action of the eicosanoids. It would thus not be necessary to invoke an effect on cerebral edema to account for the action of a variety of drugs in traumatized brain.

5. CONCLUSION

Cerebral edema, the major consequence of breakdown of BBB, can have devastating clinical consequences if the increase in ICP is not controlled. Over the years, this has provided an impetus for studies on the pathophysiology of the edematous process, in turn leading to increased understanding of the mechanisms involved. The emphasis on edema, however, may have obscured the importance of other factors that might contribute to functional abnormalities associated with various types of injury to the brain. It may be significant that despite great strides made over the years in elucidating the mechanisms of formation and resolution of vasogenic edema, no corresponding striking advances in therapy have occurred. Osmotic agents and steroids, the latter on empiric grounds, remain the backbone of measures at the disposal of the clinician.[50] Interestingly, in the case of vasogenic edema, osmotherapy results in dehydration of normal tissues without significant changes in water content of the edematous white matter,[17,65] while the effects of steroids on cerebral edema continue to be a subject of controversy.[42,63] If cerebral injury leads to functional disturbances independent of the consequences of cerebral edema, elucidation of the mechanisms underlying these disturbances would provide a solid basis for the development of new rational modes of therapy for conditions in which cerebral edema has been considered as the major cause of neurologic complications.

6. SUMMARY

As a consequence of a variety of experimental studies several types of cerebral edema have been delineated, two of which are of particular clinical interest. Vasogenic edema develops when the normal permeability characteristics of cerebral microvessels are altered by injury or when abnormally permeable vessels proliferate in tumor tissue. Three phases of vasogenic edema can be distinguished: (1) formation and spread from the area of injury, (2) a period of equilibrium when the exudation of fluid into the tissue is balanced by its removal from the affected areas, and (3) resolution when no more fluid is extravasated while the process of clearance of the edema fluid continues. In vasogenic edema, the extravasated fluid is derived from the plasma and contains serum proteins. The fluid accumulates primarily in the white matter.

Ischemic and postischemic edematous changes are a result of complex processes, including shifts of fluid between compartments, net changes in water and electrolyte content, and delayed increase in blood–brain barrier (BBB) permeability. In contrast to vasogenic edema, initially only cortical tissue is involved. With time, tissue damage also affects the microvessels and, with breakdown of normal permeability characteristics, vasogenic edema ensues.

Functional disturbances in injured brain, usually ascribed to presence of cerebral edema, may in fact be due to alterations in a variety of processes. These include the arachidonic acid cascade and several neurotransmitter systems that are affected by trauma, quite independent of the development of cerebral edema.

ACKNOWLEDGMENTS

The author's experimental work has been supported over the years by grants from the Medical Research Council of Canada and by the Canadian Donner Foundation. The assistance of Linda Michel in preparation of this manuscript is gratefully acknowledged.

REFERENCES

1. Arabi B, Long DM: Dynamics of cerebral edema. The role of an intact vascular bed in the production and propagation of vasogenic brain edema. *J Neurosurg* 51:779–784, 1979.
2. Astrup J, Siesjo BK, Symon L: Thresholds in cerebral ischemia—the ischemic penumbra. *Stroke* 12:723–725, 1981.
3. Baethmann A, Oettinger W, Rothenfusser W, et al: Brain edema factors: Current state with particular reference to plasma constituents and glutamate. *Adv Neurol* 28:171–195, 1980.
4. Bakay L, Haque IU: Morphological and chemical studies in cerebral edema. *J Neuropathol Exp Neurol* 23:393–418, 1964.
5. Baker RN, Cancilla PA, Pollock PS, et al: The movement of exogenous protein in experimental cerebral edema. *J Neuropathol Exp Neurol* 30:668–679, 1971.
6. Bell BA, Symon L, Branston NM. CBF and time thresholds for the formation of ischemic cerebral edema, and effect of reperfusion in baboons. *J Neurosurg* 62:31–41, 1985.
7. Black KL, Hoff JT: Leukotrienes increase blood–brain barrier permeability following intraparenchymal injections in rats. *Ann Neurol* 18:349–351, 1985.
8. Blasberg RG, Gazedam J, Patlak CS, et al: In Cervos-Navarro J, Ferszt R (eds): *Advances in Neurology.* Vol. 28: *Brain Edema.* Raven, New York, 1980, pp. 255–270.
9. Bodsch W, Hurter T, Hossmann K-A: Immunochemical method for quantitative evaluation of vasogenic brain edema following cold injury in rat brain. *Brain Res* 249:111–121, 1982.
10. Bothe HW, Bodsch W, Hossmann KA. Relationship between specific gravity, water content, and serum protein extravasation in various types of vasogenic brain edema. *Acta Neuropathol (Berl)* 64:37–42, 1984.
11. Bruce DA, Ter Weeme C, Kaiser G, et al: Mechanisms and time course for clearance of vasogenic cerebral edema. In Popp AJ, Bourke RS, Nelson LR (eds): *Seminars in Neurological Surgery.* Vol. 4: *Neural Trauma.* Raven, New York, 1979, pp. 155–172.
12. Bundgaard M: Vesicular transport in capillary endothelium: Does it occur? *Fed Proc* 42:2425–2430, 1983.
13. Cancilla PA, Frommes Sp, Kahn LE, et al: Regeneration of cerebral microvessels: A morphologic and histochemical study after local freeze-injury. *Lab Invest* 40:74–82, 1979.
14. Chan PH, Schmidley JW, Fishman RA, et al: Brain injury, edema, and vascular permeability changes induced by oxygen-derived free radicals. *Neurology (NY)* 34:315–320, 1984.

15. Chan PH, Longar S, Fishman RA: Oxygen-free radicals: Potential edema mediators in brain injury. In Inaba Y, Klatzo I, Spatz M (eds): *Brain Edema. Proceedings of the Sixth International Symposium.* Springer-Verlag, New York, 1985, pp. 317–323.

16. Clasen RA, Bezkorovainy A, Pandolfi S: Protein and electrolyte changes in experimental cerebral edema. *J Neuropathol Exp Neurol* 41:113–128, 1982.

17. Clasen RA, Cooke PM, Pandolfi S, et al: Hypertonic urea in experimental cerebral edema. *Arch Neurol* 12:424–434, 1965.

18. Clasen RA, Pandolfi S,Russel J, et al: Hypothermia and hypotension in experimental cerebral injury. *Arch Neurol* 19:472–486, 1968.

19. Clasen RA, Sky-Pack HH, Pandolfi S, et al: The chemistry of isolated edema fluid in experimental cerebral injury. In Klatzo I, Seitelberger F (eds): *Brain Edema.* Spinger-Verlag, New York, 1967, pp. 536–553.

20. Crockard A, Iannotti F, Hunstock AT, et al: Cerebral blood flow and edema following carotid occlusion in the gerbil. *Stroke* 11:494–498, 1980.

21. Deane BR, Greenwood J, Lantos PL, et al: The vasculature of experimental brain tumors. Part 4. The quantification of vascular permeability. *J Neurol Sci* 65:59–68, 1984.

22. Deane BR, Papp MI, Lantos PL: The vasculature of experimental brain tumors. Part 3. Permeability studies. *J Neurol Sci* 65:47–58, 1984.

23. Dila CJ, Pappius HM: Cerebral water and electrolytes. An experimental model of inappropriate secretion of antidiuretic hormone. *Arch Neurol* 26:85–90, 1972.

24. Dila CJ, Ford RM, Tatlow WFT: The neurological aspects of the inappropriate ADH syndrome. *An R Coll Physicians Surg Can* 3:40, 1970.

25. Elliott KAC, Jasper H: Measurement of experimentally induced brain swelling and shrinkage. *Am J Physiol* 157:122–129, 1949.

26. Fenske A, Sinterhauf K, Reulen H: The role of monoamines in development of cold-induced edema. In Pappius HM, Feindel W (eds): *Dynamics of Brain Edema.* Springer-Verlag, New York, 1976, pp. 150–154.

27. Fenstermacher JD: Volume regulation of the central nervous system. In Staub NC, Taylor AE (eds): *Edema.* Raven, New York, 1984, pp. 383–404.

28. Fenstermacher JD, Patlak CS: The movement of water and solutes in the brains of mammals. In Pappius HM, Feindel W (eds): *Dynamics of Brain Edema.* Springer-Verlag, New York, 1976, pp. 87–94.

29. Fishman RA: Brain edema. *N Engl J Med* 293:706–711, 1975.

30. Fujimoto T, Walker JT Jr, Spatz M, et al: Pathophysiologic aspects of ischemic edema. In Pappius HM, Feindel W (eds): *Dynamics of Brain Edema.* Springer-Verlag, New York, 1976, pp. 171–180.

31. Gazendam KM, Go KG, Van Zanten AK: Composition of isolated edema fluid in cold-induced brain edema. *J Neurosurg* 51:70–77, 1979.

32. Go KG, Baethmann A (eds): *Recent Progress in the Study and Therapy of Brain Edema.* Plenum, New York, 1984.

33. Herzog I, Levy WA, Scheinberg LC: Biochemical and morphological studies of cerebral edema associated with intracerebral tumors in rabbits. *J Neuropathol Exp Neurol* 24:244–255, 1965.

34. Hirano A: The fine structure of brain in edema. In Bourne GH (ed): *The Structure and Function of Nervous Tissue.* Vol. 2. Academic, New York, 1969, pp. 69–135.

35. Hirano A, Becker NH, Zimmerman HM: Pathological alterations in the cerebral endothelial cell barrier to peroxidase. *Arch Neurol* 20:300–308, 1969.

36. Hossmann K-A: Development and resolution of ischemic brain swelling. In Pappius HM, Feindel W (eds): *Dynamics of Brain Edema.* Springer-Verlag, New York, 1976, pp. 219–227.

37. Hossman K-A: Treatment of experimental cerebral ischemia. *J Cereb Blood Flow Metab* 2:275–297, 1982.

38. Hossman K-A: The pathophysiology of ischemic brain swelling. In Inaba Y, Klatzo I, Spatz M (eds): *Brain Edema, Proceedings of the Sixth International Symposium.* Springer-Verlag, New York, 1985, pp. 367–384.

39. Hossmann K-A, Wechsler W, Wilmes F: Experimental peritumorous edema. Morphological and pathophysiological observations. *Acta Neuropathol (Berl)* 45:195–203, 1979.

40. Inaba Y, Klatzo I, Spatz M (eds): *Brain Edema, Proceedings of the Sixth International Symposium.* Springer-Verlag, New York, 1985.

41. Ito U, Go KG, Walker JT Jr, et al: Experimental cerebral ischemia in mongolian gerbils. III. Behaviour of the blood–brain barrier. *Acta Neuropathol (Berl)* 34:1–6, 1976.

42. Katzman R, Pappius HM: *Brain Electrolytes and Fluid Metabolism.* Williams & Wilkins, Baltimore, 1973.

43. Katzman R, Gonatas N, Levine S: Electrolytes and fluids in experimental focal leukoencephalopathy. *Arch Neurol* 10:58–65, 1964.

44. Katzman R, Clasen RA, Klatzo I, et al: Report of the Joint Committee for Stroke Facilities; XV. Brain edema in stroke. Study group of brain edema in stroke. *Stroke* 8:509–540, 1977.
45. Klatzo I: Presidential address. Neuropathological aspects of brain edema. *J Neuropathol Exp Neurol* 26:1–14, 1967.
46. Klatzo I, Chui E, Fujiwara K, et al: Resolution of vasogenic brain edema. In Cervos Navarro J, Ferszt R (eds): *Advances in Neurology*. Vol 28: *Brain Edema*. Raven, New York, 1980, pp. 359–373.
47. Klatzo I, Wisniewski H, Smith DE: Observations on penetration of serum proteins into the cerebral nervous system. *Prog Brain Res* 15:73–88, 1965.
48. Klatzo I, Wisniewski H, Steinwall O, et al: Dynamics of cold injury edema. In Klatzo I, Seitelberger F (eds): *Brain Edema, Proceedings of the Symposium*. Springer-Verlag, New York, 1967, pp. 554–563.
49. Long DM: Microvascular changes in cold injury edema. In Go KG, Baethmann A (eds): *Recent Progress in the Study and Therapy of Brain Edema*. Plenum, New York, 1984, pp. 45–54.
50. Long DM: New therapies for brain edema. In Inaba Y, Klatzo I, Spatz M (eds): *Brain Edema, Proceedings of the Sixth International Symposium*. Springer-Verlag, New York, 1985, pp. 565–577.
51. Long DM, Hartmann JF, French LA: The response of experimental cerebral edema to glucocorticoid administration. *J Neurosurg* 24:843–854, 1966.
52. Marmarou A, Shulman K, Poll W: The time course of brain tissue pressure and local CBF in vasogenic edema. In Pappius HM, Feindel W (eds): *Dynamics of Brain Edema*, Springer-Verlag, New York, 1976, pp. 113–121.
53. Marmarou A, Takagi H, Shulman K: Biomechanics of brain edema and effects on local cerebral blood flow. In Cervos-Nararro J, Ferszt R (eds): *Advances in Neurology*. Vol. 28: *Brain Edema*. Raven, New York, 1980, pp. 345–358.
54. Marmarou A, Nakamura T, Tanaka K, et al: The time course and distribution of water in the resolution phase of infusion edema. In Go KG, Baethmann A (eds): *Recent Progress in the Study and Therapy of Brain Edema*. Plenum, New York, 1984, pp. 37–44.
55. Nelson SR, Mantz M-L, Maxwell JA: Use of specific gravity in the measurement of cerebral edema. *J Appl Physiol* 30:268–271, 1971.
56. Norris JW, Pappius HM: Cerebral water and electrolytes. Effect of asphyxia, hypoxia, and hypercapnia. *Arch Neurol* 23:248–258, 1970.
57. Papadopoulos SM, Black KL, Hoff JT: The role of leukocytes and lipoxygenase inhibition on experimental vasogenic edema. American Association of Neurology Surgeons, Denver meeting, Poster No. 60, 1986.
58. Pappius HM: Biochemical studies on experimental brain edema. In Klatzo I, Seitelberger F (eds): *Brain Edema, Proceedings of the Symposium*. Springer-Verlag, New York, 1967, pp. 445–460.
59. Pappius HM: Fundamental aspects of brain edema. In Vinken PJ and Bruyn GW (eds): *Handbook of Clinical Neurology*, Vol. 16. North Holland, Amsterdam, 1974, pp. 167–185.
60. Pappius HM. Normal and pathological distribution of water in brain. In Cserr HF, Fenstermacher JD, Fencl V (eds): *Fluid Environment of the Brain*. Academic, New York, 1975, pp. 183–199.
61. Pappius HM: Mapping of cerebral functional activity with radioactive deoxyglucose: Application in studies of traumatized brain. In Cervos-Navarro J, Ferszt R (eds): *Advances in Neurology*. Vol. 28: *Brain Edema*. Raven, New York, 1980, 271–279.
62. Pappius HM: Local cerebral glucose utilization in thermally traumatized rat brain. *Ann Neurol* 9:484–491, 1981.
63. Pappius HM: Dexamethasone and local cerebral glucose utilization in freeze-traumatized rat brain. *Ann Neurol* 12:157–162, 1982.
64. Pappius HM, Dadoun R: Biogenic amines in injured brain. *Trans Am Soc Neurochem* 17:298, 1986.
65. Pappius HM, Dayes LA: Hypertonicurea. Its effect on distribution of water and electrolytes in normal and edematous brain tissues. *Arch Neurol* 13:395–402, 1965.
66. Pappius HM, Gulati DR: Water and electrolyte content of cerebral tissues in experimentally induced edema. *Acta Neuropath (Berl)* 2:451–460, 1963.
67. Pappius HM, McCann WP: Effects of steroids on cerebral edema in cats. *Arch Neurol* 20:207–216, 1969.
68. Pappius HM, Wolfe LS: Some further studies on vasogenic edema. In Pappius HM, Feindel W (eds): *Dynamics of Brain Edema*. Springer-Verlag, New York, 1976, pp. 138–143.
69. Pappius HM, Wolfe LS: The effects of indomethacin and ibuprofen on cerebral metabolism and blood flow in traumatized brain. *J Cereb Blood Flow Metab* 3:448–459, 1983.
70. Pappius HM, Wolfe LS: Functional disturbances in brain following injury: Search for underlying mechanisms. *Neurochem Res* 8:63–72, 1983.
71. Pappius HM, Wolfe LS: Involvement of serotonin and catecholamines in functional depression of traumatized brain. *J Cereb Blood Flow Metab* 3(Suppl.1):S226–S227, 1983.

72. Pappius HM, Wolfe LS: Effects of drugs on local cerebral glucose utilization in traumatized brain: Mechanisms of action of steroids revisited. In Go KG, Baethmann, A (eds): *Recent Progress in the Study and Therapy of Brain Edema*. Plenum, New York, 1984, pp. 11–16.

73. Pappius HM, Oh JH, Dossetor JB: The effect of rapid hemodialysis on brain tissues and cerebrospinal fluid of dogs. *Can J Physiol Pharmacol* 45:129–147, 1967.

74. Raimondi AJ, Evand JP, Mullan S: Studies of cerebral edema. III. Alterations in the white matter; an electron microscopic study using ferritin as a labelling compound. *Acta Neuropath (Berl)* 2:177–197, 1962.

75. Rapoport SI: Roles of cerebrovascular permeability, brain compliance and brain hydraulic conductivity in vasogenic brain edema. In Popp AJ, Bourke RS, et al. (eds): *Seminars in Neurological Surgery*. Vol. 4: *Neural Trauma*. Raven, New York, 1979, pp. 51–61.

76. Reulen HJ, Graham R, Spatz M, et al: Role of pressure gradients and bulk flow in dynamics of vasogenic edema. *J Neurosurg* 46:24–35, 1977.

77. Reulen JH, Tsuyumu M, Tack A, et al: Clearance of edema fluid into cerebrospinal fluid. A mechanism for resolution of vasogenic brain edema. *J Neurosurg* 48:754–764, 1978.

78. Schuier FJ, Hossmann K-A: Experimental brain infarcts in cats. II. Ischemic brain edema. *Stroke* 11:593–601, 1980.

79. Sokoloff L: Relation between physiological function and energy metabolism in the central nervous system. *J Neurochem* 29:13–26, 1977.

80. Sokoloff L, Reivich M, Kennedy C, et al: The [^{14}C]-deoxyglucose method for measurement of local cerebral glucose utilization: Theory, procedure, and normal values in the conscious and anesthetized albino rat. *J Neurochem* 28:897–916, 1977.

81. Spatz M: Attenuated blood–brain barrier. In Lajtha A (ed): *Handbook of Neurochemistry*. Vol. 7. Plenum, New York, 1984, pp. 501–543.

82. Sutton LN, Bruce DA, Welsh FA, et al: Metabolic and electrophysiologic consequences of vasogenic edema. In Cervos-Nararro J, Ferszt R (eds): *Advances in Neurology*. Vol. 28: *Brain Edema*. Raven, New York, 1980, pp. 241–254.

83. Unterberg AW, Baethmann AJ, Wahl M: Effects of arachidonic acid, free radicals and leukotrienes on blood–brain barrier function and vasomotor reactivity. American Association Neurology Surgeons, Denver meeting, Poster No. 168, 1986.

84. Unterberg A, Dantermann C, Miller-Esterl W, et al: Inhibition of the kallikrein–kinin system in vasogenic brain edema. In Inaba Y, Klatzo I, Spatz M (eds): *Brain Edema, Proceedings of the Sixth International Symposium*. Springer-Verlag, New York, 1985, pp. 294–298.

85. Vorbrodt AW, Lossinsky AS, Wisniewski HM et al: Ultrastructural observations on the transvascular route of protein removal in vasogenic brain edema. *Acta Neuropath (Berl)* 66:265–273, 1985.

86. Waggener JD, Beggs JL: Vasculature of neural neoplasms. *Adv Neurol* 15:27–49, 1976.

87. Wolfe LS: Eicosanoids: Prostaglandins, thromboxanes, leukotrienes and other derivatives of carbon-20 unsaturated fatty acids. *J Neurochem* 38:1–14, 1982.

88. Wolfe LS, Rostoworowski K, Pappius HM: The endogenous biosynthesis of prostaglandins by brain tissue in vivo. *Can J Biochem* 54:629–640, 1976.

12

Drug Delivery to the Brain by Blood–Brain Barrier Circumvention and Drug Modification

Nigel H. Greig

1. INTRODUCTION

Recent advances in the fields of pharmacology and molecular biology have spurred the development of several interesting and highly effective therapeutic agents for the treatment of a wide variety of malignant, infectious, and genetic diseases. While many chemotherapeutic agents exhibit excellent activity in vitro, their therapeutic efficacy is often significantly diminished when administered to appropriate animal models, since they are unable to gain access to the diseased site at a sufficient concentration for an appropriate time. This is particularly common when the disease is sequestered within the CNS. As described by Neuwelt (Chapter 17, Vol. 2, Parts A and B) and Greig (Chapter 16, Vol. 2), this problem is commonly encountered by neurooncologists in the treatment of both primary and metastatic brain tumors. As a consequence, the prognosis of such patients remains extremely poor.[92] Possibly less extreme but of equal importance is the clinical observation that CNS drug-delivery problems are prevalent in the treatment of patients with acute cerebral bacterial and viral infections as well as with neurotransmitter and enzyme-deficiency diseases, such as Parkinson, Huntington, and Tay–Sachs disease.

This chapter summarizes the difficulties associated with delivering water-soluble drugs and, in particular, anticancer drugs to the brain. It discusses techniques designed (1) to optimize systemic pharmacokinetics, and (2) to circumvent the blood–brain barrier (BBB) and thereby deliver increased drug concentrations to the brain, maximizing their therapeutic effects. However, even a cursory knowledge of pharmacology confirms that many classes of drugs, prime examples of which are the anesthetics and antidepressants, freely enter the brain. The physicochemical factors that determine brain uptake are therefore outlined first.

Nigel H. Greig • Laboratory of Neurosciences, National Institute on Aging, National Institutes of Health, Bethesda, Maryland 20892.

2. FACTORS INFLUENCING DRUG CONCENTRATIONS IN BRAIN

Several factors co-determine the time course of a drug in the brain following its systemic administration: (1) the concentration versus time profile of the free unbound drug in the plasma; (2) the permeability of the compound at the BBB; (3) the rate of cerebral blood flow (of particular relevance to lipophilic compounds); and (4) the rates of metabolism and of binding of the compound within the brain.

Following the systemic administration of a drug, its plasma concentration reaches a peak and then declines as the compound is redistributed, metabolized, and eliminated. For some agents, the brain is an ideal organ for drug uptake, since it is highly perfused. However, despite a high rate of blood flow, the brain is different from other organs in that its blood vessels are lined by endothelial cells that are joined by continuous belts of tight intercellular junctions that block paracellular diffusion.[218] These endothelial cells contain no fenestrations and only a few cytoplasmic vesicles that do not appear to be involved in transport functions.[29] This BBB acts as a continuous cellular layer between blood and brain and restricts the brain entry of most nontransported hydrophilic and charged compounds. As a consequence, only compounds that have an appreciable solubility in the lipid component of the endothelial cell membranes are able to cross the barrier freely by diffusion and significantly enter the brain in significant amounts.

It is the physicochemical characteristics of a compound, related to its chemical structure, that determine both its pharmacologic action and overall pharmacokinetics. Simplistically, most drugs are organic molecules formed from carbon-containing backbones with different functional groups. The number and structural arrangement of the hydrocarbons and nature of the functional groups impart to the molecule its specificity of action (i.e., fit for a receptor) and its ability to penetrate tissues and cells. To discuss logically the absorption, distribution, and disappearance of a drug, one must first recognize the possible role of different functional groups and use this information to predict the pharmacokinetics and activity of newly developed agents.

2.1. Influence of Physicochemical Properties on Drug Organ Uptake

Several extensive reviews discuss the principles of drug design by using the analysis of quantitative structure–activity relationships (QSAR) the fundamentals of which were primarily laid by Hansch,[109–112] Hammett[108] and Taft.[243] This chapter makes no attempt to discuss these in detail. However, since these principles are fundamental to an understanding of how drugs enter the brain, and, in addition, form the basis of the rationale for either reversibly or irreversibly modifying drugs to augment their brain entry, an oversimplified view is essential. Hansch,[109-112] and Flynn and Yalkowsky[77,265] have provided a more in-depth analysis.

Probably the most important property of a drug is a balanced lipo/hydrophilicity, since transport within the body is via the hydrophilic phases of plasma and extracellular fluid, whereas passage into cells or across biologic barriers, such as those of the brain and gastrointestinal (GI) tract, is via lipid phases. Indeed, there is a linear relationship between the cerebrovascular permeability of a compound and its lipophilicity[151,205,206] (Fig. 1). Although a certain degree of lipid solubility is essential for compounds to transverse biologic membranes, agents that partition predominantly in the lipid phase are subject to four pharmacokinetic drawbacks:

1. Lipophilic compounds are difficult to solubilize in an innocuous vehicle for in vivo administration.
2. They will remain at the site of injection following intramuscular, subcutaneous, or intraperitoneal administration. This may be of value for the slow release of drug to maintain low but steady-state concentrations but is disadvantageous for optimizing brain delivery.
3. Lipophilic drugs often become heavily plasma protein bound. This may similarly be of value in maintaining low drug concentrations over an extended period but is generally not advantageous for optimizing cerebral drug uptake.
4. Lipophilic compounds may form micelles,[129] or, when administered as a high concentration, fall out of solution following intravenous (IV) injection.

It should be emphasized that many lipid-soluble drugs that freely enter the brain are readily water soluble, often at concentrations in excess of that required for CNS activity. A number of other physicochemical characteristics are of importance, e.g., depolarization or disruption of ordered water structure. Unfortunately, the discussion of these falls outside the scope of this chapter.

Almost all empiricisms and theories in current use agree that the addition of a hydrophobic group to a compound will increase its absorption, i.e., transport across membranes. This results from a direct increase in the biologic lipid-water partition coefficient by the addition of the hydrophobic substituent. Several in vitro models exist to measure the partition of compounds between immiscible aqueous and lipid phases, mimicking the plasma and the cell membrane. The octanol-water system has been the most extensively studied. Using this model, Leo et al.[149] measured the partitioning, P, of more than 5000 compounds. This study provided (1) a quantitative measure of the lipophilic nature of a wide variety of compounds; (2) an assessment of how, for analogous compounds, different substituents alter the partitioning of the compound between water and

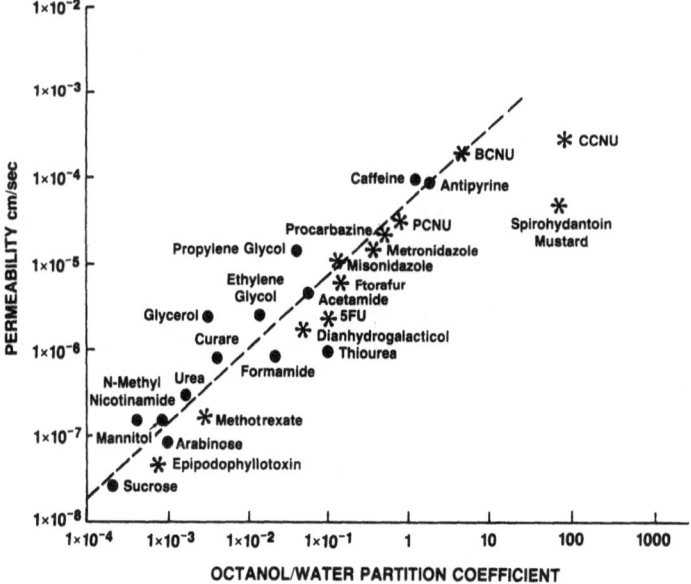

Figure 1. Relationship between octanol/water partition coefficient and cerebrovascular permeability. (∗) designates anticancer agents.

Table I. Constants for Solubility
and Electronic Effects of Various
Substituents in the Meta-Position
of Phenoxyacetic Acid[a]

R	Solubility[b] (π)	Electron effect[c] (σ)
n-C$_4$H$_9$	+1.90	−0.15
SCF$_3$	+1.58	+0.51
SF$_5$	+1.50	+0.68
n-C$_3$H$_7$	+1.43	−0.15
OCF$_3$	+1.21	+0.35
I	+1.15	+0.28
CF$_3$	+1.07	+0.55
C$_2$H$_5$	+0.97	−0.15
Br	+0.94	+0.23
SO$_2$CF$_3$	+0.93	+0.93
Cl	+0.76	+0.23
SCH$_3$	+0.62	−0.05
CH$_3$	+0.51	−0.17
OCH$_3$	+0.12	−0.27
NO$_2$	+0.11	+0.78
H	0.0	0.0
COOH	−0.15	+0.27
COCH$_3$	−0.28	+0.52
CN	−0.30	+0.63
OH	−0.49	−0.36
NHCOCH$_3$	−0.79	−0.02
SO$_2$CH$_3$	−1.26	+0.73

[a] ⬡—O—CH$_2$—COOH
[b] See Eq. (1), from Hansch.[109-111]
[c] Hammett's sigma constant for 4-substituents.[108] In general, the values for π and σ are approximately constant and additive in a variety of different aromatic systems, as long as no strong group interactions occur.

lipid phases; and (3) how the same substituent attached to different compounds similarly alters the partition coefficient of those compounds.

Measuring the partition coefficient of a series of compounds differing in a single substituent, Hansch et al.[110-113] were able to calculate and assign any substituent a numeric rating π as to whether it increased, $+\pi$ or decreased, $-\pi$ the lipid solubility of the compound relative to a parent compound (Table I). For example, the partition coefficients of p-ethyl and p-methylbenzoic acids can be measured between octanol and water. The two compounds differ chemically by a CH$_2$ substituent, the π value which can be determined by the difference in the logarithms of the partition coefficients:

$$\pi CH_2 = \log P_{ethyl} - \log P_{methyl} \tag{1}$$

where P_{ethyl} and P_{methyl} are the partition coefficients of the p-ethyl and p-methylbenzoic acids, and π is the substituent lipophilicity. Similarly πCH_2 has the same quantitative value for the difference in the logarithms of the partition coefficient between either ethyl- or methylbenzoate or p-butyl and p-propylbenzoate, or indeed between any two compounds differing structurally by a single CH_2 substituent, except when electronic or steric factors intervene.

Using the substituent values in Table I, it is possible to (1) reduce the chemical formula of a compound into a series of substituents and additively calculate the log P of that compound without experimentally measuring its partition between octanol and water; and (2) predict the log P value of an unsynthesized compound and from this predict the in vivo transport of the compound across biologic membranes.

To gain a fuller understanding of how alterations in the lipophilicity of a compound affect its transport across biologic membranes, it is necessary to construct a homologous series that covers a wide range of partition coefficients. Probably the simplest way of studying this is by the sequential addition of a CH_2 substituent into the alkyl chain of a compound. The value of a πCH_2 in octanol-water is 0.5. As a consequence, only seven consecutive homologues are required to span a 1000-fold range in partition coefficients in half-log increments. In addition, the alkyl portion of a compound is sometimes unrelated to its activity. Thus, the addition of sequential CH_2 substituents increases the lipophilicity of the compound and, within limits, its transport across biologic membranes. However, alkyl addition can in some circumstances affect drug–receptor interactions, which would reduce biologic response despite higher drug–target concentrations.

One of the best examples of the correlation between permeability and increasing alkyl chain length was demonstrated by Scheuplein and Blank,[219] who measured the transport of nine aliphatic alcohols across excised human striatum. From basic permeability theory, one would expect a linear increase in the logarithm of the transport of a compound with increasing alkyl chain length, as shown by Eq. (2), which describes the rate of transport, or flux, of a compound across a rate-limiting membrane that separates two aqueous phases.

$$F = \frac{P}{R} dC \qquad (2)$$

where, F is the flux, P is the lipid-water partition coefficient, dC is the concentration difference on either side of the membrane, and R is the membrane resistance to diffusion. For many compounds into which consecutive methyl subtituents are added, their transport across biologic membranes to their target is the rate-limiting step for activity (except when a specific drug–receptor fit is essential). It is therefore possible to construct structure–activity curves for many series of compounds, such as in Fig. 2 which relates the structure of a series of barbiturate homologues to their hypnotic activity in rodents. Such curves can be described by the parabolic equation

$$\log (1/C) = -k\pi^2 + k'\pi + k'' \qquad (3)$$

or

$$\log (1/C) = -k(\log P)^2 + k'(\log P) + k'' \qquad (4)$$

These equations relate the molar concentration C of a drug that elicits a constant biologic response (i.e., ED_{50} or LD_{50}), to the substituent lipophilicity π or to the lipid partition coefficient log P, of the drug. The values of k, k', and k'' are determined by regression analysis. The reciprocal of the concentration reflects the fact that higher potency is associated with a lower dose, and the negative sign of the π or $(\log P)^2$ term reflects

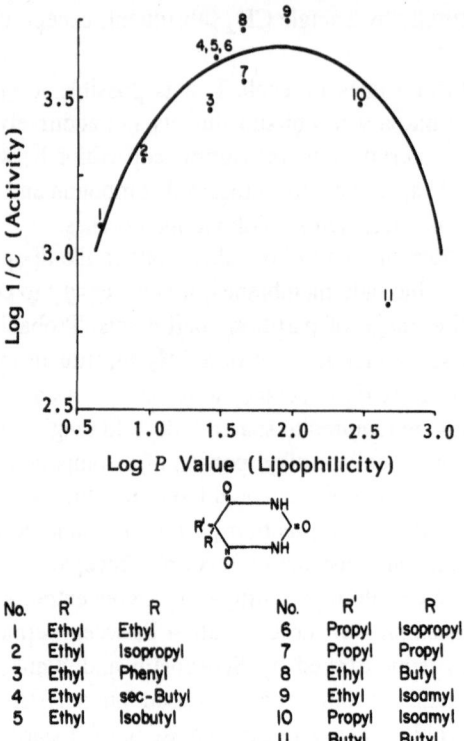

Figure 2. Quantitative structure–activity relationship between the structure of a series of barbiturate homologues and their hypnotic activity in rodents. (Adapted from Hansch et al.,[111] Leo et al.[149])

the expectation of an optimum lipophilicity, designated by π_0 or log P_0. Setting the
derivative $d \log (1/C)/d \log P$ equal to zero and solving the resulting equation for log P,
yields the value of log P_0 for that set of congeners under those specific conditions.
Assuming that Eq. (4) is a useful model for the variation in C with the hydrophobic effects
of the substituents, then regression analysis employing additional extrathermodynamically
derived substituent constants (for electronic σ and/or steric effects) can be added, when
required, to improve the fit of Eq. (4) and better delineate the various effects of a given
substituent.

Most equimolar structure–activity curves show, in addition to an initial linear phase
of slope π, a leveling off and then a reduction in activity with increasing chain length (and
hence lipophilicity) beyond a critical value. There are a variety of possible explanations
for this plateau or decline in activity, which are dependent on biologic parameters (e.g.,
receptor specificity and metabolism) and physical properties (e.g., reduced aqueous solubility, partitioning into inert phases, complex formation, and binding to sites other than
the target). The importance of these is demonstrated by the nitrosoureas (see Section 5.1
and Fig. 8).

Thus far, the effects of chemical structure on the absorption, transport, and solubility
of a compound have been limited to simple alkyl additions. The influence of adding a
polar substituent to each member of an homologous series is a complicating factor that
affects both the dissociation of the compound into ionized and un-ionized forms as well as
its octanol–water partition coefficient. As a consequence, the octanol–water partition

coefficient of each member of the new series will be lower than that of the corresponding homologue in the reference series.

The binding of drugs to receptors often requires the interaction of ionized groups. As a result, most pharmacologically active compounds are either weak acids or weak bases, coexisting in ionized and un-ionized forms. Due to the hydration of the ion and the charged nature of the lipoprotein that forms cell membranes, the ionized species of a drug is both absorbed and transported across biologic membranes to a far lesser degree than the un-ionized species. The equilibrium between the ionized and un-ionized forms is thus an additional factor determining the penetration of a compound into the brain. This equilibrium is determined by the pH of the local environment and the compound's dissociation constant K_a, from the Henderson–Hasselbalch equation [Eq. (5)], where pK_a is $-\log K_a$:

$$pK_a = pH + \log \frac{\text{(un-ionized acid)}}{\text{(ionized acid)}} \text{ or } pH + \log \frac{\text{(ionized base)}}{\text{(un-ionized base)}} \qquad (5)$$

In each aqueous phase, the concentration of the un-ionized species C_u, can be related to the total concentration of drug, C_t, the pH, and the pK_a of the weak acid or base as follows:

$$C_u = \frac{C_t}{1 + 10 \, (pH - pK_a)} \qquad (6)$$

Physiologic pH is determined by a variety of factors, including alveolar ventilation, the kidney and body buffers, and is regulated within narrow limits by a multiplicity of feedback loops. The pK_a of a compound, however, can be altered by chemical modification, since this is related to the electron-withdrawing, $\sigma+$, and electron-donating, $\sigma-$, capacities and proximities of the compound's substituent groups (Table I).

The combination of Eqs. (2) and (6) is known as the pH-partition theory and interrelates the lipid solubility of a compound with its pK_a and the local pH. How these combine to affect the penetration of drugs across biologic membranes is best demonstrated by the absorption of drugs from the GI tract into blood. It is at this interphase that the largest differences in pH are encountered within the body. The pH difference between the blood and the brain, however, is considerably smaller. Under normal conditions, the cerebrospinal fluid (CSF) of humans is more acidic than arterial blood by only 0.1 pH units or less. Since the homeostatic regulation of both is good, it is difficult to take advantage of such a small pH differential.

Despite this, small alterations in plasma pH can cause an increase in the brain uptake of compounds that possess a pK_a close to that of plasma. It can be seen from Eq. (6) that such circumstances result in relatively large changes in the ionized and un-ionized fractions of the compound. Schulman et al.[220] demonstrated that the brain uptake of morphine, pK_a 7.93, was increased by up to threefold in alkalotic rats (mean pH 7.62) as compared with acidotic rats (mean pH 7.16). Such alterations in physiologic pH might occur in surgical patients during respiratory alkalosis.[139] A similar pH-dependent brain uptake has been demonstrated to occur following the intracarotid bolus injection of nicotine (pK_a 6.16 and 10.96) or lactate (pK_a 3.83) when administered in solutions of different pH,[186] some of which were extreme. Such manipulations are not feasible in humans, however, and the pK_a of many ionized drugs is not close to physiologic pH. As a consequence, dissociation behavior is an important parameter when designing reversible derivatives (prodrugs) when the linkage involves an ionizable group.

With regard to drugs whose site of action lies within the brain, Hansch[111] has postulated a log P of 2.0 for optimal brain uptake. This value agrees with those of several successful agents that elicit activity within the CNS (log P values: chloroform 1.93, diazepam 2.82, chloretone (chlorobutanol) 2.03, diphenylhydantoin 2.47, glutethimide 1.90, and chlorodiazepoxide 2.44). For drugs that have an octanol–water partition coefficient in excess of log P_0 (Fig. 1), their rates of BBB transfer are rapid and are generally limited by cerebral blood flow. The rates of blood–brain transfer of lipid insoluble and polar drugs that have a log P value of less than −1.0 are usually limited by their permeability at the cerebral endothelium; while the rate of transfer of drugs whose log P value lies between these limits is a function of both.

2.2. Drug Interactions with Plasma Proteins

Only unbound free drug is able to cross biologic membranes such as the BBB and enter the brain to elicit a pharmacologic response. Many systemically administered therapeutic compounds, however, become bound in varying degrees with various affinities, to plasma constituents, particularly serum albumin, which constitutes in excess of 50% of plasma proteins. This binding can significantly alter a compound's distribution both to the brain as well as to other organs, and also extend its half-life by altering its rate of elimination. Protein binding may not only prevent the rapid excretion of drug but may protect it from metabolism as well, thereby maintaining free drug concentrations at almost steady-state conditions over a protracted period of time by releasing the bound drug as the free drug disappears through tissue binding, elimination, or metabolism.

In many ways, it is surprising that a wide variety of compounds of diverse chemical structure and pharmacologic action bind to serum albumin, γ-globulins, and other plasma protein fractions, since there exists a high degree of structural specificity for these interactions. This can be explained by studying the surface structure of a plasma protein such as serum albumin. The molecule is globular, with its tertiary structure stabilized by internal disulfide and hydrogen bonds. At physiologic pH, it bears a net negative charge at its surface and is covered with numerous charged groups. These include carboxylate anions from glutamic and aspartic acids; protonated nitrogen atoms from lysine, arginine, and histidine; hydroxyl groups from tyrosine, serine, and theronine; and cysteine sulfhydryls. Together, these account for the water solubility of albumin despite the fact that it contains a high proportion of hydrophobic residues, composed of aliphatic and aromatic side chains from phenylalanine, tyrosine, leucine, isoleucine, alanine, proline, valine, and methionine. An eloquent in-depth analysis concerning the molecular aspects of ligand binding to serum albumin may be found in the report by Kragh-Hansen.[143]

With such a preponderance and spacing of ionic, hydrogen bonding and hydrophobic groups, complementary binding sites, often several, for a variety of compounds exist. Nevertheless, the nature and position of specific substituents on a compound can influence the binding characteristics of that compound to plasma proteins. Although few comprehensive studies have been undertaken, a number of general conclusions can be drawn from the analysis of the plasma protein binding of a few, albeit limited, homologous series:

1. The presence of ionizing groups increases the binding affinity of the compound to plasma protein. Davison and Smith[53] demonstrated that benzoic acid binds more strongly to plasma proteins than does phenol or two benzamide derivatives.

Table II. Apparent Association Constants for
the Interaction of Human or Bovine Serum Albumin
with Certain Homologous Long-Chain Anions

Ligand	No. of carbons	High-energy association constant, $K'a$ (M^{-1})	Ref.
Saturated			
Laurate	12	1.6×10^6	
Myristate	14	4.0×10^6	
Palmitate	16	6.0×10^7	88
Stearate	18	8.0×10^7	
Oleate	18	1.1×10^8	
Unsaturated			
Octanol	8	3.0×10^3	
Decanol	10	7.0×10^4	244
Dodecanol	12	1.5×10^5	
Octylsulfonate	8	6.0×10^5	
Decylsulfonate	10	1.4×10^6	244
Dodecylsulfonate	12	1.2×10^6	

Reynolds et al.[209] likewise demonstrated that long-chain carboxylic acids and sulfates have higher affinities than their comparable alcohols.

2. The presence of polar substituents such as hydroxyl groups, as well as their position, contribute to the binding interaction.[244] Davison and Smith[53] further demonstrated that the addition of a hydroxyl group to benzoic acid increases its binding affinity. The addition of a hydroxyl or amide in the *ortho* position rather than the *para* or *meta* positions appeared to result in a further increase in the binding affinity.

3. The addition of alkyl chains, either as substituents on aromatic acids or in straight chains on alcohols, acids, or sulfates, as well as the addition of aromatic rings, greatly increases the binding affinity (Table II).[88,244] Such an enhancement is due to the interaction of the lipophilic portions of both the drug and protein rather than to hydrogen bonding.

While these rules are generally true, exceptions do exist, and some rules are superseded by others. Warfarin, for example, binds to plasma protein with a greater affinity than does hydroxywarfarin. Hydroxywarfarin contains an additional ionizable group but is less lipophilic than warfarin.[191]

There have been several extensive reviews of the binding of drugs to plasma proteins.[134,248] However, it is seldom that these are related to the brain. Rarely are the principles of restrictive and nonrestrictive organ clearance discussed. These are of particular relevance when explaining why protein binding restricts drug entry from some organs but not others. As described above, the interaction between plasma proteins and drugs is usually reversible. The extent to which binding occurs is dependent on the concentration of the drug, the concentration of the binding proteins (the former changes continuously with time while the latter is approximately steady-state), and finally the affinity of the drug for the protein. This is expressed as an association constant, $K'a$.

It is possible to measure free and bound drug concentrations experimentally by ultracentrifugation or equilibrium dialysis techniques.[102,194,196,238] By determining these over a range of total drug concentrations, while maintaining the plasma protein concentration constant, it is possible to calculate both the $K'a$ value and the number of classes of identical binding sites by Scatchard analysis of the data.[218]

Since protein binding affects both hydrophilic and lipophilic drugs, the physicochemical parameter of lipophilicity alone does not guarantee the brain uptake of a drug. This point is of particular relevance later in the analysis of prodrugs of both chlorambucil and methotrexate (see Section 5.2.2). Many lipophilic compounds, which should freely enter the brain, are restricted from doing so by high-affinity binding to plasma proteins, resulting in minimal free drug concentrations. Prime examples are bilirubin,[156] erythrocin (red dye No. 3),[19,157] and Evans blue.[78] At normal plasma concentrations, all these agents are highly and tightly bound to serum albumin, and consequently none significantly enters the brain. However, the brain uptake of these compounds does occur when their concentration in plasma saturates their plasma protein-binding sites and there is a resulting increase in the concentration of the free compounds. For example, in kernicterus, hyperbilirubinemia increases the free bilirubin concentration in plasma. As a consequence, bilirubin freely enters the brain and causes neurotoxicity.[156] For such compounds, their brain extraction is less than or equal to their unbound fraction in plasma. As a result, brain extraction is limited by the binding of the compounds to plasma proteins and alters with the unbound fraction. Diazepam, valproic acid, and phenytoin are examples of drugs that are similarly restrictively cleared by the liver.[23,161,196,215]

Conversely, despite high protein binding and an insignificant plasma concentration of unbound drug, nonrestrictive elimination of a compound by an organ can occur when the organ extraction is greater than, and independent of, the free drug concentration. In this case, the binding of a compound to plasma proteins does not protect it from elimination or organ uptake. Under such circumstances, the drug–protein complex can be viewed as a delivery system for organ uptake or clearance, since the drug is stripped off the plasma proteins as they pass through the organ. A prime example of this occurrence in the brain involves the fatty acid palmitate. Palmitate is extensively bound by plasma proteins, in excess of 99.99%, even at high concentrations.[191] Despite this, approximately 5% of the compound is unidirectionally cleared by brain during a single pass.[191] Dhopeshwarkar et al.[56] demonstrated that following intracarotid administration, labeled palmitate is incorporated into brain phospholipids within 15 sec. Similarly, despite high plasma protein binding, penicillin is nonrestrictively cleared by renal tubular secretion, and propranolol and lidocaine are eliminated nonrestrictively by the liver.[134,161,215,248]

The best explanation of this restrictive/nonrestrictive clearance phenomena is the tug-of-war model.[161] Using a lipophilic drug as an example, so that permeability problems at the BBB can be neglected, and assuming that the time required for the drug to disassociate from the plasma protein (K'off time) is short compared with the transit time of the drug through the organ of uptake, a free drug $[D]$ in blood, which passes through the cerebral vasculature, is capable of interacting with plasma proteins to form a reversible drug–protein complex $[D-P]$ [Eq. (7)]. It is also capable of interacting with a cell within the brain, where it may bind to a receptor, to form a drug-receptor complex $[D-R]$. At the cell, the drug initiates a pharmacologic response, is released, and is then metabolized to a metabolite $[M]$, according to the scheme

$$[D - P] \rightleftharpoons [D] \rightleftharpoons [D - R] \rightarrow [M] + [R] \tag{7}$$

The probability that the drug will interact with either the plasma protein or brain receptor is determined by the relative affinities of the drug for the plasma protein, $K'a$, and receptor, $1/K_m$.

For restrictively cleared drugs, the affinity between the drug and plasma protein is greater than that between the drug and receptor. Thus, the drug preferentially binds to plasma proteins and only the limited free drug concentration in plasma is at liberty to equilibrate with the brain. When affinity between the drug and receptor is larger than that between the drug and plasma protein, the drug–receptor complex is favored; despite protein binding, the drug is extracted nonrestrictively into the brain. As the drug is metabolized or enters an intracellular pool, rapid dissociation of the drug from plasma protein occurs during its transit through the organ to maintain pseudoequilibrium.[161]

It is unfortunate that for most therapeutic agents with which this chapter is concerned, their binding affinities to plasma proteins are generally greater than those to cells within the brain or, in the case of water-soluble compounds, the BBB prevents their intracerebral affinity from becoming effective. As a consequence, for brain, unlike liver and kidney, the plasma extraction of most drugs is restrictive.

In general, there is a linear relationship between the pharmacologically active free drug concentration and the total drug concentration.[121] However, for drugs that possess a high binding affinity for plasma proteins, $K'a > 1 \times 10^4$ M^{-1},[164a] there is generally a dose range, often at a high concentration, within which relatively small increases in dose produce relatively large increases in free drug.[240] This saturation of protein binding has been demonstrated to occur with therapeutic doses of salicylate,[5] valproic acid,[23] disopyramide,[170] methotrexate[238] (Fig. 3) and melphalan.[99,102] Similarly, the unbound fraction of a drug can increase as a result of competitive binding of metabolites or concomitantly administered drugs,[194,196,215] as well as during hypoalbuminemia associated with neoplastic disease, surgery, or aging.[134,248]

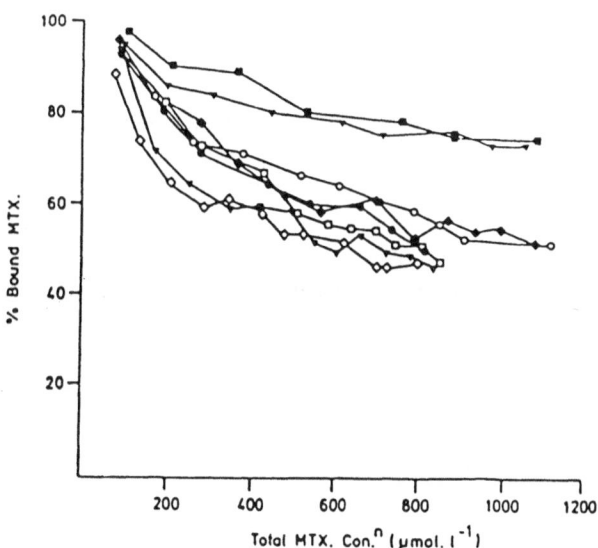

Figure 3. Decrease in the percentage of plasma protein-bound methotrexate in individual sera with increasing methotrexate concentration. (Each symbol represents individual patients.)(From Steele et al.[238])

In almost all cases, an increase in the free plasma pool of drug results in higher drug concentrations in all organs but causes toxicity. When an increase in the concentration of free drug is predicted, a smaller initial dose is administered. Notable exceptions to this exist for drugs that can be administered with rescue techniques to salvage normal tissue. For example, following the IV administration of normal-dose methotrexate to humans, a negligible amount of methotrexate is taken up by the brain and the CSF : plasma ratio is approximately 1 : 17,000.[222] On the other hand, high-dose 24-hr methotrexate infusion in combination with leucovorin rescue 24–36 hr later has been found to increase brain methotrexate levels to cytocidal concentrations and to heighten the CSF : plasma ratio to 1 : 33.[36] A combination of factors probably accounts for this increase in brain methotrexate levels: (1) peak plasma concentrations are dramatically increased during high-dose therapy, but this would not affect the CSF : plasma ratio; (2) Steele et al.[238] demonstrated that the plasma protein binding of methotrexate decreases during high-dose therapy from 99% to approximately 50%, due to saturation of the binding sites (Fig. 3), resulting in up to a 500-fold increase in free methotrexate levels; and (3) drug infusion maintains plasma methotrexate levels constant for sufficient time for the brain concentration to become uniformly high and to equilibrate completely with plasma drug levels.

As the alkylating agent melphalan causes relatively little other toxicity following its profound hematologic toxicity, it is now being used clinically in high-dose regimens together with autologous bone marrow transplantation.[43,44] Greig et al.[99,102] reported that at total plasma concentrations in excess of 33 μM, which are commonly achieved during high-dose therapy,[113] plasma protein binding becomes saturated and free melphalan concentrations increase significantly. Although no studies have yet reported brain melphalan concentrations during high-dose therapy, for reasons similar to those discussed for methotrexate, they would be expected to be increased. Whether this would be of benefit in the treatment of patients with primary or secondary brain tumors sensitive to melphalan (medulloblastoma or malignant melanoma metastases) remains to be determined. It may, however, be of benefit in the prevention of the initial development of CNS metastases.

3. OPTIMIZATION OF DRUG UPTAKE BY BRAIN

Recent advances in analytic chemistry and a fuller understanding of kinetics have improved the effective use of several classes of pharmacologic compounds, particularly anticancer agents whose unselective actions on both tumor and host cells require that they be used in concentrations as high as the host can tolerate. In vitro cell-sensitivity experiments often provide valuable information concerning the approximate dose of a drug required for a pharmacologic effect; such data can be correlated with drug-monitoring studies to assess whether minimally effective concentrations are achieved clinically. The pharmacokinetics of a wide variety of clinically used drugs were tabulated by Gilman, Goodman, and Gilman,[84] and those of the anticancer agents more extensively discussed by Ames et al.[6] and Chabner.[35] In general, most pharmacokinetic studies report plasma drug concentrations and, if related to the CNS, CSF levels as well. Since many of the intimate biochemical mechanisms of the drugs are unknown, the relationship between the administered dose, the measurable plasma and CSF concentrations, the target-tissue concentration, and, of greatest importance, the final therapeutic effect, remains complex and largely unknown. The critical drug concentration is at the target. For tumors, this includes the entire neoplastic area, because a high drug concentration in the necrotic center of a

CNS tumor, where barrier integrity is reduced,[97] will not necessarily result in cytocidal concentrations in the actively growing periphery,[92,93] (Chapter 16, Vol. 2). Although tissue drug concentrations cannot usually be measured directly in patients, a knowledge of plasma and CSF concentrations can nevertheless give valuable information.

1. Depending on the volume of distribution of the drug, plasma concentrations provide an upper estimate of the target concentration. In cases in which the volume of distribution is small, indicating that the drug may be immobilized into a storage compartment (i.e., plasma protein bound), this can be a severe overestimation. The quantity of drug available to the brain following systemic administration is related to cerebral blood flow, the compound's free un-ionized plasma concentration, and its cerebrovascular permeability. For hydrophilic drugs, plasma concentrations generally remain considerably higher than concomitant brain concentrations, except during the latter phase of drug elimination, when plasma drug levels have fallen through clearance and/or metabolism and drug is sequestered in the brain from earlier entry. Under such circumstances, CSF/plasma ratios are extremely misleading. The same ratio can, however, provide a useful estimate of brain drug entry under steady-state conditions, which for most therapeutic agents are seldom reached, or at early time points, after the initial distributive phase, when plasma drug concentrations are high and approximately steady. During such conditions, should plasma drug concentrations either not reach or quickly fall below therapeutic concentrations, it is unlikely that target tissue within the brain would be subjected to an effective drug concentration. By contrast, the brain concentration of a lipophilic drug can be substantially higher than the concomitant plasma level almost immediately after the plasma peak. The brain concentration–time profile can be related to the plasma peak height and decline, since the brain disappearance of lipophilic agents is generally through backdiffusion.

2. The rate of drug elimination from plasma provides valuable information concerning the duration of drug action. This information can be used to assess which route of drug administration best maintains therapeutic drug concentrations. With the exception of the anticancer alkylating agents, which produce irreversible complexes by covalently binding to nuclear substituents in cells,[6,61] most other pharmacologic compounds form reversible complexes with receptors. Thus a steady supply of drug must be maintained despite local clearance and metabolism to compete for the receptors and keep them saturated. This is of particular relevance to cell-cycle-dependent anticancer agents that are only effective against sensitive cells in the correct fraction of the cell cycle.

3. Plasma drug concentrations provide valuable information relating to drug bioavailability after administration and organ clearance.

4. Monitoring of plasma drug levels provides the possibility of individualizing patient treatment.

In the clinical setting, techniques to optimize drug brain entry by altering pharmacokinetics are limited. They rely on (1) altering the frequency of administration, and (2) altering the route of administration, from oral to systemic in cases of poor bioavailability, or from systemic to local (intracarotid or intrathecal).

3.1. Frequency of Drug Administration

The primary factor that determines the duration of time between successive administrations of a drug is the disappearance half-life of the compound. An appropriately long disappearance half-life is required for a compound to enter the brain. A very long half-life indicates that the compound may be trapped within an intracellular pool or that it may be heavily protein bound; under such circumstances, circulating free plasma and brain concentrations are low. However, a rapid disappearance half-life will not permit sufficient time for significant drug to enter the brain. For example, the trypanocide suramin has an excessively long disappearance half-life. It binds strongly to plasma proteins, minimally enters cells or the brain, and is excreted only slowly by the kidney. As a consequence, the drug can be detected in plasma up to 3 months after its IV administration,[84] and weekly doses are sufficient to maintain systemic therapeutic drug levels. By contrast, the half-lives of the anticancer pyrimidine analogues 5-fluorouracil (5-FU) and cytosine arabinoside are extremely short. The principal route of elimination of 5-FU is hepatic metabolism, which appears to exhibit dose dependency. The rate-limiting step in 5-FU catabolism is its initial reduction by dihydrouracil dehydrogenase, the eventual products being CO_2, urea, ammonia, and γ-fluoro-β-alanine.[6,35,61] Following the distributive phase, the initial half-life of 5-FU is only approximately 8 min.[175] Cytosine arabinoside is similarly cleared rapidly by the liver; cytidine deaminase inactivates it to uracil arabinoside, with a half-life of 12 min.[75,116,250] Rapid clearance prevents either of these compounds from reaching significant concentrations within the brain.[93] Indeed, both possess such short half-lives that IV drug infusion rather than successive IV boluses are required to maintain therapeutic drug concentrations in plasma and peripheral tissues.

While continuous drug infusion can optimize the systemic pharmacokinetics of a drug, it will not necessarily increase its brain concentration into the therapeutic window. Hornbeck and Byfield[32,123] reported that 5-FU reaches insignificant and subcytotoxic concentrations in CSF after both high-dose IV bolus and lower dose IV infusion. By contrast, Weinstein et al.[254] reported high CSF serum cytosine arabinoside ratios of up to 0.58 following the IV infusion of cytosine arabinoside and Ho et al.[116] reported a ratio of 0.4. These high ratios probably result from the saturation of the systemic metabolism of cytosine arabinoside. Indeed the prolongation of plasma drug levels by this administration approach has proved superior to four others, which have included (1) the use of depot preparations of cytosine arabinoside by the administration of the prodrug cytosine arabinoside-5'-palmitate, (2) the administration of a cytosine arabinoside analogue in which the amino group of the cytosine moiety was modified to reduce the metabolism of the compound (unfortunately this resulted in a dramatic loss in the activity of the compound)[60]; (3) the administration of tetrahydrouridine, an effective inhibitor of deaminase activity, the use of which is still under investigation.[179]; and (4) the use of the analogue anhydrocytosine arabinoside (cyclocytidine), which is a soluble depot form of cytosine arabinoside (this agent does not become phosphorylated, its activity proceeding via cytosine arabinoside-5'-triphosphate).[124,251]

3.2. Route of Drug Administration

There are essentially three routes by which a drug can be delivered to the brain: (1) by IV administration, including the oral, intramuscular, subcutaneous, and intraperitoneal routes as well as inhalation, whereby drug enters the venous system before reaching the brain; (2) by intraarterial administration, using either the carotid or vertebral arteries; and (3) intra-CSF administration, by the lumbar or intraventricular routes.

3.2.1. Intraarterial Drug Administration

The goal of intraarterial infusion is to increase the therapeutic index of a compound that already demonstrates activity following its IV administration. This can be achieved by either increasing the target tissue concentration of drug without the side effect of concomitantly increasing peripheral tissue drug levels or by decreasing the peripheral tissue concentrations without lowering those of the target.

Due to the technical problems and expense of administering drugs into the carotid and vertebral arteries, the technique is often reserved for patients with brain tumors. When assessing whether the technique would prove more advantageous than IV drug administration, several factors have to be evaluated:

1. All of the advantage must result from the first circulation of the drug through the brain. After the drug leaves the target and enters the systemic circulation its pharmacokinetics are similar to those following IV administration.
2. The brain must not be a site of drug toxicity.
3. An increase in the concentration of drug at the target site will result in an increase in activity,[229] if a maximal effect has not already been achieved.

The advantage of the route of administration occurs only during the initial passage of the drug through the brain, which is not compatible with the concentration–time concept. As a consequence, drugs that are ideal for intraarterial infusion must (1) bind covalently to the target tissue to prevent brain washout, and (2) possess a short half-life. They are thus metabolized or eliminated before reaching the systemic circulation, which reduces their peripheral toxicity. If acute systemic toxicity is the dose-limiting factor in the IV administration of the drug, a larger dose can be given intraarterially than intravenously. If cumulative systemic toxicity is dose limiting, however, a longer infusion or larger number of doses can be given intraarterially.

It is possible to calculate the relative advantage, R_d, of the intraarterial versus intravenous route:

$$R_d = \frac{\text{target concentration (IA)}}{\text{target concentration (IV)}} = 1 + \frac{Cl_{tb}}{Q_a} \qquad (8)$$

where, Cl_{tb} is the total body clearance of the drug, which can be calculated following IV administration, and Q_a is the blood flow through the infused artery. The predicted R_d of a variety of commonly used anticancer agents is shown in Table III. Although the R_d of all, with the exception of cyclophosphamide, which requires hepatic activation,[6,35,61] is increased, it should be remembered that low cerebrovascular permeability may still limit many intraarterially administered drugs from reaching therapeutic concentrations within the brain. Consequently, most investigations have utilized BCNU[118,162] for brain perfusion. Levin et al.[150,153,154] have demonstrated that the concentration of BCNU achieved in the brain was fourfold greater after intracarotid infusion than after IV administration.

The reversible osmotic disruption of the BBB (Chapter 17, Vol. 2, Parts 1 and 2) in conjunction with intracarotid drug infusion significantly increases the brain uptake of water-soluble agents and allows their brain sequestration following barrier closure.[181] The combination of these techniques—osmotic barrier opening and arterial drug administration (both using the same arterial catheter)—thus overcomes the requirement of lipophilicity for drug uptake by the brain, as well as the formation of covalent irreversible bonds to the target tissue to reduce drug washout.

3.2.2. Intra-CSF Drug Administration

The administration of drugs directly into the CSF, by either the intralumbar or intraventricular routes, bypasses the BBB to achieve immediately high CSF concentrations of drugs that are normally restricted from entering the brain. Drug administration by

Table III. Regional Drug-Delivery Advantage for the Intraarterial Brain Delivery of Selected Anticancer Agents

Drug	$Cl_{tb}{}^a$ (ml/min)	R_d value[b] Internal carotid artery (400–600 ml/min)[c]	Vertebral artery (200–400 ml/min)[c]
Thymidine	40,000	67.6–101.0	101.0–201.0
5-Fluorouracil (5-Fu)	4,000	7.7–11.0	11.0–21.0
Cytosine arabinoside	3,000	6.0–8.5	8.5–16.0
BCNU	1,000	2.6–3.5	3.5–6.0
Adriamycin[d]	900	2.5–3.3	3.3–5.5
AZQ[d]	400	1.7–2.0	2.0–3.0
Cisplatinum	400	1.7–2.0	2.0–3.0
Methotrexate	200	1.3–1.5	1.5–2.0
Cyclophosphamide[e]		1.0	1.0

[a] Cl_{tb} is the total body clearance of the drug.
[b] R_d represents the relative advantage of drug delivery by the intraarterial versus the intravenous route [Eq. (8)].
[c] Blood flow through the internal carotid and vertebral arteries in humans (Neuwelt, personal communication).
[d] Adriamycin and cisplatinum have been demonstrated by Neuwelt et al. [183] to be neurotoxic in the dog; as a consequence, their brain uptake should be avoided.
[e] Drugs that require activation at sites other than the target have no drug-delivery advantage by the intraarterial route.

these routes has proved of significant value in the treatment of tumors and infections sequestered within the CSF, such as meningeal leukemia, leptomeningeal carcinomatosis, and gram-negative bacillary meningitis.[85,92,204,216] Since the volume of distribution of a drug delivered intrathecally is considerably smaller than that following its systemic administration, a smaller initial dose will achieve a high CSF concentration without the toxicity associated with the systemic administration of the drug, although CNS toxicity may be greater and should be carefully monitored. This is of particular relevance for intrathecally administered anticancer agents. However, since all drugs are eventually cleared from the CSF into the systemic circulation via the sagittal sinus and capillaries of the brain, low, but nevertheless significant, drug concentrations will eventually appear in the plasma after several hours. A further advantage of intra-CSF drug administration is that the pharmacokinetic parameters of a compound in CSF can be significantly different and occasionally more optimal compared with those following its systemic administration, as the rate of metabolism and route of elimination of the drug are often altered. Prime examples of each are demonstrated by (1) cytosine arabinoside, and (2) methotrexate, the two most common intrathecally administered anticancer agents:

1. IV-administered cytosine arabinoside is rapidly deaminated to the inactive uracil arabinoside by cytidine deaminase,[116] located predominantly in the liver (see Section 3.2). As a consequence, the plasma half-life of the parent compound is short, approximately 12 min in humans,[75,250] and ineffective concentrations enter the brain.[92] By contrast, human brain contains only low amounts of cytosine deaminase.[116] The disappearance of cytosine arabinoside from human CSF is biphasic following its intraventricular administration with half-lives of 30 min and 3.5 hr.[82]

2. On the other hand, IV– administered methotrexate disappears from plasma bi-

phasically with an initial half-life, 45 min,[116] which is primarily dependent on renal excretion. This involves glomerular filtration and the active secretion of methotrexate from the proximal renal tubule, mediated by a transport system for weak organic acids.[126,158] The final protracted elimination half-life, 8–10 hr, reflects enterohepatic methotrexate circulation. By contrast, other factors jointly determine the disappearance of methotrexate from CSF. Indeed, the rate of disappearance of most compounds from CSF is governed primarily by CSF bulk flow, from the ventricles to the cisterns, and excretion at the sagittal sinus. High CSF methotrexate concentrations are maintained for a longer time compared with plasma. The CSF component half-lives of methotrexate are 4.5 and 14 hr following its administration by the lumbar route.[13]

Thus, for both cytosine arabinoside and methotrexate, intra-CSF rather than systemic administration appears more appropriate for the treatment of CSF-sequestered diseases. This advantage, however, does not necessarily apply to disease sites within the brain parenchyma.

Although the ependymal cells that line the ventricles, and the pial–glial membranes do not present a barrier between the CSF and the brain extracellular fluid (ECF), it is generally accepted that the brain penetration of most intrathecally administered agents is minimal. Electron micrographs of brain parenchyma demonstrate that the extracellular space between adjacent cells is extremely tortuous. The rate of penetration of a compound into brain from CSF is primarily determined by its aqueous diffusion constant. This is a physical characteristic of a drug and is a function of its molecular weight.[71] In addition, should the drug enter cells during its passage through the brain, as is the case for both cytosine arabinoside and methotrexate, or cross capillaries, then its effective diffusion distance into the parenchyma of the brain would be even shorter.[72] The measurement of the transport of materials between CSF and brain is complex but has nevertheless been studied by ventriculocisternal perfusion in both the dog and the monkey.[11,193] While few similar studies have been undertaken on the transport of compounds from the subarachnoidal CSF into brain or spinal cord, which is relevant following drug administration by the lumbar route, it is probable that the principles and permeabilities of drug penetration into the brain from the ventricles and the subarachnoid space are similar. However, it should be stressed that a major morphologic difference exists between these two systems. At the pial–glial interface, the subarachnoid space extends into the brain in the form of the Virchow–Robin space. These subarachnoidal extensions are major sites of tumor infiltration in leptomeningeal carcinomatosis[188] and may act as channels for drugs following their intra-CSF administration. Despite this, it is generally believed that the brain penetration of most CSF-administered drugs is minimal, whether they are administered via the ventricular or lumbar route.[93] Even sustained high CSF drug concentrations will only achieve a significant quantity of drug in those portions of the CNS within a few millimeters from the CSF circulation. Using a mathematical model, Blasberg et al.[12] calculated the brain concentration of methotrexate at various distances into the parenchyma from the ependymal surface at 1, 18, 24, and 48 hr following methotrexate ventriculocisternal perfusion in humans. Methotrexate penetration into the brain was slow and, although significant concentrations were calculated to be present in periventricular areas, at distances of 8 mm and more the concentration was probably nontherapeutic. Hryniuk and Bertino[125] reported that $> 1 \times 10^{-6}$ M of methotrexate is required for cytotoxicity. Levin[150] calculated that for a tumor present at a depth of 2 cm from the ependymal surface, a piece of brain 4 cm in ventriculosubarachnoidal breadth, which is approximately average in humans, required perfusion on both surfaces for 48 hr to achieve 8% of

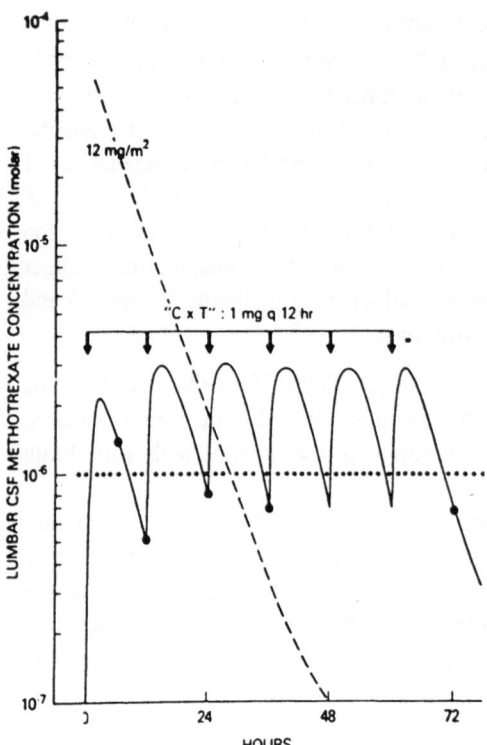

Figure 4. Ventricular cerebrospinal fluid (CSF) concentrations of methotrexate following a "concentration × time" dose schedule delivery of 1 mg methotrexate by Ommaya reservoir (●—●) at 6 intervals of 12 hr. The dashed curve (---) reflects mean methotrexate levels achieved following a conventional dose regimen. The horizontal dotted line (. . . .) at 1×10^6 molar estimates the maximum cytocidal concentration for most human leukemic cells. (From Bleyer et al.[14])

the CSF concentration within the tumor. From a practical point of view, it is difficult to maintain high steady-state CSF concentrations of methotrexate, or indeed any drug, over such a protracted period. A minimal time interval exists between concurrent drug administrations, whether by the intralumbar or the intraventricular route, and protracted high CSF methotrexate concentrations have been associated with the onset of methotrexate CNS toxicity, particularly leukoencephalopathy.[203]

Interestingly, Bleyer et al.[14] have demonstrated that CSF methotrexate levels can be maintained above cytocidal concentrations for 72 hr by intraventricular administration of methotrexate (1 mg) at 12-hr intervals (Fig. 4). This regimen proved superior to the intraventricular administration of a single higher dose (12 mg/m²) in the treatment of meningeal leukemia. The latter regimen causes a higher incidence of leukoencephalopathy, particularly in combination with radiotherapy, and maintained CSF methotrexate concentrations at cytocidal levels for approximately 24 hr. The concentrations of methotrexate within the brain, however, were not calculated following either regimen.

Whether or not cytocidal concentrations of methotrexate are attained within brain, it is unlikely that intrathecal chemotherapy would be of value in the treatment of intraparenchymal tumors because (1) the prohibitively high CSF drainage of drugs and the depth of the tumor within the brain would result in inconsistent and low drug concentrations at the tumor; and (2) intra-CSF drug administration, particularly by the lumbar route, is often contraindicated in the presence of solid intraparenchymal tumors due to the likelihood of herniation.

Greig (unpublished results) administered a variety of anticancer agents, including methotrexate, singly via the intraventricular route to rats implanted with Walker carcinosarcoma in the cerebral cortex. None survived longer than the untreated controls despite the proven sensitivity of the tumor to the drugs. This confirms the inadequate access of intraventricularly administered drugs, even as small as 5-FU (130 M_r), to intraparenchymal tumors.

For water-soluble compounds larger than methotrexate (454 M_r), such as gentamicin (approximately 477 M_r), amphotericin B (924 M_r), or proteins such as interferon (approximately 19,000 M_r), diffusion from CSF into brain would be expected to be even more limited. Although intraventricular amphotericin B and gentamicin are of value in the treatment of gram-negative bacillary meningitis,[204,262,263] they are ineffective against intraparenchymal abscesses (Neuwelt, personal communication). Similarly intra-CSF administered interferon has been reported to be minimally effective in extending the survival of rabbits, or indeed, of humans infected with rabies.[117,171] Billiau et al.[10] injected interferon intrathecally into monkeys and were unable to detect it except in minute concentrations in the pia mater and most superficial cortex. Similarly, Greig et al.[100] injected 1×10^4 units of interferon into the cisterna magna of rats. Brain interferon levels were below assay sensitivity (8 units/g), even in those areas adjacent to CSF. All brain areas were carefully rinsed with sterile tissue culture medium to remove any contaminating CSF prior to interferon analysis.

The route of intra-CSF drug administration is an additional factor that determines the distribution of a drug throughout the CSF compartments and secondarily within both the brain and spinal cord. Clinically, the intraventricular route is commonly used during the treatment of leptomeningeal disease. The intralumbar route is commonly used in the prophylactic treatment of acute lymphocytic leukemia, since it does not require the surgical placement of an Ommaya reservoir,[189,216] which is essential for intraventricular drug administration. This device consists of a surgically implanted subcutaneous reservoir with a connection to a lateral ventricle (Fig. 5).

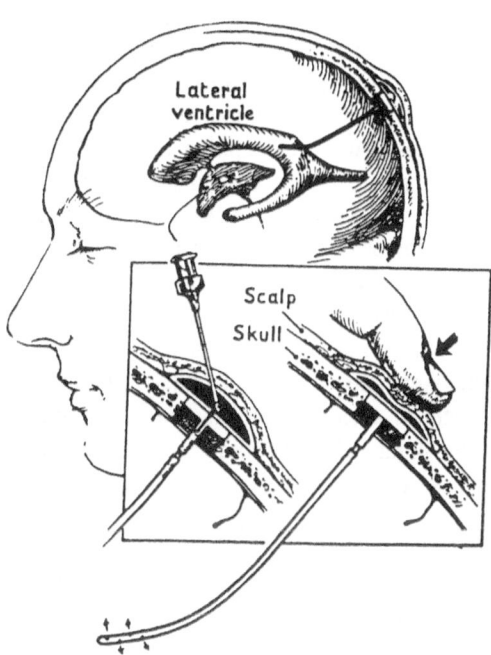

Figure 5. Schematic representation of an Ommaya reservoir. The reservoir is implanted subcutaneously and attached to a catheter, the tip of which extends into the lateral ventricle. (From Ommaya.[189])

Table IV. Estimated Total Cerebrospinal
Fluid Volume in Man

Subjects	CSF (ml + SD)
Adults	130 + 30
Older children	100 + 20
Young children	80 + 20
Infants	50 + 10

From Rieselbach et al.[210]

While intralumbar drug administration is initially less complicated than the intraventricular route, the approach has a variety of disadvantages:

1. Optimal drug delivery to basal cisterns and ventricles is frequently not achieved. As described, the direction of CSF flow is from the ventricles to the basal cisterns and then in both cephalad and caudal directions to the subarachnoid spaces. Drugs administered via the intralumbar route must therefore ascend the spinal subarachnoid space to reach the cisterns. From there, in order to equilibrate with the ventricular system, they must overcome a descending unidirectional current of newly formed CSF that constantly dilutes the subarachnoid CSF. Three factors that can optimize intralumbar drug administration, albeit in a limited manner, are (a) the volume, (b) the specific gravity of the drug solution, and (c) the position of the patient during and immediately after drug instillation. For example, (a) Reiselbach et al.[210] demonstrated in the monkey that, for a compound injected into the lumbar space, an injection volume of 10% and 25% of the total CSF was required for it to reach the cisterns and ventricles, respectively. Table IV shows the total CSF volume in humans. Since the lumbar subarachnoid space is small, it is not possible to inject into it volumes even approaching 25% of the total CSF volume. As an aside, it should be noted that the total CSF volume is not proportional to age, weight, or total body surface area; a factor that should be considered when calculating intra-CSF drug doses for adults and infants for the chemotherapy of CSF-sequestered diseases. (b and c) Vehicles with specific gravities greater than human CSF, 1.004–1.009 at 37°C,[86] are considered hyperbaric and have been reported to increase the rate of cephalad movement of lumbar injected drugs. Alazraki et al.[2,3] demonstrated that, compared to the normal lumbar administration of the drug, the addition of 10% dextrose in water containing amphotericin B, specific gravity 1.034, increased both the rate of appearance and the final concentration of drug in the cisterna magna of monkeys placed in the Trendelenburg position during intralumbar administration. The effects of these maneuvers, however, are probably lost within a matter of minutes, as the sugar solution rapidly equilibrates with the CSF. It remains undetermined whether these two maneuvers would augment the delivery of drug to the ventricles in humans. Not only are nests of malignant cells harbored in the choroid plexuses throughout the ventricles of patients with leptomeningeal carcinomatosis and meningeal leukemia,[188a] but the basal cisterns and ventricles act as bacterial reservoirs, representing major sites of infection in most cases of coccidioidal and cryptococcal meningitis.[105,262]

2. The lumbar subarachnoid CSF volume is small compared with that in the ventricles. As a consequence, arachnoiditis frequently follows lumbar drug administration,

particularly of amphotericin B. The addition of hydrocortisone sodium succinate, 20 mg/dose, has been recommended to reduce irritation.[262]

3. Shapiro et al.[222] demonstrated with radiolabeled albumin that inadequate drug distribution and CSF leakage sometimes occurs after repeated lumbar injections.

4. The drug may inadvertently be injected into the subdural or epidural spaces, causing toxicity. Likewise, neurotoxicity has been associated with the poor placement or obstruction of Ommaya devices.[221,222]

5. Intralumbar administration is limited by the total number and frequency of injections that are feasible. As a consequence, a small number of high concentration boluses are administered. This increases the possibility of neurotoxicity, which for many drugs, such as methotrexate, is related to the peak dose. In addition, this does not permit protracted maintenance of cytocidal drug concentrations.[14]

By contrast, the Ommaya reservoir administers drugs directly into the lateral ventricles, which, in the opinion of Bleyer et al.[13,14] for the treatment of meningeal leukemia, Glass[85] for the treatment of meningeal carcinomatosis, and Diamond and Bennet[57] and Rahal[204] for the treatment of fungal and bacillary meningitis, is superior to lumbar drug administration. The reasons for this have been discussed, how they eventually affect the CSF pharmacokinetics of an agent can be demonstrated by methotrexate. As previously described, methotrexate disappears from CSF following a biphasic curve, with half-lives of 4.5 and 14 hr, after its lumbar administration in humans.[13] After a 6.25-mg/m^2 methotrexate intralumbar dose, peak ventricular CSF methotrexate concentrations varied from 2×10^{-5} to 6×10^{-7} M (the latter level being below the cytocidal concentration.[114]) Bode et al.[15] have reported that the CSF disappearance of methotrexate after ventricular administration to humans is biphasic with component half-lives of 1.7 and 6.6 hr. Following a 6.25-mg/m^2 methotrexate dose, via an Ommaya reservoir, however, peak ventricular concentrations reached 2×10^{-4} M.[222] The drug appeared in lumbar CSF at 1 hr and reached a peak lumbar concentration of 8×10^{-5} M at 4 hr.

Other interesting techniques that have used both the technology and experience gained from the Ommaya reservoir are the ventriculolumbar CSF perfusion technique and the intratumor Ommaya reservoir. In the former, high concentrations of drugs can be administered into ventricular CSF via an indwelling subcutaneous Ommaya reservoir. The agents perfuse throughout the CSF and are finally removed via an outflow cannula placed in the lumbar subarachnoid space.[201] During the infusion, CSF pressure is controlled by changing the height of the outflow column at the lumbar end of the system. By increasing the pressure within the system, the rate of bulk flow is increased proportionally. Using this technique, it is potentially possible to perfuse the entire CSF and maintain high drug concentrations over a period of several hours. The technique has been used with some success on a small number of patients with refractory CNS leukemia.[201,216] However, due to the required medical support and technical complexity, it is unlikely that the technique will ever achieve widespread clinical use.

The second approach, the intratumor Ommaya reservoir, was developed on the premise of delivering high concentrations of water-soluble anticancer agents directly into the tumor. Rubin et al.[216] inserted an Ommaya device directly into the tumor bed of glioma patients during surgical resection and subsequently treated them with methotrexate. Only seven patients were studied—too small a sample to adequately assess the technique's potential.

As an aside, a recent and interesting variation of this technique is the surgical placement of radioactive rods into brain tumors during their resection,[173],[174] thereby delivering radiation, as opposed to drugs, locally over an extended period of time.

4. REVERSIBLE BLOOD–BRAIN BARRIER MODIFICATION

The BBB maintains the homeostatic environment of the brain so that it can function irrespective of large systemic fluctuations in essential compounds (see Chapter 17, Vol. 2, Parts 1 and 2). In addition, it protects the brain from toxic agents and degradation products present in the systemic circulation. A variety of maneuvers have been demonstrated to increase the permeability of the barrier to intravascular markers (see Chapters 22 and 27, Vol. 2). These have included hypercapnia, hypoxia, hypervolemia, ischemia, portacaval anastomosis, several heavy metals, and a variety of solvents.[92],[192],[205] Virtually all cause structural damage and irreversible or severely prolonged barrier breakdown and are deleterious. Two techniques, however—osmotic opening and Metrazol opening—have proved successful in consistently and innocuously opening the BBB. Consequently, both have been employed to increase the brain uptake of therapeutically relevant water-soluble agents, thereby demonstrating that absolute and continuous BBB integrity is not necessarily essential for the maintenance of normal brain function.

4.1. Osmotic Opening

Neuwelt has extensively described the underlying mechanisms and the experimental and clinical states in which the osmotic opening technique has been successfully applied (see Chapter 17, Vol. 2, Parts 1 and 2).

4.2. Metrazol Opening

Several studies have demonstrated that electroconvulsive shock and chemically induced seizures result in an increase in cerebrovascular permeability to intravascular markers[148],[159],[176] (Chapter 27, Vol. 2). Greig[94] and Hellmann[95],[98] demonstrated in the rat that Metrazol, a CNS stimulant, which acts as an analeptic agent when administered in high concentrations, significantly increased the brain uptake of [^{125}I]serum albumin and horseradish peroxidase (HRP) by threefold. The phenomenon proved innocuous and was completely reversed within 4 hr. Greig et al.[98] used the technique, together with Na pentobarbital (to reduce peripheral seizure activity), to increase the brain uptake of the anticancer agents razoxane and melphalan (Fig. 6). Neither enters normal brain in significant or therapeutic concentrations.[92],[96],[98],[103a] This Metrazol-enhanced brain uptake led to an increase in the efficacy of razoxane against intracerebrally sequestered L1210 leukemia cells in mice.[98] (Fig. 7).

The mechanisms underlying this phenomenon remain undetermined. Lorenzo et al.[159],[161] reported that Metrazol-induced seizures caused opening of tight junctions in immobilized and artificially ventilated cats and significantly increased cerebral blood flow. However, there was no correlation between the regional distribution of permeability increase and the regional distribution of increased blood flow. Although hypertension, which would cause both vasodilatation and the recruitment of redundant capillaries, has been associated with the Metrazol opening technique, it is unlikely that the demonstrated permeability increase would result from this alone. Greig and Hellmann[95] reported that increases in cerebrovascular permeability only occurred at Metrazol concentrations that

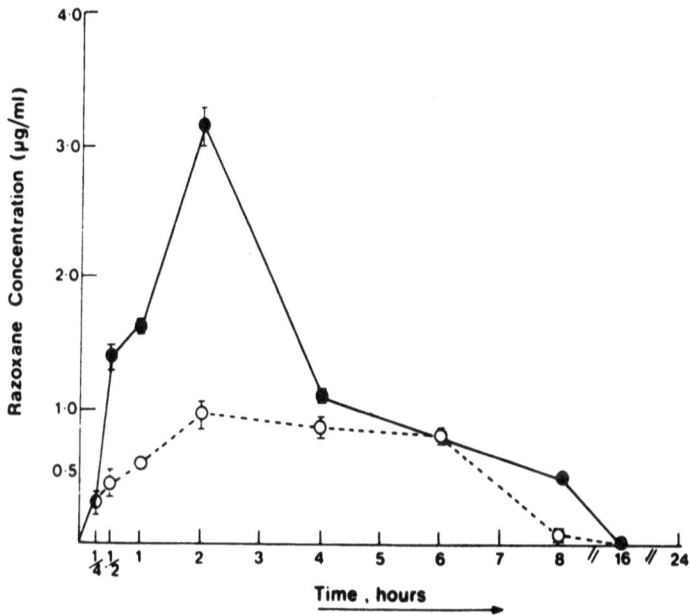

Figure 6. Cerebrospinal fluid (CSF) concentration of razoxane (+SEM), following its intravenous (IV) administration in the rat, alone and in combination with Metrazol. (○) CSF razoxane, following administration of razoxane 100 mg/kg IP. (●) CSF razoxane, following the administration of razoxane 100 mg/kg IP, and Metrazol 60 mg/kg IV. (From Greig et al.[95])

caused CNS seizure activity, as detected by the appearance of paroxysmal discharges during electroencephalographic (EEG) recording. Subthreshold doses that caused similar increases in arterial blood pressure caused no increase in permeability. It could prove interesting to extend this study and disassociate seizure activity from the other Metrazol-induced phenomena with anticonvulsive agents, such as ethosuximide, trimethadione, or valproic acid, and then assess whether cerebrovascular permeability alterations occur.

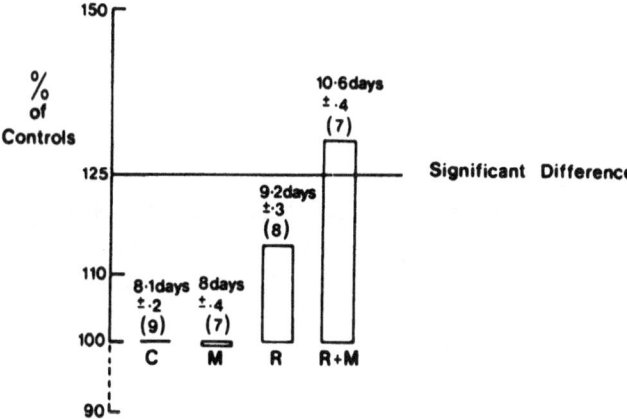

Figure 7. The effect of administration of razoxane, Metrazol, and razoxane together with Metrazol on the survival of mice intracerebrally implanted with L1210 leukemia. C, controls (no treatment); M, Metrazol (60 mg/kg IV); R, razoxane (40 mg/kg IP); M + R, Metrazol plus razoxane (60 mg/kg IV and 40 mg/kg IP, respectively). All administered on days 1–4. (From Greig et al.[98])

While some degree of vasodilatation may be essential for Metrazol-induced barrier opening, it is probable that some complex combination of neuronal excitation, coupled with an increased blood flow and vasodilatation, is responsible for the Metrazol-enhanced cerebrovascular permeability. Any permeability differences between brain areas could then be explained by regional variations in the susceptibility of local capillaries to tension and the local effect of Metrazol on neuronal activity.[95]

4.3. Miscellaneous Techniques

A wide variety of other physical and chemical insults have been reported to increase cerebrovascular permeability both innocuously and reversibly. Some have been suggested and indeed may eventually be of clinical use. However, the results of many remain unsubstantiated, and several have been refuted.[92,100,182] The more interesting ones are reviewed briefly.

4.3.1. Arterial Hypertension

Johansson and colleagues[132,133] have extensively studied the effects of chemically induced acute hypertension on cerebrovascular permeability (Chapter 22). These workers clearly demonstrated that intravascular pressure in the small vessels is the major determinant of barrier breakdown in hypertension. Autoregulation, involving progressive arteriolar vasoconstriction with increasing pressure, maintains cerebral blood flow at a constant rate over a wide range of arterial pressure.[65,205] A breakdown occurs, however, when arterial pressure rises above 180–200 mm Hg.[207] Above this limit, cerebral blood flow rises steeply with further increases in pressure, and barrier opening and structural damage occur.[133,207] Johansson et al.[133] reported that high $Paco_2$ and X-irradiation increase the susceptibility of cerebral vessels to hypertension, lowering the arterial threshold at which barrier reduction occurs. It is possible that tumor capillaries, which often lack structural support[97] (Chapter 16, Vol. 2), are more susceptible to increases in arterial pressure than is normal brain. Suzuki et al.[242] recently tested this hypothesis and demonstrated an increase in the efficacy of adriamycin with acute hypertension over adriamycin alone in the treatment of rats with intracerebral implants of DBLA-6 leukemia. Whether this highly experimental technique could ever be safely and successfully transferred to the clinic remains undetermined.

4.3.2. Hypercapnia

Clemedson et al.[42] were the first to demonstrate an increase in cerebrovascular permeability in the presence of a raised $Paco_2$. Barrier reduction occurs at a threshold of approximately 10% CO_2, the degree of opening being related to both the concentration of CO_2 and duration of exposure.[47] It is probable that the phenomenon is augmented by both vasodilatation and hypertension. Nevertheless, it is unlikely that hypercapnia-induced barrier opening, although apparently reversible,[88] would ever be of therapeutic value, since it causes multiple small cerebral hemorrhages, edema, and alterations in cerebral metabolism.[42,47,48,92,205]

4.3.3. Hypoxia and Ischemia

Although, from a pathophysiologic view, hypoxia and ischemia are distinct conditions, they are discussed together, since experimentally they are inseparable. In general, the BBB is resistant to hypoxia and ischemia, as evidenced by its impermeability to Trypan blue several hours after death.[26] The induction of barrier opening by hypoxia and ischemia in animals has required extreme conditions, the clamping of the common carotid artery, and/or hypoxia for hours. Conversely, both neurons and glial cells are exquisitely sensitive to the lack of oxygen, which causes cell death in a matter of minutes.[27] Tumor cells contain anaerobic mechanisms that might protect them from reductions in oxygen tension that would cause severe damage to normal brain tissue. This technique is wholly unsuitable for the enhancement of brain drug delivery for the treatment of any human disease.

4.3.4. X-Irradiation and Fast Neutron Radiation Therapy

Nair and Roth[176,177] reported that X-rays, in a single dose of 10,000 rad, caused a subtle alteration in BBB permeability that was not as great as that caused by either Metrazol or osmotic imbalance. Nevertheless, it increased the brain uptake and efficacy of [^{35}S]acetazolamide, an anticonvulsant, in rats.[177] Clinically, radiation to the brain is administered in considerably lower doses, generally in 160–300 rad fractions over a protracted period.[33] Single fractions of up to 1000 rad have been administered,[140] but this is rare. When these lower doses are administered to animals, however, they produce no alteration in cerebrovascular permeability.[63,90,155] Thus, it appears that at clinically relevant doses, X-irradiation causes no acute reduction in the permeability of the BBB. While a synergism between radiation therapy and several anticancer agents has been demonstrated,[169,233] this is more likely to be due to the synchronization of the cells in mitosis or to an interaction of the molecular mechanisms involved in cytotoxicity, rather than to a radiation-induced increase in the brain uptake of water-soluble anticancer agents.

Chronically, however, radiation therapy may indeed cause permeability changes within the brain, particularly in white matter. The tolerance of the normal brain to radiation is generally the limiting factor in the successful treatment of a CNS tumor by radiotherapy. Radiation necrosis of the normal neural tissue has been reported following the conventional radiotherapy of both CNS tumors and tumors of the scalp.[128,165] Neurologic symptoms related to this generally occur from 9 months up to 2 years following the cessation of treatment with doses as low as 4500 rad over 30 days or 9000 rad over 80 days. This radiation-induced necrosis is characterized by capillary alterations, causing areas of enhancement in CT scans that are difficult to differentiate from tumor regrowth; fine structural vascular alterations such as thickened vessel walls, endothelial proliferation, perivascular fibrosis, fibrinoid necrosis, and thrombosis have been reported.[30] In addition, calcification, astrocyte proliferation, and demyelination, focal or diffuse, are commonly encountered. It has been suggested that myelin-producing oligodendrocytes are more radiosensitive than other glial cells.[146] This area has been extensively reviewed by Fike et al.[74]

Hypoxia has been suggested as one of the chief factors that limits the effect of X-irradiation on tumor tissue, although it should be emphasized that oxic tumor cells are often difficult to kill as well. Hypoxic malignant cells comprise a large proportion of cells in necrotic areas within primary brain tumors. Several attempts have been made to increase the radiosensitivity of these cells. These have included the administration of hyperbaric oxygen during radiotherapy and the use of the electron-affinic agent metronidazole. The effect of both on patient survival has generally proved disappointing. Another approach to the treatment of hypoxic tumor cells involves their irradiation by high-linear energy-transfer particles, such as fast neutrons, which, unlike conventional photons, are directly ionizing. Fast neutrons are uncharged particles with a mass of approximately 2000 times that of an electron. These particles have three significant advantages over conventional photons: (1) they require a lower oxygen enhancement ratio (i.e., the ratio of hypoxic to aerated doses needed to achieve the same biologic effect is approximately similar at all survival levels); (2) fast neutrons have a directly lethal effect on tumor cells, and there is minimal or no repair of sublethal or potentially lethal damage; and (3) there is minimal variation in the sensitivity of tumor cells to fast neutrons with cell cycle.

The acute and chronic effects of fast neutrons on the brain and, more specifically, on cerebrovascular permeability are less well documented. It is probable, however, that vascular damage is a major causative factor in the delayed brain injury related to the fast neutron radiation therapy of brain tumors. Zook et al.[266] irradiated the brains of healthy dogs with total doses of 1333, 2000, 3000, or 4500 rad of fast neutrons over 7 weeks. Those that received 2000, 3000, and 4500 rad developed severe neurologic symptoms and either died or were killed when moribund at 62–720, 47–96, and 32–98 days postirradiation, respectively. All, irrespective of radiation dose, possessed marked vascular damage, characterized by foci of multiple hemorrhages. In addition, the integrity of the BBB was reduced to IV-administered Trypan blue in each.

While fast neutron radiation therapy in patients with brain tumors has culminated in total tumor eradication, no clinical trials have demonstrated a concomitant improvement in either the quality or length of patient survival.[34,91,223] Death has resulted from the deleterious effects of the treatment on normal brain. Both fast-neutron and photon radiation therapy appear to alter cerebrovascular permeability chronically, and the former may do so acutely. These alterations, however, cannot be separated from the deleterious actions of both treatments on normal neuronal tissue, and appear to be irreversible. While radiation therapy is both essential and a mainstay in the treatment of brain tumors, its use is not applicable to the enhancement of brain drug delivery.

4.3.5. Chemical Solvents

A variety of chemical solvents, of which dimethyl sulfoxide (DMSO) and ethanol are the most often cited, have been reported to act as carriers for water-soluble drugs into brain. It has been suggested that they increase barrier permeability by acting to enhance the solubility or fluidity of the membranes of the cerebral capillary endothelial cells.

Broadwell et al.[24,25] and Wright et al.[264] recently reported that DMSO reversibly and innocuously opens the BBB to intravascular HRP in the mouse. Clinical trials were initiated on the basis of these experiments, in which DMSO plus cyclophosphamide or adriamycin were administered to patients with brain tumors.[67,138] Both Greig et al.[100] and Neuwelt et al.[182] undertook similar studies to those of Broadwell, in rat and mouse, clearly demonstrating that DMSO, in a variety of doses administered by several routes, had no effect on the brain penetration of HRP, [^{125}I]serum albumin, Evans blue–albumin, hexosaminidase A, or the anticancer agents methotrexate and melphalan. The only consistent effect of DMSO was that it caused neurotoxicity at high concentrations[100,182,264] and increased serum osmolality to levels incompatible with life.[213] Indeed, DMSO-induced changes in skin and bladder permeability have been related to its toxic actions, resulting in histologically identifiable structural alterations and lesions.[4,21] Clinical trials confirmed the inability of DMSO to increase the brain uptake of water-soluble drugs.[65,134] In fact, DMSO may inhibit the alkylating activity of drugs such as chlorambucil.[259]

The intracarotid administration of ethanol in concentrations above 30% has been demonstrated to increase the permeability of the BBB to intravascular markers. Such concentrations of ethanol, however, are not compatible with continued life. Interestingly, the intrathecal administration of benzyl alcohol, commonly used as an antibacterial agent, has likewise been associated with neurotoxicity.[55] Philips and Cragg[197–199] studied the effect of acute and chronic 6-month administration of ethanol, at the highest tolerable dose, on the permeability of [^{14}C]sucrose into the brain of rats. Even at anesthetic levels, it proved to have no effect on the integrity of the BBB. Hillbom and Tervo[115] confirmed this and also demonstrated that ethanol and acetaldehyde have no effect on the integrity of the BBB. The combination of ethanol and an alcohol-related stress, such as starvation, may, however, cause a weakening of barrier integrity, resulting in a small but significant increase in its permeability.[198]

4.3.6. Drugs

Several reports have indicated that the intracarotid administration of various anticancer agents, including etoposide[235] (VP-16), 5-FU[168] and hydroxyurea,[73] reversibly open the BBB.[92] Spigelman et al.[119,235,236] reported that the intracarotid infusion of etoposide in rat caused a threefold increase in cerebrovascular permeability. The effect was dose dependent and reversible. Unlike the osmotic and Metrazol-opening techniques, which were reversed within 4 hr,[92,95,181] the integrity of the BBB was reduced for 4–5 days after etoposide administration. The mechanism of etoposide-induced barrier opening is obviously very different from that induced by either osmotic imbalance or Metrazol. Since the duration of barrier reduction was so long, a direct toxic effect of etoposide on the cerebrovascular endothelium cannot be ruled out, especially as no histopathologic studies have yet been undertaken to assess the acute or long-term neurotoxicity of the compound. It is interesting to note that Greig et al.,[97] Edvinsson et al.,[67] and Cserr[46] reported that the duration of time that the integrity of BBB is reduced following the damage caused by a direct intracerebral injection is approximately 1 day. This is a considerably shorter time than that caused by etoposide. Indeed, the only other technique that causes barrier reduction for this duration involves the intracarotid administration of sodium dehydrocholate, a synthetic bile salt, which was associated with neurotoxicity.[234] Many questions remain concerning the ultimate significance of the etoposide opening technique, the mechanisms involved, its reproducibility (not all animals are affected), and possibly of greatest importance, whether or not it demonstrates an acceptable risk/benefit ratio. Further investigations are obviously warranted. Whatever their outcome, however, the technique would have severe limitations in the clinic. Barrier reduction over such a prolonged period would probably result in severe cerebral edema, as well as the brain entry of potentially neurotoxic drugs and metabolites if the technique was combined with the co-administration of antineoplastic agents.[137,183]

MacDonell et al.[168] have reported that 5-FU caused an increase in cerebrovascular permeability which was reversed within 7 hr. Neuwelt et al.[182] were unable to confirm this in a quantitative study using 5-FU concentrations twice those of the former study. Neuwelt et al.[183] did, however, demonstrate that the antineoplastic agent cisplatinum caused barrier reduction in the dog following its intracarotid administration. This, however, was associated with severe neurotoxicity and produced multiple hemorrhagic and necrotic foci. Interestingly, intracarotid cisplatinum has demonstrated notable activity in the treatment of brain tumors in children.[58,141] Heavy metals such as mercuric or nickel chloride have similarly been demonstrated to cause irreversible barrier damage and neurotoxicity.[192]

5. CHEMICAL MODIFICATION OF DRUGS

It is apparent from the preceding sections that the systemic pharmacokinetics of a compound can be altered dramatically by a variety of biologic approaches (i.e., schedule, concentration, and route of administration). By understanding the factors that govern the penetration of drugs into the CNS, it is possible to optimize systemic pharmacokinetics, albeit in a limited manner, to maximize brain drug uptake. For compounds with the correct physicochemical properties, biologic manipulation can make it possible to either increase and/or maintain the brain drug concentration within the therapeutic window for longer (see Sections 2 and 3). For other drugs, however, intraparenchymal drug delivery is inadequate despite these manipulations. The polarity and functional groups present on the drug, in many cases essential to drug–receptor interactions, make the Galenic formulation unsuitable for activity at the target by virtue of a transport deficiency. It is possible to overcome such pharmacokinetic limitations of an otherwise active compound by chemical modification.

In theory, there are three approaches by which this can be achieved: (1) by synthesizing active analogues that are more lipophilic than the parent compound; (2) by synthesizing prodrugs to link an active compound temporarily to a transport moiety (the transport moiety is generally lipophilic in nature and often masks a water-soluble substituent on the parent compound); and (3) by synthesizing active compounds that are structurally similar to endogenous agents, which are transported into the brain by a carrier-mediated transport system.

In general, the first technique, the synthesis of active analogues, is most often undertaken to increase the brain uptake of a compound. The fundamentals underlying the chemical modification of a drug to increase its lipophilicity are outlined in Section 2.1. The major difference between the design and synthesis of analogues and prodrugs is that analogues are irreversible derivatives of active parent compounds in which a substituent, often not essential to activity, is chemically modified or replaced. Prodrugs, however, are reversible derivatives in which any functional group, irrespective of its involvement in drug–receptor interactions, is linked to a temporary transport moiety.

The principle common to both analogues and prodrugs is that the improved physicochemical characteristics of the compound will enable it to readily reach its site of action. However, unless there is some form of retention, most often as receptor binding, the compound will reach its target, produce a momentary burst of activity, and then rapidly equilibrate within the rest of the body. This can cause systemic side effects unless there is a selectivity of drug action in the target tissue alone. The additional concept of prodrugs, which is seldom achieved for those that are brain directed, is that they should be cleaved from their transport moiety within the target organ, and thereby become sequestered there.

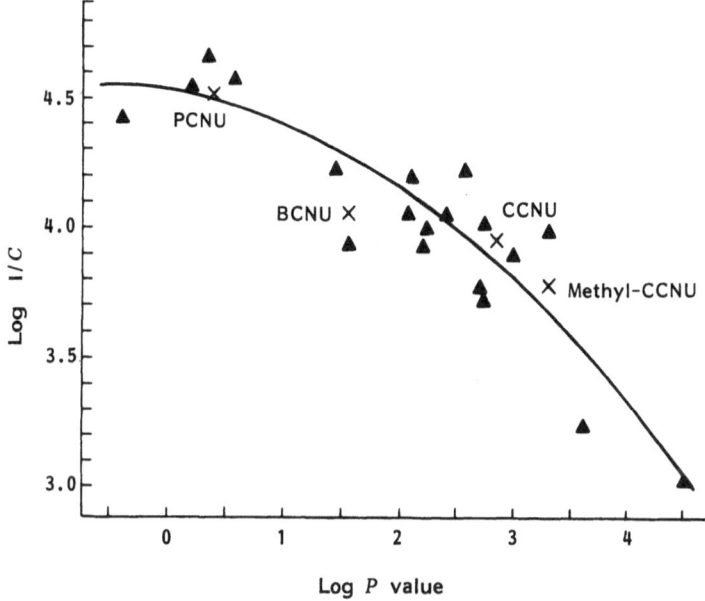

Figure 8. Quantitative structure–activity relationship between the lipid partition coefficient of 23 nitrosourea derivatives and their activity against intracerebrally implanted L1210 leukemia in the mouse. $1/C$ = activity. (From Hansch et al.[112])

5.1. Analogues

The principles governing the development of drug analogues for improved brain penetration have been outlined in Section 2. Quantitative structure–activity relationships have been undertaken on a variety of drug classes to optimize their brain delivery and activity. The most active depressant drugs of all classes have partition coefficients of approximately log P value 2.0, in the octanol/water system. The most extensive similar studies undertaken in the field of cancer chemotherapy have involved the nitrosoureas. Calculations of the log P values of 23 nitrosoureas were undertaken by Hansch et al.[112] and were related to their activity against intracerebrally implanted L1210 leukemia in the mouse. Figure 8 demonstrates the results, where C is the molar concentration producing a 175% increase in the survival of the treated animals compared with the controls. The negative slope of the plot of log $1/C$ versus log P indicates that increasing the lipophilicity of the nitrosoureas beyond an optimal of approximately log P of value 0.3 yields compounds less active against intracerebral L1210 leukemia in mice. In studies involving the commonly used nitrosoureas, BCNU, CCNU, and methyl-CCNU (log P values 1.53, 2.83, and 3.3, respectively) (Fig. 9), Greig[92] and Levin and Kabra[152] confirmed that their activities decrease progressively from BCNU to CCNU to methyl-CCNU against intracerebral Walker carcinosarcoma and 9L gliosarcoma in the rat. Levin and Kabra[153] further reported that the rate and extent of brain RNA, DNA, and protein binding by BCNU was more than 10 times higher than that of CCNU following the IV administration of similar concentrations. Clinical trials using these compounds in the treatment of primary brain tumors have in general confirmed these studies.[64,92]

It is interesting that while the nitrosourea study by Hansch et al.[109] demonstrated that

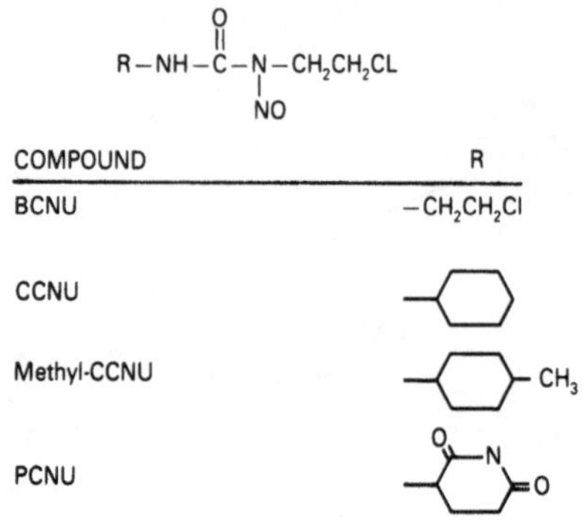

Figure 9. Chemical structures of the commonly used nitrosoureas, BCNU, CCNU, methyl-CCNU, and PCNU.

highly lipophilic derivatives are less active than their moderately lipophilic analogues, unlike similar studies on a variety of depressant and hypnotic drugs, no parabolic relationship between activity and log P was demonstrated. Furthermore, the optimal log P value was on the order of -0.4 to 0.4 rather than 2.0. Similar to many other classes of drugs, a combination of factors is involved in the final activity of the nitrosoureas. They rapidly and spontaneously decompose to form chloroethyl moieties[237] that can alkylate purine and pyrimidine bases of DNA and cross-link double-stranded DNA.[188] This disruption in nucleic acid function accounts for their antitumor activity. The reduced aqueous solubility and increased binding of the analogues to serum albumin in direct proportion to their log P values[248] probably results in the alkylation, and hence inactivity of the more lipophilic derivatives before they reach the brain. Interestingly, Weinkam et al.[253] reported that human serum albumin-mediated catalysis of the nitrosoureas occurs.

Hansch's[112] and Levin's[152] studies prompted investigations of the moderately lipophilic nitrosourea, PCNU (log P value 0.37) (Fig. 9), which eventually culminated in clinical trials. As expected, this compound demonstrated significant activity against intracerebral tumors in rodents and appeared to enter the brain readily.[178] Its clinical activity in patients with brain tumors has proved equal but not superior to that of BCNU.[92]

Two further interesting compounds specifically designed for intracerebral uptake and activity that did not use the Hansch approach, are spirohydantoin mustard (SHM) and the aziridinylbenzoquinone (AZQ). The observation that the centrally active anticonvulsive agent 5,5-diphenylhydantoin (phenytoin) penetrated the BBB in significant quantities[76] and preferentially localized in brain tumor tissue in concentrations of up to 4 times that in normal brain,[212] prompted the investigation of hydantoin derivatives as carriers for nitrogen mustard for brain tumor therapy. Peng et al.[195] synthesized and tested a number of nitrogen mustard hydantoin derivatives against a variety of murine tumors. The moderately lipophilic compound SHM was one of the most promising. 5,5-Diphenylhydantoin has a log P value of 2.47 in the octanol–water system. The addition of a nitrogen mustard alkylating group to the molecule, to confer antitumor activity, significantly increased its partition coefficient to log P value 4.69. Such compounds are too lipophilic, being minimally soluble in the aqueous phases of plasma and ECF. The removal of an aromatic ring, to yield the drug SHM (Fig. 10), reduced the lipophilicity of the compound to approximately log P value 2.5, within the calculated optimal range of lipophilicity for

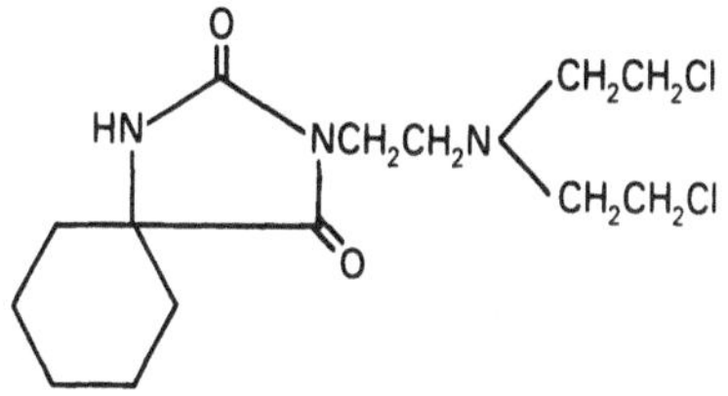

Figure 10. Chemical structure of spirohydantoin mustard (SHM).

brain uptake. Following IV administration, SHM was found to penetrate the CSF of the dog.[200] However, the CSF concentrations achieved were lower than expected, which may have been due to plasma protein binding. Despite this, the compound has demonstrated significant activity against intracerebral ependymoblastoma in the mouse. However, the assessment of the clinical activity of SHM in patients with CNS tumors has thus far been hampered by dose-limiting neurotoxicity, which resembles anticholinergic overdrive.[92] Recent studies have indicated that the co-administration of either physostigmine or haloperidol reduces this side effect, permitting the dose escalation of SHM.[225] Ongoing clinical trials are reassessing the effectiveness of the compound in brain tumor therapy. It is interesting to note that Mauger and Ross[166] previously used the hydantoin ring structure as a carrier in a series of active bis(2-chloroethyl)aminoarylhydantoins. Unfortunately no brain activity experiments were reported.

The quinone structure is a common chemical feature in a number of active anticancer agents: adriamycin, daunorubicin, streptonigrin, and mitomycin C, all derived from fermentation of plant products.[6,61] Among the hundreds of purely synthetic quinone analogues evaluated, *p*-benzoquinone derivatives containing multiple aziridinyl groups proved to possess significant antitumor activity. As a consequence, Chou et al.[39] synthesized a series of aziridinyl quinones of differing solubilities and tested their activity against a variety of intraperitoneal and intracerebral tumors in the mouse. The most apparent qualitative structure–activity relationship was that the most polar water-soluble derivatives demonstrated the greatest activity against the intraperitoneal ascitic tumors, whereas the more lipophilic compounds induced the greatest inhibition of the intracerebrally sequestered tumors.[62,142] AZQ (2,5-diaziridinyl-3,6-bis-carboethoxy-amino-1,4-benzoquinone, (Fig. 11) log *P* value 0.05 (J. Driscoll, personal communication) was eventually chosen for pharmacokinetic studies and clinical trials. Being a small un-ionized compound, it was found to enter the brain readily, and reached approximately 30% of its corresponding plasma concentration in the monkey.[89] AZQ has demonstrated

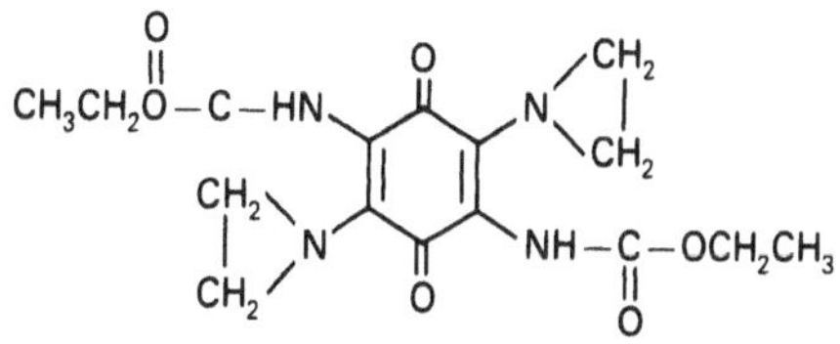

Figure 11. Chemical structure of aziridinylbenzoquinone (AZQ).

some limited activity, which was inferior to BCNU, against primary brain tumors.[92] It is unlikely that AZQ would ever be of therapeutic value as a single agent, but it may prove useful in combination chemotherapy.

Analysis of the design of the three described compounds, PCNU, SHM, and AZQ, demonstrates several interesting points that are reiterated in the design of prodrugs. First, it is possible to develop a rational series of active compounds from a lead agent to produce compounds that may have greater activity or more optimal pharmacokinetics for the treatment of a specific disease or for activity in a specific location. The discovery of the lead compound is most often a combination of both luck and knowledge. Second, an optimal lipophilicity, log P value 2.0, has been predicted for the transport of drugs into the brain. This value, however, is not universal. A combination of factors, such as plasma protein binding and rate of systemic disappearance, determines the final concentration of a drug in the brain. These additional factors can substantially lower the optimal log P value for a specific series of compounds.

5.2. Prodrugs

The prodrug principle consists of the covalent attachment of a carrier moiety to an active drug to alter its physicochemical properties. This is followed by the enzymatic cleavage of the carrier moiety from the prodrug, preferably within the target organ, to release the active drug (Fig. 12). Depending on the nature of the transport moiety, the regeneration of the parent compound can be either enzymatically or chemically mediated.

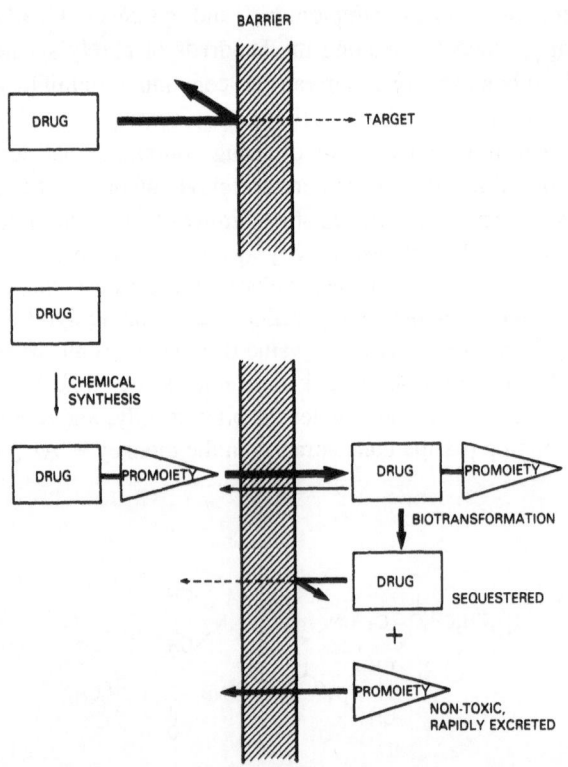

Figure 12. The concept of brain-directed prodrugs.

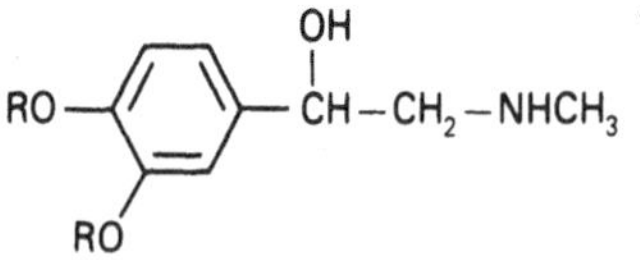

Figure 13. Chemical structures of epinephrine and the epinephrine prodrug dipivaloylepinephrine (DPE).

The conditions that trigger chemically reconverting prodrugs are somewhat limited. It is therefore difficult to achieve any form of targeting with them, as there is little opportunity for their selective activation within a specific site. In addition, such prodrugs are often unstable prior to administration. As a consequence, this review is limited to a discussion of enzymatically activated prodrugs.

As reviewed by Higuchi and Stella,[114] Stella,[239] Notari,[184] Sinkula and Yalkowsky,[228] and Wermuth,[256] well-designed prodrugs satisfy the following criteria:

1. The linkage between the active parent compound and the transport moiety must be broken in vivo.
2. The generation of the active form must occur with sufficiently rapid kinetics to ensure effective drug concentrations at the site of action before systemic prodrug metabolism occurs.
3. The prodrug, as well as the released transport moiety, must be nontoxic.
4. The prodrug is preferably inactive or less active than the parent compound.

The chemical manipulation of a drug requires a handle from which derivatives can be synthesized. The two commonly used starting groups are hydroxyl and carboxylic acid residues.

5.2.1. Hydroxyl-Linked Prodrugs

Starting from hydroxylic derivatives, increased lipophilicity can be obtained simply by esterification or sometimes by etherification. For example, dipivaloylepinephrine (DPE) is a dipivalate ester of epinephrine (Fig. 13), that crosses the cornea; it is used in the treatment of glaucoma.[167] McClure[167] reported that DPE is 100 times more potent than epinephrine in lowering intraocular pressure. In addition, the associated systemic toxicities of epinephrine, related to its untoward cardiac effects, are greatly reduced. The greater effectiveness of DPE compared with epinephrine, and subsequent decreased dose requirement, are due to the increased corneal penetration of the ester, following topical administration, and a decreased rate of epinephrine metabolism. The prodrug DPE is significantly more lipophilic than epinephrine by virtue of its low polarity pivalate moieties that mask the phenolic hydroxyl groups of epinephrine. In addition, the metabolism of epinephrine, which involves the methylation of the *meta*-hydroxyl moiety of the phenolic ring, is delayed until DPE undergoes enzymatic hydrolysis to epinephrine. This prolongs the duration of action of the prodrug. DPE is relatively stable in the absence of enzymes, making it ideal for topical application. Following its absorption, however, DPE rapidly regenerates epinephrine after its hydrolysis by unspecific esterases present in the plasma and lens.

The site direction of prodrugs to the brain is more complicated, since the compounds are subjected to enzyme catalysis in the plasma and during their first passage through the liver before they reach the brain. Crevelling et al.[45] have reported an increased brain uptake of the norepinephrine derivatives, 3,4,β-triacetyl and 3,4,β-trimethylsilyl nor-epinephrine, relative to norepinephrine itself. Unfortunately, neither regenerated nor-epinephrine in brain tissue at a rate sufficient to produce activity.

More successful experiments were undertaken by Horn et al.[122] on the potent, selective, and long-lasting dopaminergic agonist 2-amino-6,7-dihydroxytetrahydronaph-thalene (ADTN) (Fig. 14).[121] Although many dopamine-receptor antagonists are available clinically, the number of useful dopamine agonists is small. It is therefore unfortunate that the dopamine analogue ADTN, like dopamine, is restricted from crossing the BBB.[120] Furthermore, previous studies on dopamine esters have demonstrated them to be of limited value, as they were extensively metabolized by monoamine oxidase (MAO), a problem not encountered with ADTN. Horn et al.[120,122] demonstrated in the rat that the intraperitoneal administration of dibenzoyl-ADTN (Fig. 14) led to the maintenance of plateau like concentrations of ADTN in brain over several hours. There appeared to be a selectivity of ADTN accumulation in dopamine-rich brain areas such as the nucleus accumbens, olfactory tubercle, striatum, cortex, and cerebellum, in descending order. In addition, there was an apparent selectivity of action on presynaptic rather than on post-synaptic dopamine receptors. Further investigations involving the diacetyl and diisobuty-ryl esters of ADTN gave peak brain concentrations of ADTN at 5–15 min after admin-istration. The enzyme hydrolysis of these esters was more rapid than that of the dibenzoyl ester, or indeed than that of the dipivaloyl ester. These produced maximal ADTN con-centrations within the brain at approximately 3 hr which were still measurable at 10 hr postadministration.

Finally, possibly the most successful example of a hydroxyl-linked brain-directed prodrug is the diacetyl derivative of morphine, heroin. By virtue of its lipid solubility, heroin readily enters the brain. It is then rapidly hydrolyzed to monoacetylmorphine and morphine, which are responsible for its pharmacologic actions.

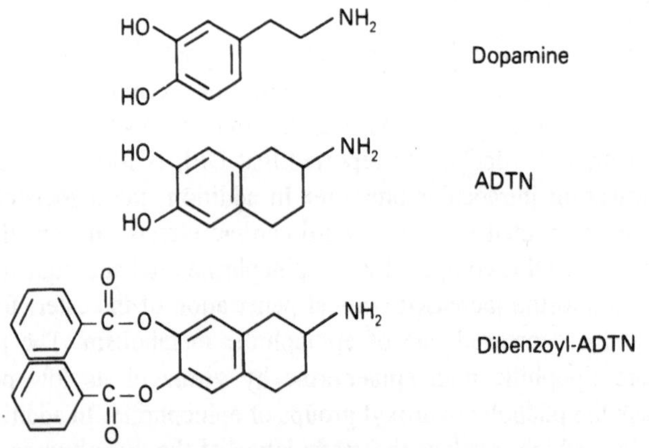

Figure 14. Chemical structures of dopamine, the dopaminergic agonist ADTN, and the ADTN prodrug dibenzoyl-ADTN.

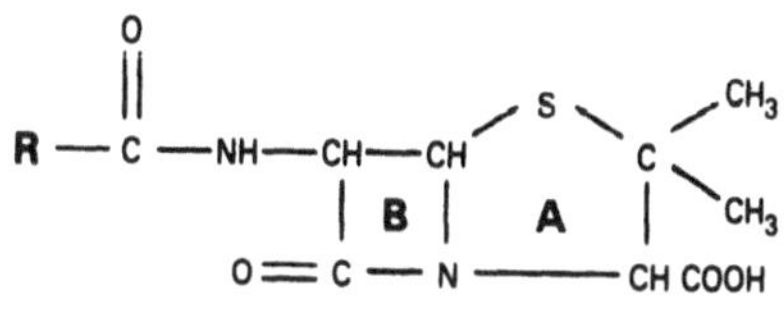

Figure 15. The basic chemical structure of the penicillins, consisting of a thiazolidine ring (A), a β-lactam ring (B), and a side chain (R).

5.2.2. Carboxyl-Linked Prodrugs

Lipophilic prodrugs can also be derived from a carboxylic function, the most commonly used derivatives being carboxylic esters. Possibly the most successful application of prodrug design in medicine has involved the penicillin β-lactam antibiotics. The basic structure of the penicillins is shown in Fig. 15; it consists of a thiazolidine ring (A) connected to a β-lactam ring (B), which together form the primary structural requirements for biologic activity. A side chain (R) is attached to the β-lactam ring, alterations in which determine many of the antibacterial and pharmacologic characteristics of the penicillin analogues. A variety of analogues have been synthesized to overcome one or more shortcomings of penicillin G.[84,107] These are related to (1) the spectrum of activity of penicillin G, (2) its resistance to penicillinases, and (3) its poor absorption from the GI tract. Penicillin G, in common with all penicillins, has a relatively acidic thiazolidine carboxyl group, pK_a 2.7. The compound is therefore highly ionized at both physiologic and intestinal pH.

Simple alkyl and aryl esters of the thiazolidine carboxylic acid were synthesized to increase the lipophilicity of the molecule to improve its oral absorption. Although such esters were both rapidly absorbed and hydrolyzed to penicillin G in rodents, they proved too stable in humans.[9,107] The thiazolidine carboxyl group is essential for antibacterial activity, therefore, ester hydrolysis must occur.

Jansen and Russell[131] resolved this problem by the synthesis of a series of double esters

$$R-CO-O-CH_2-O-CO-CH_3,$$

where R is the penicillin residue, in which the second hydroxyl group of the first ester was esterified by a simple carboxylic acid. Such acyloxymethyl esters rapidly regenerated penicillin in the blood and tissues of humans. Enzyme hydrolysis proceeds via a chemically unstable hydroxymethyl ester that rapidly collapses to release the parent compound and formaldehyde.

The synthesis of acyloxymethyl esters was applied to ampicillin,[48] an aminobenzyl penicillin derivative (Fig. 16), that is the most widely used broad spectrum antibacterial penicillin. Ampicillin contains an amino group on the side chain. The compound therefore exists as a zwitterion in the pH range of the GI tract and is poorly absorbed. The pK_a of the amine and acid functions are 7.2 and 2.6, respectively.

Several acyloxymethyl ampicillin esters, bacampicillin, pivampicillin and talampicillin (Fig. 16), have been synthesized.[16,41,226] All are more lipophilic than ampicillin and yet can be administered orally as their water-soluble amine salts. The esters penetrate, greater than 98%, the GI tract and are rapidly hydrolyzed, in less than 15 min, to ampicillin in the plasma. Simon et al.[227] reported that serum levels of ampicillin are equal following the oral and IV administrations of equimolar concentrations of bacampicillin

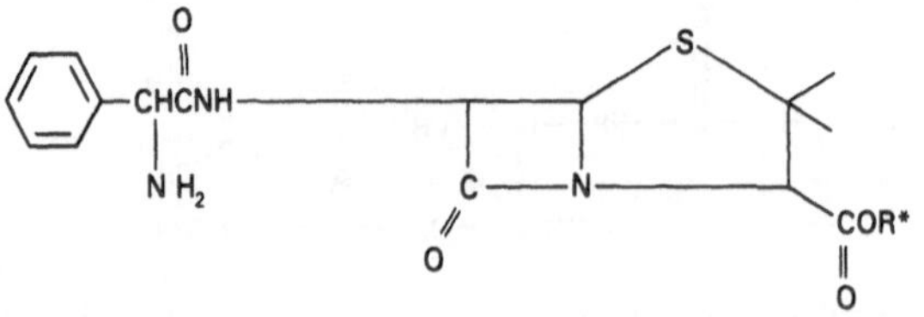

COMPOUND	R*
Ampicillin	—H or ⊖
Pivampicillin	—CH₂OCOC (CH₃)₃
Bacampicillin	—CH (CH₃) OCOOCH₂CH₃
Talampicillin	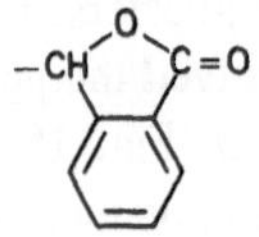 —CH⟨...⟩C=O

Figure 16. Chemical structures of ampicillin and the ampicillin prodrugs pivampicillin, bacampicillin, and talampicillin.

and ampicillin, respectively. Furthermore, the oral administration of bacampicillin resulted in fivefold higher serum ampicillin levels than equimolar oral ampicillin. Clinical trials have confirmed the efficacy and safety of these acyloxymethyl ampicillin ester prodrugs. The recommended daily oral dose of bacampicillin is 0.8 g for the treatment of common infections. This results in less toxicity to intestinal flora but provides equal activity to 2.0 g of oral ampicillin per day.

An interesting further development in the double ester concept is the probenecid ester of ampicillin.[38] This represented an attempt to design a prodrug in which the liberated metabolite of ester hydrolysis would further improve the pharmacokinetics of the regenerated parent compound. It was postulated that the released probenecid would extend the plasma half-life of ampicillin by reducing its active secretion in the kidney. In preliminary studies in the dog and monkey, peak serum ampicillin concentrations following the oral administration of the probenecid prodrug were lower than those following the similar administration of bacampicillin, pivampicillin, or talampicillin. There was, however, an indication that the half-life of ampicillin was extended following the administration of the probenecid prodrug in the dog. The analysis of further studies is awaited with interest.

The prodrug principle has been successfully extended to other penicillin analogues. The esterification of the side-chain carboxyl group of carbenicillin, a dicarboxylic penicillin (Fig. 17), has resulted in two orally absorbed carbenicillin prodrugs.[31,38] These are the indanyl ester, geocillin or cardinocillin, and the phenyl ester, carfecillin (Fig. 17). Interestingly, the hydrolysis of these simple aryl esters of the side chain carboxyl group, rather than those of the thiazolidine ring, proceeds fairly rapidly in human blood. Both prodrugs have proved active in the treatment of urinary tract infections. Finally, the pivaloyloxymethyl ester of mecillinam, pivmecillinam, an antibiotic possessing an amidino rather than an amido link between the β-lactam nucleus and side chain (Fig. 18), was successfully developed to increase the GI absorption of mecillinam.[50,211]

The development of carboxyl-linked prodrugs for brain delivery has met with mixed success. Considerable effort has been channeled into the evolution of GABA-mimetics that are capable of entering the brain. The central inhibitory neurotransmitter GABA (Fig.

COMPOUND	R*
Carbenicillin	−H or ⊖
Carfecillin	
Geocillin	

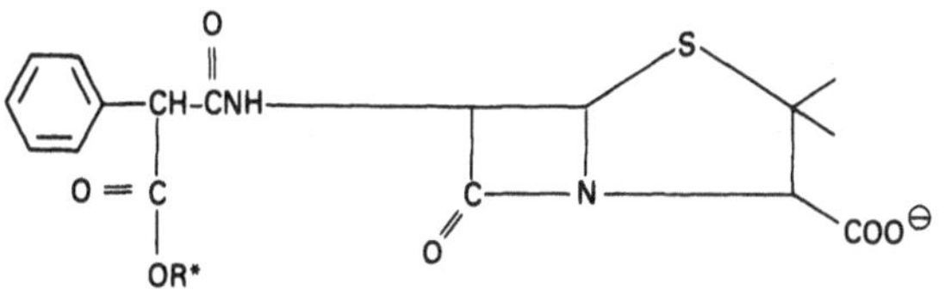

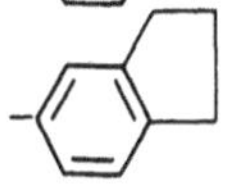

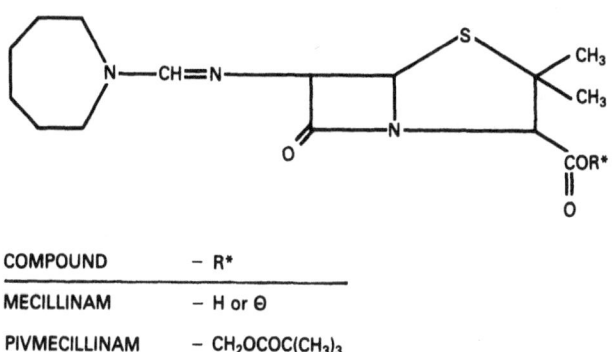

Figure 17. Chemical structures of carbenicillin and the carbenicillin prodrugs carfecillin and geocillin.

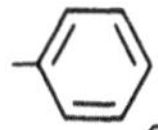

19) is a zwitterion at physiologic pH, pK_a 4.0 and 10.7, and minimally penetrates the BBB. A variety of potential GABA prodrugs have been synthesized to mask the amino group by acylation[79]; these have included *N*-pivaloyl and *N*-benzoyl-GABA.[82] Both compounds appeared to enter the brains of rats in significant amounts, reaching peak concentrations at 30 and 5 min, respectively. However, the in vivo enzymatic cleavage of such amides has proved difficult in humans. In addition, a variety of GABA analogues have been synthesized. The most interesting of these are muscimol and baclofen (Fig. 19). Muscimol, unfortunately, is rapidly metabolized systemically so that only a small proportion, 0.02%, of the unchanged compound reaches the brain.[8,163] Furthermore, its octanol–water partition coefficient, log *P* value 2.0, is not significantly different from that of GABA, log *P* value 2.1. Conversely, baclofen appears to exhibit brain activity but this is probably not related to its GABA-like structure.[52,55,202]

A further GABA analogue, which appears to be a potent and specific GABA-agonist, is isoguvacine (Figs. 19 and 20).[136] Unfortunately, like GABA, the compound is a zwitterion at physiologic pH, pK_a 3.6 and 9.8, and minimally enters the brain. Krosgaard-Larsen et al.[145] synthesized a variety of carboxyl-linked esters of isoguvacine. Simple

COMPOUND	− R*
MECILLINAM	− H or ⊖
PIVMECILLINAM	− $CH_2OCOC(CH_3)_3$

Figure 18. Chemical structures of mecillinam and the mecillinam prodrug pivmecillinam.

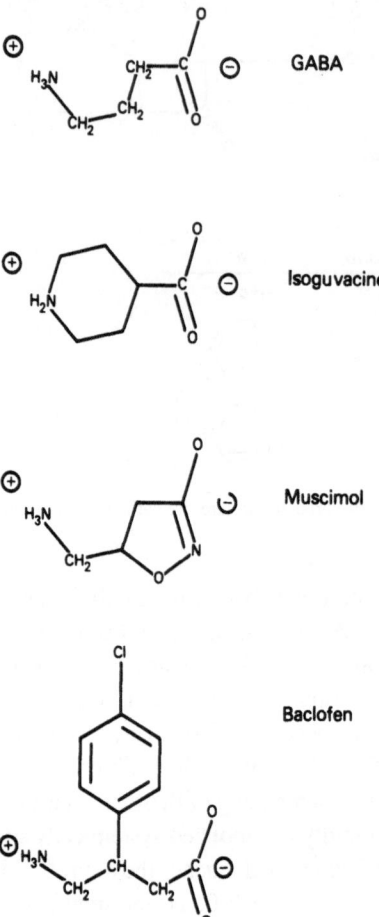

GABA

Isoguvacine

Muscimol

Baclofen

Figure 19. Chemical structures of GABA, and the GABA analogues isoguvacine, muscimol, and baclofen.

alkyl esters, such as the butyl ester (Fig. 20, I), proved to be too stable. As a consequence, a series of acyloxymethyl isoguvacine esters were prepared (Fig. 20, II–V).[70] The hydrolysis half-lives of II–V, to release isoguvacine under physiologic conditions, were more rapid and were strongly dependent on the structure of the acyl moieties in the ester groups. The rates of decomposition correlated well with their anticonvulsant activity in mice. Compound I demonstrated no activity until 17 hr postadministration, while compounds II and III demonstrated activity after 5 hr and 1 hr, respectively. Compounds IV and V were inactive. It is probable that both were hydrolyzed too rapidly in the blood to reach significant concentrations in the brain. Further studies of these compounds are awaited with interest.

In the field of cancer chemotherapy, several investigators have attempted to design brain-directed prodrugs for the treatment of CNS tumors. Two anticancer agents that have been extensively studied are methotrexate and chlorambucil. Over short periods of time, neither appear to cause CNS toxicity following their administration after reversible BBB disruption.

Methotrexate, log P value 1.85, is restricted from entering the brain by the presence of two amine and two carboxyl residues, which become charged at physiologic pH (Fig.

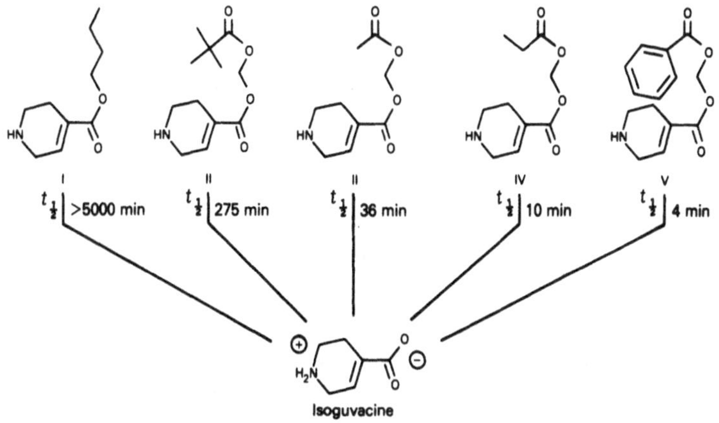

Figure 20. Chemical structures of isoguvacine and five isoguvacine prodrugs, together with the conversion half-lives of the prodrugs to isoguvacine in the presence of human serum. (From Falch et al.[70])

21). As a result, less than 0.1% of the drug exists in the un-ionized form. Rosowsky et al.[213,214] have undertaken extensive studies on the structural alteration of the glutamine moiety of the molecule, and in particular of the two ionizable carboxyl groups. The chemical modification of the pteroyl moiety or the C-O group in the amide link that separates the glutamine and pteroyl moieties of methotrexate causes a dramatic decrease in the activity of the compound.

The primary purpose of Rosowsky's manipulations of methotrexate was to produce lipophilic derivatives that would freely enter tumor cells by passive diffusion. This would overcome resistance pathways associated with alterations in the normal carrier-mediated transport of methotrexate into cells. Since the α-carboxyl group proved to be essential for activity, it was important that the derivatives regenerated methotrexate. Rosowsky synthesized a variety of alkyl esters, up to 16 carbons chain-length, as well as amide and L-homocysteic (SO_3H) analogues. Their activities were evaluated by their ability to inhibit the enzyme dihydrofolate reductase, in a cell-free system, and to kill CEM lymphoblasts in vitro.

Since similar physicochemical properties were required to enhance the transport of methotrexate into tumor cells and across the BBB, pharmacokinetic experiments were undertaken to assess the brain uptake of the lipophilic γ-monobutylester (Fig. 21) in the rhesus monkey.[213] Preliminary experiments had demonstrated that it broke down to

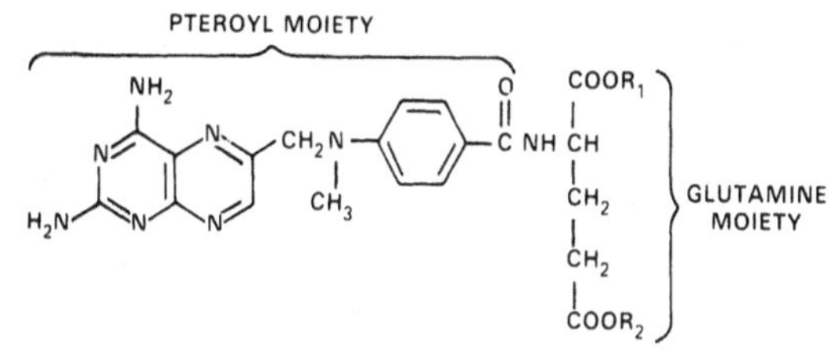

COMPOUND	R_1	R_2
METHOTREXATE	H	H
γ MONOBUTYL METHOTREXATE	H	C_4H_9

Figure 21. Chemical structures of methotrexate and the methotrexate prodrug γ-monobutyl methotrexate.

methotrexate alone in vivo. Following IV administration, however, the compound became highly protein bound, greater than 99%. As a consequence, the combined CSF levels of the ester and methotrexate were not significantly greater than those following the IV administration of equimolar methotrexate.

An interesting variation of these methotrexate prodrug studies was the synthesis of a series of nonclassic antifolates by Elslager and Davoll[68] which produced trimetrexate, a lipophilic analogue of methotrexate that lacks a glutamine residue. Like methotrexate, trimetrexate was a potent inhibitor of the enzyme dihydrofolate reductase.[130] However, as trimetrexate is lipophilic, log P value 0.88 versus methotrexate log P value -1.85, it was found to inhibit the growth of tumor cells that were resistant to methotrexate by virtue of an intracellular transport defect.[135] Like methotrexate, its in vivo antitumor activity in murine models was found to be schedule-dependent. However, due to its different physicochemical characteristics, its pharmacokinetics were different from methotrexate. It had a biphasic clearance, 40 min and 10 hr in humans,[130] that was longer than methotrexate and primarily dependent on biotransformation as opposed to renal secretion. Despite the extended plasma half-life and lipophilicity of the compound, however, minimal trimetrexate has been found to enter the brain.[156] Balis et al.[7] reported that following IV drug administration to rhesus monkey, the CSF/plasma ratios of trimetrexate and methotrexate were 0.034 and 0.021, respectively, which were not significantly different. Similar to the lipophilic methotrexate γ-monobutyl ester of Rosowsky et al.[213] trimetrexate was found to bind heavily to plasma proteins.[7]

Greig and Rapoport[104] have undertaken extensive systemic and brain pharmacokinetic and activity studies on a variety of esters of the alkylating agent chlorambucil (Fig. 22). Chlorambucil is restricted from entering the brain by a single carboxyl group, which becomes ionized at physiologic pH. The in vitro measurement of the rate of hydrolysis of the esters in rat and human plasma demonstrated that all regenerated chlorambucil. However, the half-lives of the short-chain alkyl esters as well as the simple and more complex aryl esters, including the prednisolone ester prednimustine, were extremely rapid. Following their IV administration in the rat, none was sufficiently stable to accumulate in the brain prior to systemic hydrolysis. The longer-chain hexyl and octyl esters (Fig. 22) proved to be more stable in plasma, with half-lives in excess of 10 min prior to their hydrolysis to chlorambucil. Unfortunately, these highly lipophilic compounds, log P value > 4, became heavily bound to plasma proteins following their IV administration in the rat. As a consequence, their total alkylating concentration in the brain was not significantly greater than that following the IV administration of equimolar chlorambucil. All the esters demonstrated activity similar to chlorambucil against intracerebrally implanted Walker carcinosarcoma in the rat. Recently, Greig and Rapoport[104a] have designed and tested compounds which by steric hindrance, have sufficient half-lives, without extensive plasma protein binding, to enter the brain. One of these has a brain/plasma ratio of between 0.45 and 0.7 compared to that of 0.017 for chlorambucil[103a], and has demonstrated activity against a metastatic brain tumor model in rats.

Finally, Bodor et al.[17-19] have attempted to use a dihydropyridine–pyridinium redox system to deliver a variety of quaternary ammonium and quaternary ammonium-linked compounds to the brain. The chemical reduction of a quaternary pyridinium salt to its tertiary dihydropyridine makes it considerably more lipophilic. N-substituted dihydropyridines are relatively unstable and are rapidly oxidized to the parent quaternary compound. The concept was first used to increase the brain uptake of the highly polar quaternary ammonium anticholinesterase inhibitor, 2-PAM (pralidoxime), by administer-

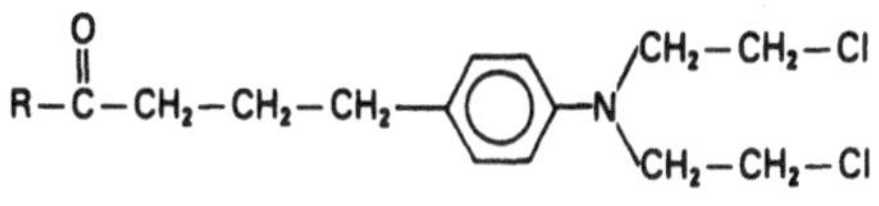

COMPOUND	R
Chlorambucil	H–O
Methyl Derivative	CH_3–O
Propyl Derivative	$CH_3\ (CH_2)_2$–O
Hexyl Derivative	$CH_3\ (CH_2)_5$–O
Octyl Derivative	$CH_3\ (CH_2)_7$–O
Methyl Benzyl Derivative	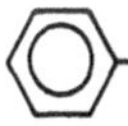–CH_2–O
Ethyl Benzyl Derivative	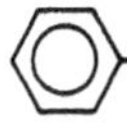–$(CH_2)_2$–O
Prednimustine	Prednisolone–O
Amide	NH_2
Glycinamide	$NH_2C(O)CH_2$ NH

Figure 22. Chemical structures of chlorambucil and a variety of chlorambucil derivatives (prodrugs).

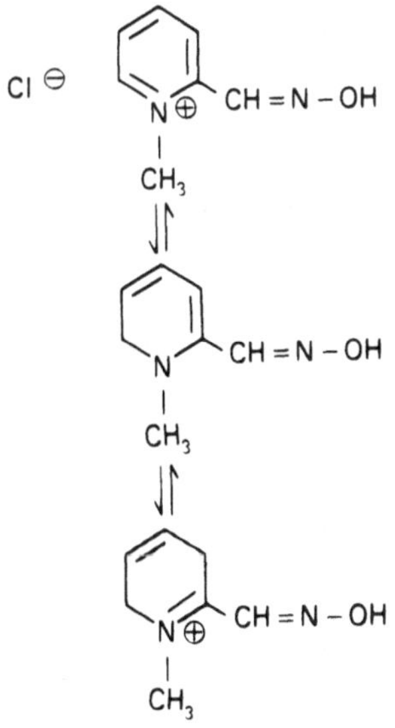

(1)

(2)

(3)

Figure 23. Chemical structures of the dihydro derivatives: (1) 2–PAM, (2) Pro–2–PAM, and (3) the stable immonium salt.

ing it as its 5,6-dihydropyridine derivative, Pro-2-PAM (Fig. 23).[19] Pro-2-PAM is stable in vitro as is its immonium salt. Although a dramatic 13-fold increase in the brain concentration of 2-PAM was achieved in the mouse, the prodrug was not brain specific. Most of the administered Pro-2-PAM was converted to 2-PAM in peripheral tissues.

This concept was taken a step further by the development of a pyridinium carrier to which drugs could be attached by a carboxyl, or more commonly, by an amide link. Studies involving a dopamine-linked compound demonstrated that some of the dihydropyridine dopamine derivative was delivered to the brain, where it became sequestered as the quaternary derivative.[18] The majority, however, was oxidized in peripheral tissues. Furthermore, there were few data to support the suggestion that dopamine was then released from the quaternary derivative, within the brain, at a sufficient rate to be of therapeutic value. The brain delivery of a variety of other dihydropyridine-linked compounds is being assessed; these include GABA as well as several anticancer agents. The success or failure of the technique ultimately relies on stabilizing the highly unstable dihydropyridine-linked compounds prior to their administration and controlling the rate of enzymatic cleavage of the drug from the carrier within the brain. The latter is particularly important since quaternary ammonium salts have been demonstrated to be actively transported out of the brain by the choroid plexus.[246,247]

5.2.3. Analysis of Prodrug Design

The reason that site specificity has thus far proved such an elusive goal for brain-directed prodrugs is that most enzymes involved in metabolizing prodrug bonds are nonspecific and ubiquitous. This is of benefit for prodrugs designed for increased corneal or GI uptake, such as dipivaloylepinephrine or bacampicillin, as they are absorbed from environments of low enzyme activity. Following their penetration into the body, they both rapidly regenerate their parent active drug in local tissue and plasma. The delivery of prodrugs to the brain is infinitely more complex. The compounds must be transported via the blood and through the liver, sites of high enzyme activity, before reaching their target. The preferential localization of highly specific enzyme systems in organs other than the liver and kidney is rare. Until one is characterized within the brain, however, true site specificity is unachievable. Thus far, brain-directed prodrugs have generally relied on designing lipophilic esters that are sufficiently stable in plasma and liver so that a significant proportion will eventually reach the brain. Here, it is hoped that they will become hydrolyzed and sequestered. Prodrugs that are too stable, such as the 3,4,β-triacetyl and 3,4,β-trimethylsilyl esters of epinephrine, reach significant concentrations within the brain but are inactive.[45] Conversely, prodrugs that are too labile, such as the short-chain alkyl esters of chlorambucil, are hydrolyzed to their minimally lipophilic parent compound before they reach the brain.[104]

An analysis of the kinetics of the hydrolysis of various ester species of drugs is beyond the scope of this chapter but has been extensively reviewed by Testa and Jenner[245] and Charton.[37] Simplistically, the affinity and reactivity of esterases for prodrugs are dependent on the lengths and shapes of the groups on either side of the ester link.[59,144] The effects of changes in the alkyl or acyl (parent drug) parts of the substrate are largely independent.[59] As a consequence, information gained from the measurement of the rate of enzyme hydrolysis of a series of ampicillin esters can be used to predict the lability of other ampicillin prodrugs by correlation analysis.[37] The same information, however, is not directly applicable to similar esters of chlorambucil, GABA or methotrexate. As demonstrated by analogues of carboxyl penicillin (Fig. 17), the rate of enzyme hydrolysis of the same simple aryl ester attached to different positions on the same compound, to the carboxyl group on either the side chain or thiazolidine ring, proceeds at very different rates. The former is more rapid.

It is nevertheless possible to make some broad generalizations. As demonstrated by prodrugs of isoguvacine and penicillin, when the regeneration of the parent drug proceeds too slowly, the rate of the ester hydrolysis of the prodrug can be increased by the synthesis of double esters. Conversely, when the ester hydrolysis of the prodrug proceeds too rapidly, a decrease in esterase affinity and reactivity to the prodrug can be induced by (1) increasing the chain length of the ester to beyond approximately 6 carbons, and (2) adding bulky substituents for steric hindrance. As demonstrated by chlorambucil, however, long-chain alkyl esters are often too lipophilic. They bind heavily to plasma proteins, distribute to liver, bile, and fat deposits, and reach minimal concentrations within the brain. In addition, they are minimally water-soluble and are therefore difficult to administer. Despite these drawbacks, heavily protein bound compounds can act as useful prodrugs to maintain the systemic drug concentrations of therapeutically valuable agents over a protracted period of time after a single IV administration.

A major complicating factor in the design of clinically useful prodrugs involves the different relative abundances of specific and nonspecific esterases in different animal species and their wide and often unspecific action.[59,260] In general, the rate of nonspecific esterase activity becomes progressively lower from mouse to rat to human.[59,104,144] Finding an appropriate animal model to assess the potential value of a series of prodrugs is therefore difficult. Although human plasma can be used to measure the rate of hydrolysis of a prodrug by plasma esterases in vitro, some prodrugs are preferentially metabolized by enzyme systems within the liver. In addition, in vivo brain uptake and activity studies necessitate the use of animal models, in which the pharmacokinetics of the prodrugs may be very different from humans and not optimal for brain uptake in that model.

5.3. Carrier-Mediated Transport at the Blood–Brain Barrier

Stereospecific, saturable carrier-mediated transport systems exist at the level of the cerebral capillary endothelium to regulate and facilitate the brain uptake of essential water-soluble compounds such as D-glucose, L-amino acids, and ions. Using an isolated brain perfusion technique (see Chapter 4, Section 3.2), Greig et al.[101,172] recently demonstrated that melphalan, a nitrogen mustard derivative of the large neutral amino acid L-phenylalanine (Fig. 24), shares the large neutral amino acid carrier system at the BBB. Melphalan brain uptake demonstrated classical concentration-dependent transport, saturation, and inhibition by L-phenylalanine. Interestingly, melphalan has been reported to be actively taken up by tumor cells by two separate amino acid transport systems. An authoritative review of this has been written by Vistica.[249] Furthermore, experiments by Adair and McElnay[1] and Bosanquet and Gilbey,[22] concerning the absorption of melphalan following its oral administration, strongly suggest that, in addition to simple diffusion, melphalan uses an amino acid carrier system at the level of the GI epithelium to cross the gut wall.

The affinity of melphalan for the BBB amino acid carrier system (the reciprocal of the Michaelis–Menten parameter K_m) (Table V) was significantly less than that of the endogenous large neutral amino acids, physiologic concentrations of which inhibited the facilitated uptake of melphalan.[101,172] Its affinity for the carrier at the BBB was also less than its affinity for the transport on tumor cells.[99] Nevertheless, this is the first quantitative study to demonstrate that exogenous drugs that are not neurotransmitter precursors

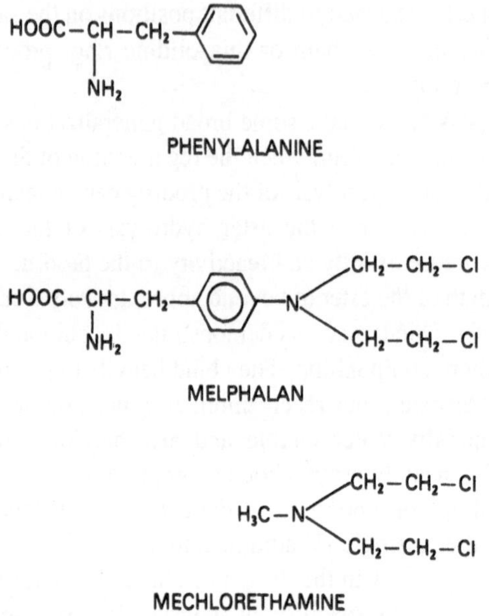

HOOC—CH—CH₂—⟨benzene ring⟩
   |
   NH₂

PHENYLALANINE

HOOC—CH—CH₂—⟨benzene ring⟩—N⟨ CH₂—CH₂—Cl / CH₂—CH₂—Cl ⟩
   |
   NH₂

MELPHALAN

H₃C—N⟨ CH₂—CH₂—Cl / CH₂—CH₂—Cl ⟩

MECHLORETHAMINE

Figure 24. Chemical structures of phenylalanine, melphalan (the nitrogen mustard derivative of phenylalanine), and mechlorethamine.

can make use of carrier-mediated transport systems to gain entry into the brain. It is probable that other amino acid drug analogues are similarly transported. One possible candidate is the glutamine antimetabolite acivicin (Fig. 25). Although melphalan is not a perfect substrate for the large neutral amino acid carrier system, it demonstrates a further technique to increase the brain uptake of drugs. It may be possible, through drug design, to synthesize further compounds that compete more readily for this system.

   Without doubt, the greatest success in making use of a BBB facilitated transport system to deliver a therapeutically useful compound into the brain has been the use of L-dopa in the alleviation of Parkinson disease. L-dopa is an endogenous large neutral

Table V. Permeability of Tracer
Concentrations of the Large Neutral
Amino Acids and Melphalan at the BBB,
Measured by the in Situ Brain
Perfusion Technique

	Saline perfusion	
Compound	$PS \times 10^4$, s^{-1}	$K_m{}^a$
Tryptophan	966.7	0.009
Phenylalanine	790.0	0.010
Leucine	411.5	0.026
Tyrosine	356.0	0.050
Isoleucine	209.8	0.051
Methionine	77.3	0.075
Melphalan	10.8	0.150

$^a K_m = \mu$moles/ml.

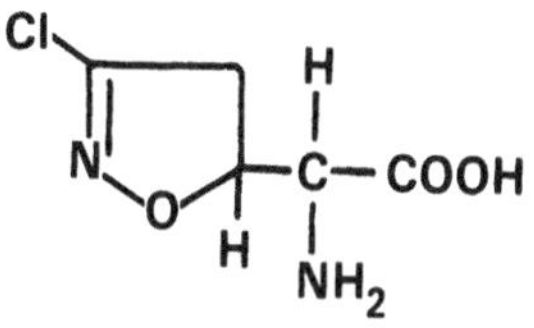

Figure 25. Chemical structure of the glutamine antimetabolite acivicin.

amino acid; it is transported by the large neutral amino acid transport system and is a precursor of the neurotransmitter dopamine. Following the oral administration of exogenous L-dopa, however, the compound is extensively decarboxylated in the GI tract and liver. This massive peripheral conversion of L-dopa to dopamine, which minimally enters the brain, and to other related metabolites is responsible for the dose-limiting side effects of L-dopa administration; chiefly hypotension, arrhythmia, nausea, and emesis. Although these characteristic side effects can be substantially reduced by the concomitant administration of L-amino acid decarboxylase inhibitor, which is restricted from entering the brain (benserazide and carbidopa) the alarming CNS side effects, particularly dyskinesia and on–off phenomena, associated with L-dopa therapy, remain unaffected. Although the co-administration of a decarboxylase inhibitor permits a reduction in the L-dopa dose, it has been suggested that the described CNS side effects are related to, and could be alleviated by, reducing the peak concentration and rapid fluctuations in the level of L-dopa in plasma.[51,106]

As previously described, Bodor and Brewster[17] have attempted to increase the brain uptake of dopamine by linking it to a pyridinium carrier. In addition, they synthesized a series of L-dopa prodrugs to increase the bioavailability of the parent compound.[20] The main sites of metabolism of L-dopa, the carboxylic, amino, and catechol functions, were protected both individually and in combination, with ester, amide, and peptide moieties. Several of the synthesized prodrugs reduced the metabolism of L-dopa both prior to and after its absorption, and this resulted in an increased bioavailability and plasma L-dopa/dopamine ratio of the prodrugs compared with the administration of L-dopa.[20] However, the shape of the concentration versus time profile of free L-dopa was similar following prodrug and L-dopa administration, and an even higher initial L-dopa plasma peak was observed following the administration of the prodrug.

Recently, Garzon-Aburbeh and colleagues[83] have developed an interesting and novel prodrug approach that has perhaps optimized the plasma concentration versus time profile of L-dopa to its facilitated transport at the BBB. To maintain a steady, high and extended concentration of L-dopa in plasma, they synthesized a glyceride derivative of L-dopa, 1,3-dihexadecanoyl-2-[(S)-2-amino-3-(3,4-dihydroxyphenyl)propanoyl]propane-1,2,3-triol, which conferred lymphotropic properties on L-dopa. Owing to its similarity to natural lipids, following oral administration, the prodrug was absorbed via the enteral route through the intestinal lymphatic system. This protected the L-dopa moiety from the usual first-pass metabolism in the GI tract and liver. Once in the bloodstream in the mouse, rat, and monkey, the prodrug then slowly released L-dopa by ester hydrolysis. Compared with the administration of L-dopa, the prodrug produced higher L-dopa plasma levels, for a longer period of time and with a more favorable L-dopa/dopamine ratio. Although no prodrug entered the brain, extended L-dopa brain concentrations were achieved; these resulted in a longer antiparkinsonian activity against oxotremorine and reserpine tests in the mouse. In addition, in both plasma and brain, the typical initial transient L-dopa peak was attenuated, thereby reducing the toxicity of L-dopa. Further studies from this group are awaited with interest.

6. CEREBRAL BLOOD FLOW

Regional cerebral blood flow is heterogeneous,[166] and, under normal circumstances, is closely linked to the local rate of energy utilization.[213] Since the rate of blood flow is a determinant of drug delivery and a limiting factor in the brain uptake of lipophilic compounds, systemically administered drugs should be available for uptake in greater quantities by those brain areas that have higher metabolic rates. This concept is fundamental to the measurement of regional cerebral blood flow by the [14C]iodoantipyrine technique.[166] In this, regional blood flow is determined by the initial heterogeneous accumulation of the lipophilic compound in the brain.

Extensive studies by Soncrant et al.[234] using the [14C]-2-deoxy-D-glucose technique to measure regional cerebral metabolism,[232,233] have demonstrated that neurotransmitter agonists and antagonists, as well as barbiturates will consistently increase and/or decrease glucose utilization, and thus blood flow, in specific and highly localized brain areas in the rat. The regions affected and the degree and duration of the effect depended on (1) the mechanism of action of the drug, (2) its initial concentration, and (3) the time elapsed following its administration.

From these studies, Soncrant et al.[232] has posed an interesting series of questions, which are directly relevant to the optimization of brain drug delivery. Can drugs by their regionally specific action on metabolism, and thus blood flow, significantly alter either their own regional delivery or the regional delivery of other compounds within the brain? Is it possible to achieve a higher concentration of a drug selectively in a confined target area of brain by either pretreating or simultaneously administering an agent that increases metabolism in that region?

The answers to these questions are under investigation; the potential of the concept, however, is more clear. In the future, it may be possible to direct drugs preferentially, such as neurotransmitter precursors, to their target neurons within the brain by increasing the cerebral metabolism of those neurons. Conversely, it may be possible to use barbiturates to distinguish or differentiate normal neuronal tissue from tissue displaying metabolic activity that is insensitive to barbiturate anesthesia. Frey and Agranoff[80] have recently demonstrated that phenobarbital uniformly reduced cerebral neuronal metabolism but not metabolism associated with astrocytic and phagocytic activity within chemically induced brain lesions in the rat. It may therefore be possible to (1) delineate more accurately brain tumors or strokes by reducing the metabolism of the surrounding neural tissue during positron emission tomographic (PET) scanning of patients; or (2) protect the normal brain from neurotoxicity by reducing its metabolism and blood flow during brain tumor chemotherapy.

7. CONCLUSIONS

The blood–brain barrier (BBB) restricts the brain uptake of a variety of clinically relevant water-soluble drugs. Even in disease states in which the integrity of the barrier is partially reduced, blood-to-brain drug transport remains a major factor in the poor therapeutic response of patients to otherwise pharmacologically active drugs (see Chapter 17, Vol. 2, Parts 1 and 2). Quantitative structure–activity relationship studies have demonstrated that the physicochemical parameter of lipophilicity as well as electronic and steric effects (which are not as well defined) are key determinants of the final CNS activities of drugs. Indeed, the free energy model of Hansch and its elaborations, in addition to other models, have rationalized the field of drug design. While quantitative correlations have made a significant contribution toward optimization of the physicochemical properties of a variety of drugs, the initial discovery of a lead compound is still very much a combination of knowledge and luck. In addition, since the physicochemical parameters of a compound affect a variety of biologic parameters, such as the permeability of the compound at biologic membranes, its plasma protein binding, metabolism, and excretion, it is evident that, within a series of compounds targeted for activity within the brain, increased potency will not necessarily be achieved by simply varying one property alone.

For several clinically proven compounds, simple alterations in the scheduling, route of administration, or sometimes the dosage of drug (if combined with a rescue technique) are sufficient to optimize the systemic pharmacokinetics of a compound to permit the brain uptake of effective concentrations. For most compounds, however, such manipulations still result in subtherapeutic drug levels within the brain. Research has thus progressed along two major avenues to overcome this. The brain uptake of certain compounds can be improved by one of several chemical manipulations. As a consequence, interest is growing in the concept of prodrugs. Selective enzyme localization is essential, however, before true targeting to brain is likely to be achieved. The brain uptake of compounds that are not amenable to chemical modification can be increased by reversible barrier modification. It should be remembered, however, that the BBB acts as a protective as well as a regulatory structure, and the brain uptake of many compounds results in neurotoxicity; thus, for these compounds, barrier modification should be avoided. The future integration of the chemical and barrier modification approaches would therefore be fruitful in initially assessing which compounds would be useful and safe to deliver to the brain. Perhaps the greatest potential for the future lies in accepting the fact that the BBB severely compromises the delivery of drugs in the treatment of all CNS sequestered diseases. The transport of a drug to a well-defined target would then be considered during the initial stages of drug design rather than as an afterthought.

ACKNOWLEDGMENTS

I wish to thank Dr. Deena Shapiro, Dr. Timothy Soncrant, and Dr. Krystyna Wozniak for their invaluable advice and help during the preparation of this article. I also wish to thank Dr. Stanley Rapoport for his support within the Laboratory of Neurosciences, National Institute on Aging, and finally, Professor Kurt Hellmann, who kindled my initial interest in the field of cancer chemotherapy.

REFERENCES

1. Adair C, McElnay J: Studies of the mechanism of gastrointestinal absorption of melphalan and chlorambucin. *Cancer Chemother Pharmacology* 17:95–98, 1986.
2. Alazraki N, Halpern S, Ashburn W, et al: Hyperbaric cisternography: Experience in humans. *J Nuc Med* 14:226–229, 1973.

3. Alazraki N, Fierer J, Halpern S, et al: Use of hyperbaric solution for administration of intrathecal amphotericin B. *N Engl J Med* 290:641–646, 1974.
4. Allenby A, Fletcher J, Schock, et al: The effect of heat, pH, and organic solvents on the electrical impedance and permeability of excised human skin. *Br J Dermatol* 81:31–39, 1969.
5. Alvan G, Bergman U, Gustafsson L: High unbound fraction of salicylate in plasma during intoxication. *Br J Clin Pharmacol* 26:145–160, 1979.
6. Ames M, Powis G, Kovach J: *Pharmacokinetics of Anticancer Agents in Humans*. Elsevier, Amsterdam, 1983.
7. Balis F, Lester C, Poplak D: Pharmacokinetics of trimetrexate (NSC 352122) in monkeys. *Cancer Res* 46:169–174, 1986.
8. Baraldi M, Grandison L, Guidotti A: Distribution and metabolism of muscimol in the brain and other tissue of the rat. *Neuropharmacology* 18:57–62, 1979.
9. Barnden R, Evans R, Hamlet J, et al: Some preparative uses of benzyl–penicillinic ethoxyformic anhydride. *J Chem Soc* 3733–3739, 1953.
10. Billiau A, Heremans H, Ververken D, et al: Tissue distribution of human interferon after exogenous administration in rabbits, monkeys and mice. *Arch Virology* 68:19–25, 1981.
11. Blasberg R, Patlak C, Fenstermacher J: Intrathecal chemotherapy: brain tissue profiles after ventriculocisternal perfusion. *J Pharmacology Exp Ther* 195:73–83, 1975.
12. Blasberg R, Patlak C, Shapiro W: Distribution of methotrexate in the cerebrospinal fluid and brain after intraventricular administration. *Cancer Treatm Rep* 61:633–641, 1977.
13. Bleyer W, Dedrick R. Clinical pharmacology of intrathecal methotrexate. I. Pharmacokinetics in nontoxic patients after lumbar injection. *Cancer Treatm Rep* 61:703–708, 1977.
14. Bleyer W, Poplack D, Simon R: Concentration × time methotrexate via a subcutaneous reservoir: A less toxic regimen for intraventricular chemotherapy of central nervous system neoplasms. *Blood* 51:835–842, 1978.
15. Bode U, Magrath I, Bleyer W, et al: Mechanism for methotrexate efflux from cerebrospinal fluid in man. *Cancer Res* 40:2184–2187, 1980.
16. Bodin N, Ekstrom B, Forsgren U, et al: Bacampicillin: A new orally well-absorbed derivative of ampicillin. *Antimicrobial Agents Chemother* 8:518–525, 1975.
17. Bodor N, Brewster M: Problems of drug delivery to the brain. *Pharmacology Ther* 19:337–386, 1982.
18. Bodor N, Roller R, Selk S: Elimination of a quaternary pyridinium salt delivered as its dihydropyridine derivative from the brain of mice. *J Pharmacol Sci* 67:685–687, 1978.
19. Bodor N, Shek E, Higuchi T: Delivery of a quaternary pyridinium salt across the blood–brain barrier as its dihydropyridine derivative. *Science* 190:155–156, 1975.
20. Bodor N, Sloan K, Higuchi T, et al: Improved delivery through biological membranes. 4. Prodrugs of L-Dopa. *J Med Chem* 20:1435–1445, 1977.
21. Borzellca J, Harris T, Bernstein S: The effect of dimethyl sulfoxide on the permeability of the urinary bladder. *Invest Urology* 6:43–52, 1968.
22. Bosanquet A, Gilbey E: Comparison of fed and fasting states on the absorption of melphalan in multiple myeloma. *Cancer Chemother Pharmacology* 12:183–186, 1984.
23. Bowdle T, Patel I, Levy R, et al: Valproic acid dosage and plasma protein binding and clearance. *Clin Pharmacol Exp Ther* 28:486–492, 1980.
24. Broadwell R, Kaplan R, Salcman M: Potential use of dimethyl sulfoxide to open the blood–brain barrier. *Soc Neurosci* 7:244, 1981.
25. Broadwell R, Salcman M, Kaplan R: Morphological effect of dimethyl sulfoxide on the blood–brain barrier. *Science* 217:164–166, 1982.
26. Broman T: *The Permeability of Cerebrospinal Vessels in Normal and Pathological Conditions*. Munksgaard, Copenhagen, 1949.
27. Brown A, Brierly J: The earliest alteration in rat neurons and astrocytes after anoxia–ischaemia. *Acta Neuropathol* 23:9–22,1973.
28. Brown T, Ettinger D, Rice A, et al: A phase I clinical trial of spirohydantoin mustard (SHM). *Proc Am Soc Clin Oncol* 3:33, 1984.
29. Bundgaard M, Frokjaer-Jensen J, Crone C: Endothelial plasmalemmal vesicles as elements in a system of branching invaginations from the cell surface. *Proc Natl Acad Sci USA* 76:6439–6442, 1979.
30. Burger P, Mahaley M, Dudka L, et al: The morphological effects of radiation administered therapeutically for intracranial gliomas. A postmortem study of 25 cases. *Cancer* 44:1256–1272, 1979.

31. Butler K, English A, Knirsch A, et al: Metabolism and laboratory studies with indanyl carbenicillin. *Del Med J* 43:366–375, 1971.

32. Byfield J, Hornbeck C: Correspondence re: I. Kerr et al. Effect of intravenous dose and schedule on cerebrospinal fluid pharmacokinetics of 5-fluorouracil in the monkey. *Cancer Res* 44:4929–4932, 1984. *Cancer Res* 45:3398–3399, 1985.

33. Cairncross J, Kim J, Posner J: Radiation therapy for brain metastases. *Ann Neurol* 7:529–541, 1980.

34. Calterall M, Bloom H, Ash D, et al: Fast neutron compared with megavoltage X rays in the treatment of patients with supratentorial glioblastoma: A controlled pilot study. *Int J Radiat Oncol Biol Phys* 6:261–266, 1980.

35. Chabner B: Antimetabolites. In Pinedo H (ed): *Cancer Chemotherapy*. Excerpta Medica, Amsterdam, 1979.

36. Chabner B, Meyers C, Oliverio V: Clinical pharmacology of anticancer drugs. *Semin Oncol* 4:165–191, 1977.

37. Charton M: Prodrug lability prediction through the use of substituent effects. *Methods Enzymol* 112:323–340, 1985.

38. Christensen B, Leanza W: USA patent #3,931,150, 1976.

39. Chou F, Khan H, Driscoll J: Potential central nervous system anticancer agents: Aziridinylbenzo quinones-2. *J Med Chem* 19:1302–1308, 1976.

40. Clayton J, Cole M, Elson S, et al: Preparation, hydrolysis, and oral absorption of the α-carboxyl ester of carbenicillin. *J Med Chem* 18:172–177, 1975.

41. Clayton J, Cole M, Elson S, et al: Preparation, hydrolysis, and oral absorption of lactonyl esters of penicillins. *J Med Chem* 19: 1385–1391, 1976.

42. Clemedson C, Hartelius H, Holmberg A: The influence of carbon dioxide inhalation on the cerebral vascular permeability to Trypan blue (the blood–brain barrier). *Acta Pathol Microbiology Scand* 42:137–149, 1958.

43. Cornbleet M, McElwain T, Kumar P, et al: Treatment of advanced melanoma with high-dose melphalan and autologous bone marrow transplantation. *Br J Cancer* 48:329–334, 1983.

44. Cornbleet M, Leonard R, Smyth J: High-dose agent therapy: A review of clinical experience. *Cancer Drug Deliv* 1:227–238, 1984.

45. Crevelling C, Daly J, Tokuyama T, et al: Labile lipophilic derivatives of norepinephrine capable of crossing the blood–brain barrier. *Experientia* 25:26–27, 1969.

46. Cserr H: Bulk flow of cerebral extracellular fluid as a possible mechanism of cerebrospinal fluid–brain exchange. In Cserr H, Fenstermacher J, Fencl V (eds): *Fluid Environment of the Brain*. Academic, London, 1975, pp. 215–224.

47. Cutler R, Barlow C: The effect of hypercapnia on brain permeability to protein. *Arch Neurol* 14:54–63, 1966.

48. Cutler R, Lorenzo A: Transport of L-aminocyclopeptane carboxylic acid from the feline cerebrospinal fluid. *Science* 161:1363–1364, 1968.

49. Daehne W, Fredriksen E, Gundersen E, et al: Acyloxymethyl esters of ampicillin. *J Med Chem* 13:607–612, 1970.

50. Daehne W, Fredriksen E, Gotfredsen W, et al: British Patent 1,290,787, 1972.

51. DaPrada M, Keller H, Pieri L, et al: The pharmacology of Parkinson's disease: Basic aspects and recent advances. *Experientia* 40:1165–1172, 1984.

52. Davies J, Watkins J: The action of β-phenyl-GABA derivatives on neurons of the cat cerebral cortex. *Brain Res* 70:501–505, 1974.

53. Davison L, Smith P: The binding of salicylic acid and related substances to plasma proteins. *J Pharmacology Exp Ther* 133:161–170, 1961.

54. DeLand F: Intrathecal toxicity with bezyl alcohol. *Toxicol Appl Pharmacol* 25:153–156, 1973.

55. Delini-Stula A: Differential effects of baclofen and muscimol on behavioral responses implicating GABA-ergic transmission. In Krogsgaard-Larsen P, Scheel-Kruger J, Kofod H (eds): *GABA Neurotransmitters*. Munksgaard, Copenhagen, 1979, pp. 482–499.

56. Dhopeshwarkar G, Subramanisian C, McConnell D, et al: Fatty acid transport into the brain. *Biochim Biophys Acta* 255:572–579, 1972.

57. Diamond R, Bennett J: A subcutaneous reservoir for intrathecal therapy of fungal meningitis. *N Engl J Med* 288:186–189, 1973.

58. Diez B, Monges J, Muriel F: Evaluation of cisplatinum in children with recurrent brain tumors. *Cancer Treatm Rep* 69:911–913, 1985.

59. Dixon M, Webb E: *Enzymes*, 3rd Ed. Academic, New York, 1979.
60. Dollinger M, Burchenal J, Kreis W, et al: Analogues of I-β-D arabino-furanosylcytosine. Studies on mechanism of action in Burkitt's cell culture and mouse leukemia, and in vivo deamination studies. *Biochem Pharmacol* 16:689–706, 1967.
61. Dorr R, Fritz W: *Cancer Chemotherapy Handbook*. Elsevier, New York, 1980.
62. Driscoll J, Dudeck L, Congleton G, et al: Potential CNS antitumor agents. VI. Aziridinylbenzoquinones. III. *J Pharm Sci* 68:185–188, 1979.
63. Edwards M, Levin V, Byrd A: Quantitative observations of the subacute effects of x-irradiation on brain capillary permeability: part II. *Int J Radiat Oncol Biol Phys* 5:1633–1635, 1979.
64. Edwards M, Levin V, Wilson C: Brain tumor chemotherapy: An evaluation of agents in current use for phase II and III trials. *Cancer Treatm Rep* 64:1179–1205, 1980.
65. Edvinsson L, MacKenzie E: Amine mechanisms in the cerebral circulation. *Pharmacol Rev* 28:275–348, 1977.
66. Edvinsson L, Nielsen K, Owman C, et al: Alterations in intracranial pressure, blood–brain barrier after subchronic implantation of a cannula into the brain of conscious animals. *Acta Physiol Scand* 82:527–531, 1971.
67. Egorin M, Kaplan R, Salcman M, et al: Cyclophosphamide plasma and cerebrospinal fluid kinetics with and without dimethyl sulfoxide. *Pharmacology Ther* 32:122–128, 1982.
68. Elslager E, Davoll J: Synthesis of fused pyrimidines as folate antagonists. In Castle R, Townsent L (eds): *Lectures in Heterocyclic Chemistry*, Vol. 2. Hetero Corp., Orem, Utah, 1974.
69. Eriksson T, Lilijequist S, Carlsson A: Ethanol-induced increase in the penetration of exogenously administered L-dopa through the blood–brain barrier. *J Pharm Pharmacol* 31:636–637, 1979.
70. Falch E, Krogsgaard-Larson P, Christensen A:. Esters of isoguvacine as potential prodrugs. *J Med Chem* 24:285–289, 1981.
71. Fenstermacher J, Blasberg R: Methods of quantifying the transport of drugs across brain barrier systems. *Pharmacol Ther* 14:217–248, 1981.
72. Fenstermacher J, Patlak C: The exchange of material between cerebrospinal fluid and brain. In Cserr H, Fenstermacher J, Fencl V (eds): *Fluid Environment of the Brain*. Academic, New York, 1975, pp. 201–214.
73. Feun L, Savaraj N, Lu, et al: Disruption of the blood–brain barrier with intracarotid hydroxyurea. *Proc Am Assoc Cancer Res* 25:364, 1984.
74. Fike J, Sheline G, Cann C, et al: Radiation necrosis. *Prog Exp Tumor Res* 28:136–151, 1984.
75. Finklestein J, Shern J, Chabner B: Pharmacological studies of tritiated cytosine arabinoside (NSC 63878) in children. *Cancer Chemother Rep* 54:35–41, 1970.
76. Firemark H, Barlow C, Roth L: The entry, accumulation and binding of diphenylhydantoin-[14C] in brain. *Int J Neuropharmacol* 2:25–38, 1963.
77. Flynn G, Yalkowsky S: Correlation and prediction of mass transport across membranes. 1. Influence of alkyl chain length on flux-determining properties of barrier and diffusant. *J Pharm Sci* 61:838–857, 1972.
78. Freedman F, Johnson J: Equilibrium and kinetic properties of the Evans blue albumin system. *Am J Physiol* 216:675–681, 1969.
79. Frey H, Loscher W: Cetyl GABA: Effect on convulsant thresholds in mice and acute toxicity. *Neuropharmacology* 19:217–220, 1980.
80. Frey K, Agranoff B: Barbiturate-enhanced detection of brain lesions by carbon-14-labeled 2-deoxyglucose autoradiography. *Science* 219:879–881, 1983.
81. Fulton D, Levin V, Gutin P, et al: Intrathecal cytosine arabinoside for the treatment of meningeal metastases from malignant brain tumors and systemic tumors. *Cancer Chemother Pharmacol* 8:285–291, 1982.
82. Galzigna L, Garbin L, Bianchi M et al: Properties of two derivatives of γ-aminobutyric acid (GABA) capable of abolishing cardiazol and bicuculline-induced convulsions in the rat. *Arch Int Pharmacodyn* 235:73–85, 1978.
83. Garzon-Aburbeh A, Poupaert J, Calesen M, et al: A lymphotropic prodrug of L-Dopa: Synthesis, pharmocological properties, and pharmacokinetic behavior of 1,3-dihexadecanoyl-2-[(S)-2-amino-3-(3,4-dihydroxyphenyl)proponyl]propane-1,2,3,-triol. *J Med Chem* 29:687–691, 1986.
84. Gilman A, Goodman L, Gilman A: *The Pharmacological Basis of Therapeutics*. Macmillan, New York, 1980.
85. Glass J, Shapiro W, Posner J: Treatment of leptomeningeal metastases. *Neurology (NY)* 25:350–354, 1978.

86. Glynn K, Alazraki N, Waltz T: Coccidioidal meningitis: Intrathecal treatment with hyperbaric amphotericin B. *Calif Med* 119:6–9, 1973.

87. Golberg M, Barlow C, Roth L: Abnormal brain permeability in carbon dioxide narcosis. *Arch Neurol* 9:498–507, 1963.

88. Goodman D: The interaction of human serum albumin with long chain fatty acid anions. *J Am Chem Soc* 80:3892–3898, 1958.

89. Gormley P, Wood J, Poplack D: Ability of a new anticancer agent AZQ to penetrate to cerebrospinal fluid. *Pharmacology* 22:196–198, 1981

90. Griffin T, Rasey J, Bleyer W: The effect of photon irradiation on blood–brain barrier permeability to methotrexate in mice. *Cancer* 40:1109–1111, 1977.

91. Griffin T, Davis R, Laramore G, et al: Fast neutron radiation therapy for glioblastoma multiforme. *Am J Clin Oncol* 6:661–667, 1983.

92. Greig N: Chemotherapy of brain metastases: Current status. *Cancer Treatm Rev* 11:157–186, 1984.

93. Greig N: Optimizing drug delivery to brain tumors. *Cancer Treatm Rev* 14:1–28, 1987.

94. Greig N, Cavanagh J: Quantitative aspects of reversible opening of the blood–brain barrier by pentylenetetrazol. *J Neuropathol Appl Neurobiol* 8:245, 1982.

95. Greig N, Hellmann K: Enhanced cerebrovascular permeability by Metrazol: Significance for brain metastases. *Clin Exp Metast* 1:83–95, 1983.

96. Greig N, Fu X, Hellmann K: Razoxane penetration into the cerebrospinal fluid of rats. *Cancer Chemother Pharmacol* 8:251–252, 1982.

97. Greig N, Jones H, Cavanagh J: Blood–brain barrier integrity and host responses in experimental metastatic brain tumors. *Clin Exp Metast* 1:229–246, 1983.

98. Greig N, Newell D, Hellmann K: Metrazol enhances brain penetration and therapeutic efficacy of some anticancer agents: Implications for brain metastases. *Clin Exp Metast* 2:55–59, 1984.

99. Greig N, Sweeney D, Rapoport S: Concentration dependent binding of melphalan to plasma proteins in healthy humans and rats. *Proc Am Assoc Cancer Res* 26:357, 1985.

100. Greig N, Sweeney D, Rapoport S: Inability of dimethyl sulfoxide (DMSO) to increase the brain uptake of water-soluble compounds: Implications to chemotherapy of brain tumors. *Cancer Treatm Rep* 69:305–312, 1985.

101. Greig N, Momma S, Sweeney D, et al: Facilitated transport of melphalan at the rat blood–brain barrier by the large neutral amino acid carrier system. *Cancer Res* 47:1571–1576, 1987.

102. Greig N, Sweeney D, Rapoport S: Melphalan concentration dependent plasma protein binding in healthy humans and rats. *Eur J Clin Pharmacol* 32:179–185, 1987.

103. Greig N, Sweeney D, Rapoport S: Brain delivery of interferon in the rat by reversible osmotic blood–brain barrier modification. *Proc Am Assoc Cancer Res* 28:1742, 1987.

103a. Greig N, Rapoport SI: Comparative brain and plasma pharmacokinetics and anticancer activities of chlorambucil and melphalan in the rat. *Cancer Chemother Pharmacol* 31:1–8, 1988.

104. Greig N, Rapoport S: Brain and peripheral pharmacokinetics of chlorambucil and esters of chlorambucil. *Cancer Chemother Pharmacol* (in press), 1988.

104a. Greig N, Rapoport SI: *Enhancing Drug Delivery to the Brain*. U.S. Patent file no. 088982, 1987.

105. Gulati A, Nath C, Shanker K, et al: Effect of alcohol on the permeability of the blood–brain barrier. *Pharm Res Commun* 17:85–93, 1985.

106. Gundert-Remy U, Hildebrandt R, Stiehl A, et al: Intestinal absorption of levodopa in man. *Eur J Clin Pharmacol* 23:69–74, 1983.

107. Hamilton-Miller J: Chemical manipulations of the penicillin nucleus: A review. *Chemotherapia* 12:73–88, 1967.

108. Hammett L: *Physical Organic Chemistry*. 2nd Ed. McGraw-Hill, New York, 1970.

109. Hansch C: A quantitative approach to biochemical structure–activity relationships. *Acc Chem Res* 2:232–239, 1969.

110. Hansch C: Strategy in drug design. *Cancer Chemother Rep* 56:433–441, 1972.

111. Hansch C, Steward A, Anderson S, et al: The parabolic dependence of drug action upon lipophilic character as revealed by a study of hypnotics. *J Med Chem* 11:1–11, 1968.

112. Hansch C, Smith N, Engle R, et al: Quantitative structure–activity relationships of antineoplastic drugs: Nitrosoureas and triazenoimidazoles. *Cancer Chemother Rep* 56:443–456, 1972.

113. Hersh M, Ludden T, Kuhn J, et al: Pharmacokinetics of high-dose melphalan. *Invest New Drugs* 1:331–334, 1983.

114. Higuchi T, Stella V: *Prodrugs as Novel Drug Delivery Systems*. American Chemical Society Symposium Series 14, American Chemical Society, Washington, DC, 1975.

115. Hillbom M, Tervo T: Ethanol and acetaldehyde do not increase the blood–brain barrier permeability to sodium fluoroscein. *Experientia* 37:936–938, 1981.

116. Ho D, Frie E: Clinical pharmacology of 1-β-D-arabinofuransylcytosine. *Clin Pharmacol Ther* 12:944–954, 1971.

117. Ho M, Nash C, Morgan C, et al: Interferon administered into the cerebrospinal space and its effect on rabies in rabbits. *Infection* 9:286–293, 1974.

118. Hochberg F, Pruitt A, Beck D, et al: The rationale and methodology for intraarterial chemotherapy with BCNU as treatment for glioblastoma. In Howell S (ed): *Intraarterial and Intracavity Cancer Chemotherapy*. Martinus Nijhoff, Boston, 1984, pp. 97–109.

119. Hollis P, Zappulla R, Spigelman M, et al: Physiological and electro-physiological consequences of etopside-induced blood–brain barrier disruption. *Neurosurgery* 18:581–586, 1986.

120. Horn A, DeKaste D, Dijkstra D, et al: A new dopaminergic prodrug. *Nature (Lond)* 276:405–407, 1978.

121. Horn A, Grol C, Dijkstra D: Facile syntheses of potent dopaminergic agonists and their effects on neurotransmitter release. *J Med Chem* 21:825–828, 1978.

122. Horn A, Kelly P, Westerink B, et al: A prodrug of ADTN: Selectivity of dopaminergic action and brain levels of ADTN. *Eur J Pharmacol* 60:95–99, 1979.

123. Hornbeck C, Floyd R, Byfield J, et al: Cerebrospinal fluid versus serum concentrations of 5-FU, allopurinol, oxypurinol, and radiation. *Cancer Treatm Rep* 66:571–573, 1982.

124. Hoshi A, Yoshida M, Kanzawa, R, et al: Specific inhibition of DNA synthesis by cyclocytidine. *Chem Pharm Bull* 20:2286–2287, 1972.

125. Hyrniuk W, Bertino J: Treatment of leukemia with large doses of methotrexate and folinic acid. Clinical-biochemical correlates. *J Clin Invest* 48:2140–2155, 1969.

126. Huang K, Wenczak B, Liu Y: Renal tubular transport of methotrexate in the rhesus monkey and dog. *Cancer Res* 39:4843–4848, 1979.

127. Huffman D, Wan S, Azarnoff D, et al: Pharmacokinetics of methotrexate. *Clin Pharmacol Ther* 14:572–579, 1973.

128. Husain M, Garcia J: Cerebral radiation necrosis: Vascular and glial features. *Acta Neuropathol (Berl)* 36:381–385, 1976.

129. Israelachvili J, Mitchell J, Ninham B: Theory of self–assembly of lipid bilayers and vesicles. *Biochim Biophys Acta* 470:185–201, 1977.

130. Jackson R, Fry D, Boritzki T, et al: Biochemical pharmacology of the lipophilic antifolate, trimetrexate. *Adv Enzyme Reg* 22:187–206, 1984.

131. Jansen A, Russell T: Some novel penicillin derivatives. *J Chem Soc* 2127–2132, 1965.

132. Johansson B, Li C, Olsson Y, et al: The effect of acute arterial hypertension on the blood–brain barrier to protein tracers. *Acta Neuropathol (Berl)* 16:117–124, 1970.

133. Johansson B, Strandgaard S, Lassen N: On the pathogenesis of hypertensive encephalopathy. The hypertensive breakthrough of autoregulation of cerebral blood flow with forced vasodilatation, flow increases and blood–brain barrier damage. *Circ Res* 34:167–171, 1974.

134. Jusko W, Gretch M: Plasma and tissue protein binding of drugs in pharmacokinetics. *Drug Metab Rev* 5:43–140, 1976.

135. Kamer B, Eibl B, Cashmore A, et al: Efficacy and transport of a new lipid-soluble antifolate, 2,4-diamino-5-methyl-6-[2,4,5-trimethyloxyanilino)methyl]quinazoline in methotrexate resistant cells. *Proc Am Assoc Cancer Res* 22:26, 1981.

136. Kaplan J, Raizon B, Desarmenien P, et al: New anticonvulsants: Schiff bases of γ-aminobutyric acid and aminobutyramide. *J Med Chem* 23:702–704, 1980.

137. Kaplan R, Wiernik P: Neurotoxicity of antineoplastic drugs. *Semin Oncol* 9:103–130, 1982.

138. Kaplan R, Riggs L, Miler C, et al: Preliminary observations on the effects of dimethyl sulfoxide on the metabolism and distribution of adriamycin. *Proc Am Assoc Cancer Res* 22:367, 1981.

139. Kauffman J, Koski W, Benson D, et al: Narcotic and narcotic antagonist pka's and partition coefficients and their significance in clinical practice. *Drug Alcohol Depend* 1:103–114, 1976.

140. Kaufmann H: Radiation therapy of metastases of the brain and spinal cord. *Adv Neurosurg* 12:68–70, 1984.

141. Khan A, D'Souza B, Wharam M, et al: Cisplatin therapy in recurrent childhood brain tumors. *Cancer Treatm Rep* 66:2013–2020, 1982.

142. Khan A, Driscoll J: Potential central nervous system antitumour agents: Aziridinylbenzoquinones. *J Med Chem* 19:313–317, 1976.

143. Kragh-Hansen U: Molecular aspects of ligand binding to serum albumin. *Pharmacol Rev* 33:17–53, 1981.

144. Krisch K: Carboxylic ester hydrolases. In Boyer P (ed): *The Enzymes*, Vol. 5. Academic, London, 1971, pp. 43–69.

145. Krogsgaard-Larsen P, Christensen A: GABA agonists. Synthesis and structure–activity studies on analogues of isoguvacine and THIP. *Eur J Med Chem* 14:157–164, 1979.

146. Lampert P, Tom M, Rider W: Disseminated demyelination of the brain following CO^{60} (gamma) radiation. *Arch Pathol Lab Med* 68:322–330, 1959.

147. Lee J: Effect of alcohol injections on the blood–brain barrier. *J Stud Alcohol* 23:4–15, 1962.

148. Lee J, Olszewski J: Increased cerebral permeability after repeated electroshocks. *Neurology (NY)* 11:515–519, 1961.

149. Leo A, Hansch C, Elkins D: Partition coefficients and their uses. *Chem Rev* 71:525–616, 1971.

150. Levin V: Pharmacological considerations in brain tumor chemotherapy. In Fewer D, Wilson C, Levin V (eds): *Brain Tumor Chemotherapy*. Charles C. Thomas, Springfield, Illinois, 1976, pp 42–74.

151. Levin V: Relationship of octanol/water partition coefficient and moleuclar weight to rat brain capillary permeability. *J Med Chem* 23:682–684, 1980.

152. Levin V, Kabra P: Effectiveness of the nitrosoureas as a function of their lipid solubility in the chemotherapy of experimental rat brain tumors. *Cancer Chemother Rep* 58:787–792, 1974.

153. Levin V, Kabra P: Brain and tumor pharmacokinetics of BCNU and CCNU following IV and intracarotid artery administration. *Proc Am Cancer Res* 16:19, 1975.

154. Levin V, Kabra P, Freeman-Dove M: Pharmacokinetics of intracarotid artery [^{14}]-BCNU in the squirrel monkey. *J Neurosurg* 48:587–593, 1978.

155. Levin V, Edwards M, Byrd A: Quantitative observations of the acute effects of x-irradiation on capillary permeability: Part I. *J Radiol Oncol Biol Phys* 5:1627–1631, 1979.

156. Levine R, Fredericks W, Rapoport S: Entry of bilirubin into the brain due to opening of the blood–brain barrier. *Pediatrics* 69:255–259, 1982.

157. Levitan H, Ziylan Z, Smith Q, et al: Brain uptake of food dye, erythrosin B, prevented by plasma protein binding. *Brain Res* 322:131–134, 1984.

158. Liegler D, Henderson E, Hahn M, et al: The effect of organic acids on renal clearance of methotrexate in man. *Clin Pharmacol Ther* 10:849–857, 1969.

159. Lorenzo A, Shiranige I, Liang M, et al: Temporary alteration of cerebrovascular permeability to plasma proteins during drug-induced seizures. *Am J Physiol* 223:268–277, 1972.

160. Lorenzo A, Hedley-White T, Eisenberg H, et al: Increased penetration of horseradish peroxidase across the blood–brain barrier induced by Metrazol seizures. *Brain Res* 88:136–140, 1975.

161. MacKichan J: Pharmacokinetic consequences of drug displacement from blood and tissue proteins. *Clin Pharmacokinet* 9:32–41, 1984.

162. Madajewicz S, West C, Park H, et al: Phase II study—Intraarterial BCNU therapy for metastatic brain tumors. *Cancer* 47:653–657, 1980.

163. Maggi A, Enna S: Characteristics of muscimol accumulation in mouse brain after systemic administration. *Neuropharmacology* 18:361–366, 1979.

164. Manz H, Woolley P, Ornitz R: Delayed radiation necorsis of brain stem related to fast neutron irradiation. A case report and literature review. *Cancer* 44:473–479, 1979.

164a. Martin B: Potential effect of plasma proteinson drug distribution. *Nature (Lond)* 207:274–276, 1965.

165. Martins A, Johnston J, Henry J, et al: Delayed radiation necrosis of the brain. *J Neurosurg* 47:336–345, 1977.

166. Mauger A, Ross W: Aryl-2-halogenalkylamines XX. The preparation and properties of some di-2-chloroethylaminoaryl substituted hydantoins and related amino acids. *Biochem Pharmacol* 11:847–858, 1962.

167. McClure D: The effects of a prodrug of epinephrine (dipivalylepinephrine) in glaucoma—General pharmacology, toxicology and clinical experience. In Higuchi T, Stella V (eds): *Prodrugs as Novel Drug Delivery Systems*. American Chemical Society Symposium Series 14. American Chemical Society, Washington DC, 1975, pp. 224–235.

168. MacDonell L, Potter P, Leslie R: Localized changes in blood–brain barrier permeability following the administration of antineoplastic drugs. *Cancer Res* 38:2930–2934, 1978.

169. Mealey J Jr, Chen T, Shupe R: Response of human glioblastomas to radiation and BCNU chemotherapy. *J Neurosurg* 41:339–349, 1974.

170. Meffin P, Robert E, Winkle R, et al: Role of concentration dependent plasma protein binding of disopyramide disposition. *J Pharm Biopharm* 7:29–46, 1979.

171. Merigan T, Baer G, Winkler W, et al: Human leukocyte interferon administration to patients with symptomatic and suspected rabies. *Ann Neurol* 16:82–87, 1984.

172. Momma S, Greig N, Smith Q, et al: Facilitated transport of melphalan at the blood–brain barrier by the large neutral amino acid carrier system. *Proc Am Assoc Cancer Res* 26:357, 1985.

173. Mundinger F: Stereotoxic interstitial therapy of nonresectable intracranial tumors with iridium-192 and iodine-125. *Prog Radio-Oncol* 2:371–380, 1982.

174. Mundinger F, Weigel K: CT-stereotactic interstitial irradiation therapy of nonresectable and recurrent intracranial tumor in children and adolescents. In Voth D, Krauseneck P (eds): *Chemotherapy of Gliomas*. de Gruyter, Berlin, 1985, pp. 241–259.

175. Myers C, Diasio R, Eliot H, et al: Pharmacokinetics of the fluoropyrimidines: Implications for their clinical use. *Cancer Treatm Rev* 3:175–183, 1976.

176. Nair V, Roth L: Effect of x-irradiation and certain other treatments on blood–brain barrier permeability. *Radiat Res* 23:249–264, 1964

177. Nair V, Sugano H, Roth L: Enhancement of anticonvulsant action of acetazolamide after head x-irradiation and its relation to blood–brain barrier changes. *Radiat Res* 23:265–281, 1964.

178. National Cancer Institute: *PCNU*. Clinical brochure. National Institutes of Health, Bethesda, Maryland 20892, July 1978.

179. Neil G, Moxley T, Manak R: Enhancement of tetrohydrouridine of 1-β-D-arabinofuranosylcytosine (cytarabine) oral activity in L1210 leukemic mice. *Cancer Res* 30:2166–2172, 1970.

180. Neil G, Buskirk H, Moxley T, et al: Biochemical pharmacological studies with 1-βD-arabinosycytosine-5′-adamantoate (NSC 117614), a depot form of cytarabine. *Biochem Pharmacol* 20:3295–3308, 1971.

181. Neuwelt E, Rapoport S: Modification of the blood–brain barrier in the chemotherapy of malignant brain tumors. *Fed Proc* 43:214–219, 1983.

182. Neuwelt E, Barnett P, Barranger J: The inability of dimethylsulfoxide (DMSO) and 5-fluorouracil (5-FU) to open the blood–brain barrier. *Neurosurgery* 12:29–34, 1983.

183. Neuwelt E, Glasberg M, Frenkel E, et al: Neurotoxicity of chemotherapeutic agents after blood–brain barrier modification: Neuropathological studies. *Ann Neurol* 14:316–324, 1983.

184. Notari E: Prodrug design. *Pharmacology Ther* 14:25–53, 1981.

185. Ohno K, Pettigrew K, Rapoport S: Local cerebral blood flow in the conscious rat as measured with [^{14}C]-antipyrine, [^{14}C]-iodoantipyrine and [^{3}H]-nicotine. *Stroke* 10:62–67, 1979.

186. Oldendorf W, Braun L, Cornford E: PH dependence of blood–brain barrier permeability to lactate and nicotine. *Stroke* 10:577–581, 1979.

187. Oldendorf W, Hyman S, Braun L, et al: Blood–brain barrier: Penetration of morphine, codeine, heroin and methadone after carotid injection. *Science* 178:984–986, 1972.

188. Oliverio V: Pharmacology of the nitrosoureas: An overview. *Cancer Treatm Rep* 60:703–707, 1976.

188a. Olson M, Chernik N, Posner J: Infiltration of the leptomeninges by systemic cancer. A clinical and pathological study. *Arch Neurol* 30:122–137, 1974.

189. Ommaya A: A subcutaneous reservoir and pump for sterile access to ventricular cerebrospinal fluid. *Lancet* 2:983–984, 1963.

190. O'Reilly R: Mechanism for pharmacodynamics of warfarin in man. *Clin Res* 16:311, 1968.

191. Pardridge W, Mietus L: Palmitate and cholesterol transport through the blood–brain barrier. *J Neurochem* 34:463–466, 1980.

192. Pardridge W, Connor J, Crawford I: Permeability changes in the blood–brain barrier: Causes and consequences. *CRC Crit Rev Toxicol* 3:159–199, 1975.

193. Patlak C, Fenstermacher J: Measurement of dog blood–brain transfer constants by ventriculocisternal perfusion. *Am J Physiol* 229:877–884, 1975.

194. Paxton J: Effects of aspirin on salivary and serum phenytoin kinetics in healthy subjects. *Clin Pharmacol Exp Ther* 27:170–178, 1980.

195. Peng G, Marquex V, Driscoll J: Potential central nervous system antitumor agents: Spirohydantoin mustard. *J Med Chem* 18:846–849, 1975.

196. Perucca E, Hebdige S, Frigo G, et al: Interaction between phenytoin and valproic acid: Plasma protein binding and metabolic effects. *Clin Pharmacol Exp Ther* 28:779–780, 1980.

197. Philips S: Does ethanol damage the blood–brain barrier? *J Neurol Sci* 50:81–87, 1981.

198. Philips S, Cragg B: Weakening of the blood–brain barrier by alcohol related stresses in the rat. *J Neurol Sci* 54:271–278, 1982.

199. Philips S, Cragg B: Blood–brain barrier dysfunction in thiamine deficient, alcohol treated rats. *Acta Neuropathol (Berl)* 62:235–241, 1984.

200. Plowman J, Lakings D, Owens E, et al: Initial studies on the penetration of spirohydantoin mustard into the cerebrospinal fluid of dogs. *Pharmacology* 15:359–366, 1977.

201. Poplack D, Bleyer A, Pizzo P: Experimental approaches to the treatment of CNS leukemia. *Am J Pediatr Hematol Oncol* 1:141–149, 1979.

202. Potashner S: Baclofen effects on amino acid release. *Can J Physiol Pharmacol* 56:150–154, 1978.

203. Price A, Jamieson P: The central nervous system in childhood leukemia II. Subacute leukoencephalopathy. *Cancer* 35:306–310, 1975.

204. Rahal J: Treatment of gram–negative bacillary meningitis in adults. *Ann Intern Med* 77:295–302, 1972.

205. Rapoport S: *Blood–Brain Barrier in Physiology and Medicine*. Raven, New York, 1976.

206. Rapoport S, Levitan H: Neurotoxicity of x-ray contrast media: Relation to lipid solubility and blood–brain barrier permeability. *AJR* 122:186–193, 1974.

207. Rapoport S, Thompson H: Opening of the blood–brain barrier by a pulse of hydrostatic pressure. *Biophys J* 15:326a, 1975

208. Reese T, Karnovsky M: Fine structural localization of a blood–brain barrier to exogenous peroxidase. *J Cell Biol* 34:207–217, 1967.

209. Reynolds J, Herbert S, Steinhardt J: The binding of some long chain fatty acid anions and alcohols by bovine serum albumin. *Biochemistry* 7:1357–1361, 1968.

210. Rieselbach R, DiChiro G, Mahgrefte B, et al: Subarachnoid distribution of drugs after lumbar injection. *N Engl J Med* 267:1273–1278 , 1962.

211. Roholt K: Pharmacokinetic studies with mecillinam and pivmecillinam. *J Antimicrob Chemother* 3(suppl B):71–81, 1977.

212. Rosenblum T, Stein A: Preferential distribution of diphenylydantoin in primary brain tumors. *Biochem Pharmacol* 12:453, 1963.

213. Rosowsky A, Abelson H, Beardsley G, et al: Pharmacological studies on the dibutyl and γ-monobutyl esters of methotrexate in the rhesus monkey. *Cancer Chemother Pharmacol* 10:55–61, 1982.

214. Rosowsky A, Forsch R, Yu C, et al: Methotrexate analogues. 21. Divergent influence of alkyl chain length on the dihydrofolate reductase affinity and cytotoxicity of methotrexate monoesters. *J Med Chem* 27:605–609, 1984.

215. Rowland M: Protein binding and clearance. *Clin Pharmacokinet* 9:10–17, 1984.

216. Rubin R, Ommaya A, Henderson E, et al: Cerebrospinal fluid perfusion for central nervous system neoplasms. *Neurology (NY)* 16:680–692,1966.

217. Runckel D, Swanson J: Effect of dimethyl sulfoxide on serum osmolarity. *Clin Chem* 26:1745–1747, 1980.

218. Scatchard G: The attractions of proteins for small molecules and ions. *Acad Sci (NY)* 51:660–692, 1949.

219. Scheuplein R, Blank I: Permeability of skin. *Physiol Rev* 5:702–747, 1971.

220. Schulman D, Kaufman J, Eisenstein M, et al: Blood pH and brain uptake of [^{14}C]-morphine. *Anesthesiology* 61:540–543, 1984.

221. Shapiro W, Chernik N, Posner J: Necrotizing encephalopathy following intraventricular installation of methotrexate. *Arch Neurol* 28:96–102, 1973.

222. Shapiro W, Young D, Mehta B, et al: Methotrexate: Distribution in cerebrospinal fluid after intravenous ventricular and lumbar injections. *N Engl J Med* 293:161–166, 1975.

223. Shaw C, Sumi M, Alvord E, et al: Fast–neutron irradiation of glioblastoma multiforme: Neuropathological analysis. *J Neurosurg* 49:1–12, 1978.

224. Shen D, Azarnoff D: Clinical pharmacokinetics of methotrexate. *Clin Pharmacokinet* 3:1–13, 1978.

225. Sigman L, Van Echo D, Egorin M, et al: Phase I trial of spiromustine. *Proc Am Soc Clin Oncol* 3:31, 1984.

226. Shiobara Y, Tachibana A, Sasaki H, et al: Phthalidyl D-α-aminobenzyl-penicillinate hydrochloride (PC-183), a new orally active ampicillin ester. 1. Absorption, excretion and metabolism of PC-183 and ampicillin. *J Antibiot* 27:665–673, 1974.

227. Simon C, Malerczyk V, Klans M: Absorption of bacampicillin and ampicillin and penetration into body fluids in healthy volunteers. *Scand J Dis* 14:228–235, 1978.

228. Sinkula A, Yalkowsky S: Rationale for design of biologically reversible drug derivatives: Prodrugs. *J Pharm Sci* 64:181–210, 1975.

229. Skipper H, Schabel F, Wilcox W: Experimental evaluation of anticancer agents. VII. On the criteria and kinetics associated with curability of experimental leukemia. *Cancer Chemother Rep* 35:1–111, 1964.

230. Sokoloff L: The (^{14}C)-deoxyglucose method for the measurement of local cerebral glucose utilization, procedure, and normal values in conscious and anesthetized albino rats. *J Neurochem* 28:897–916, 1977.

231. Sokoloff L: Relationship between physiological function and energy metabolism in the central nervous system. *J Neurochem* 29:13–26, 1977.

232. Soncrant T, Pizzolato G, Battistin L: The use of drugs as probes of cerebral function. In Battistin L (ed): *PET and NMR: New Perspectives in Neuroimaging and Clinical Neurochemistry*. Liss, New York, 1986, pp. 131–149.

233. Spence A, Geraci J: Combined cyclotron fast-neutron and BCNU therapy in a rat brain-tumor model. *J Neurosurg* 54:461–467, 1981.

234. Spigelman M, Zappulla R, Malis L, et al: Intracarotid dehydrocholate infusion: A new method for prolonged reversible blood–brain barrier disruption. *Neurosurgery* 12:606–612, 1983.

235. Spigelman M, Zappulla R, Holland J, et al: Characterization of etopside-induced blood–brain barrier disruption. *Proc Am Assoc Cancer Res* 25:383, 1984.

236. Spigelman M, Zappulla R, Johnson J, et al: Etopside-induced blood–brain barrier disruption. *J Neurosurg* 61:674–678, 1984.

237. Sponzo R, DeVita V, Oliverio V: Physiological disposition of 1-(2-chloroethyl)-3-cyclohexyl-1-nitro-sourea (CCNU) and 1-(2-chloroethyl)-3-(4-methyl cyclohexyl)-1-nitrosourea (MeCCNU) in man. *Cancer* 31:1154–1159, 1973.

238. Steele W, Lawrence J, Stuart J, et al: The protein binding of methotrexate by the serum of normal subjects. *Eur J Clin Pharmacol* 15:363–366, 1979.

239. Stella V: The control of drug delivery via bioreversible modification. In Juliano R (ed): *Drug Delivery Systems, Characteristics and Biomedical Applications*. Oxford University Press, New York, 1980, pp. 112–175.

240. Stewart D, Leavens M, Moore M, et al: Cis DDP: human CNS distribution and use as a radiosensitizer in malignant brain tumors. *Cancer Res* 42:2474–2479, 1982.

241. Stoller R, Jacobs S, Drake J, et al: Pharmacokinetics of high-dose methotrexate. *Cancer Treatm Rep* 6:19–24, 1975.

242. Suzuki M, Abe I, Sato H: Changes in drug delivery (by blood–brain barrier dysfunction) on arachnoid leukemia: Implications for CNS leukemic dissemination. *Clin Exp Metast* 1:163–171, 1983.

243. Taft R: Separation of polar, steric and resonance effects in reactivity. In Newman M (ed): *Steric Effects in Organic Chemistry*. Wiley, New York, 1956, pp.556–675.

244. Teresi D, Luck J: The combination of organic anions with serum albumin. *J Biol Chem* 194:823–834, 1952.

245. Testa B, Jenner P: Drug metabolism: Chemical and biochemical aspects. In Jenner P, Testa B (eds): *Drugs and the Pharmaceutical Sciences*. Vol. 4. Dekker, New York, 1976, pp. 3–455.

246. Tochino Y, Schanker L: Active transport of biological amine compounds by the rabbit choroid plexus. *Pharmacologist* 6:177, 1964.

247. Tochino Y, Schanker L: Active transport of quaternary ammonium compounds by the choroid plexus in vitro. *Am J Physiol* 208:666–673, 1965.

248. Valner J: Binding of drugs by albumin and plasma protein. *J Pharm Sci* 66:447–465, 1977.

249. Vistica D: Cellular pharmacokinetics of the phenylalanine mustards. *Pharm Ther* 22:379–405, 1983.

250. Wan S, Huffman D, Azarnoff D, et al: Pharmacokinetics of 1-β-D-arabino-furanosylcytosine in humans. *Cancer Res* 34:392–397, 1974.

251. Wang M, Sharma R, Bloch A: Studies on the mode of action of 2,2′anhydro-1-β-D-arabinosylcytosine. *Cancer Res* 33:1265–1271, 1973.

252. Way E, Kemp J, Young J, et al: The pharmacological effects of heroin in relationship to its rate of biotransformation. *J Pharmacol* 9:144–154, 1960.

253. Weinkam R, Liu T, Lin H: Protein mediated chemical reactions of the chloroethylnitrosoureas. *Chem Biol Interact* 31:167–178, 1980.

254. Weinstein H, Griffin T, Feeney J, et al: Pharmacokinetics of continuous intravenous and subcutaneous infusion of cytosine arabinoside. *Blood* 59:1351–1353, 1982.

255. Weir E, Cashmore A, Dreyer R, et al: Pharmacology and toxicity of a potent non-classical 2,4-amino quinazoline folate antagonist, trimetrexate, in normal dogs. *Cancer Res* 42:1696–1702, 1982.

256. Wermuth C: Designing prodrugs and bioprecursors. In Jolles G, Wooldridge K (eds): *Drug Design: Fact or Fantasy?* Academic, New York, 1984, pp. 47–72.

257. Westerink B, Dijkstra D, Feenstra M, et al: Dopaminergic prodrugs: Brain concentrations and neu-rochemical effects of 5,6- and 6,7-ADTN after administration as dibenzoyl esters. *Eur J Pharmacol* 61:7–15, 1980.

258. Wheeler G, Bowden B, Grimsley J: Interrelationships of some chemical, physiochemical, and biological activities of several 1-(2-haloethyl)-1-nitrosoureas. *Cancer Res* 34:194–200, 1974.

259. Wickstrom E: Chlorambucil inhibition of dimethyl sulfoxide and thiosulfate: Implications for chlorambucil chemotherapy. *Med Hypoth* 6:1035–1041, 1980.

260. Williams F: Clinical significance of esterases in man. *Clin Pharmacokinet* 10:392–403, 1985.
261. Williams T: Meningitis: Special techniques in treatment. *Mod Treatm* 7:606–617, 1970.
262. Winn W: The treatment of coccidioidal meningitis: The use of amphotericin B in a group of 25 patients. *Calif Med* 101:78–89, 1964.
263. Witorsch P, William T, Ommaya A, et al: Intraventricular amphotericin B: Use of a subcutaneous reservoir in four patients with mycotic meningitis. *JAMA* 194:699–702, 1965.
264. Wright D, McCormick P, Lawrence P, et al: Infusions of dimethyl sulfoxide (DMSO): Blood–brain barrier effects. *J Cereb Blood Flow Metab* 5:s75–s76, 1985.
265. Yalkowsky S, Flynn A: Transport of alkyl homologs across synthetic and biological membranes: A new model for chain length: Activity relationships. *J Pharm Sci* 62:210–217, 1973.
266. Zook B, Bradley E, Casarett A, et al: Pathological findings in canine brain irradiated with fractionated fast neutrons or photons. *Radiat Res* 84:562–578, 1980.

13

The Blood–Ocular Barrier

Mark J. Kupersmith and Manoucher Shakib

No treatise on the blood–brain barrier (BBB) is complete without a discussion of the barriers between the blood and the most specialized portion of the central nervous system (CNS), the eye. The eye is a complex organ with distinct anatomic structures and blood supplies that form several nonuniform blood–ocular barriers (Fig. 1). Analysis of the passage of substances into the anterior and posterior chambers, the vitreous, the optic nerve, the choroid, and the retina reveals different levels of penetration into each. The vitreous has no direct vascular input and lies between the aqueous and retinal barriers. The breakdown of the barrier between the blood and the retina plays a significant role in the onset of numerous retinal diseases. A barrier similar to the BBB occurs at the level of the optic nerve.

1. BLOOD–AQUEOUS BARRIER

Although the blood–aqueous barrier is less significant than the blood–retinal barrier, it plays an important role in regulating the exchange between blood and intraocular fluid. Aqueous humor is secreted by the ciliary processes into the posterior chamber, behind the lens, and passes through the pupil to the anterior chamber in front of the lens. Free diffusional exchange between the aqueous and the vitreous in the posterior chamber is limited only by hyaluronic acid in the formed vitreous.[19] Active transport, possibly by pinocytosis, may carry drugs into the posterior chamber.[13,43]

The blood–aqueous barrier occurs at the blood vessels and the epithelium of the iris and ciliary body. The arterial blood supply to the iris arises from the arterial circle formed by seven anterior ciliary arteries arising from the recti muscles and two long posterior ciliary arteries.[29] The blood flow to the iris seems to be regulated by sympathetic innervation.[29] Branches from the arterial circle course from the iris root through the anterior stroma to form an incomplete ring of capillaries that supplies the dilator muscle and

Mark J. Kupersmith • Departments of Neurology and Ophthalmology, New York University, New York, New York 10016.   Manoucher Shakib • Department of Ophthalmology, New York University, New York, New York 10016.

collarette area. The capillaries are nonfenestrated and have continuous basement membrane and some pericytes (Fig. 1). Tight junctions, which are impermeable to circulating horseradish peroxidase (HRP), are found in monkeys.[20,48] In other animals and humans these junctions are not as significant. The zonula occludentes of these tight junctions do not occupy most of the interendothelial space, hence they are less stable and open with

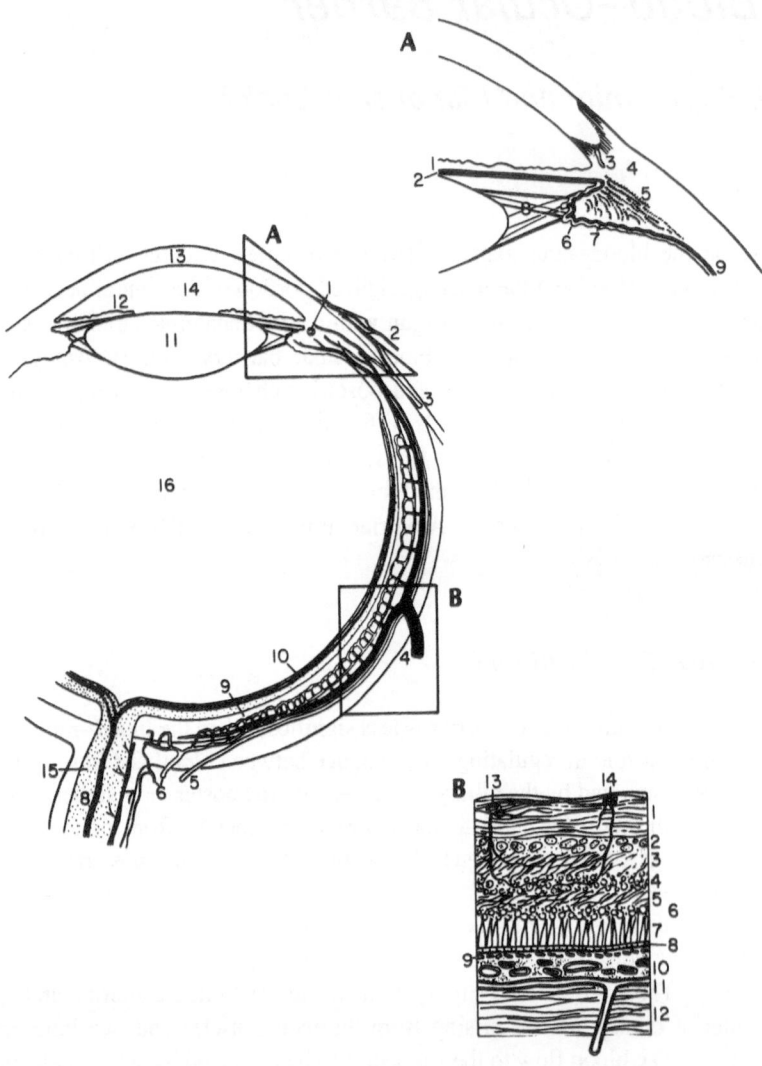

Figure 1. The eye and its blood supply. (1) Major arterial circle of iris; (2) conjunctival vessels; (3) anterior ciliary vessels; (4) vortex vein; (5) long posterior ciliary artery; (6) short posterior ciliary arteries; (7) pial vessels; (8) central vessels of retina; (9) choroid vessels; (10) retina vessels; (11) lens; (12) iris; (13) cornea; (14) anterior chamber; (15) optic nerve; (16) posterior chamber. (A) Angle of the eye (triangle insert). (1) Iris stroma; (2) iris pigment epithelium; (3) outflow meshwork; (4) canal of Schlemm; (5) ciliary muscle; (6) ciliary processes; (7) ciliary epithelium; (8) zonules; (9) para plana. (B) Retina–Choroid (rectangle insert). (1) Optic nerve fibers; (2) ganglion layers; (3) inner plexiform layer; (4) inner nuclear layer; (5) outer plexiform layer; (6) outer nuclear layer; (7) rod and cone layer; (8) pigment epithelium; (9) choriocapillaris; (10) layer of large choroidal vessels; (11) suprachoroid; (12) sclera; (13) arteriole; (14) venule.

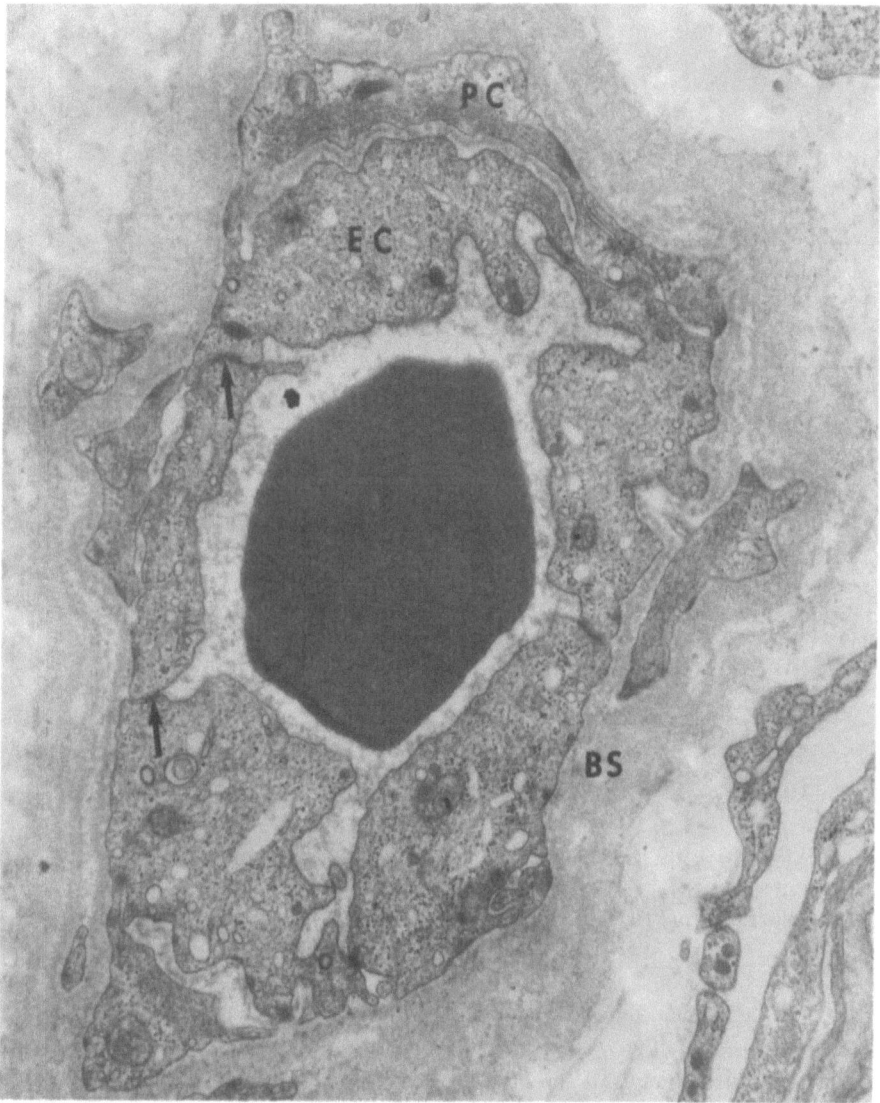

Figure 2. Electron micrograph of human iris capillary showing nonfenestrated endothelial cells (EC) with their junctional complexes (arrows), a continuous basement membrane (BS), and part of a pericyte (PC) are seen. (× 21,000)

locally administered histamine or prostaglandin.[12,40] The permeability of the iris vessels to protein can be altered by mechanical stimulation, and paracentesis can cause a breakdown in the blood–aqueous barrier.[29] The late leakage of intravenously administered fluorescein from the pupillary margin in normal subjects results from passage of fluorescein through the iris vessels and iris tissues into the aqueous.[23] The iris vessels seem to have no role in producing the aqueous humor. The anterior surface of the iris is lined by discontinuous cells, similar to those found in the iris stroma, so no barrier exists at this level. The aqueous fluid and stroma have a free exchange of liquids and solutes.

By contrast, the epithelium of the posterior iris is continuous with the nonpigmented layer of the ciliary epithelium and has tight junctions (Fig. 2). The posterior iris has a blood aqueous barrier formed by the posterior epithelium in addition to the iris vessels.

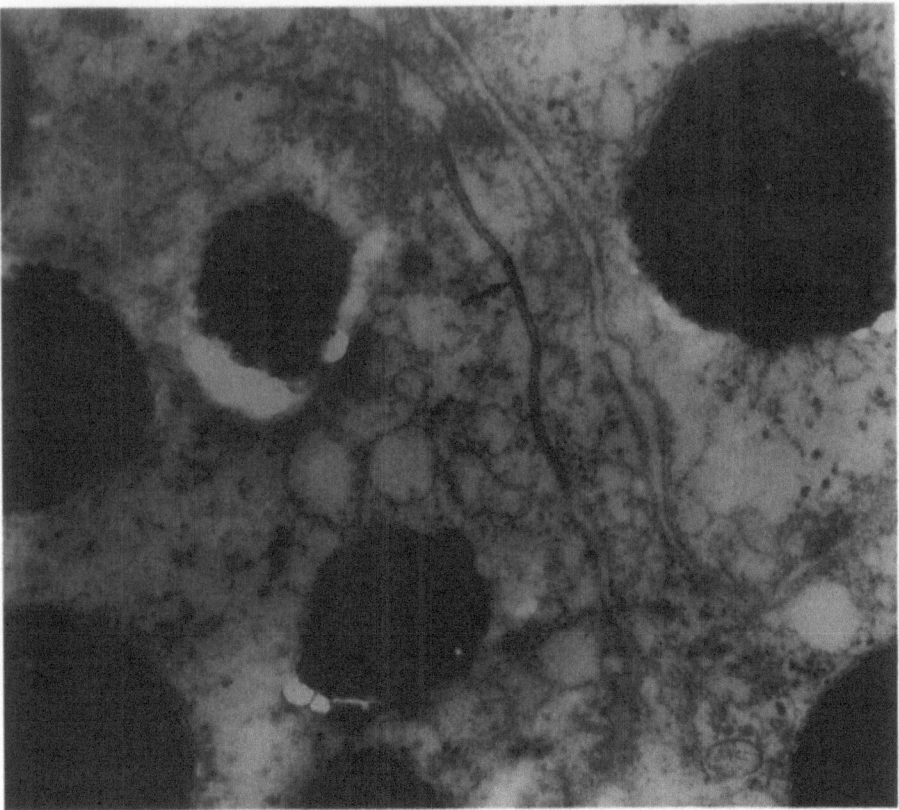

Figure 3. Electron micrograph of human pigmented epithelium of posterior iris showing a large area of tight junction (arrow). Note the outer leaflets of adjacent cell membrane fused to form an intermediate layer, zonula occulent. See also Fig. 6. (× 60,000)

Aqueous humor is produced by the ciliary body. Large-diameter arteries branching from the long posterior ciliary arteries, anterior ciliary arteries, and the major circle of the iris supply the ciliary body. The ciliary muscle is supplied by thick-walled arteries with tight junction endothelium. Fenestrations ($\leq 10 \, \mu$m) in the capillary vessel walls (Fig. 3) permit plasma proteins to escape into the loose connective tissue of the ciliary stroma, increasing osmotic pressure of the stroma.[1] The pigmented epithelium, which lies on the stroma, permits the passage of macromolecules, such as HRP, through intercellular clefts or gap junctions[38] (Fig. 4). The nonpigmented epithelium, which lies between the pigmented epithelium and the posterior chamber, has selective permeability to specific ions, decreasing the stroma osmotic pressure.[3] The tight junctions between the pigmented and nonpigmented epithelium are the site of blood aqueous barrier in the ciliary body (Fig. 5). Desmosomal junctions between the nonpigmented epithelial cell are found near their apices by sealing strands. Although these junctions bear a morphologic resemblance to other types of epithelium with leaky junctions under normal circumstances, little contribution to the aqueous arises here. The nonpigmented epithelium performs osmotic work by transporting certain ions and molecules across the cells into the aqueous. The aqueous has higher concentrations of bicarbonate and ascorbate and lower concentrations of calcium and urea than in the plasma.

The aqueous fills the anterior (0.25 ml) and posterior (0.06 ml) chambers of the eye with an osmotic pressure 3–4 mOsm greater than that of plasma.[14] The aqueous flows at about 2 μl/min. The protein concentration is less than 50 mg/dl. Approximately 70% of the sodium enters the aqueous by cellular transport, which can be inhibited by ouabain. Smaller amounts of chloride are similarly transported. Bicarbonate enters the aqueous by either diffusion or exchange with carbon dioxide, which is inhibited by carbonic anhydrase inhibitors, or both. Amino acids are transported by three carriers, for acidic, basic, and neutral molecules. Cyanide, ouabain, and probenecid depress the transport of amino acids.[45]

Aqueous humor production can be altered experimentally by several methods. Elevation in intraocular pressure (IOP) decreases its production. Paracentesis leads to an enlargement of the intercellular spaces between the limiting ciliary epithelium, permitting passive diffusion.[42] An injection of prostaglandins or histamine into the anterior chamber also causes uncontrolled leaking with elevation of the protein concentration in the aqueous humor. Intraarterial, but not intravenous, urea also separates the ciliary epithelial and vascular endothelial intercellular tight junctions without cell damage.[34,36] Intravenous urea lowers IOP, HRP passes into the anterior chamber by way of Schlemm's canal (the normal nonvascular drainage of aqueous from the eye) as a result of backdiffusion from episcleral veins.[37] Other hyperosmotic agents, such as mannitol and glycerol, lower IOP as well. The formation of aqueous continues, even with lowered systemic arterial blood pressure. Carbonic anhydrase inhibitors, such as acetazolamide and methazolamide, decrease aqueous production by approximately 50%.[26] Activating adenylate cyclase also decreases aqueous flow.[21]

The cornea is a physical barrier to penetration of substances into the eye. Diffusion of drugs through the cornea depends on how long the drug remains in contact with the

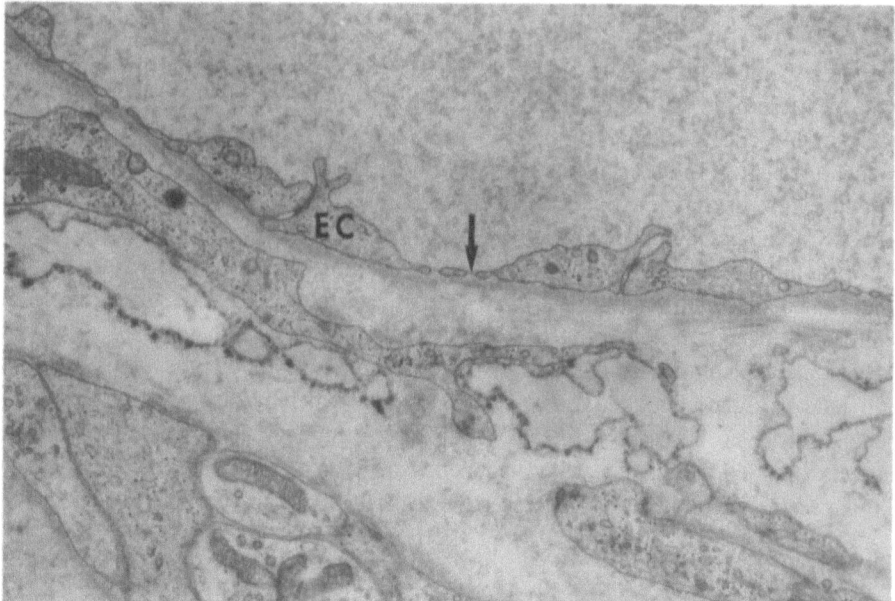

Figure 4. Electron micrograph of human ciliary body capillary showing fenestration in the endothelial cell of the vessel wall (arrow). (× 20,500)

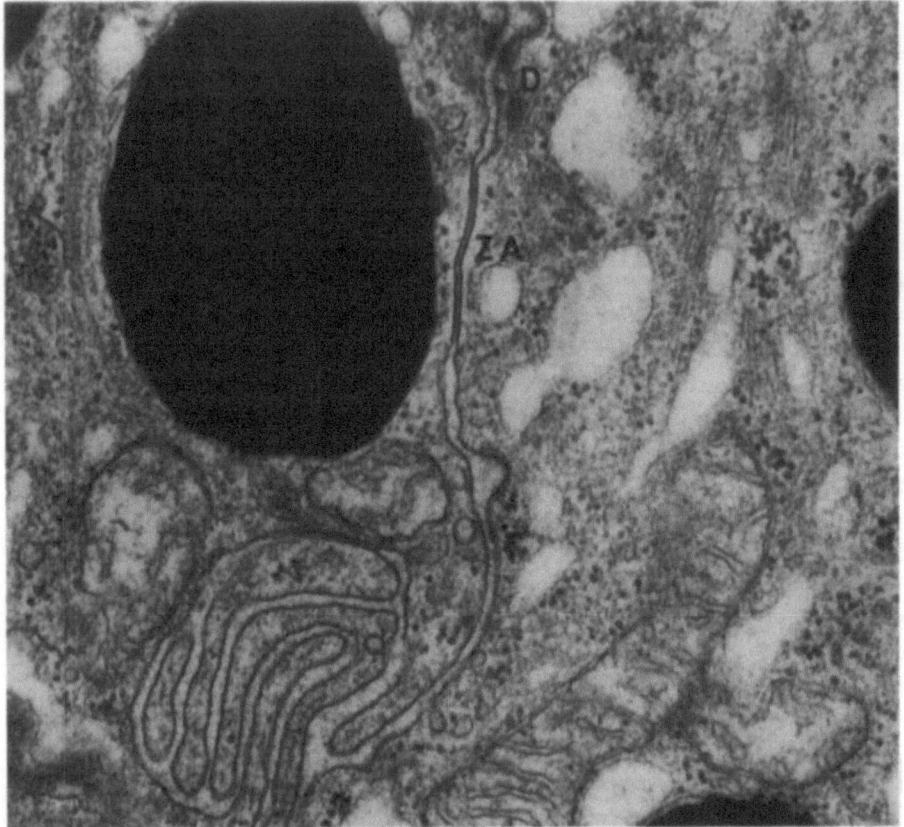

Figure 5. Electron micrograph of junctional complex between two pigmented cells of a human ciliary body showing a zonula adherens (ZA) or gap junction and desmosome (D). (× 60,000)

corneal epithelium, drug lipid solubility, and the condition of the tear film and various layers of the cornea. Drugs that alter the aqueous humor can be topically administered.[26] Epinephrine reduces aqueous formation by 30% without altering carbonic anhydrase. Epinephrine also facilitates aqueous outflow. Miotics increase the outflow of the aqueous from the eye by opening the trabecular meshwork, the site from which the aqueous leaves the eye. Cholinergic agonists (e.g., pilocarpine, carbachol, methacholine, acetylcholine) or anticholinesterase agents (e.g., eserine, neostigmine, isoflurophate, phospholine iodide, and demecarium bromide) have all been used as miotics in glaucoma patients. The β-adrenergic blocker timolol reduces aqueous production.[26]

Surgery of the anterior segment of the eye alters the blood–aqueous barrier. Slit-lamp aqueous fluorophotometry is a sensitive method whereby disruption of the blood–aqueous barrier in vivo can be detected in humans.[49] Intraocular lens fixation following cataract extraction disrupts the blood–aqueous barrier as measured by increased leakage of fluorescein.[33] Indomethacin decreases the leakage of fluorescein in these subjects.[39]

2. BLOOD–RETINAL BARRIER

The retinal vessels and retinal pigment epithelium are the two sites of blood–retinal barrier. The vessels of the choriocapillaris have lumens of ≤ 40 μm with numerous fenestrations in their endothelium of ≤ 80 mm wide (Fig. 7). Macromolecules such as albumin, myoglobin, and gammaglobulin escape easily from the choroidal capillaries.[2] The permeability of the choroidal vessels to gammaglobulin and albumin is 5 and 30 times greater than permeabilities of the vessels in the kidney and muscle, respectively.[5] HRP passes easily through the vessel walls into the connective tissue of the choroid (Fig. 8). The choriocapillaris is supplied by the posterior ciliary arteries.[22] The normally high rate of blood flow through the choroid exhibits no autoregulation and is diminished by eleva- tion in IOP and stimulation of the cervical sympathetic ganglion.[4] Blood flow may be increased by stimulation of the sphenopalatine ganglion. Bruch's membrane, located between the retinal pigment epithelium and the choriocapillaris, is a diffusion barrier to large molecules (Figs. 8 and 9). HRP and fluorescein, which escape from the choriocapil- laris, pass easily through Bruch membrane (Fig. 8). The pigmented epithelium of the

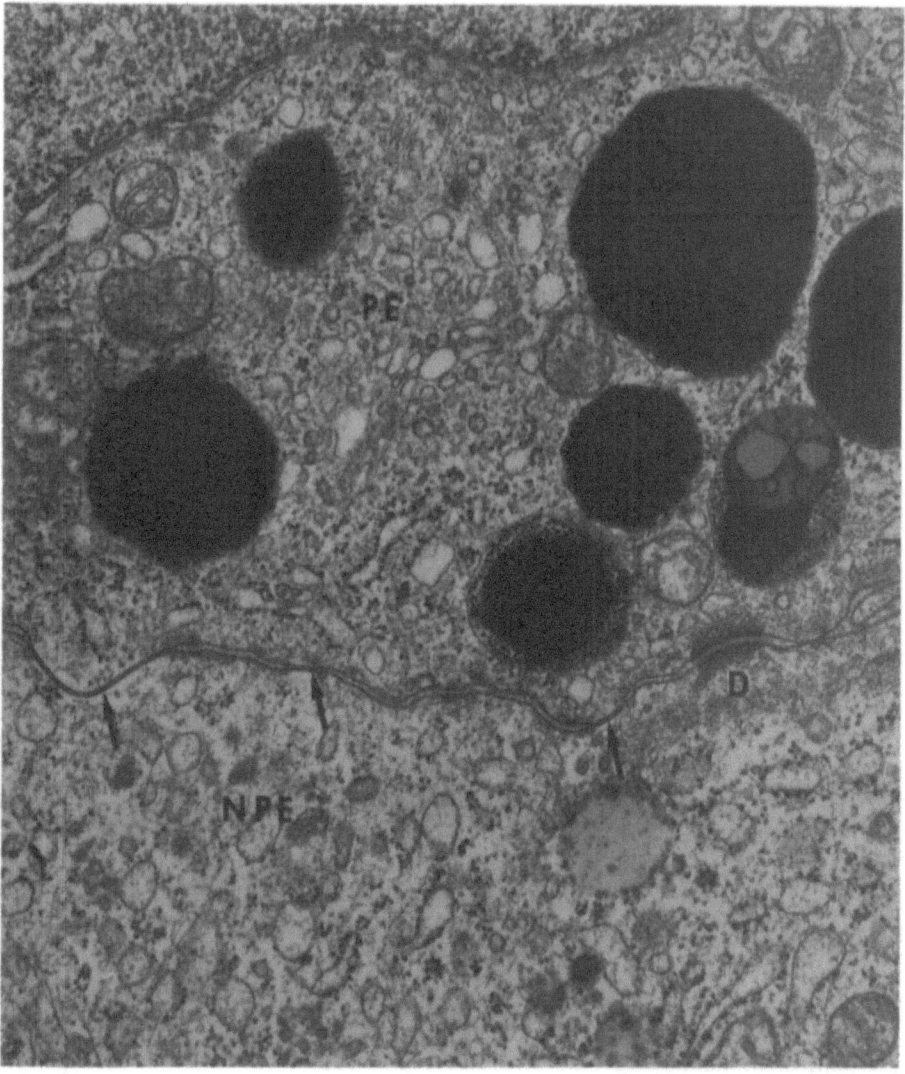

Figure 6. Electron micrograph of junctional complexes between two adjacent pigmented epithelium (PE) and nonpigmented epithelium (NPE) of the human ciliary body. Note the area of tight junction (arrows) and desmosome (D). (× 36,000)

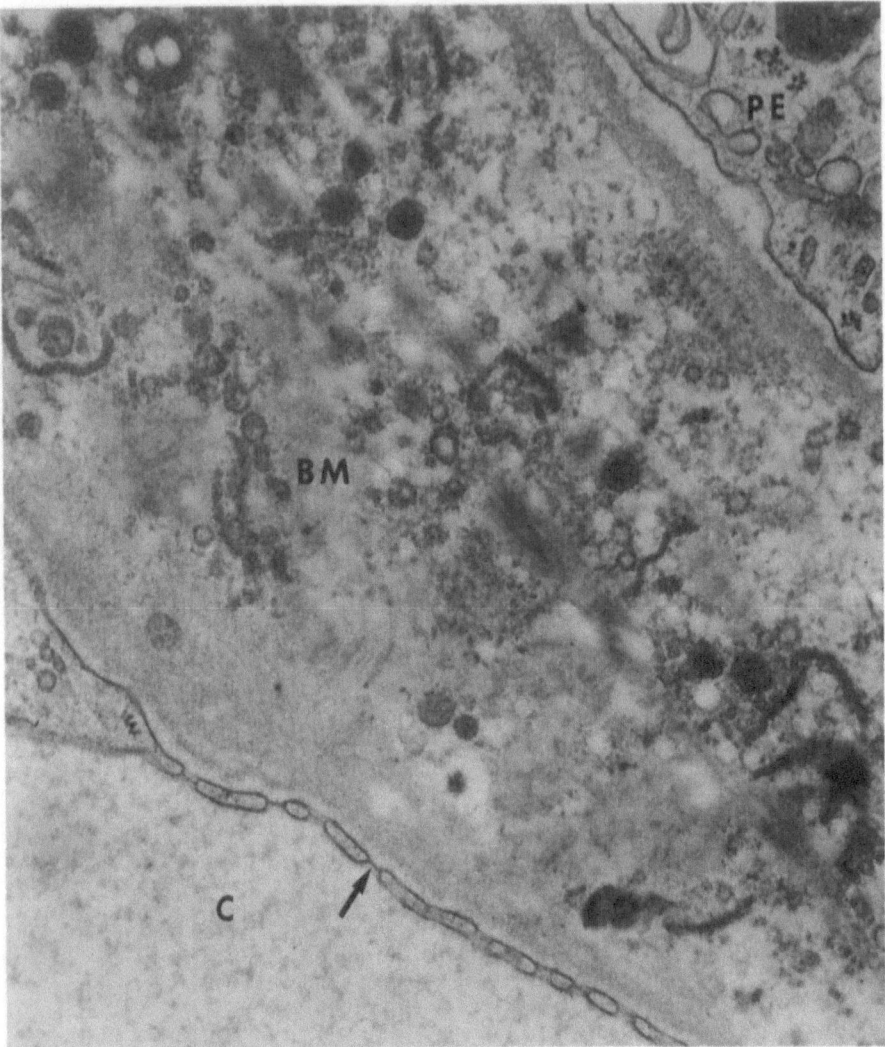

Figure 7. Electron micrograph of choriocapillaris (c), Bruch membrane (BM), and pigmented epithelium of the rat retina pigment epithelium (PE). The endothelium cells of the capillary have fenestrations (arrow). (× 25,000)

retina is the significant barrier between the retina and choroid (Figs. 9 and 10). Such macromolecules as HRP, trypan blue, and fluorescein penetrate the intercellular spaces between adjacent retinal pigment epithelium but are blocked by tight junctions between the lateral surfaces of the cells (Fig. 10). The application of histamine does not open these tight junctions (Figs. 11 and 12). Intraarterial injection of hypertonic solutions open the tight junctions without damage to the pigment epithelium, so that fluid from the choriocapillaris leaks into the retina.[36] The epithelial cell membrane is highly permeable to potassium and poorly permeable to sodium and chloride. D-glucose is actively transported by a stereospecific carrier across the retinal pigment cell layer into the retina. Amino acids, some of which may serve as false neurotransmitters in the retina, are transported from the vitreous to the blood vessels across the retinal pigment. Prostaglandins in the vitreous are similarly cleared by the choroid.[6] The choriocapillaris is vital in providing the proper metabolic milieu for the outer layers of the retina.

Except for the pigment epithelium, the 10 layers of the retina do not significantly prevent diffusion of various macromolecules (Fig. 13). The outer and inner limiting membranes have wide gaps and fail to block HRP from reaching the pigment epithelium when injected into the vitreous. The glia (Muller cells) offer no major barrier in the retina. The retinal vessels are in direct contact with glia and are the other site of the blood–retinal barrier. Muller cells, like brain glia, maintain the homeostasis of electrolytes and of the neuronal extracellular environment.

The inner retinal layers receive their blood supply via branches of the central retinal artery, which arises from the ophthalmic artery, entering the optic nerve just posterior to the globe. The retinal circulation demonstrates autoregulation to maintain constant blood flow despite alterations in perfusion pressure, IOP, and changes in oxygen or carbon dioxide concentrations. Arteriolar constriction is seen with 100% oxygen inspiration, and vasodilatation is demonstrated with hypoxia, but retinal blood flow remains stable under these conditions.[16]

The retinal vessels have interendothelial junctions similar to those in the brain but different from those found in vessels of other organs with continuous endothelium (Fig.

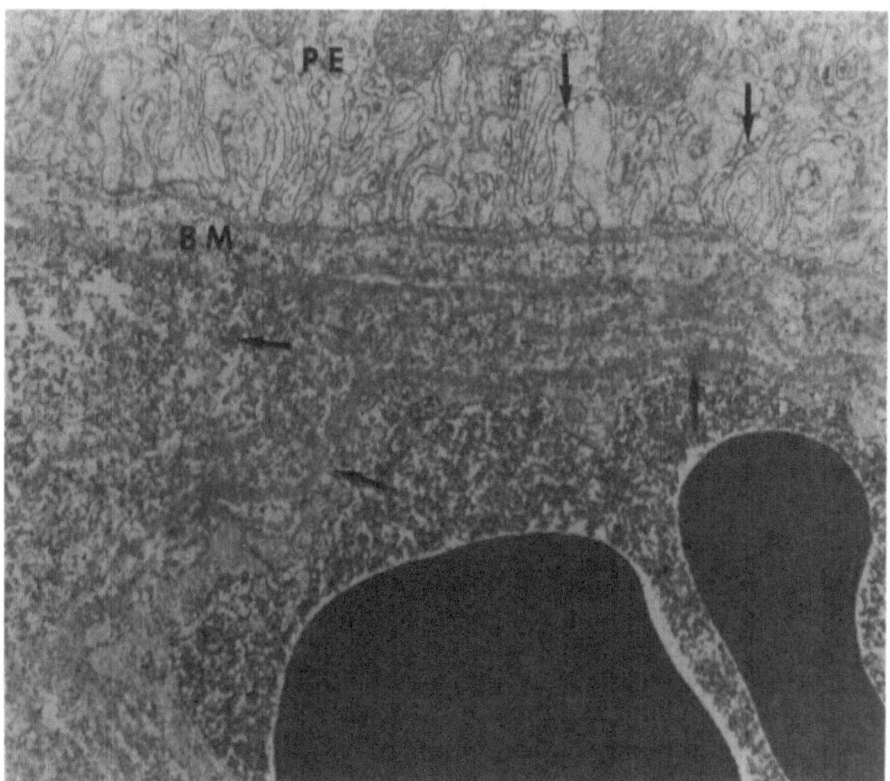

Figure 8. Electron micrograph of rat choriocapillaris after injection of horseradish peroxidase showing penetration of the tracer (arrows) into the Bruch membrane, choroidal tissue, and into the pigmented epithelium. (× 21,000)

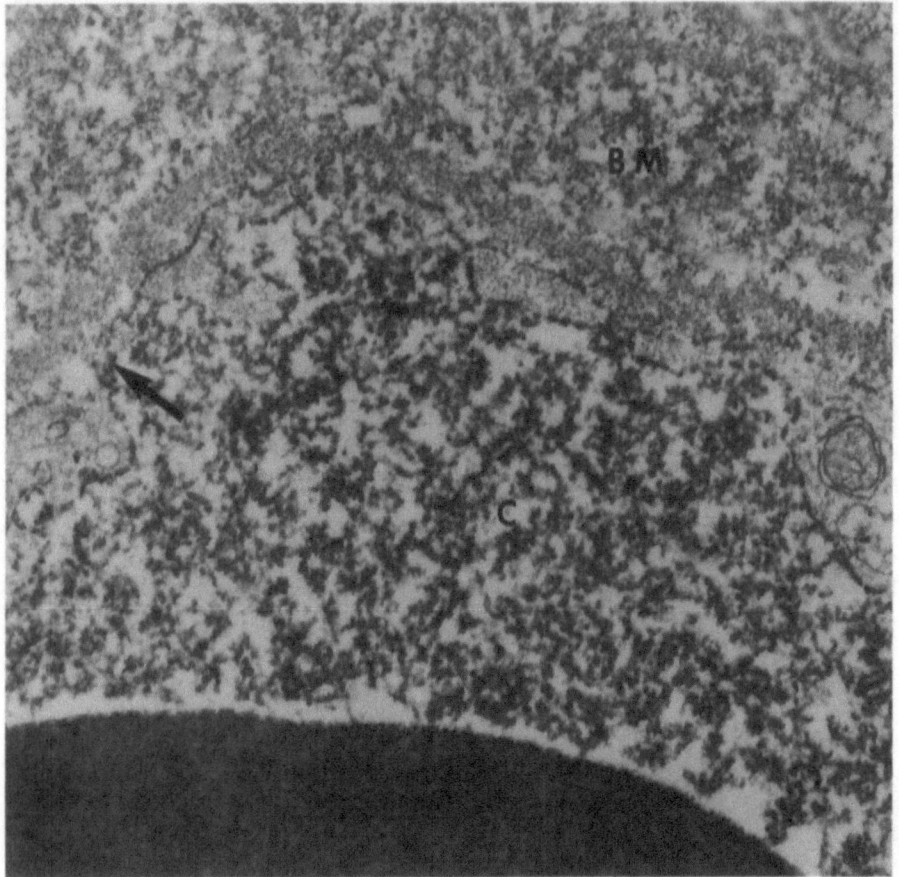

Figure 9. Electron micrograph of the rat choriocapillaris (C) showing the pores or fenestrations, which are permeable to peroxidase (arrow). (BM) Bruch membrane. (× 64,000)

14). Histamine increases the permeability of all the vessels in the eye by opening up the interendothelial junctions except those of the retinal vessels (Fig. 16). Zonula occludentes seal the intercellular spaces around most of the interface between the endothelial cells.[12,40] These tight junctions, the main sites of the blood–retinal barrier, block the escape of fluorescein, HRP, trypan blue, and microperoxidase[44] (Fig. 16). The retinal endothelial cells, like those in the brain, have elevated alkaline phosphatase activity,[9] monoamine oxidase, and dopa-decarboxylase activities.[10]

The barrier at the retinal arteriole remains intact even when the cerebral vessels become permeable with acute elevation of carotid arterial pressure.[27] Disruption of the intercellular junctions is rarely seen with elevated blood pressure except when frank retinal arteriole necrosis is present. Hypertonic perfusion of the carotid circulation shrinks the endothelial cells reversibly and opens the tight junctions of the retinal and cerebral vessels. Intravascular fluorescein and Evans blue–albumin leak into the vitreous from the opened blood–retinal barrier at the level of the retina and optic nerve.[36]

The retinal capillaries differ from other capillaries in that nonspecific exchange of molecules does not occur and pinocytotic activity is rarely seen in the vessel walls. Steep gradients of specific molecules are established across the blood–retinal barrier. Active transport, particularly of amino acids, is necessary to supply the retina. Carrier-mediated transport of amino acids and D-glucose occurs. Magnesium is actively transported to higher concentrations in the extracellular space of the retina than in the plasma. Potassium is actively transported from the vitreous and the retina into the blood. The concentration of sodium is similar in plasma and the retinal extracellular fluids. Substances with high lipid/water partition coefficients pass freely from the blood into the vitreous. Various antibiotics penetrate into the vitreous in relationship to this coefficient.[6]

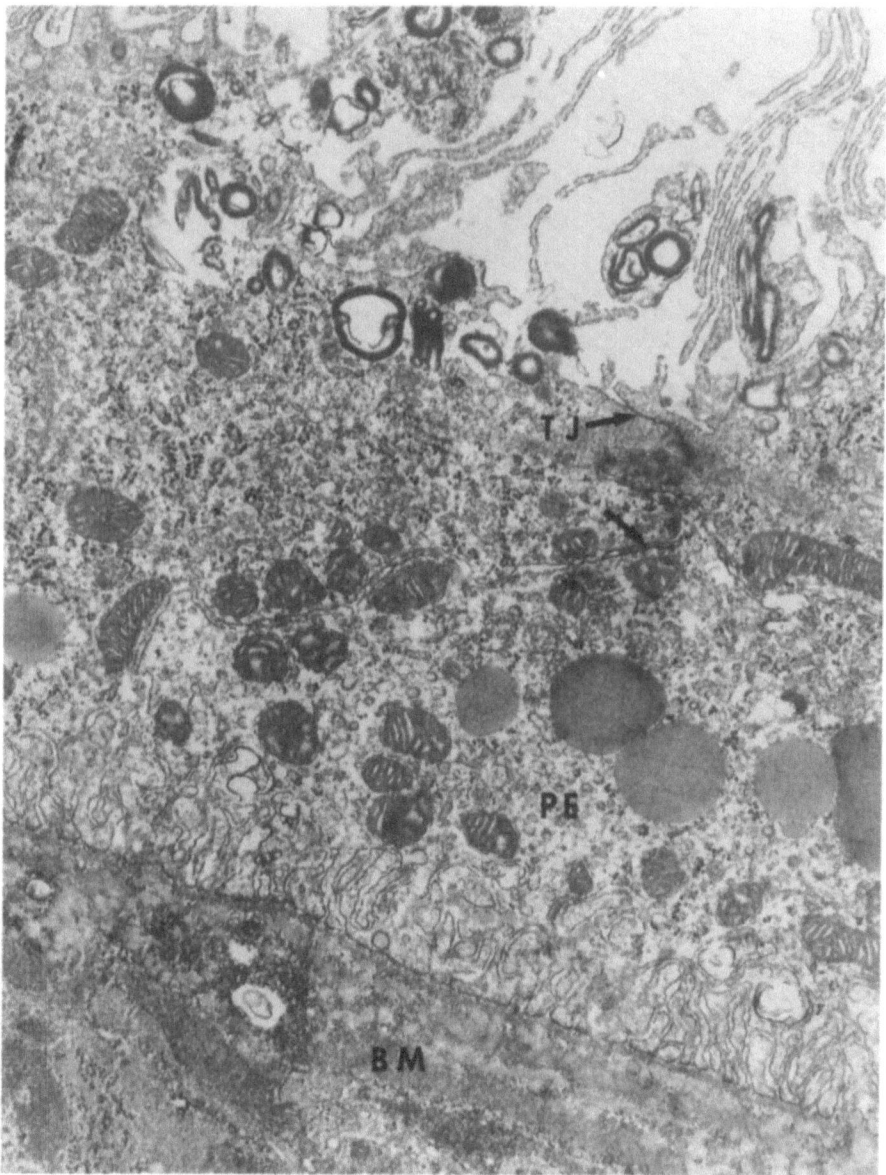

Figure 10. Electron micrograph of Bruch membrane (BM) and pigmented epithelium (PE) of the retina after injection of horseradish peroxidase in the blood circulation of the rat. The tracer appears in the intercellular space (arrow) but not beyond the interepithelial tight junctions (TJ). (× 21,000)

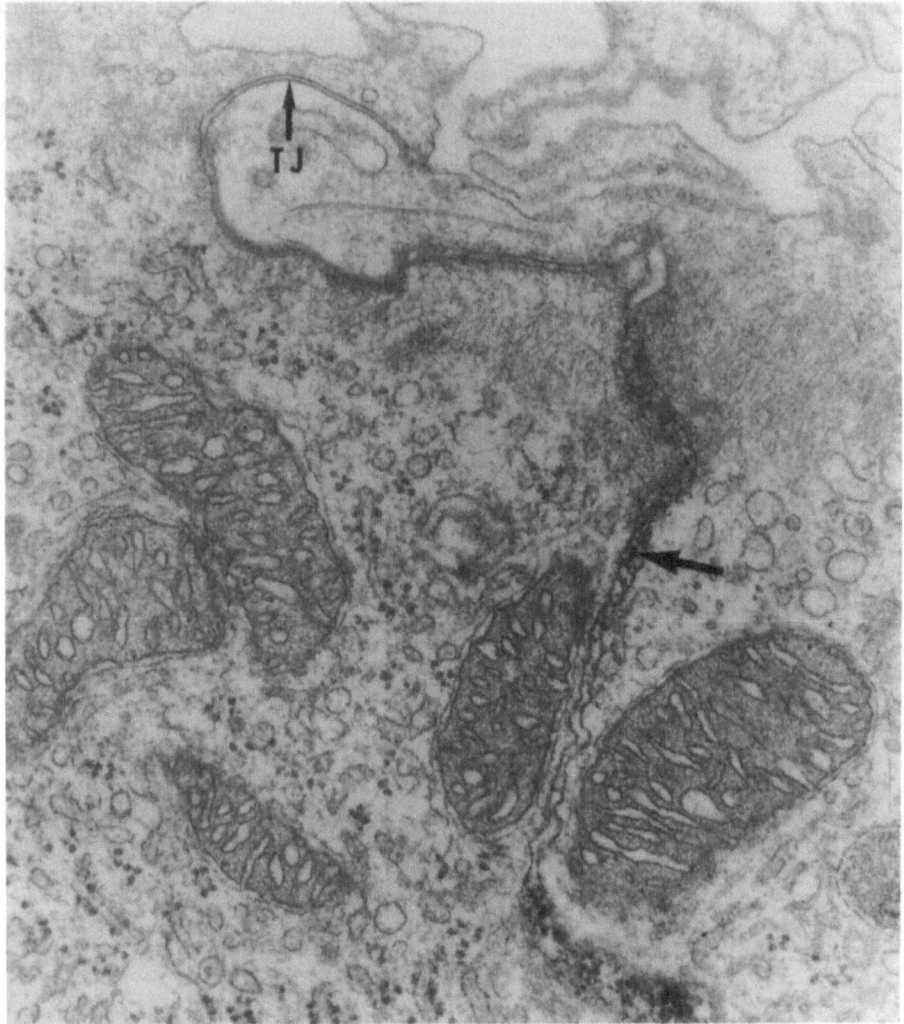

Figure 11. Electron micrograph of the intercellular space between two retina pigmented epithelial cells. The migration of tracer (arrow) is stopped at the site of the tight junction (TJ). (× 55,000)

Fluorescein Angiography and Fluorophotometry

The movement of intravascular fluorescein, a dibasic acid, into the eye is considered individually because of its use in fluorescein angiography and vitreous fluorophotometry. Fluorescein (diameter of 11 Å) has a blood–retinal permeability coefficient at the inner layer of the retina of 0.14×10^{-5} cm/sec. Passive diffusion of fluorescein is related to lipid solubility and blood and retinal pH.[7] Fluorescein is actively transported across the retinal vessels from the vitreous even against a high intravascular fluorescein concentration. By contrast, intravenous fluorescein will minimally enter the anterior portion of the vitreous body via the ciliary body in a normal eye. Metabolic inhibitors, such as probenecid or penicillin, change the direction of fluorescein toward free exchange across the retina. Fluorescein movement out of the eye is blocked and leakage into the retina occurs over the entire retinal surface after administration of these drugs. Fluorescein transport across the blood–retinal barrier may be similar to organic anion transport in other organs.

Fluorescein angiography is a dynamic method of studying the retinal and choroidal circulations and the blood–retinal barriers at the retinal vessels and the retinal pigment epithelium (Fig. 17) (there are many books on this subject; short reviews are found in refs. 1 and 15) . Briefly, fluorescein (3 ml of 25% or 5 ml of 10% fluorescein) is injected intravenously and seconds later rapid sequential photography at 1 sec is performed. Excitor and barrier filters permit the transmission of light emitted by fluorescein excited by blue light. Approximately 80% of intravascular fluorescein is albumin bound and the free 20% is the most fluorescent. Choroidal arteries do not leak fluorescein and are clearly

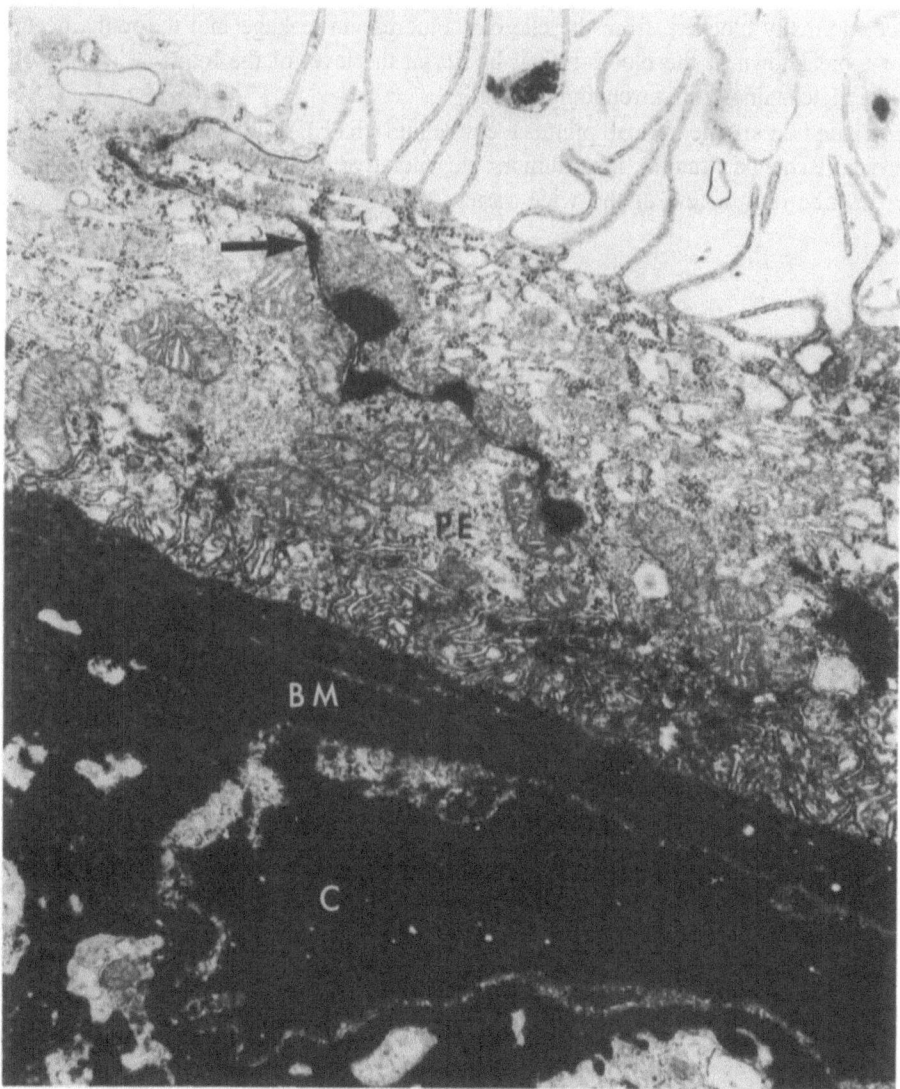

Figure 12. Electron micrograph of horseradish-peroxidase-injected, histamine-treated rat retina showing a large amount of the tracer in the choriocapillaris (C), Bruch membrane (BM), and intracellular space between pigmented epithelium (PE). The tight junctions between pigment epithelium remained closed and no tracer could pass beyond them (arrow). (× 19,000)

visible prior to filling of the choroidal capillaris. The fenestrated capillaries of the choroid freely permit extravasation of dye, so that the extravascular hyperfluorescence prevents detailed study of the choroidal circulation. The large choroidal vessels appear hypo-fluorescent in contrast to the remaining extravascular fluorescein late in the angiogram. The retinal pigment blocks fluorescein transmission relative to the degree of pigmentation of the cells. Retinal arterioles and the vessels on the optic nerve surface are sharply delineated. Vessels as small as precapillary arterioles can be resolved. The normal retina, which does not contain fluorescein, is contrasted with the intravascular fluorescein. The vessels of the optic disc have a blood–optic nerve barrier similar to the BBB so that no fluorescein leaks into the neural tissue. By contrast, fluorescein leaks into the prelaminar region of the optic nerve from the choriocapillaris causing late hyperfluorescence at the disc border. Defects in the retinal pigment epithelium permit the transmission of fluores-cence, normally blocked, from the choroid. Fluorescein leakage into the retina occurs from a breakdown of the blood–retinal barrier at the level of the retinal vessels. Both result in late staining of surrounding retina.[50]

Disruption of the retinal pigment epithelium (RPE) results from inflammation, choroidal ischemia, trauma, local tumors, degenerative diseases, and toxic substances. We describe only a few of the wide variety of diseases that demonstrate fluorescein

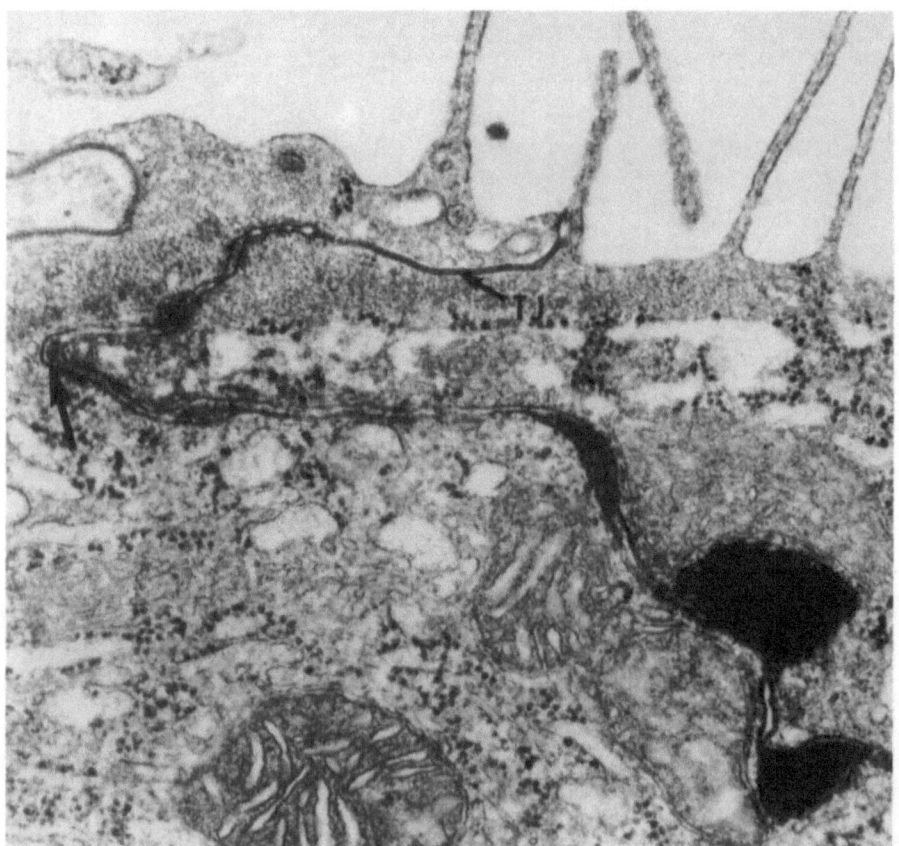

Figure 13. Higher-magnification electron micrograph of the tight junction (TJ) blocking passage of horseradish peroxidase (arrow) between pigment epithelium in the rat retina in Fig. 11. (× 49,000)

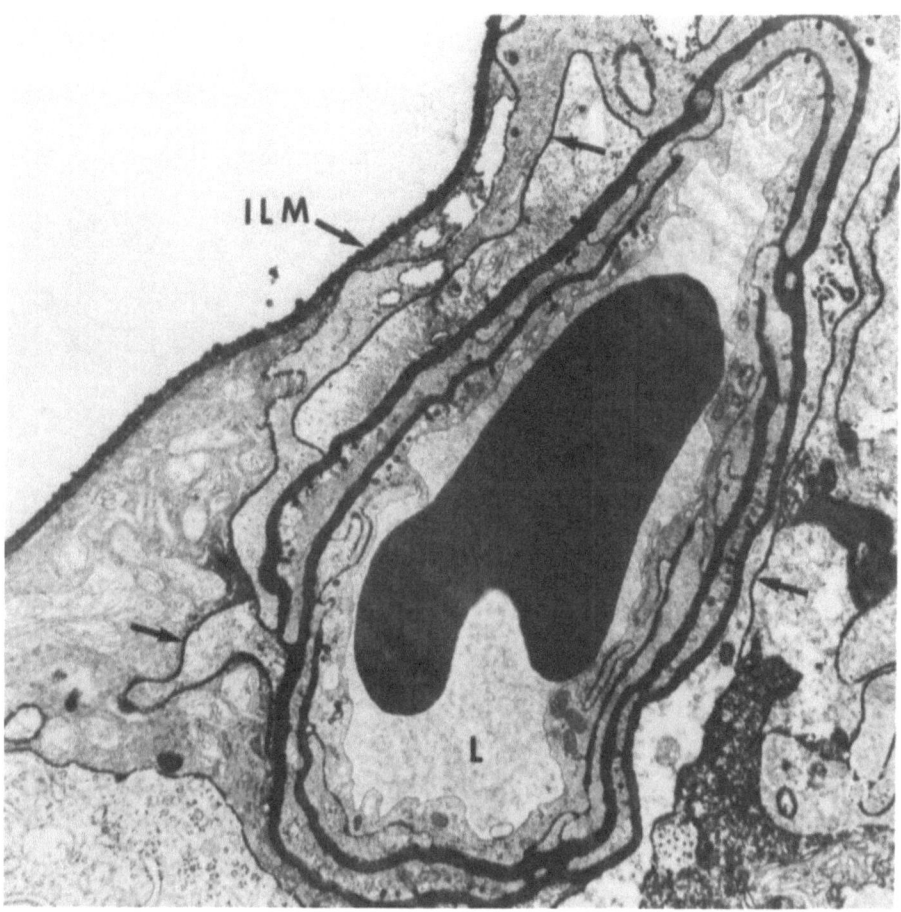

Figure 14. Electron micrograph of the retina after injection of horseradish peroxidase into the rat vitreous shows the tracer in the intracellular spaces of the retina (arrows). The tracer passes through the internal limiting membrane (ILM) into the retina but does not pass into the capillary lumen (L) despite intense staining of the basement membrane. (× 15,600)

angiographic evidence of a breakdown of this blood–retinal barrier. An acute bilateral inflammation of the RPE and the choriocapillaris, acute multifocal posterior placoid pigment epitheliopathy, occurs in young subjects following viral infections. Fluorescein angiography demonstrates occlusion in the choroidal vessels and staining of the RPE without significant leakage into the retina. This disorder may be associated with cerebral vasculitis. Inflammation of the uvea (associated with meningoencephalitis, poliosis, and vitiligo) in Vogt–Koyanagi–Harada syndrome or in other causes of posterior uveitis causes leakage of fluorescein from the choroid into the retina. Serpiginous choroidopathy, acute retinal pigment epithelitis, and presumed ocular histoplasmosis also alter the fluorescein angiographic pattern of the RPE.

Local accumulation of fluid and fluorescein occurs under the RPE from detachment of RPE in the macula of subjects over the age of 50 as a part of macula degeneration. Leakage from the RPE or retinal capillaries with accumulation of fluid between the RPE and the sensory retina occurs in younger patients with disorders such as central serous retinopathy[41] (Fig. 18). Mechanical trauma, drugs such as chloroquine and Mellaril, solar

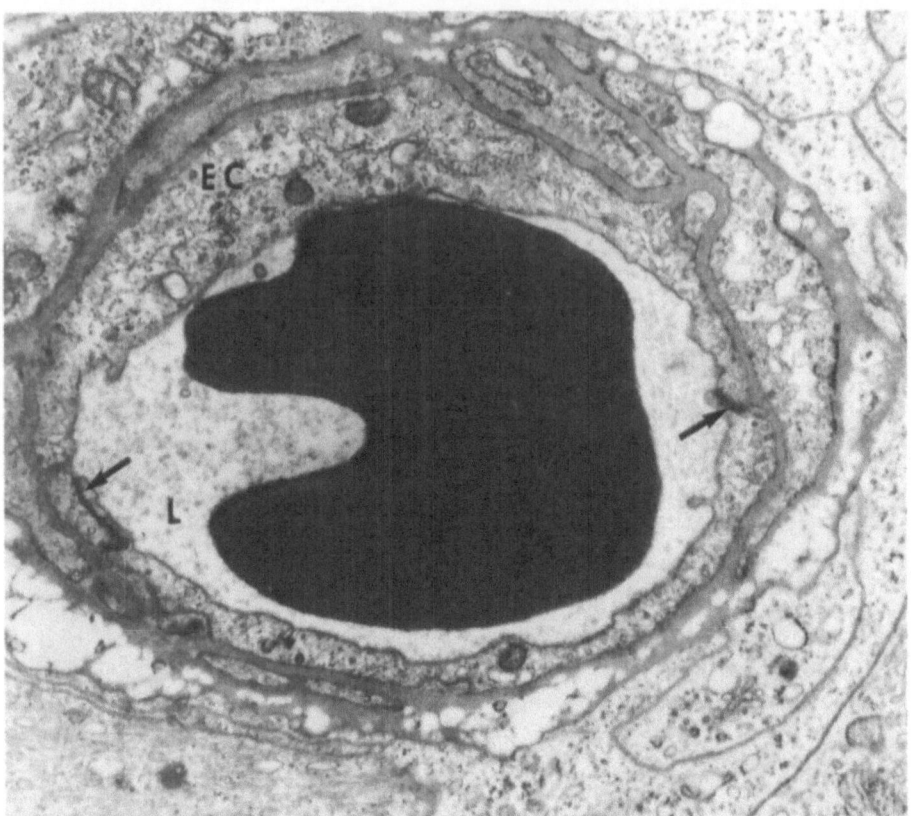

Figure 15. Electron micrograph of human retinal capillary showing nonfenestrated endothelium (EC), tight junctions (arrows), and lumen (L). (× 16,500)

and laser thermal burns, and various retinal degenerations also result in a loss of RPE.[19] Subretinal neovascularization (new vessels with incompetent endothelium) grow from the choriocapillaris through the defects in the RPE in many of the above disorders. Protein, lipid, and fluorescein leak profusely into the retina in these areas of neovascularization because the vessels do not have the morphologic characteristics of the normal retinal vessels. Tumors disrupt the RPE retinal barrier in proportion to the degree of RPE damage from local pressure and associated subretinal neovascularization.

Edema of the optic nerve head occurs from many causes (Fig. 19). Papilledema of increased intracranial pressure results in fluorescein leakage if the edema causes capillary telangiectasia or significantly dilated veins. Ischemia of the optic nerve in subjects with giant cell arteritis or anterior ischemic optic neuropathy also causes fluorescein leakage. The papillitis of multiple sclerosis, sarcoidosis, or collagen vascular diseases demonstrates significant fluorescein leakage from the disc. Inflammation of the adjacent retina or vitreous can cause secondary inflammation of the optic disc vessels with increased permeability. Optic disc edema is found in acute hypertension, uremia, toxemia, and retinal vein thrombosis.

Retinal diseases almost always disrupt the blood–retinal barrier in the retina.[50] Congenital retinal vascular defects[24] (arteriovenous malformations, angiomas, retinal telangiectasia, and macroaneurysms) lead to edema, hemorrhages, and lipoid and proteinaceous exudates into the surrounding retina because of defective endothelial cells and increased permeability in these lesions (demonstrable on fluorescein angiography). Neovascularization does not occur in these congenital malformations.

Diabetic retinopathy disrupts the blood–retinal barrier by several mechanisms. Early findings include intraretinal microvascular abnormalities with capillary occlusions and microaneurysms causing hemorrhages, exudates, and intraretinal edema. Retinal vascular endothelial proliferation in response to retinal ischemia is seen later. As in other causes of neovascularization, the vessels are highly permeable because of fenestrations in the endothelium and the lack of true tight junctions. Leakage of blood, exudates, edema, and fluorescein result (Fig. 19). Hypertension, pregnancy, and uremia aggravate the blood–retinal barrier breakdown in diabetics.

Occlusion of the central retinal vein or branch vein, hypoperfusion of the retina in arteriole occlusive disease, retinal phlebitis, and hyperviscosity syndromes can cause distention of retinal veins, with venous stasis, retinal edema and hemorrhages, and fluorescein leakage. Endothelial cells that are stretched by venous dilation are damaged by stagnant blood flow and resulting hypoxia. Leakage from retinal arterioles occurs in collagen vascular disease or infectious (syphilis) arteritis, hypertensive encephalopathy, sarcoidosis, and severe arteriosclerotic ischemia. The retinal endothelium can be irreversi-

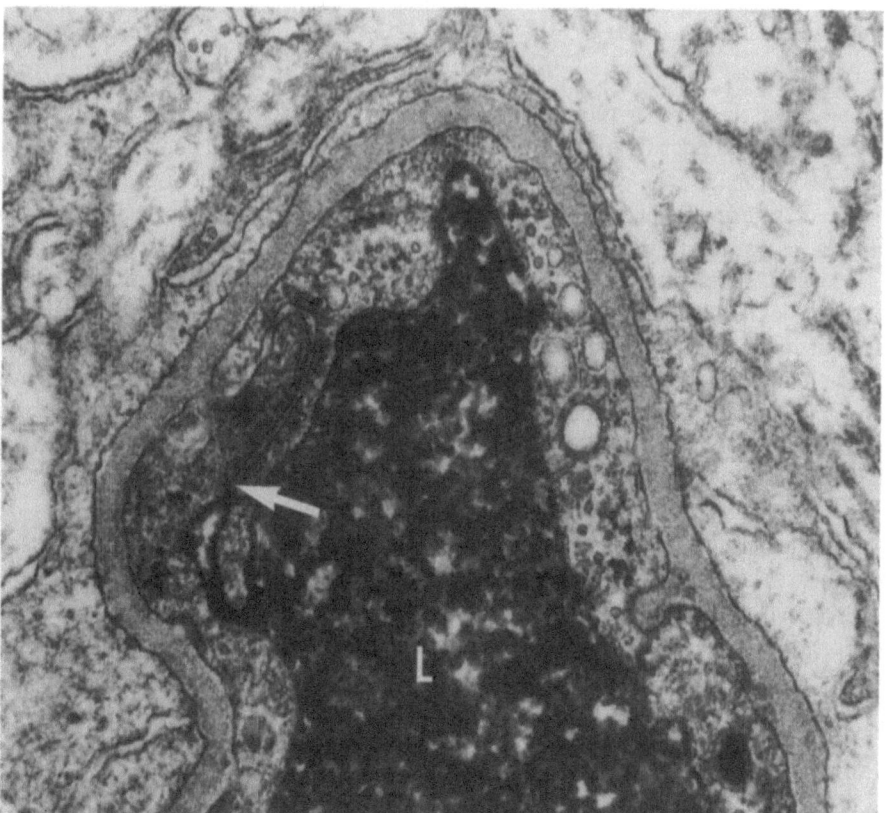

Figure 16. Electron micrograph of rat retinal capillary after injection of peroxidase into the circulation and application of histamine. No tracer can be seen beyond the tight junction (arrow). A large amount of tracer is present in the capillary lumen (L). (× 71,000)

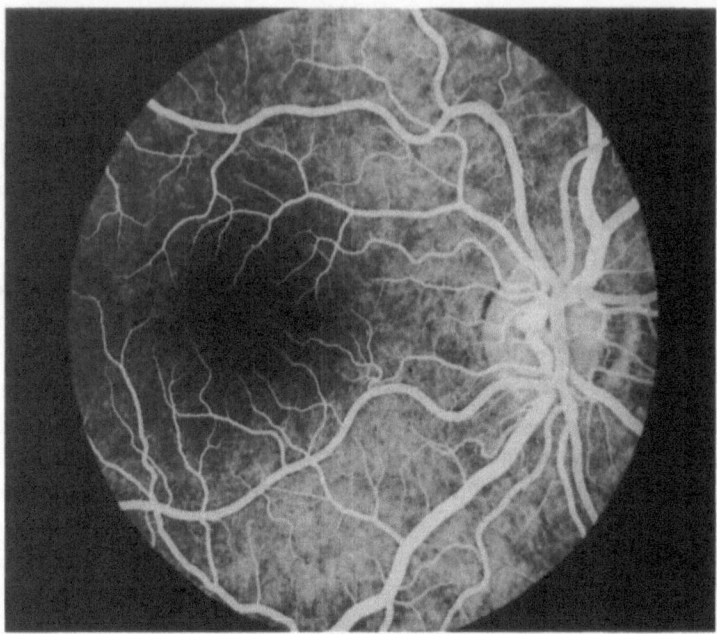

Figure 17. A normal fluorescein angiogram showing filling of retinal arteries, venules, and capillaries. A capillary-free zone is seen in the macula.

bly injured by ischemia or hypoxia. Inflammation of the arteries or veins also results when the vessels are secondarily affected by vitritis, uveitis, retinal inflammation (as in toxoplasmosis), or in immune-complex or infectious emboli involvement of the vessel walls. Infection of the endothelial cells by cytomegalovirus can also cause an acute breakdown in the barrier as part of a fulminant vasoocclusive disorder.

Neovascularization with fluorescein leakage is seen in retinal arteriole occlusive diseases of sickle cell anemia, oxygen toxicity in the premature (retinal lental fibroplasia), and idiopathic peripheral retinal occlusions (Eale disease).

The retinal endothelium is altered following cataract extraction.[46] Retinal capillaries demonstrate increased pinocytotic activity for HRP if vitreous is lost during the procedure. After lens removal, the open tight junctions of the retinal capillaries cause edema, with leakage into the macula and optic disc and from there into the vitreous. The edema in the macula is worsened by topical administration of epinephrine.

Mild defects in the blood–retinal barrier may only be apparent on vitreous fluorophotometry. One hr following intravenous injection of fluorescein, the fluorescence in the vitreous is measured in 0.25-mm steps from the retina to the cornea.[30,35,51] The posterior, middle, and anterior vitreous concentrations of fluorescein are determined and compared against established normative concentrations to determine whether the blood–retinal barrier breakdown has permitted higher than normal levels of fluorescein to enter the vitreous. Localized blood–retinal barrier breakdown is differentiated from a generalized blood–retinal barrier disruption by sampling fluorescein concentration in the vitreous adjacent to multiple retinal sites.

Vitreous fluorophotometry was first used to demonstrate blood–retinal breakdown in adult-onset diabetes prior to funduscopic or angiographic abnormalities.[11,12] More severe diabetic retinopathy causes more extensive blood–retinal barrier breakdown as measured by this method. In experimental models of diabetes, blood–retinal barrier restoration may be helped by insulin.

Patients with retinitis pigmentosa have RPE and retinal photoreceptor loss, resulting in elevated vitreous fluorescein concentration. Higher concentrations are found in most patients with retinitis pigmentosa who also have macula edema and fluorescein angiographic evidence of capillary leakage.[17] By contrast, other forms of nightblindness of retinal origin[32] or cases with RPE defects (Stargardt maculopathy)[18] may have normal vitreous fluorophotometry.

Hypertensive retinopathy that has progressed to cause hemorrhages, cotton-wool spots, edema, or exudates has obvious vitreous fluorophotometry evidence of blood–retinal breakdown. By contrast, hypertensive patients, only with narrowing of vessels, do not demonstrate blood–retinal barrier disruption.[25] Despite the relative resilience in the retinal arterioles, acute severe rises in blood pressure can disrupt the retinal endothelial blood–retinal barrier.

Ocular inflammatory diseases demonstrate vitreous fluorophotometric evidence of blood–retinal barrier defects. Peripheral retinal or choroidal inflammation, as in pars planitis, cause leakage from many areas of the retina.[28] Cases with retrobulbar neuritis, without funduscopic abnormalities, have fluorescein leakage into the vitreous which is not found in cases with noninflammatory optic neuropathies.[8]

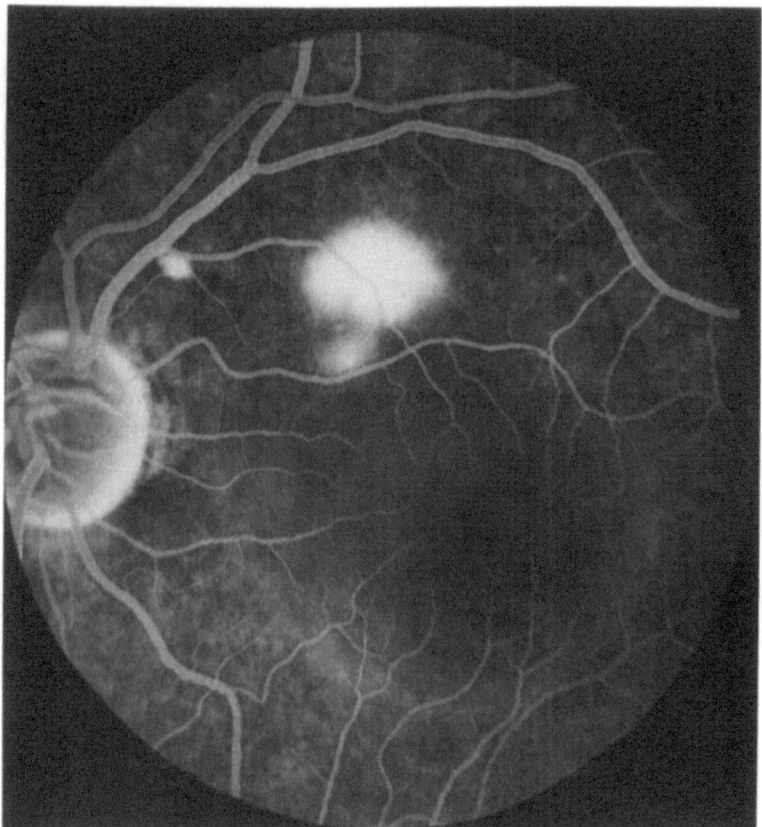

Figure 18. An abnormal later phase of fluorescein angiogram showing leakage from vessels along the superior temporal arcade in a patient with central serous choroidopathy.

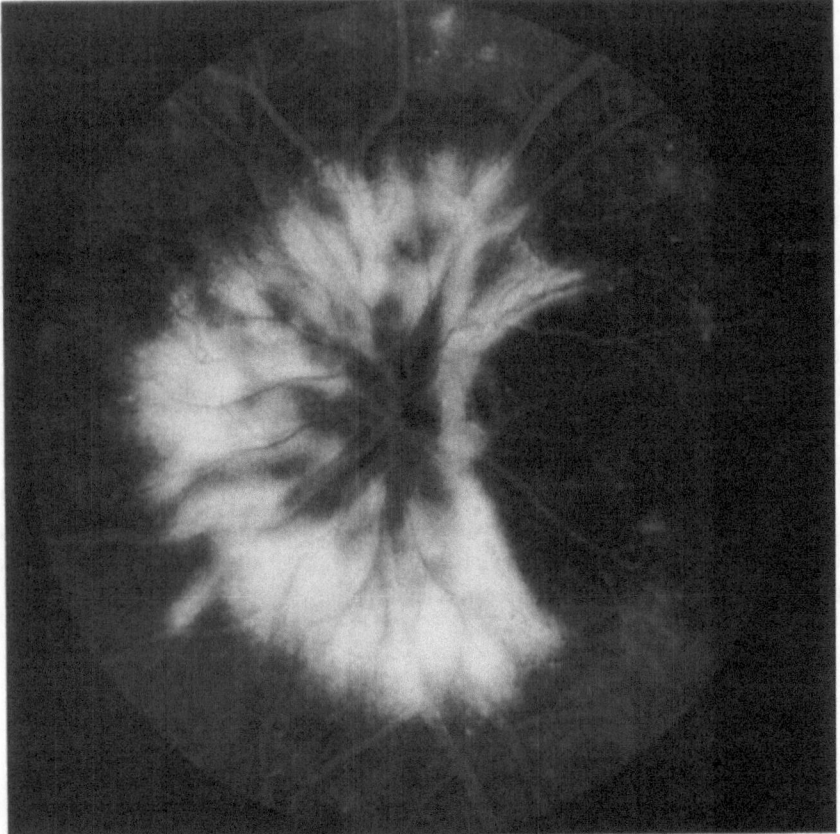

Figure 19. Fluorescein angiogram in a diabetic with neovascularization of the optic disc demonstrating diffuse severe leakage.

Vitreous fluorophotometry also reveals blood–retinal breakdown at the retinal endothelium and RPE following cataract extraction.[47] The older the surgery, the higher the probability of demonstrating the blood–retinal barrier defect.[31]

Fluorophotometry and angiography permit the in vivo study of the barriers of different areas in the eye in the living human. The effects of ocular and systemic diseases, blood pressure and flow, and various drugs can be easily demonstrated. In particular, the investigation of the barrier at the level of the retinal endothelial cells provides insight into the function of the BBB.

ACKNOWLEDGMENTS

We would like to thank Douglas MacDonald for electron microscopic work and Jill Kressel for the preparation of the manuscript.

REFERENCES

1. Archer DA: Fluorescein angiography. In Krill AE (ed): *Hereditary Retinal and Choroidal Diseases*. Harper & Row, New York, 1972, pp. 73–187.
2. Bill A: Capillary permeability to and extravascular dynamics of myoglobin albumin and gammaglobulin in the uvea. *Acta Physiol Scand* 73:204–219, 1968.
3. Bill A: Blood circulation and fluid draining in the eye. *Physiol Rev* 55:383–417, 1975.
4. Bill A: Some aspects of the ocular circulation. *Invest Ophthalmol Vis Sci* 26:410–422, 1985.
5. Bill A, Tornquist P, Alm A: Permeability of the intraocular blood vessels. *Trans Ophthalmol Soc UK* 100:332–336, 1980.
6. Bito L, deRousseau CJ: Transport functions of the blood–retinal barrier system and the microenvironment of the retina. In Cunha-Vaz JG (ed): *The Blood–Retinal Barriers*. Plenum, New York, 1979, pp. 133–164.
7. Blair N, Rusin M, Shakin E: The effect of pH on the transfer of fluorescein across the blood–retinal barrier. *Invest Ophthal Vis Sci* 2:1133–1139, 1985.
8. Braude LS, Cunha-Vaz JG, Goldberg MF, et al: Diagnosing acute retrobulbar neuritis by vitreous fluorophotometry. *Am J Ophthalmol* 91:764–773, 1981.
9. Cunha-Vaz JG: The blood–ocular barriers. *Surv Ophthalmol* 23:779–796, 1979.
10. Cunha-Vaz JG: Sites and functions of the blood–retinal barriers. In Cunha-Vaz JG (ed): *The Blood–Retinal Barriers*. Plenum, New York, 1979, pp. 101–118.
11. Cunha-Vaz JG: Vitreous fluorophotometry. In Cunha-Vaz JG (ed): *The Blood–Retinal Barriers*. Plenum, New York, 1979, pp. 195–210.
12. Cunha-Vaz JG, Shakib M, Ashton N. Studies on the permeability of the blood–retinal barrier. I. On the existence, development and site of a blood–retinal barrier. *Br J Ophthalmol* 50:441–453, 1966.
13. Cunha-Vaz JG, Abreu JR, Campos JA, et al: Early breakdown of the blood–retinal barrier in diabetes. *Br J Ophthalmol* 59:649–656, 1975.
14. Davson H, Matchett PA: The kinetics of penetration of the blood–aqueous barrier. *J Physiol (Lond)* 122:11–32, 1953.
15. Ernest T: Recent correlates of ocular physiology and fluorescein angioscopy, angiography, and fluorophotometry. *J Contin Educ Ophthalmol* 41:13–27, 1979.
16. Ernest TJ: Blood flow in normal and abnormal retinal vessels. *Trans Ophthalmol Soc UK* 100:343–345, 1980.
17. Fishman G, Cunha-Vaz JG, Salzano T: Vitreous fluorophotometry in patients with retinitis pigmentosa. *Arch Ophthalmol* 99:1202–1207, 1981.
18. Fishman G, Cunha-Vaz JG, Travassos AC: Vitreous fluorophotometry in patients with fundus flavimaculatus. *Arch Ophthalmol* 100:1086–1088, 1982.
19. Foulds WS, Moseley H, Eadie A, et al: Vitreal, retinal, and pigment epithelial contributions to the posterior blood–ocular barrier. *Trans Ophthalmol Soc UK* 100:341–342, 1980.
20. Freddo TF, Raviola G: Freeze-fracture analysis of the interendothelial junctions in the blood vessels of the iris in *Macaca mulatta*. *Invest Ophthalmol Vis Sci* 23:154–167, 1982.
21. Gregory D, Sears M, Bausher L, et al: Intraocular pressure and aqueous flow are decreased by cholera toxin. *Invest Ophthalmol Vis Sci* 20:371–381, 1981.
22. Hayreh SS: Pathogenesis of visual field defects. Role of auxiliary circulation. *Br J Ophthalmol* 54:289–311, 1970.
23. Hayreh SS, Scott WE: Fluorescein iris angiography. *Arch Ophthalmol* 96:1383–1389, 1978.
24. Henkind P, Walsh JB: Retinal vascular anomalies, pathogenesis, appearance and history. *Trans Ophthal Soc UK* 100:425–433, 1980.
25. Jampol LM, White S, Cunha-Vaz JG: Vitreous fluorophotometry in patients with hypertension. *Arch Ophthalmol* 101:888–890, 1983.
26. Kolker AE, Hetherington J: *Becker-Schaffer's Diagnosis and Therapy of the Glaucomas*. CV Mosby, St Louis, 1976, pp. 325–330.
27. Laties AM, Rapoport SI, McGlinn A: Hypertensive breakdown of cerebral but not of retinal blood vessels in rhesus monkey. *Arch Ophthalmol* 97:1511–1514, 1979.
28. Mahlberg PA, Cunha-Vaz JG, Tessler HH: Vitreous fluorophotometry in pars planitis. *Am J Ophthalmol* 95:189–196, 1983.
29. Marsh RJ, Ford SM: Blood flow in the anterior segment of the eye. *Trans Ophthalmol Soc UK* 100:388–396, 1980.

30. Maurice DM: A new objective fluorophotometer. *Exp Eye Res* 2:33–38, 1963.
31. Miyake K: Blood–retinal barrier in eyes with long-standing aphakia with apparently normal fundi. *Arch Ophthalmol* 100:1437–1439, 1982.
32. Miyake Y, Gto S, Ando F, et al: Vitreous fluorophotometry in congenital stationary night blindness. *Arch Ophthalmol* 101:574–579, 1983.
33. Miyake K, Asakura M, Kobayashi H: Effect of intraocular lens fixation on the blood–aqueous barrier. *Am J Ophthalmol* 98:451–455, 1984.
34. Okisaka S, Kuwabara T, Rapoport SI: Selective destruction of the pigmented epithelium in the ciliary body of the eye. *Science* 184:1298–1299, 1974.
35. Palestine AG, Brubaker RF: Pharmacokinetics of fluorescein in the vitreous. *Invest Ophthalmol Vis Sci* 21:542–549, 1981.
36. Rapoport S: Osmotic opening of blood–brain and blood–ocular barriers. *Exp Eye Res* 25 (suppl): 499–509, 1977.
37. Raviola G: Blood–aqueous barrier can be circumvented by lowering intraocular pressure. *Proc Natl Acad Sci USA* 73:638–642, 1976.
38. Raviola G: The structural basis of the blood–ocular barriers. *Exp Eye Res* 25 (suppl): 27–63, 1977.
39. Sanders RD, Manus CK, Lieberman HL, et al: Breakdown and reestablishment of blood–aqueous barrier with implant surgery. *Arch Ophthalmol* 100:588–590, 1982.
40. Shakib M, Cunha-Vaz JG: Studies on the permeability of the blood–retinal barrier. IV. Role of the junctional complexes of the retinal vessels on the permeability on the blood–retinal barrier. *Exp Eye Res* 5:229–234, 1966.
41. Shakib M, Rutkowski P, Wise GN: Fluorescein angiography and the retinal pigment epithelium. *Ophthalmology* 74:206–218, 1972.
42. Smelser G, Pei YFL: Cytological basis of protein leakage into the eye following paracentesis. *Invest Ophthalmol* 4:249–263, 1965.
43. Smith RS, Rundt LA: Ultrastructural studies of the blood–aqueous barrier: II. The barrier to horseradish peroxidase in primates. *Am J Ophthalmol* 76:937–947, 1973.
44. Smith RS, Rundt LA: Ocular vascular and epithelial barriers to microperoxidase. *Invest Ophthalmol* 14:556–560, 1975.
45. Stone R. The transport of para-aminohippuric acid by the ciliary body and by the iris of the primate eye. *Invest Ophthalmol Vis Sci* 18:807–818, 1979.
46. Tso M: Pathological study of cystoid macular edema. *Trans Ophthalmol Soc UK* 100:408–413, 1980.
47. Tso M, Shih CY: Experimental macular edema after lens extraction. *Invest Ophthalmol Vis Sci* 16:381–392, 1977.
48. Vegge T: A study of the ultrastructure of the small iris vessels in the vervet monkey (*Cercopithecus acthiops*). *Z Zellforsch* 123:195, 1972.
49. Waltman SR, Oestrick C, Krupin T, et al: Quantitative vitreous fluorophotometry: A sensitive technique for measuring early breakdown of the blood–retinal barrier in young diabetic patients. *Diabetes* 27:85–87, 1978.
50. Yannuzzi LA, Gitter KA, Schatz H (eds): *The Macula: A Comprehensive Text and Atlas.* Williams & Wilkins, Baltimore, 1979.
51. Zeimer RC, Blair NP, Cunha-Vaz JG: Vitreous fluorophotometry for clinical research. II. Methodology of data acquisition and processing. *Arch Ophthalmol* 101:1757–1761, 1983.

Index

391